Introduction to
BIOFUELS

Mechanical Engineering Series
Frank Kreith, Series Editor

Introduction to
BIOFUELS

David M. Mousdale

CRC Press
Taylor & Francis Group
Boca Raton   London   New York

CRC Press is an imprint of the
Taylor & Francis Group, an **informa** business

Library of Congress Cataloging-in-Publication Data

Mousdale, David M.
  Introduction to biofuels / David M. Mousdale.
    p. cm. -- (Mechanical engineering series)
  Includes bibliographical references and index.
  ISBN 978-1-4398-1207-5 (hardcover : alk. paper)
  1. Biomass energy. I. Title. II. Series.

TP339.M693 2011
662'.88--dc22   2010022941

Contents

Preface
How to Get the Most from This Book

There are many excellent reasons for wanting to know more about biofuels. You may be—and I hope many readers are—a chemist who sees biological methodologies as natural extensions to established synthetic methods that can offer radically new perspectives and opportunities for chiral molecules and "smart" catalysts. But you also may *not* know that chemistry offers unique routes to biofuels.

From a life sciences background, biofuels are a signpost to the future of *bio*-based industries—with *bio*manufacturing and *bio*production—and these represent new industrial sectors for the student of contemporary economics. The shifting energy landscape of the twenty-first century is both worrying and intellectually challenging. Renewable energy technologies are in high demand; are biofuels part of that scientific portfolio that might bring "energy independence" and "energy security," revitalize rural infrastructures, and wean the developed world off its "addiction" to oil?

Chemical engineers will already appreciate that industrial-scale products require a massive infrastructure of hardware that is seen in petrochemical refineries. But what will be used as inputs to future biorefineries? What will emerge—and how can those outputs be maximized for yield and be successfully purified?

The genetic engineer naturally has interests in the manipulation and construction of microbial cells as biocatalysts as well as in the expression of desirable traits in plants that are common to future bioenergy crops and to the world's major food-bearing plant species. Can evolution be accelerated in the laboratory and thence into new fields on marginal land or in aquatic systems, possibly on the shores of seas and oceans?

"Sustainability" is a key issue that runs as a central thread through contemporary college courses in subjects covering a diversity of topics: geography, energy studies, transportation, and mechanical engineering. Any "bio" element in energy-related issues will inevitably require an assessment of impacts on agronomy, land use, soil erosion, and water availability. Brazil is a remarkable example of a major transformation from a neocolonial rural economy to a nation closing on its admission into the "G8 club" of modern industrial giants. Brazil has used biofuels to revolutionize its use of climate and natural resources. Is that development a sustainable process? Can it be transplanted elsewhere?

The burgeoning interest in biofuels in "self-help" areas as well as in rural employment has sparked community college and informal courses in biofuels science and technology. I hope that students will find parts of this book technically useful and stimulating to a wider interest in biofuels.

WHAT ARE THE CRUCIAL SCIENTIFIC QUESTIONS ABOUT BIOFUELS?

At the core of the subject lie two groups of technologies—chemical and biological—that transform plant materials such as cereal grains, sugarcane juice, vegetable seed oil, or grass cuttings into liquid or gaseous fuels. There are implications—biological, chemical, physical, and engineering—for the upstream agenda of plant biomass provision and the downstream distribution and use of biofuels. But are biofuels really that useful? Obtaining renewable energy generation is *easy:* Think of power systems functioning today from winds, waves, waterfalls, tides, solar, and geothermal energy. With sufficient renewable electricity, hydrogen can be produced instantly from the electrolysis of water and can substitute for *all* fossil fuels as the perfect *zero* carbon power source—forever (or as far into the future as anyone can see).

But *no* renewable can generate carbon or, as combined in biological molecules, hydrogen, nitrogen, oxygen, sulfur, phosphorus, etc. Oil supplies the mix of hydrocarbons with which petrochemistry generates the myriad chemical intermediates that underpin modern life. Plastics are the most obvious example, but petrochemicals give rise to the chemical entities in many medicines, synthetic fabrics, myriad commodity goods, and—importantly for biofuels—fertilizers, pesticides, and herbicides. What will become of supporting the *material* demands of an eventual global population in excess of 9 billion? The answer may be: *not* petrochemicals.

An article in the journal *Energy Policy* that was published in January 2009[1] concluded:

> The new formula…computes fossil fuel reserve depletion times for oil, coal and gas of approximately 35, 107 and 37 years, respectively. This means that coal reserves are available up to 2112, and will be the only fossil fuel remaining after 2042.

Energy Policy is unlikely to be on the reading lists for biology, chemistry, or chemical engineering courses but may be accessed by students of politics, economics, and energy issues of contemporary society. The estimates in that journal article may be wildly inaccurate, but they suggest that within decades a complete industrial sector is required: designed, constructed, and fully functional to replace petrochemicals. If the world is very fortunate—or, perhaps, very unlucky—oil reserves might last for a century or even more. Global warming will continue unabated and, as many experts predict, the tipping point of climatic catastrophe will be reached eventually. Only terrestrial and/or marine plant biomass can supply the carbon (and hydrogen) that petroleum hydrocarbons already provide—as an extender to dwindling oil supplies, an alternative to oil in mitigating climate change, or an outright replacement for petrochemicals.

Biofuels are manufactured, distributed, and sold; schemes to manufacture new biofuels are under discussion or are being practically implemented around the world. They may only be stepping stones before electric vehicles and the hydrogen economy become affordable worldwide. They may help, hinder, or have no overall effect on efforts to minimize greenhouse gas emissions. They are certainly contentious, but understanding the actual and potential impact of biofuels on society requires serious study of a broad sweep of technological, sociological, and energy-policy topics that intermingle and intertwine.

HOW CAN STUDENTS OF DIFFERING ACADEMIC BACKGROUNDS AND INTERESTS USE THIS BOOK?

For the (molecular) biologist and biotechnologist, Chapters 1–4 and 7 will occupy familiar ground, and Chapters 5, 8, and 10 will broaden understanding of biofuels and their production methods to biochemical engineering and industrial chemistry. The following are useful textbooks:

- Perez-Blanco, H. 2009. *The Dynamics of Energy. Supply, Conversion, and Utilization.* Boca Raton, FL: CRC Press.
- Blanch, H. W., and Clark, D. S. 1997. *Biochemical Engineering.* Boca Raton, FL: CRC Press.
- Rogers, P. P., Jalal, K. F., and Boyd, J. A. 2007. *An Introduction to Sustainable Development.* London: Earthscan.

For chemists and chemical engineers, most of Chapters 1, 5, and 8 will be fairly close to a core syllabus. However, for the more "bio" topics, the following are good introductions to new languages and vocabularies:

- Moat, A. G., Foster, J. W., and Spector, M. P. 2002. *Microbial Physiology,* 4th ed. New York: John Wiley & Sons, Inc.
- Anastas, P. T., and Warner, J. C. 2000. *Green Chemistry: Theory and Practice.* Oxford, England: Oxford University Press.

For students interested in biofuels but without a formal training in modern chemistry, biochemistry, or biology, I recommend the following set of articles from *Scientific American* and *National Geographic:*

- Huber, G. W., and Dale, B. E. 2009. Grassoline at the pump. *Scientific American* 301:52.
- Wald, M. L. 2009. The power of renewables. *Scientific American* 300:50.
- Zweibel, K., Mason, J., and Fthenakis, V. 2008. A solar grand plan. *Scientific American* 298:64.
- Satyapal, S., Petrovic, J., and Thomas, G. 2007. Gassing up with hydrogen. *Scientific American* 296:80.
- Ashley, S. 2007. Diesels come clean. *Scientific American* 296:80.
- Wald, M. L. 2007. Is ethanol for the long haul? *Scientific American* 296:42.
- Kammen, D. M. 2006. The rise of renewable energy. *Scientific American* 295:82.
- Gibbs, W. W. 2006. Plan B for energy. *Scientific American* 295:84.
- Grant, P. M., Starr, C., and Overbye, T. J. 2006. A power grid for the hydrogen economy. *Scientific American* 295:58.

- Romm, J. J., and Frank, A. A. 2006. Hybrid vehicles gain traction. *Scientific American* 294:72.
- Bourne, J. K., Jr. 2007. Green dreams. Making fuel from crops could be good for the planet—after a breakthrough or two. *National Geographic* October:38.

ROAD MAPS FOR FOLLOWING DIFFERENT COURSES OF STUDY AND TEACHING RESOURCES

The sequence of chapters provides a narrative offering answers to key questions on modern biofuels, but various routes can be followed for different interests and study aims:

- the spectrum of products claimed as "bio," their properties, and state of development toward commercialization: Chapters 1, 2, 7, 8, and 10
- the emergence of major new "white" (industrial) biotechnology: Chapters 1–5, 7, and 10
- biofuels as important elements in contemporary energy studies: Chapters 1 and 2 (early sections) and 6–10
- biofuels and the "greening" of industrial chemistry: Chapters 1, 2, 5, 6, and 7 (parts), 8, and 10
- biofuels as commodities and industrial products: Chapters 1 and 2 (early sections), 6–8, and 10
- sustainable development issues: Chapters 1, 2, 6, and 8–10
- "green" (agricultural) biotechnology: Chapters 1, 2, and 5 (parts), and 7–10

Each chapter begins with questions whose answers are later summarized in key informational points. Embedded sections termed "STEM" (science, technology, engineering, and math) will provide detailed derivations and (where appropriate) equations for a subset of topics that are essential but often completely missed (or taken for granted). Together, the STEM topics form a thread of essential technologies and also a guide to how some quantitative parameters of crucial importance to the debates on biofuels have been reached. Some mathematical models—or the information with which to build them—for energy computations and biochemical systems are now freely available in Excel formats, and I have indicated downloading sites for these.

Every chapter is extensively referenced so that an individual reader can pursue a topic in depth using the indicated primary literature (many publications have 2008 and 2009 entry dates), up-to-date reviews, and Internet sources.

Science is a participatory experience—*not* a spectator sport. That truism also applies to the commentator and the consultant, who must be as up-to-date as the research scientist and knowledgeable about important new data and ideas appearing this week and next.

Finally, although it may seem quixotic to attempt to pull together information and ideas from so many disparate areas—biotechnology, microbial and plant physiology, industrial biomanufacture, organic chemistry, chemical engineering, genetic engineering, ecology, economics, oil industry studies, social development, and thermodynamics—the quest has its rationale. The public debate inevitably overlaps many of these topics; sometimes, the topics are confused or misrepresented. None of the issues can be usefully studied in isolation from *all* of the others.

But they can all be studied.

REFERENCE

1. Shafieea, S., and Topal, S. 2009. When will fossil fuel reserves be diminished? *Energy Policy* 37:181.

Acknowledgments

This book could not have been written without the continuously updated supply of statistical data provided by many international and national agencies and associations. In particular, I acknowledge my gratitude to UNICA and ANFAVEA in Brazil, the U.S. National Biodiesel Board and the Renewable Fuel Association, and British Petroleum plc ("Beyond Petroleum").

My sincere thanks are also due to all open-access journals and to the U.S. National Institutes of Health for devising and maintaining the digital archive of journal literature (PubMed Central). We live in an age when national funding bodies are beginning to demand of scientists they fund that they make their work freely and openly available to the public; only some career librarians might feel threatened by this.

Wide access to cutting-edge information and data is the only means of ensuring an informed debate that can leave the closed world of dedicated scientific practitioners and influence the often hostile responses experienced by science when it touches—as biofuels certainly do—the lives of billions of people.

David Mousdale
(dmousdale@beocarta.com)

The Author

David Mousdale was educated at Oxford (BA, biochemistry, 1974) and Cambridge (PhD, 1979). He researched growth control and integration mechanisms in plants and plant cell cultures before turning to enzyme responses to xenobiotics, including the first isolation of a glyphosate-sensitive enzyme from a higher plant.

In the microbial physiology and biochemistry of industrial fermentations, he developed metabolic analysis to analyze changes in producing strains developed by serendipity (i.e., classical strain improvement) or by rational genetic engineering. He became managing director of beòcarta Ltd in 1997. Although much of the work of beòcarta Ltd has been focused on antibiotics and other secondary metabolites elaborated by streptomycetes, it has been extended to vitamins, enzymes, amino acids, carboxylic acids, heterologous protein expression, novel bioactives, and animal cell bioreactors for the manufacture of biopharmaceuticals.

Recent projects have included enzyme production for the food industry, enzymes for processing of lignocellulose substrates for biorefineries, the recycling of glycerol from biodiesel manufacture, the metabolic analysis of extracellular polysaccharide production by fungi, and the metabolomics of industrial fermentations.

Units and Conversion Factors

ENERGY

The joule (J) is the force of 1 newton (N) acting over a distance of 1 meter (m).

The calorie is the amount of heat required to raise the temperature of 1 gram (g) of water from 14.5°C to 15.5°C.

The British thermal unit (BTU) is the amount of heat energy needed to raise the temperature of 1 pound (lb) of water by 1 degree Fahrenheit (°F).

1 joule = 0.239 calories

1 calorie = 4.187 J

1 British thermal unit (Btu) = 1.055 joules (1.055 kJ)

The kilowatt-hour (kWh) is 1 kilowatt (kW) of power expended for 1 hour (h); 1 kWh = 3.6×10^6 J.

1 gigajoule (GJ) = 10^9 J = [278] kWh

1 exajoule (EJ) = 10^{18} J

1 quad = 1 quadrillion Btu (10^{15} Btu) = 1.055 EJ

1 quad = 172 million barrels of oil equivalent (boe)

$1.00/GJ = $1.055 per million Btu, or $1.00 per million Btu = $0.948/GJ

ENERGY INDUSTRIES

In the United States, 1 barrel (bbl) = 42 gallons = 159 liters (L)

In Europe, 1 metric ton (tonne) = 7.3 bbl

1 barrel oil per day (often bpd) = 50 tonnes per year

Gas is measured in the United States and United Kingdom in cubic feet (cf) and in Europe in cubic meters (m^3); 1 m^3 = 35.3 cf.

For gas conversion, 1 barrel of oil equivalent (boe) = 6,000 cf.

1 tonne of coal = 4.879 boe crude oil

MASS, VOLUME, CONCENTRATION

1 U.S. ton (short ton) = 2,000 lb

1 imperial ton = 2,240 lb

1 tonne = 1,000 kilograms (kg) = 2,205 lb

1 U.S. gallon = 0.833 imperial gallon

1 imperial gallon = 4.55 liters = 1.20 U.S. gallon

1 liter = 0.264 U.S. gallon = 0.220 imperial gallon

1 U.S. bushel = 0.0352 m^3

The mole is the mass (in grams) of 6.022×10^{23} molecules of a compound or 6.022×10^{23} atoms of an uncombined element.

1 mol per liter = M = 1,000 millimolar (mM) = 10^6 micromolar (μM)
1 m^3 = 1,000 liters
1 liter = 1,000 milliliters (mL)

AREA UNITS

1 hectare (ha) = 10,000 m^2 = 2.47 acres; 1 acre = 0.405 hectares
1 U.S. ton/acre = 2.24 ton/ha
1 ton/hectare = 0.446 ton/acre

1 Ethanol as the Leading "First-Generation" Biofuel

1.1 INTRODUCTION

How, when, and why did biofuels emerge to form a vibrant, modern industrial sector, spanning biotechnology, biochemical engineering, and large-scale chemistry? In contemporary society, industrially manufactured biofuels, such as ethanol, deriving from food crops (sugarcane and cereal grains) are not simply analogous to traditional chemical fuel molecules elaborated in petrochemical refineries but also involve issues fiercely (and globally) debated by energy analysts, environmentalists, politicians, agronomists, and economists. Do biofuels make thermodynamic sense? Have they any sustainable future? Only by considering both the science and the history of biofuels production can we accurately position the (bio)technologies inside contemporary debates on world energy and environmental issues

1.2 HISTORICAL DEVELOPMENT OF ETHANOL AS A FUEL FROM NEOLITHIC TIMES TO THE TWENTIETH CENTURY

Biotechnology—the application of living cells to elaborate products of medical or other uses in human society—has a history almost as long as that of agriculture. The development of molecular archaeology (i.e., the trace chemical analysis of residues on pottery sherds and other artifacts recovered from archaeological strata) has begun to specify discrete chemical compounds as markers for early biotechnological activities—in particular the production of ethanol.[1] Molecular radiocarbon dating and dendrochronology, archaeobotany, and archaeology have combined to establish an ancient lineage:

- In western Asia, wine making can be dated as early as 5400–5000 BC in what today is northern Iran and, farther south in Iran, at a site from 3500–3000 BC.[1]
- In Egypt, predynastic wine production began at approximately 3150 BC, and a royal wine-making industry had been established at the beginning of the Old Kingdom (2700 BC).[2]
- Wild or domesticated grape (*Vitis vinifera* L. subsp. *sylvestris*) can be traced back to before 3000 BC at sites across the western Mediterranean region, Egypt, Armenia, and along the valleys of the Tigris and Euphrates rivers.

This is similar to the modern distribution of the wild grape (used for 99% of today's wines) from the Adriatic coast, at sites around the Black Sea and southern Caspian Sea, Turkey, the Caucasus and Taurus Mountains, Lebanon, and the islands of Cyprus and Crete.[3]

• Partial DNA sequence data identify a yeast similar to the modern *Saccharomyces cerevisiae* as the biological agent used for the production of wine, beer, and bread in ancient Egypt, ca. 3150 BC.[2]

The occurrence of *V. vinifera* in regions in or bordering on the Fertile Crescent that stretched from Egypt though the western Mediterranean region and to the lower reaches of the Tigris and Euphrates is crucial to the understanding of Neolithic biotechnology (i.e., wine making). When ripe, grapes supply not only abundant sugar but also other nutrients (organic and inorganic) necessary for rapid microbial fermentations as well as the causative yeasts themselves—usually as passengers on the skins of the fruit. Simply crushing (pressing) grapes initiates the fermentation process that, in unstirred vessels (i.e., in conditions that soon deplete oxygen levels) produces ethanol at 5–10% by volume (approximately 50–100 g/L).

In China, molecular archaeological methodologies such as mass spectroscopy and Fourier transform infrared spectrometry have placed "wine" (i.e., a fermented mixture of rice, honey, and grape as well as, possibly, other fruit) as being produced in an early Neolithic site in Henan Province from 6500 to 7000 BC.[4] Geographically, China lies well outside the accepted natural range of the Eurasian *V. vinifera* grape but is home to many other natural types of grape. Throughout the Old World, the earliest known examples of wine making, separated by more than 2,000 km and occurring between 7,000 and 9,000 years ago, were probably independent events, perhaps an example on the social scale of the "convergent evolution" well known in biological systems at the genetic level.

Grape wines, beers from cereals (einkorn wheat, one of the founder plants in the Neolithic revolution in agriculture, was domesticated in southeastern Turkey, ca. 8000 BC), and alcoholic drinks made from honey, dates, and other fruits grown in the Fertile Crescent are likely to have had ethanol concentrations below 10% by volume. The physicochemical concentration of the ethanol in such liquids by distillation results in a wide spectrum of potable beverages known collectively as "spirits." The evolution of this chemical technology follows a second line of technical evolution[5,6]:

• Chinese texts from ca. 1000 BC warned against overindulgence in distilled spirits.
• Whisky (or whiskey) was widely known in Ireland by the time of the Norman invasion of 1170–1172.
• Arnold de Villeneuve, a French chemist, wrote the first treatise on distillation, ca. 1310, and a comprehensive text on distilling was published in Frankfurt-am-Main (Germany) in 1556.
• The production of brandies by the distillation of grape wines became widespread in France in the seventeenth century.
• The first recorded production of grain spirits in North America was that by the director general of the colony of New Netherland in 1640 (on Staten Island).

- In 1779, 1,152 stills had been registered in Ireland—although this number had fallen drastically to 246 by 1790 as illicit moonshine pot stills flourished.
- In 1826, a continuously operating still was patented by Robert Stein of Clackmannanshire, Scotland, but the Bureau of Excise of the United Kingdom accepted the twin-column distillation in 1830; this apparatus, with many variations and improvements to the basic design, continues to yield high-proof ethanol (ca. 95% by volume, v/v).

Distillation yields "rectified spirit" (96.4% v/v), a binary azeotrope with a boiling point of 78.15°C. "Absolute" alcohol, prepared by the physical removal of the residual water, has the empirical formula of C_2H_6O, molecular weight 46.07; it is a clear and colorless liquid with a boiling point of 78.5°C and a density (at 20°C) of 0.789 kg/L. Absolute alcohol absorbs water vapor rapidly from the air and is entirely miscible with liquid water. As a chemical known to alchemists and medicinal chemists in Europe and Asia, it found many uses as a solvent for materials insoluble or poorly soluble in water, more recently as a topical antiseptic, and as a general anesthetic (although pharmacologically highly difficult to dose accurately).

For the explicit topic of this volume, however, its key property is its inflammability: Absolute alcohol has a flash point of 13°C. Any volatile, inflammable alcohol or related organic compound of low molecular mass (aldehydes, ketones, and ethers) could have been the obvious choice for the internal combustion engine, but ethanol had the unique advantage of already being produced on a global scale at the end of the nineteenth century.

STEM TOPIC 1.1: AZEOTROPES AND ETHANOL PURIFICATION

Azeotropes result from the deviation from ideality of "real" liquids (i.e., liquids do not behave exactly as assumed by classical thermodynamics). Any liquid has a vapor pressure; the higher the volatility of a liquid, the higher is its vapor pressure. Mixtures of two liquids with different boiling points might be expected to exhibit behavior where the vapor pressure of the mixture lies between the individual vapor pressures of the two individual liquids, but deviations from this ideal are well known. Azeotropes are formed when the mixture has a vapor pressure lower than that of either binary component; for example, water boils at 100°C, pure ethanol at 78.5°C, and the azeotrope at 78.2°C. Intuitively, the azeotrope will be predominantly ethanol, and the actual result is 96.4% ethanol by volume, or 95.6% by mass (see Table STEM 1.1).

The technology of distillation is *fractional* distillation, in which the volatilized mix is repeatedly condensed and reboiled so that its composition approaches the maximum content of the more volatile component. When an azeotrope is not formed, eventually a pure vapor (100% of the more volatile binary component) will emerge. Starting from an ethanol–water mixture *below* 96.4% (v/v) ethanol, however, fractional distillation will produce the azeotropic mixture and leave beyond pure water in the heated vessel. Starting from an

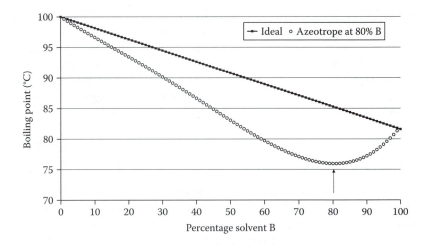

FIGURE STEM 1.1 Variation in boiling points of ideal and azeotrope-forming (↑) binary liquid mixtures.

ethanol–water mixture *above* 96.4% (v/v) ethanol, fractional distillation will produce the azeotropic mixture and leave behind pure ethanol in the heated vessel (see Figure STEM 1.1).

Azeotropic distillation adds a third solvent to prevent the undesired azeotrope from forming; benzene was one such third solvent used, but it is toxic and contaminates the final product. Most modern fuel ethanol plants use molecular sieving, where water molecules are stripped out of the liquid mixture because they can penetrate into the molecular-scale pores in the beads of the adsorbent material, leaving pure ethanol to pass through. The Praj Industries (Pune, India) Web site provides a useful guide to the chemical engineering of ethanol purification in the context of fuel ethanol production (www.praj.net).

TABLE STEM 1.1
Azeotropes Formed between Water and Organic Solvents

Component A	Component B	Azeotrope (% B, v/v)	Azeotrope bp (°C)
Water	Dichloromethane	98.5	38.1
Water	Ethanol	96.4	78.2
Water	Tetrahydrofuran	95.0	65.0
Water	Ethyl acetate	91.5	70.3
Water	Propan-2-ol	87.4	80.3
Water	Acetonitrile	84.0	76.1

By 1905, ethanol had indeed emerged as the fuel of choice for automobiles among engineers and motorists, and the Automobile Club of America sponsored a competition for alcohol-powered vehicles in 1906. Opinion was heavily swayed by fears about oil being a rare commodity, by rising gasoline prices, and the monopolistic practices of Standard Oil.[7] Henry Ford planned to use ethanol as the primary fuel for his Model T (introduced in 1908) but soon opted for the suddenly cheaper alternative of gasoline. Price competition between ethanol and gasoline had proved crucial; the removal of excise duty from denatured ethanol in 1907 came too late to stimulate investment in fuel ethanol production and develop a distribution infrastructure in what proved to be a narrow window of opportunity for fuel ethanol.

Ford was not alone in considering a variety of possible fuels for new engines. Rudolf Diesel (who obtained his patent in 1893) developed the first prototypes of the high-compression, thermally efficient engine that still bears his name with powdered coal in mind (a commodity that was both cheap and readily available in nineteenth century Germany). Via kerosene, he later arrived at the use of crude oil fractions, the marked variability of which later caused immense practical difficulties in the initial commercialization of diesel engines.[8]

The modern oil industry had already begun in Titusville, Pennsylvania, in the summer of 1859 with a drilled extraction rate of 30 barrels a day—equivalent to a daily income of $600.[9] By 1888, tsarist Russia had allowed Western European entrepreneurs to open up oil fields in Baku (in modern Azerbaijan) with a productive capacity of 50,000 barrels a day. On January 10, 1901, the Spindletop well in Texas began gushing, reaching a maximum flow of 62,000 barrels a day. Immediately prior to the outbreak of World War I, the main oil-producing countries could achieve outputs in excess of 51 million tons per year, or 1 million barrels a day. In 1902, 20,000 vehicles drove along American roads, but this number had reached over a million by 1912. These changes were highly welcome to oil producers, including (at least, until its forced breakup in 1911) the Standard Oil conglomerate. Kerosene intended for lighting domestic homes had been a major use of oil but, from the turn of the century, electricity had increasingly become both available and preferable (or fashionable).

Greatly aiding the convergence of interest between vehicle manufacturing and the oil industry was the dominance of U.S. domestic production of oil: In 1913, the oil produced in the United States amounted to in excess of 60% of the world's total (Figure 1.1). The proximity within national boundaries of production lines for automobiles (in Detroit) and oil refining capacities firmly cast the die for the remainder of the twentieth century and led to the emergence of oil exploration, extraction, and processing, and the related petrochemical industry, as the dominant features of the interlinked global energy and industrial feedstock markets.

Nevertheless, Henry Ford continued his interest in alternative fuels, sponsoring conferences on the industrial uses of agricultural mass products (grain, soybeans, etc.) in 1935–1937; the Model A was often equipped with an adjustable carburetor designed to allow gasoline, alcohol, or a mixture of the two—an early example of what today is termed the flexible fuel vehicle (FFV).

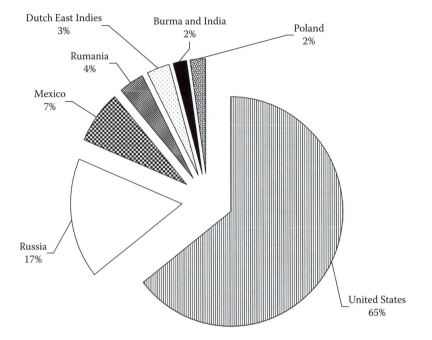

FIGURE 1.1 Geographical breakdown of world oil production in 1913. (Data from Tugendhat, C., and Hamilton, A. 1975. *Oil: The Biggest Business.* London: Eyre Methuen.)

1.3 OIL SUPPLY AND OIL PRICE IN THE TWENTIETH CENTURY: NECESSITY AND ALTERNATIVE FUEL PROGRAMS

Many commentators take the oil crisis of 1973, following the Yom Kippur War, as the pivotal moment that catalyzed the interest in and sustained the development of biofuels. The twentieth century was marked by wars from beginning to end, and one or more of these earlier conflicts were crucial drivers in severely shaking confidence in the assumed continuity of oil supplies for transportation fuels.

Even a cursory glance at Figure 1.1 will show how disadvantaged were the German, Austro-Hungarian, and Ottoman empires in comparison with the Allied powers in World War I, especially after the entry of the United States in 1917. With only minor eastern European oil fields beyond the vagaries of naval blockade and interception, the ingenuity of the German chemical industry was severely stretched by the effort to substitute imports (including fuel oils) by innovations with synthetic, ersatz products. Since then and throughout the twentieth and early twenty-first centuries, any state entering into global or regional wars has faced the same strategic imperatives: how to ensure continued oil supplies and how (if possible) to control access to them. Oil refineries and storage tanks are to be targeted, sea lanes interdicted, and, if possible, foreign oil fields secured by invasion.

In the past 90 years, wars and economic depressions often demanded attempts to substitute ethanol for gasoline. In the 1920s and 1930s, several countries (Argentina, Australia, Cuba, Japan, New Zealand, the Philippines, South Africa, and Sweden)

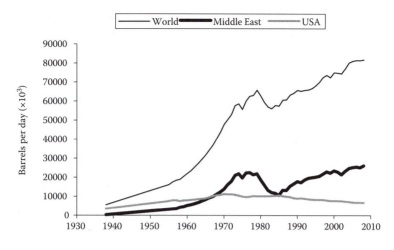

FIGURE 1.2　Oil production. (Data from BP statistical reviews.)

used ethanol blends in gasoline; alcohol-fueled vehicles became predominant in Germany during World War II, and by 1944 the U.S. Army had developed a nascent biomass-derived alcohol industry.[10] Such programs, however, were mostly of a contingency (or emergency) nature, highly subsidized, and, once oil began flowing in increasingly large amounts after 1945, generally abandoned.

In the decade immediately preceding 1973, the United States had lost its dominance of world oil production (Figure 1.2). Other major players were expanding (e.g., the Middle East reached 30% of world oil production) and new producers were appearing—notably, Africa (Libya, Algeria, and Nigeria), which accounted for 13% of world oil.[11] Allowing for inflation, world oil prices slowly *decreased* throughout the 1960s (Figure 1.3); at the time, this was perceived as a natural response to increasing oil production, especially with relative newcomers such as Libya and Nigeria contributing significantly.

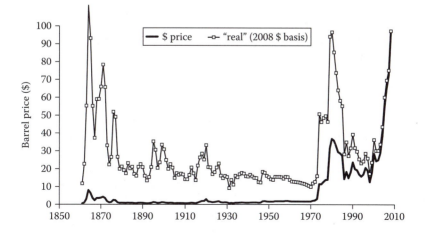

FIGURE 1.3　Historical oil price, yearly averages. (Data from BP statistical reviews.)

Global production after World War II followed an exponential rate of increase. However, political changes and a growing cooperation between oil-producing states in the Organization of Petroleum Exporting Countries (OPEC) and the Organization of Arab Petroleum Exporting Countries (OAPEC) led to new agreements that reversed the real oil price erosion and restricted oil extraction. In Libya, average production was reduced from a peak of 3.6 million barrels per day before June 1970 to approximately 2.2 million barrels per day in 1972 and early 1973; the Kuwaiti government enforced a ceiling of 3 million barrels per day in early 1972 from the previous 3.8 million barrels per day.[9]

When war did eventually break out in October 1973, OAPEC proposed with immediate effect to cut back output by 5% with a further 5% each month until a settlement in accord with United Nations (UN) resolutions was effected. In addition, the gulf states of OPEC, together with Iran, imposed unilateral price increases of up to 100%. The immediate effect on world oil prices was severe (Figure 1.3). More importantly, although prices decreased from the initial peaks in 1973 and 1974, prices began a second wave of rapid increase in 1979, after the Iranian revolution, to reach a new maximum in 1981.

From more than $50 a barrel in 1981, prices then confounded industry analysts again, despite the subsequent conflict between Iran and Iraq, and crashed down to $20 by the late 1980s. However, for over a decade real oil prices had been continuously threefold (or greater) than those paid in 1970. Although not reaching the real prices recorded in the 1860s during the American Civil War (when industrialization was a new phenomenon for most of the world), the oil price inflation between 1973 and 1981 represented a markedly different scenario from any experienced during the twentieth century—in dollar or real terms—despite world wars and major economic depressions (Figure 1.3).

Across the industrially developed states of the Organization for Economic Cooperation and Development (OECD—the United States, Japan, Germany, France, United Kingdom, Italy, and Canada), although the real price of imported crude oil had decreased between 1960 and 1973 by an average of 1% per annum, the inflation-adjusted price increased by 24.5% per annum between 1973 and 1980. The result was that the oil crisis soon developed into a deep economic crisis even in those economically and technically advanced OECD nations.[11] Because gasoline prices were buffered by the (frequently high) taxes included in the at-pump prices in the OECD countries, gasoline prices to motorists increased by only two- or threefold between 1970 and 1980 while crude oil prices rose over eightfold. In contrast, industrial and domestic oil prices increased approximately fivefold.[12]

The future for oil supplies to net oil importers was highly problematic, but purely for geopolitical treasons. Although known oil reserves in 1973 amounted to 8.8 × 10[10] tons, over 55% was in the Middle East, mostly in OAPEC countries (Figure 1.4). In the days of the cold war, the Soviet Union (USSR), Eastern Europe, and China accounted for only 16.3% of world oil production but were net exporters of both crude oil and oil products, whereas the United States was a net importer of both (Figure 1.5). In the United States, oil represented 47% of total primary energy consumption; other OECD countries were even more dependent on oil: 64% in Western Europe and 80% in Japan.[13]

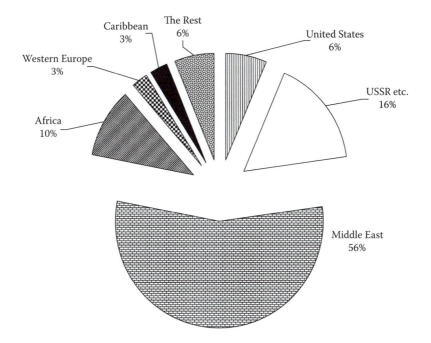

FIGURE 1.4 Known oil reserves at end of 1973. (Data from *BP Statistical Review of the World Oil Industry, 1973.* 1974. London: British Petroleum Co. Ltd.)

The developed economies of the OECD countries responded to the oil price shocks of the 1970s by becoming more oil efficient. Although total OECD gross domestic product (GDP) increased by 19% between 1973 and 1980, total oil imports fell by 14%, and the oil used to produce each unit of GDP fell by 20%. To offset the reduced use of oil, however, coal and especially nuclear energy source utilization

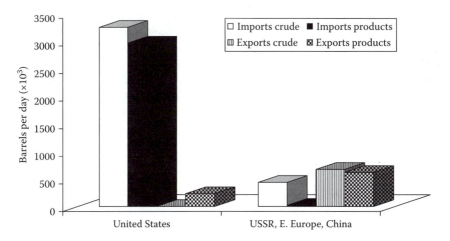

FIGURE 1.5 Oil imports/exports 1973. (Data from *BP Statistical Review of the World Oil Industry, 1973.* 1974. London: British Petroleum Co. Ltd.)

increased greatly.[14] Energy conservation became a priority ("energy demand management" measures) and technologies for the improved efficiency of energy use were much developed, advertised, and retrofitted to both domestic and industrial premises. "Fuel switching" was much less obvious in the strategies adopted by OECD countries. The substitution of gasoline for road transport by alcohol, liquefied gas, and other alternative fuels was widely advocated; however, by 1980, Canada was unique in having adopted a comprehensive policy covering all aspects of oil use and providing oil reduction targets as well as financial incentives.

1.4   CASE STUDY 1: BRAZIL AND SUGARCANE ETHANOL

For an emerging economy like Brazil's, the economic dislocation posed by sustained oil price rises was potentially catastrophic. In November 1973, Brazil relied on imports for over 80% of the country's oil consumption; in the course of the following year, the total import bill rose from $6.2 billion to $12.6 billion, and the trade balance collapsed (Figure 1.6). For the preceding decade, the Brazilian economy had enjoyed high growth rates (Figure 1.6). Industrialization had proceeded well and the inflation rate had reached its lowest level since the 1950s.[15] The Brazilian government opted against economic stagnation; rather, it aimed to pay for the higher oil bills by achieving continued growth. To meet the challenges of energy costs, the Second National Development Plan (1975–1979) decreed the rapid expansion of indigenous energy infrastructure (hydroelectricity) as well as nuclear power and alcohol production as a major means of import substituting for gasoline.

Cane sugar was the key substrate and input for Brazil's national fuel alcohol program. Sucrose production from sugarcane (*Saccharum* sp.) in Brazil has a long history from its days as a colony of Portugal. Brazil had become the world's leading sugar supplier by the early seventeenth century, but sugar production was based initially on slave labor and remained (by the twentieth century) inefficient. This represented a potential for rapid growth after 1975 because large monoculture plantations had been long established in the coastal regions of the northeast and southeast of the country. Expansion of cultivated land was greatly encouraged for the modern export crops—sugarcane, cotton, rice, corn, soybeans, and wheat—at the expense of the more traditional crops, including manioc, bananas, peanuts, and coffee. Sugarcane cultivation increased by 143% between 1970 and 1989, when expressed as land use, but production increased by 229% as Brazil's historically low use of fertilizer began to be rectified.[15]

Brazil is also the southernmost producer of rum as an alcoholic spirit, but *cachaça* is the oldest and most widely consumed national spirit beverage with a yearly production in 2009 in excess of 1.5 billion L. The primary fermentation for cachaça uses juice extracted from sugarcane, and large industrial plants had been established after the end of World War II; a variety of yeasts had been developed, suitable for continuous or discontinuous fermentations, the former reusing and recycling the yeast cells.[16]

Before distillation, the fermentation is allowed to become quiescent (as in all traditional potable alcohol processes), the yeast cells settling and then being removed (along with other residual solids) in technologically more advanced facilities by centrifugation.

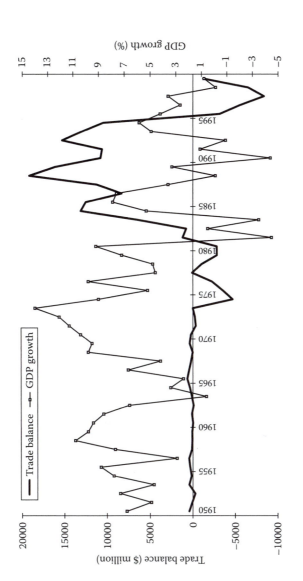

FIGURE 1.6 Brazil's economy, 1950–1999. (Data from Baer, W. 2001. *The Brazilian Economy: Growth and Development*, 5th ed. Westport, CT: Praeger Publishers.)

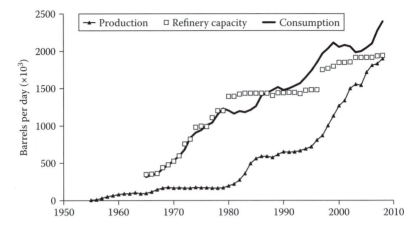

FIGURE 1.7 The Brazilian oil economy up to 2006. (Data from BP statistical reviews.)

Batch ("pot still") and continuous distillation are both used, and final alcohol concentrations are in the range of 38–48% (by volume). Predating the oil crises of the 1970s and 1980s, the first moves toward using cane sugar as a substrate for industrial ethanol production independent of beverages dated from 1930, when the Sugar and Alcohol Institute (Instituto do Açúcar e do Álcool, IAA) was set up; in 1931, a decree imposed the compulsory addition of 5% ethanol to gasoline, and the blending was increased to 10% in 1936. Four decades of experience had therefore been garnered in Brazil before fuel substitution became a priority on the political agenda.[17]

The final element in Brazil's developing strategy to produce "gasohol" was, ironically, petroleum itself. Brazil had produced oil at a low rate from at least 1955, but the offshore deposits discovered by the state-owned company PETROBRÁS were so large that, by 1998, domestic oil production equaled 69% of domestic consumption.[15] Production continued to increase (Figure 1.7), and by 2005 Brazil had become a significant global producer, accounting for 2.2% of world oil production. This was equivalent to that of the United Kingdom, considerably higher than either Malaysia or India (both 0.9%), and approaching half that of China (4.6%).[18] Indigenous refining capacity also increased during the 1970s and again after 1996 (Figure 1.7). The ability to produce alcohol as a fuel or (when mixed with gasoline) as a fuel additive became an ongoing feature of Brazilian economic life.

STEM TOPIC 1.2: THE ENERGY CONTENT OF ETHANOL

As a volatile chemical compound viewed as a gasoline substitute, pure ethanol has one major drawback: In comparison with the typical hydrocarbon components of refined oils, ethanol is more oxygenated and its combustion in oxygen generates less energy compared with either a pure hydrocarbon or a typical gasoline.

The enthalpies of combustion are indicated for the two comparable reactions:

$$C_2H_5OH + 3O_2 \rightarrow 2CO_2 + 3H_2O, \Delta H = -1,371 \text{ kJ/mol}$$

$$C_8H_{18} + 12\tfrac{1}{2}O_2 \rightarrow 8CO_2 + 9H_2O, \Delta H = -5,103 \text{ kJ/mol}$$

Ethanol is denser than octane: 785 g/L in comparison with 700 g/L for a typical octane-type hydrocarbon. The molecular masses of the two fuels are 46 and 114, respectively. Multiplying enthalpy of combustion by the molarity of the (assumed pure) fuel gives:

For ethanol, energy density = $1,371 \times (785/46) = 23,396$ kJ/L

For octane, energy density = $5,103 \times (700/114) = 31,334$ kJ/L

Ethanol, therefore, has only 75% of the energy density of a typical gasoline hydrocarbon, and this is not mitigated by the higher density of ethanol because liquid volumes are dispensed *volumetrically* and higher weights in fuel tanks represent higher loads in moving vehicles. Offsetting this, ethanol has a higher octane number (leading to higher engine efficiencies) and generates an increased volume of combustion products (gases) per energy unit burned; if these factors are optimized, the differential advantages of gasoline are significantly eroded.[19]

With slightly different input data, figures of ethanol's relative energy density of 70% of that of conventional gasoline are widely quoted. Other potential biofuels usually suffer from the same comparison; for example, methanol has only 50% of the energy density of gasoline.

The high miscibility of refined oil products allows the use of low-ethanol additions to standard gasoline (e.g., E10: 90% gasoline, 10% ethanol) that require no modifications to standard gasoline-burning vehicles. Dedicated ethanol-fueled cars were the initial favorite of the Brazilian Alcohol Program (PROÁLCOOL); sales of alcohol-powered vehicles reached 96% of total sales in 1980 and over 4 million such vehicles were estimated to be in the alcohol "fleet" by 1989 (Figure 1.8).[20] In retrospect, the choice of engines dedicated to ethanol as a fuel proved to be disastrous because the high initial market penetration was not maintained and sales of alcohol-powered vehicles had almost ceased by 1996.

The major reason for this reversal of fortune for ethanol-fueled vehicles was the collapse in oil prices during the late 1980s and 1990s; by 1998, the real price of crude oil was very similar to that before November 1973 (Figure 1.3). By 1986, ethanol production from sugarcane in Brazil had increased nearly 20-fold from a low and declining production level in 1983 (Figure 1.9). Faced with this mismatching of technologies and oil prices, the Brazilian government took drastic action[19]:

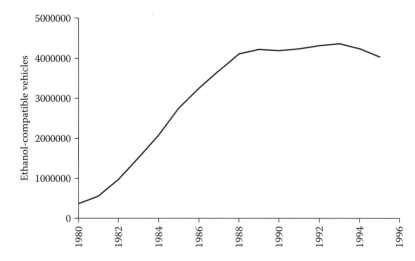

FIGURE 1.8 Ethanol-compatible vehicles in Brazil, 1980–1996. (Data from Melges de Andrade, A., Morata de Andrade, C. A., and Bodinaud, J. A. 1998. In *Biomass Energy: Data, Analysis and Trends,* 87. Paris: International Energy Agency, and ANFAVEA [Associação Nacional dos Fabricantes de Veículos Automotores]. http://www.anfavea.com.br.)

- In 1990, the IAA, the body through which governmental policy for ethanol production had been exercised, was abolished.
- In 1993, a law was passed that all gasoline sold in Brazil would have a minimum of 20% ethanol by volume.
- Prices of sugarcane and ethanol were deregulated and tariffs on sugar exports were abolished in 1997.

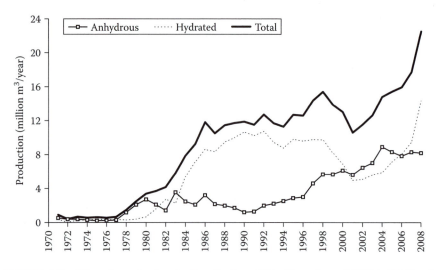

FIGURE 1.9 Ethanol production in Brazil after 1970. (Data from UNICA [União da Agroindústria Canvieira do São Paolo]. http://www.unica.com.br.)

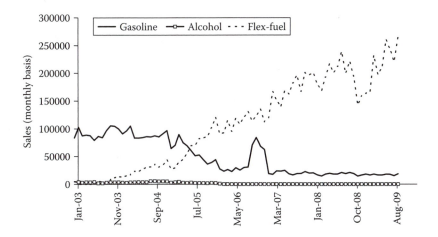

FIGURE 1.10   Sales of flexibly fueled vehicles in Brazil. (Data from ANFAVEA [Associação Nacional dos Fabricantes de Veículos Automotores]. http://www.anfavea.com.br.)

- In January 2006, the tax rate for gasoline was set to be 58% higher than that for hydrated ethanol (93% ethanol, 7% water), and tax rates were made advantageous for any blend of gasoline and anhydrous ethanol with ethanol content in excess of 13%.

In 2003, Brazilian automobile producers introduced true FFVs with engines capable of being powered by gasoline, 93% aqueous ethanol, or a blend of gasoline and anhydrous ethanol.[21] In 2004, FFVs sold in Brazil were 16% of the total market, but during 2005 their sales overtook those of conventional gasoline vehicles and have continued to dominate the automobile market, outselling conventional vehicles in 2009 by 14:1 (Figure 1.10). This turned out to be a highly prescient development as crude oil prices, which had been only slowly increasing during 2003 and early 2004, surged to new highs in 2005 (Figure 1.3).

Despite seasonal peaks and troughs, ethanol production has doubled since the 1980s (Figure 1.9). Domestic demand for ethanol-containing fuels became so great that the ethanol percentage was reduced from 25 to 20% in March 2006; this occurred despite the increased production of anhydrous ethanol for blending.[22] Brazil had evolved a competitive, consumer-led dual-fuel economy where motorists made rational choices based on the relative prices of gasoline, ethanol, and blends. Astute consumers have been observed to buy ethanol only when the pump price is 30% below gasoline blends; equal volumes of ethanol and gasoline are still, as noted before, divergent on their total energy (and, therefore, mileage) equivalents.

By 2005, Brazil had become the world's producer of ethanol with 37% of the total and exporting 15% of that indigenously produced ethanol. PETROBRÁS began the construction of a 1,000-mile pipeline from the rural interior of the state to the coast for export purposes.[20]

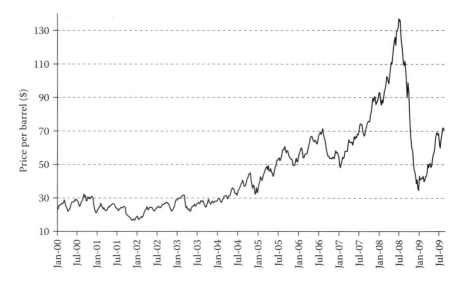

FIGURE 1.11   Crude oil price after 2000. (Data from U.S. Department of Energy, Energy Information Administration.)

In 2005, Brazilian authors published a survey that summarized many official statistics in Portuguese language publications; the following major impact factors were claimed for fuel ethanol production in Brazil[23]:

- After 1975, fuel ethanol substituted for 240 billion L of gasoline, equivalent to $56 billion in direct importation costs and $94 billion if costs of international debt servicing are included. After 2004, the severe increases in oil prices clearly acted to augment the benefits of oil substitution (Figure 1.11).
- The sugar/ethanol sector represented 3.5% of the gross national product and had a gross turnover of $12 billion, employed (directly and indirectly) 3.6 million people, and contributed $1.5 billion in taxation revenues. Approximately half of the total sugarcane grown in Brazil in 2003 was dedicated to ethanol production.
- In 2004, sugarcane production required 5.6 million ha and represented only 8.6% of the total harvested land. However, over 120 million of low-productivity pasture, natural pastures, and low-density savannas could potentially be dedicated to sugarcane production for ethanol, with a potential ethanol yield of more than 300 billion L per year.

Ethanol has continued to be a major exported commodity from Brazil: Between 1998 and 2005, exports of ethanol increased more than 17-fold, while sugar exports increased less than twofold even though price volatility was evident with both commodities (Figure 1.12). As a report for the International Bank for Reconstruction and Development and World Bank first published in October 2005 noted, average wages in the sugar-ethanol sector were higher than the mean for all sectors in Brazil.[24] As

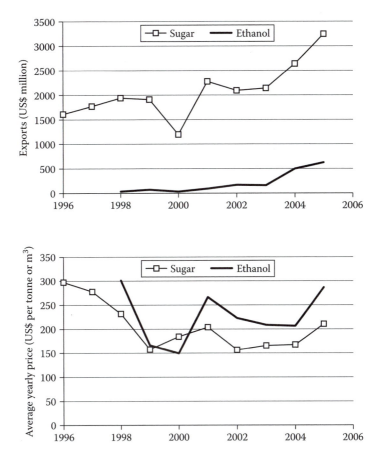

FIGURE 1.12 Exports of sugar and ethanol from Brazil, 1996–2005. (Data from UNICA (União da Agroindústria Canvieira do São Paolo). http://www.unica.com.br.)

a source of employment, sugarcane ethanol production directly employs more than 1 million people and is far more labor intensive than is the petrochemical industry: 152 times more jobs are estimated to have been created than would have been the case from an equivalent amount of petroleum products.[25]

Viewed from the perspectives of fermentation technology and biochemical engineering, ethanol production in Brazil improved markedly after 1975; fermentation productivity (expressed as cubic meters of ethanol per cubic meter of fermentation tank capacity volume per day) increased by 130% between 1975 and 2000.[23] This was due to continuous incremental developments and innovations; no reports of radically new fermentor designs in Brazil have been published (although very large fermentors, up to 2 million L in capacity, are used) and ethanol concentrations in batch fermentations are in the range of 6–12% (v/v).

The control of bacterial infection of fermentations has been of paramount importance, and selection of robust wild strains of the yeast *Saccharomyces cerevisiae* has systematized the traditional experience that wild strains frequently overgrow laboratory starter cultures.[26] The uses of flocculent yeast strains and continuous cultivation have also been technologies adopted in Brazil in response to the increased production of sugarcane ethanol.[27] Technical development of downstream technologies has been made in the largest Brazilian provider of distillation plants (Dedini S/A Indústrias de Base, www.dedini.com.br). Conventional (bubble cap trays), sieve trays, and azeotropic distillation methods/dehydration (cyclohexane, monoethylene glycol, and molecular sieving) processes operate at over 800 sites—up from 300 sites prior to 2000.[27]

On a longer term basis, genomic analysis of sugarcane promises to identify plant genes for programs to improve sugar plant growth and productivity by genetic engineering.[28,29]

1.5   CASE STUDY 2: STARCH-BASED ETHANOL IN THE UNITED STATES

If ethanol production in Brazil exemplified the extrapolation of a mature technology for sugar-based fermentation and subsequent distillation, the development of the second major ethanol fuel market—from corn in the United States—adopted a different approach to alcohol production, adaptation, and development that employed starchy seeds in the production of malt and grain spirits (bourbon, rye, whiskey, whisky, etc.). The biological difference from sugar-based ethanol fermentations lies in the carbon substrate (i.e., starch glucan polymers; see Figure 1.13).

Historically, seeds and grains have been partially germinated by brewers to generate the enzymes capable of depolymerizing "storage" polysaccharides. With whiskey, for example, barley (*Hordeum vulgare* L.) seeds are germinated and specialized cells in the seed produce hydrolytic enzymes for the degradation of polysaccharides, cell walls, and proteins. The malted barley can be used as a source of enzyme activities to break down the components of starch in cooked cereal grains (e.g., maize [*Zea mays* L.]) solubilized in sequential hot-water extractions, which are combined before the yeast cells are added but not sterilized so as to maintain the enzyme activities into the fermentation stage.[30,31]

Starch is usually a mixture of linear (amylose) and branched (amylopectin) polyglucans. For starch hydrolysis, the key enzyme is α-amylase, active on α-1,4 but not α-1,6 linkages (in amylopectin); consequently, amylose is broken down to maltose and maltotriose and (upon prolonged incubation) to free glucose and maltose. However, amylopectin is only reduced to a mixture of maltose, glucose, and oligosaccharides containing α-1,6 linked glucose residues, thus limiting the amount of fermentable sugars liberated (Figure 1.14). Cereal-based ethanol production plants use the same biochemical operations but replace malted grains with α-amylase and other polysaccharide-degrading enzymes added as purified products.

For much of the twentieth century, ethanol production as a feedstock in the formation of a large number of chemical intermediates and products was dominated in the United States by synthetic routes from ethylene as a product of the petrochemical industry, reaching 8.8×10^5 tonnes per year in 1970.[32] The oil price shocks of the early 1970s certainly focused attention on ethanol as an "extender" to gasoline but a mix of legislation and economic initiatives, starting in the 1970s, was required to

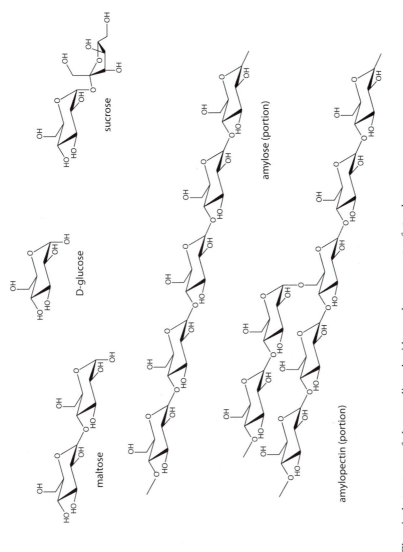

FIGURE 1.13 Chemical structures of glucose, disaccharides, and components of starch.

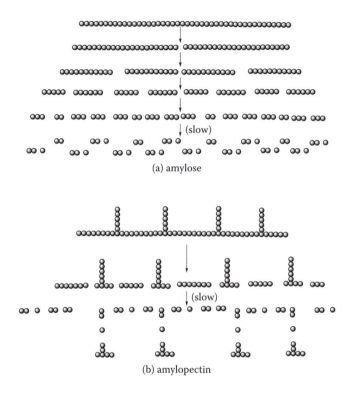

(a) amylose

(b) amylopectin

FIGURE 1.14   Prolonged degradation of starch components by amylase.

engender a large-scale bioprocessing industry; in particular, three federal environmental regulations were important[33,34]:

- The 1970 Clean Air Act (CAA, amended in 1977 and 1990) began the requirement for cleaner burning gasoline and (eventually) the mandatory inclusion of oxygenates (i.e., oxygen-rich additives).
- The 1988 Alternative Motor Fuels Act promoted the development of ethanol and other alternative fuels and alternative fuel vehicles (AFVs).
- The 1992 Energy Policy Act defined a broad range of alternative fuels but, more urgently, required that the federal vehicle fleet include an increasing number of AFVs and that they be powered by domestically produced alternative fuels.

Although ethanol was always a good candidate oxygenate for gasoline, the compound first approved by the Environmental Protection Agency was methyl tertiary butyl ether (MTBE), a petrochemical industry product. Use of MTBE increased until 1999, but reports then appeared of environmental pollution incidents caused by MTBE spillage; state bans on MTBE came into force during 2002, and its consumption began to decline (Figure 1.15).[35] Ethanol was established by then as a corn-derived, value-added product; when the tide turned against MTBE use,

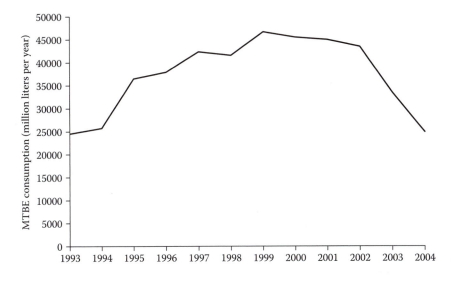

FIGURE 1.15   MTBE consumption in the United States. (Data from the U.S. Department of Energy, Energy Information Administration.)

ethanol production increased rapidly after showing little sustained growth for most of the 1990s (Figure 1.16).

In the 7 years after January 1999, the number of ethanol refineries in the United States nearly doubled and production capacity increased 2.5-fold (Figure 1.17).[36] By 2005, the United States had become the largest ethanol producer nation; Brazil and

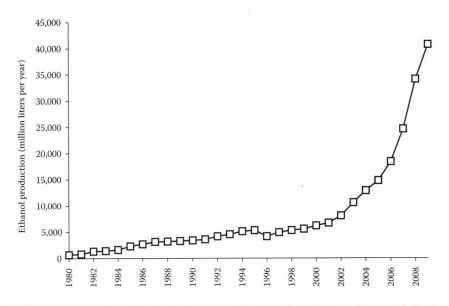

FIGURE 1.16   Ethanol production in the United States. (Data from the Renewable Fuels Association, including an estimated figure for 2008.)

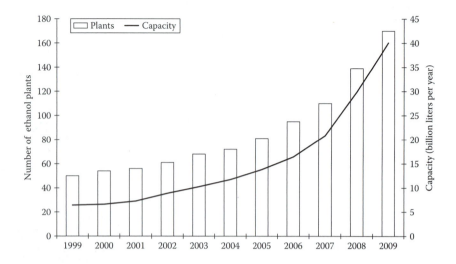

FIGURE 1.17 Growth of U.S. ethanol production, January 1999–January 2009. (Data from the Renewable Fuels Association.)

the United States accounted for 70% of global production and, apart from China, India, France, and Russia, no other nation accounted for >1% of the total ethanol produced. To further underline the perceived contribution of renewable fuels to national energy use, the 2005 Energy Policy Act created a renewable fuels standard that envisioned renewable fuel use increasing from 4 billion gallons per year in 2006 to 7.5 billion gallons per year in 2012. This implies a further expansion of ethanol production because of the dominant position of E85 (85% ethanol, 15% gasoline) vehicles in the AFV market place (Figure 1.18).

U.S. fuel ethanol production has been almost exclusively from corn, although sorghum (*Sorghum bicolor* L.), barley, wheat, cheese whey, and brewery waste have made small contributions. A detailed study of sugar sources for ethanol production concluded that only sugarcane molasses offered competitive feedstock and processing costs to established corn-based technologies (Figure 1.19), although annual capital cost investments could be comparable for corn, sugarcane and sugarbeet molasses, and juice as rival feedstocks.[37] Corn ethanol production developed from wet milling of corn; data compiled in the mid-1990s indicate that more than 70% of the large ethanol facilities then used wet milling.[34] Wet milling, schematized in Figure 1.20, produces four important liquid or solid by-products:

- corn steep liquor—a lactic acid bacterial fermentation product (starting from ca. 5% of the total dry weight of the grain extracted with warm water) with uses in the fermentation industry as a nitrogen source
- corn oil (with industrial and domestic markets)
- corn gluten feed (a low-value animal feed)
- corn gluten meal (a higher value, high-protein animal feed)

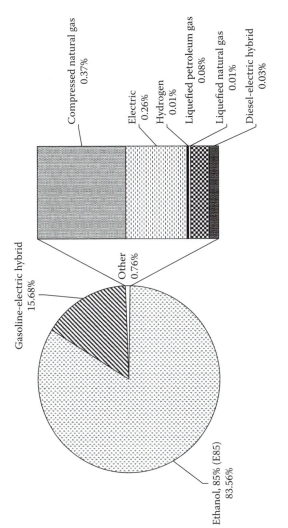

FIGURE 1.18 Alternative- and hybrid-fuel vehicles. (Data from U.S. Department of Energy, Energy Information Administration.)

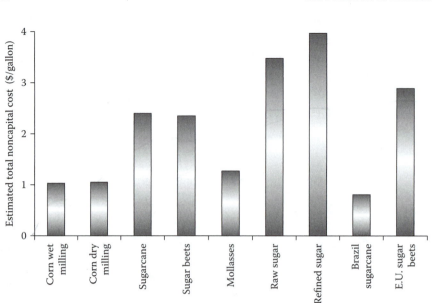

FIGURE 1.19 Estimated ethanol production costs. (Data from Shapouri, H., Salassi, M., and Fairbanks, J. N. 2006. *The Economic Feasibility of Ethanol Production from Sugar in the United States*. Washington, D.C.: U.S. Department of Agriculture.)

Together with the possibility of collecting CO_2 from the fermentation step as a saleable commodity, this multiplicity of products gave wet milling flexibility in times of variable input and output prices, although requiring a higher initial capital investment.[34] Other sources of flexibility and variation in the wet milling procedure arise at the starch processing stage. Although α-amylase is used to liquefy the starch, saccharification (using glucoamylase) can be differently controlled—at one extreme, producing a high-glucose, low-solids substrate for fermentation and, at the other, producing a low-glucose concentration that is continually replenished during the fermentation by the ongoing activity of the glucoamylase in the broth.

In contrast to wet milling, dry milling produces only CO_2 and distiller's dried grains with solubles (DDGS) as by-products; however, it has become the favored approach for corn ethanol production because of lower start-up costs.[38] Dry milling should conserve more of the nutrients for yeast growth in the fermentation step—in particular, nitrogenous inputs (free amino acids, peptides, and protein), inorganic and organic phosphates, and some other inorganic ions (potassium, sodium, magnesium, etc.)—but this has little, if any, impact on overall process economics (Table 1.1). Figure 1.21 is a representative example of the complete bioprocess for ethanol and DDGS.[39]

Unlike Brazilian sucrose-based ethanol, corn-derived ethanol has been technology driven (especially in the field of enzymes and improved yeast strains with

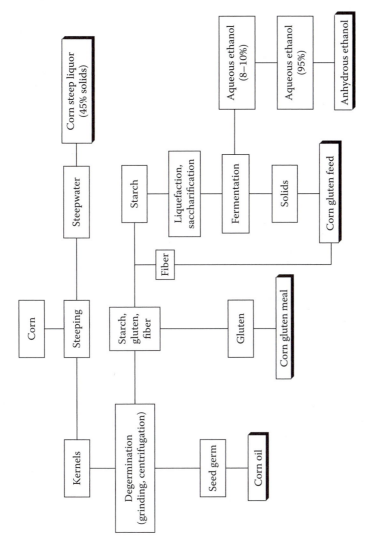

FIGURE 1.20 Outline of corn wet milling and ethanol production.

TABLE 1.1

Estimated Ethanol Production Costs[a] from Corn Milling Technologies

	Wet Milling 2005[b]	Dry Milling 2005[b]	Dry Milling Model Data[c] (40 mgy)[d]	Dry Milling Model Data[c] (80 mgy)
Feedstock costs	0.712	0.707	0.877	0.840
By-product credits	0.411	0.223	0.309	0.286
Net feedstock costs	0.301	0.484	0.568	0.554
Operating costs:				
Electricity	0.061	0.058	0.040	0.039
Fuels	0.145	0.211	0.160	0.112
Waste management	0.031	0.007		
Water	0.015	0.003	0.004	0.004
Enzymes	0.067	0.042	0.040	0.040
Yeast	0.031	0.005	0.010	0.006
Chemicals	0.055	0.036	0.010	0.013
Denaturant	0.059	0.054	0.072	0.062
Maintenance	0.088	0.062	0.020	0.052
Labor	0.093	0.058	0.010	0.020
Administrative	0.055	0.042		
Other	0.000	0.004		
Total variable cost	1.002	1.065	0.934	0.902

[a] Dollars per gallon.

[b] Shapouri, H., Salassi, M., and Fairbanks, J. N. 2006. *The economic feasibility of ethanol production from sugar in the United States.* Washington, D.C.: U.S. Department of Agriculture.

[c] Dale, R. T., and Tyner, W. E. 2006. Purdue University staff paper #06-04, West Lafayette, IN.

[d] mgy = million gallons per year output.

high ethanol tolerance) and may be or become capable of yielding up to 23% by volume of ethanol in batch fermentations within 60 hours.[39,40] The typical commercially available enzymes used to liberate the sugars present in starches and their properties are summarized in Table 1.2. Innovations in biocatalysts and fermentation engineering for corn ethanol facilities are covered at greater length in Chapter 3.

The availability of enzyme preparations with higher activities for starch degradation to maltooligosaccharides and glucose has been complemented by the use of proteases that can degrade corn kernel proteins to liberate amino acids and peptides to accelerate the early growth of yeast cells in the fermentor. Protein digestion also aids the access of amylases to difficult-to-digest starch residues, thus enhancing overall process efficiency and starch-to-ethanol conversion. Table 1.3 contains indicative patents and patent applications awarded or filed for corn ethanol technologies.

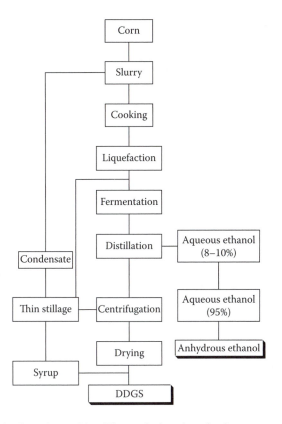

FIGURE 1.21   Outline of corn dry milling and ethanol production.

STEM TOPIC 1.3: PREDICTING ETHANOL YIELD FROM CORN STARCH CONTENT

From a dry grinding process, an ethanol content of 2.8 gallons/bushel has been quoted, leading to another frequent quote of 450 gallons/acre (at 160 bushels/acre, U.S. average). How much of the theoretical maximum yield does this represent?

With a bushel of corn grains weighing in at 56 pounds and containing 15% water by weight and starch at 70% (dry weight basis), 1 bushel = 56×0.454 kg, giving $56 \times 0.454 \times 0.7$ kg starch = 17.8 kg. Starch is polymerized glucose with a unit glucan molecular mass of $180 - 18$ (i.e., 162 g/mol). The bushel contains 17,800/162, or 109.9, mol glucose equivalent.

The ethanol formed is 2.8×3.785, or 10.6, L; this is equivalent to $10,600 \times 0.795$ g ethanol and $10,600 \times (0.795/46)$, or 183, mol. Because [1 mol] of glucose can form 2 mol of ethanol if the biochemical pathways function at maximum efficiency, the actual yield is $183 \times 100/(109.9 \times 2)$—that is, 83%.

Not all the starch in the milled grain is solubilized and hydrolyzed, and not all the available glucose may be fermented to ethanol. Grain companies are breeding corn varieties with higher starch content, biotech companies are inserting amylase genes into corn, and analytical chemists seek easy, on-site instrumentation with which to measure and predict ethanol yields from deliveries of grain.

As the multiplicity of U.S. corn ethanol producers has increased, the relative contributions of large and small facilities have shifted: In 1996, Archer Daniels Midland accounted for more than 70% of the total ethanol production, but by late 2006 this had fallen to just 21%.[34,41] The multiplication of ethanol producers has continued and, although the largest three producers still (in mid-2009) account for 28%, over 90

TABLE 1.2

Typical Enzymes for Fuel Ethanol Production from Cereals

Manufacturer and Enzyme	Type of Enzyme	Use	Properties
Novozymes			
BAN® (Thermozyme®)	α-Amylase	Starch liquefaction	
Termamyl®	α-Amylase	Starch liquefaction	Heat stable
Liquozyme®	α-Amylase	Starch liquefaction	Heat stable, broad Ph tolerance, low calcium requirement
Viscozyme®	α-Amylase	Starch liquefaction	Optimized for wheat, barley, and rye mashes
Spirizyme®	Glucoamylase	Saccharification	Heat stable
Alcalase®	Protease	Fermentation	
Genencor International			
Spezyme®	α-Amylase	Starch liquefaction	Heat stable
Distillase®	Glucoamylase	Saccharification	
G-Zyme®	Glucoamylase	Saccharification	Also added before saccharification
STARGEN™	α-Amylase + glucoamylase	Saccharification and fermentation	Enzyme blend
FERMGEN™	Protease	Fermentation	
Fermenzyme®	Glucoamylase + protease	Saccharification and fermentation	Enzyme blend
Alltech[a]			
Allcoholase I™	α-Amylase	Starch liquefaction	
High T™	α-Amylase	Starch liquefaction	Heat stable
Allcoholase II™	Glucoamylase	Saccharification	

[a] Now marketed by Enzyme Technology.

TABLE 1.3

Patent Applications and Patents Awarded for Corn Ethanol Technologies

Date of Filing or Award	Title	Applicant/Assignee	Patent/Application
January 21, 2003	Process for producing ethanol	ZeaChem, Inc. (Golden, CO)	US 6509180
October 1, 2004	Improved process for the preparation of ethanol from cereals	Etea S.r.l. (Savigliano, Italy)	EP 1 536 016 A!
March 9, 2004	Method for producing fermentation-based products from high oil corn	Renesson LLC (Bannockburn, IL)	US 6703227
May 25, 2004	Fermentation-based products from corn and method	Renesson LLC (Bannockburn, IL)	US 6740508
October 20, 2005	Methods and systems for producing ethanol using raw starch and fractionation	Bruin and Associates, Inc. (Sioux Falls, SD)	US 2005/0233030 A1
October 27, 2005	Continuous process for producing ethanol using raw starch	Bruin and Associates, Inc. (Sioux Falls, SD)	US 2005/0239181 A1
February 23, 2006	Removal of fiber from grain products including distiller's dried grains with solubles	R. Srinivasan and V. Singh	US 2006/0040024 A1
May 2, 2006	Heterologous expression of an *Aspergillus kawachi* acid-stable alpha amylase	Genencor International (Palo Alto, CA)	US 7037704
July 11, 2006	Process for producing ethanol from corn dry milling	ZeaChem, Inc. (Golden, CO)	US 7074603
August 1, 2006	Method for producing fermentation-based products from corn	Renesson LLC (Deerfield, IL)	US 7083954
September 5, 2006	Alcohol production using sonication	UltraForce Technology LLC (Ames, IA)	US 7101691
November 16, 2006	Hybrid enzymes	Novozymes A/S (Bagsvaerd, Denmark)	US 2006/0257984 A1
February 11, 2008	Fermentation process for the preparation of ethanol form a corn fraction	Renessen LLC (MO) and Cargill, Inc.(MN)	US2009/0017164 A1
January 8, 2009	Dry-mill ethanol plant extraction enhancement	IntegroExtraction, Inc. (Worthing, SD)	US2009/0181153 A1
April 28, 2009	Process for producing ethanol and for energy recovery	Stanley Consultantrs, Inc.(Muscadine, IA)	US7524418

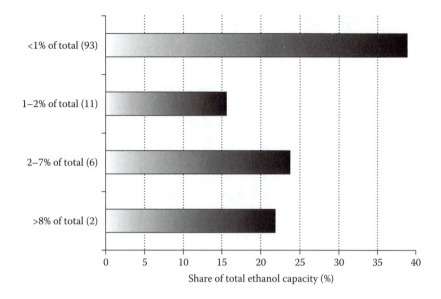

FIGURE 1.22 Contributors to U.S. fuel ethanol production. (Data from the Renewable Fuels Association.)

much smaller companies contribute nearly 40% of the total capacity (Figure 1.22). The ethanol production capacity in the United States amounted to 13 billion gallons per year, of which 11.5 billion gallons per year were operational and a further 1.5 billion gallons per year were under construction.[41]

Outside North America, construction of the first ethanol facility in Europe to utilize corn as the feedstock commenced in May 2006 in France; AB Bioenergy France began ethanol production in 2007. The parent company, Abengoa Bioenergy (www.abengoa.com), operates three facilities in Spain producing 5,550 million L of ethanol a year from wheat and barley grain. Also, a plant in Norrköping, Sweden, began producing 50 million L of ethanol annually from wheat in 2001; the product is blended with conventional imported gasoline at up to 5% by volume. These and other representative ethanol facilities in Europe and Asia are listed in Table 1.4.

TABLE 1.4

Industrial Sites for Ethanol Production from Cereals and Sugar in Europe and Asia

Manufacturer	Location	Substrate	Capacity (million liters per year)
Lantmännen Agroetanol	Norrkoping, Sweden	Wheat grain	210
CropEnergies Group	Germany, Belgium, France	Wheat grain, sugar beet	700
Abengoa Bioenergia	Spain, France	Wheat and barley grain	800
HSB Agro Industries Ltd.	Ringus, Rajasthan, India	Rice and sorghum grain	11
Jilin Fuel Ethanol Co. Ltd.	Jilin, China	Corn	900

1.6 THERMODYNAMIC AND ENVIRONMENTAL ASPECTS OF ETHANOL AS A BIOFUEL

Arguments in favor of ethanol and other biofuels tend to mix four key but logically distinct points:

1. Fossil energy resources are finite and may become seriously depleted during the twenty-first century.
2. Biofuels avoid dependency on oil imports.
3. Biofuels augment sustainable development across the globe, offering new industries to developing nations.
4. Biofuels can reduce global CO_2 emissions.

Economists and socioeconomists tend to concentrate on the first two cases; biological scientists are far more comfortable with the latter two arguments. Much attention has been given, therefore, to the questions of whether ethanol—as the first mass-produced biofuel—can help solve either the global energy supply problem or the ecological crisis of global warming caused by greenhouse gas emissions.

1.6.1 NET ENERGY BALANCE

Even before the end of the 1970s, serious scientific discussion had commenced regarding the thermodynamics of fuel ethanol production from biological sources. The first in-depth study that compiled energy expenditures for ethanol production from sugarcane, sorghum, and cassava under Brazilian conditions concluded that the net energy balance—that is, the ratio between the energy produced (as ethanol) to the total energy consumed (in growing the plants, processing the harvest, and all the various stages of the ethanol production process)—was positive.[42]

Thermodynamically, energy can be neither created nor destroyed, but can be converted from one form into another. Energy is unavoidable for the production of ethanol, and the summation could be made across the entire production process including human labor that must be replenished (i.e., energy goes into supplying the food for the work force), machinery, fuel, seeds, irrigation, and all agrochemicals (many of which are derived from petrochemicals). These can be described as direct or mechanical, as distinct from essential environmental energy inputs such as sunlight; climatic factors, including seasonal hours of sunlight, precipitation, average temperature, etc., certainly affect crop yield and will therefore have an impact on the economics of ethanol production as the cost of corn grain varies from year to year. Ethanol has a measurable energy yield in internal combustion engines; the energy inputs can be (although not unavoidably) fossil fuel consumptions (such as diesel, coal, natural gas, etc.). The ratio between energy in the ethanol produced and the energy consumed in the conversion debits is the net energy balance (NEB).

Viewed from the broadest perspective, four distinct cases can be distinguished with far-reaching implications for macroeconomics and energy policy:

1. Ethanol production has a net yield of energy and has no absolute dependency on nonrenewable energy inputs.
2. Ethanol production has a net yield of energy but is dependent on nonrenewable energy inputs.
3. Ethanol production has a net loss of energy (a yield of <1) but has no dependency on nonrenewable energy inputs.
4. Ethanol production has a net loss of energy as well as being dependent on nonrenewable energy inputs.

The most acceptable conclusion economically and environmentally is case 1. Although case 2 would pose problems to the short-term adoption of biofuels, however, eventually, there would be no insuperable problem if alternative energy sources were to become more widely available or to support the required agricultural base with agrochemicals derived exclusively from petrochemicals or other nonrenewable sources (especially if fermentation-derived bioprocesses supplanted purely synthetic routes). Moreover, cases 3 and 4 are not in themselves intrinsically unacceptable: As was pointed out nearly 20 years ago, power stations burning oil or coal all show a net energy loss but that is not used as an argument to shut them down.[43]

Domestic and industrial devices that are electrically driven require power stations as an interface for energy conversion, with the associated economic and thermodynamic costs. Filling gasoline tanks with sugar or cereal grains cannot fuel automobiles; converting these biological substrates to ethanol can—but again at economic and energetic costs. Cases 3 and, especially, 4 are equivalent to twentieth century wartime scenarios where fuel was produced synthetically; case 3 can be one of sustainable, long-term development *if* nonfossil energy sources such as hydroelectric, geothermal, wind, wave, and solar energy can eventually be used to meet the net energy costs inherent in ethanol production from agricultural inputs.

To carry on this important thread of discussion, when analysts tacitly assume that all (or most) of the energy required for biofuel production is inevitably derived from fossil fuels, this is equivalent to what the International Energy Agency (IEA) terms its reference scenario for future energy demands (i.e., that, up until 2030, coal, gas, and oil will be required for 75% of the world's power generation, with nuclear, hydro, biomass, and other renewables accounting for the remaining 25%).[44] The IEA's alternative scenario for 2030 postulates fossil fuel usage for power generation decreasing to 65% of the total. In the future, therefore, the *fossil fuel* inputs and requirements for corn- and biomass-derived ethanol may significantly decrease, thus affecting the quantitative assessment of energy balance calculations for biofuels. Realistically, in 2009 major ethanol producers (e.g., China and the United States) used electrical power generated effectively entirely from fossil fuels.

A review of a broad portfolio of renewable energy technologies in the 1970s that covered methane production from algae and livestock waste as well as sugarcane, cassava, timber, and straw concluded that *only* sugarcane-derived ethanol was capable of yielding a net energy gain.[45] Data from early Brazilian and U.S. studies of sugarcane-derived ethanol are summarized in Table 1.5. It was evident even in these early studies that the net energy balance was highly influenced by the utilization of bagasse (i.e., the combustion of the waste product to supply steam generation

TABLE 1.5

Energy Balances for Ethanol Production from Sugarcane

Publication	Energy Conversion Debits[a]			Energy Outputs[a]		
	Sugarcane Growing	Ethanol Production	Total	Ethanol	Residue	Total
Da Silva et al. (1978)						
Only ethanol considered	4,138	10,814	14,952	18,747		18,747
All bagasse converted to steam	4,138	10,814	14,952	18,747	17,500	36,247
Hopkinson and Day (1980)						
Only ethanol considered	8,500	10,800	19,300	18,400		18,400
All bagasse converted to steam	8,500	10,800	19,300	18,400	17,200	35,600

[a] 106 kcal per hectare per year.

and other energy requirements on site or, although not considered at that time) as a saleable power commodity to the broader community. The energy balance was, however, much lower than the comparative quoted case of gasoline produced from oil extracted in the Gulf of Mexico oil fields (i.e., 6:1).[46]

Subsequently, the energetics of corn-derived and sucrose-derived ethanol processes have been the subjects of scrutiny and increasingly elaborate data acquisition and modeling exercises. Economic analyses agree that the production of ethanol from sugarcane is energetically favorable, with net energy gains that rival or equal those for the production of gasoline from crude oil in subterranean deposits.[47,48]

For corn-derived fuel alcohol the energy position is highly ambiguous. One major area of disagreement is in the inclusion of coproduct credits. As we have seen in Figures 1.20 and 1.21, corn-derived ethanol has coproducts that have economic value (as saleable commodities) but also affect the energy balance if their energy contents can be assessed. In a 1978 study of corn-derived ethanol, an overall energy balance could be computed to be small (approximately 1.01); even attaining the energy breakeven point required the efficient utilization of wastes for on-site energy generation and the sale and agricultural use of by-products counted as energy outputs (see Figures 1.21 and 1.22).[49]

A crucial parameter is therefore the ratio between the energy retrieved from ethanol and coproducts and the *fossil fuel* energy inputs involved in its production—a quotient that has been termed the "fossil energy ratio" or "bioenergy ratio"[50]:

$$(E_{net} + E_{coproduct})/(E_A + E_B + E_C + E_D)$$

where
E_A is the fossil fuel energy required for the production of the plant inputs
E_B is the fossil fuel energy required during crop growth and harvesting

TABLE 1.6

Energy Balances for Ethanol Production from Corn

HHV Energy Balance[a]	LHV Energy Balance[b]	Ref.
	0.96	51
1.25		52
	0.74	53
	0.92	54
1.38		55
1.20		56
	1.32	57
	1.33	58
	0.74	59
0.79		61
1.27		60

[a] High heat value for ethanol (83,961 Btu per gallon).
[b] Low heat value for ethanol (76,000 Btu per gallon).

E_C is the fossil fuel required during transport of the harvested crop

E_D is that expended during the conversion process

By 2002, conflicting estimates had appeared for corn ethanol, but the published accounts displayed very different methodological approaches and quantitative assumptions. Even as crucial a parameter as the energy content of ethanol used in the computations varied in the range of 74,680–84,100 Btu per gallon. The choice of lower or higher heat values (measured by reference to water or steam, respectively) is also required for the energy inputs; if the choice is used consistently in the calculations, there will be no effect on the net energy gained or lost. Ten of the studies were discussed in a 2002 review.[51–60] Table 1.6 collects the data, adding to the comparison an extra publication from 2001.[61]

The obvious spread of energy balance values has generated a sustained argument over the choice of relevant input parameters but a more pertinent conclusion is that none of the balances greatly exceeded 1.00 (the arithmetical average is 1.08). It could have been concluded before 2002, therefore, that only a highly efficient production process (including the maximal utilization of whatever by-products were generated) could deliver a net energy gain, although changes in agricultural practices, higher crop yields, increased fermentation productivity, etc. might all be anticipated to contribute to a gradual trend of piecemeal improvement.

Since 2002, the polemics have continued, with NEBs calculated with various assumptions:

- Patzek (2004): 0.92 (with no by-product energy credits allowed)[62]
- Pimentel and Patzek (2005): 0.85 (after adjustment for the energy in by-products)[63]

- Dias de Oliveira, Vaughan, and Rykiel (2005): 1.10 (compared with 3.7 for sugar-derived ethanol in Brazil)[64]
- Farrell et al. (2006): 1.20[65]
- Hill et al. (2006): 1.25, although this depended mostly on counting the energy represented by the DDGS by-product (Figure 1.21) as an output.[66]

STEM TOPIC 1.4: NET ENERGY BALANCE COMPUTATIONS

The ERG Biofuel Analysis Meta-Model (EBAMM) was developed by students and faculty of the Energy and Resources Group and Richard & Rhoda Goldman School of Public Policy at UC Berkeley and originally published in 2006. The model can be downloaded from http://rael.berkeley.edu/ebamm/.

The originators of the EBAMM model disliked NEB (or net energy ratio, NER) computations because of the then variable practice in either subtracting coproduct energies from input totals or including them in energy outputs. They considered that the net energy value (NEV)—the arithmetic difference between total energy inputs and outputs per liter—was a more reliable measure. How are NEVs calculated?

Energy inputs are divided into agricultural (the fossil energy inputs for growing the crop) and biorefinery (converting corn into ethanol). Each total is the result of multiple inputs that include fertilizers and pesticides, human labor, the means of shipping farm produce to the ethanol facility, and even the issue of wastewater cleanup from the process. For the 2006 status of corn ethanol production, agricultural energy totals 5.5 MJ/L ethanol and ethanol conversion energy at 15 MJ/L; the output energy in ethanol is 21 MJ/L and coproducts amount to 4.1 MJ/L. Therefore:

$$NEV = (21 + 4.1) - (5.5 + 15) = 4.6 \text{ MJ/L}$$

$$NER = (21 + 4.1)/(5.5 + 15) = 1.22$$

The single largest energy input for the agricultural phase is nitrogenous fertilizer (45% of the total); direct fossil fuel uses have less impact: for example, 7% (farm gasoline) and 14% (farm diesel). In contrast, 91% of the total ethanol conversion input energy is represented by fossil fuel usage—in this case, calculated as coal and natural gas for power generation, etc.

Explore the EBAMM spreadsheets to define how sensitive the final NEV and NER ratios are to different assumptions made for the input parameters!

The continued failure to demonstrate overall net energy balances much in excess of 1.00 is striking. Put into the perspective of the historical measures of energy balance from the oil industry, corn-derived ethanol remains relatively inefficient.[66–68] Assuming a net energy balance of 1.2, the notional expenditure of five units of ethanol would be required to generate each net unit, or five barrels of oil would be required to generate the energy equivalent of six barrels of oil.

In Europe, where ethanol production is much less developed than in North America and Brazil, attention has focused on wheat starch and beet sugar. Studies of energy balances with wheat have consistently shown negative energy balances (averaging 0.74) while values for sugarbeet range from 0.71 to 1.36 and average 1.02. However, projections made by the International Energy Agency suggest that both feedstocks will show positive energy balances as combined fertilizer and pesticide usage drops and biotechnological conversion efficiencies improve.[69] Portuguese analysts defined yet another term, the energy renewability efficiency (ErenEf), as

$$ErenEf = (FEC - E_{in, fossil, prim}) \times 100/FEC$$

where *FEC* is the fuel energy content and $E_{in, fossil, prim}$ represents the unavoidable fossil energy input required for the production of the biofuel.[70]

Under French conditions, sugarbeet-derived ethanol was renewable even without taking into account any coproduct credits, but was maximal with an allocation of the energy inputs based on the mass of the ethanol and coproducts (*ErenEf* = 37%); wheat grain-derived ethanol was entirely dependent on this allocation calculated for the DDGS (*ErenEf* = 48%). These values were equivalent to positive net energy balances of 1.59 (sugarbeet) and 1.92 (wheat).

When a wider group of example processes from the global fuel ethanol industry was surveyed, widely divergent bioenergy ratios were computed (Figure 1.23). The most obvious inconsistency is that between two molasses-derived ethanol processes. One (in India) was derived from the case of a distillery fully integrated into a sugar mill, where excess low-pressure steam was used for ethanol distillation,

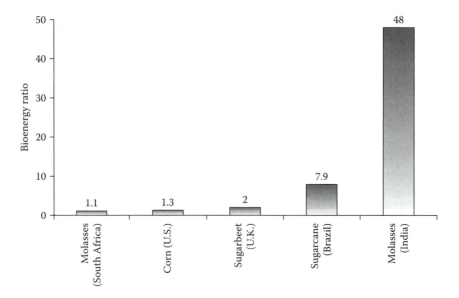

FIGURE 1.23 Ratio of ethanol energy content to fossil fuel energy input for ethanol production systems. (Data from von Blottnitz, H., and Curran, M. A. 2007. *Journal of Cleaner Production* 15:607.)

while a South African example was for a distillery distant from sugar mills and reliant on coal and grid electricity for its energy needs.[48]

Leaving behind scientific textbooks and journals, an article in the October 2007 issue of *National Geographic* magazine accepted the marginal energy gain in corn ethanol production and the disappointingly small reduction of total CO_2 emissions in the total production/use cycle, but noted the much clearer benefits of sugarcane ethanol.[71] Two quotes encapsulate the increasingly ambiguous reactions that are increasingly evident as the public debate over biofuels spreads away from scientific interest groups to embrace environmentalist and other lobbies:

> It's easy to lose faith in biofuels if corn ethanol is all you know.
> If alcohol is a "clean" fuel, the process of making it is very dirty...

It is worth repeating the energy balance in the petroleum industry because the critics of biofuels often do so: In the 1940s, 1 barrel of oil energy expended extracted 100 barrels of oil. The ratio has decreased with item and now approximates "only" 1 barrel expended for every 10 generated, although there are massive energy uses in subsequently refining the crude oil.[72]

1.6.2 EFFECTS ON EMISSIONS OF GREENHOUSE GASES AND OTHER POLLUTANTS

The three major greenhouse gases (GHGs) are carbon dioxide (CO_2), methane (CH_4), and nitrous oxide (N_2O); CH_4 and N_2O are 20–300 times more powerful than CO_2 in global warming impact; fortunately, they are much less abundant—even in the twenty-first century atmosphere. Public perception centers on CO_2 and the related concept of carbon footprint—a measurement of all greenhouse gases produced by individuals or businesses or societies, *which has units of tons, tonnes, or kilograms of CO_2 equivalent (CO_2e)*. Every production process also has its carbon footprint.

STEM TOPIC 1.5: CO_2 EMISSIONS FROM CORN AND SUGARCANE ETHANOL PRODUCTION

Returning to the EBAMM model, CO_2e values can be computed for agricultural and ethanol conversion phases and compared with total emissions from gasoline manufacture. How do the two first-generation ethanol processes compare?

The values are first expressed per liter ethanol before conversion on an energy basis (21.2 MJ/L ethanol):

Corn, agricultural phase = 780 g CO_2e/L,
ethanol conversion = 1,353 g CO_2e/L

Sugarcane, agricultural phase = 858 g CO_2e/L,
ethanol conversion = 1,254 g CO_2e/L

Both processes can be awarded coproduct credits of 525 g CO_2e/L, giving net GHG emissions of

Corn = 780 + 1353 − 525, or 1608, g CO_2e/L (i.e., 75.8 g CO_2e/MJ)

Sugarcane = 858 + 1254 − 525, or 1586, g CO_2e/L (i.e., 74.8 g CO_2e/MJ)

In practice, small additions are then made to account for distribution (e.g., by tanker) of the ethanol from the production facility: 1.4 g CO_2e/MJ for each. With a GHG total for conventional gasoline of 94 g CO_2e/MJ, the GHG reductions then become

Corn = 94 − (75.8 + 1.4) = 16.8 g CO_2e/MJ

Sugarcane = 94 − (74.8 + 1.4) = 17.8 g CO_2e/MJ

The two processes are quite comparable, offering 18 or 19% ameliorations in GHG emissions. In addition, the agricultural phases of both processes are dominated by GHG emissions associated with nitrogen fertilizer production and the release of N_2O from the fertilizers on the fields after application: 60.6% for corn and 61.3% for sugarcane.

Beyond the primary production phase, direct comparisons of GHG emissions resulting from the combustion of anhydrous ethanol, ethanol–gasoline blends, and gasoline are straightforward to perform but are poor indicators of the overall consequences of substituting ethanol for gasoline. Instead, from the early 1990s, full fuel cycle analyses were performed to estimate GHG emissions (projected beyond 2000) *throughout* the production process for ethanol to gauge the direct and indirect consequences of gasoline replacement, including:

- changes in land use and the replacement of native species by energy crops
- agricultural practices and the potential for utilizing agricultural wastes
- materials manufacture and the construction of facilities
- ethanol production
- transportation of feedstocks and ethanol
- fuel usage per mile driven

Non-carbon-based fuels (i.e., electric vehicles powered using electricity generated by nuclear and solar options) were superior, while corn ethanol showed no net advantage (Figure 1.24).[73] More recent estimates place corn ethanol production as giving modest reductions in GHG emissions, in the range of 12–14%.[65,66] The GREET model of the Argonne National Laboratory indicates steeper reductions (Figure 1.25).[50] An unavoidable complication in any such calculations, however, is that they assume that harvested crops are from land already under cultivation; converting "native" ecosystems to biofuel production would inevitably alter the natural carbon balance and reduce the potential savings in GHG emissions.[66]

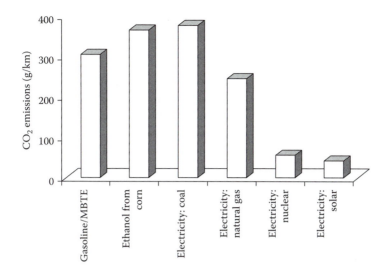

FIGURE 1.24 Total fuel cycle carbon dioxide emissions. (Data from Bergeron, P. 1996. In *Handbook on Bioethanol: Production and Utilization,* ed. Wyman, C. E., chap. 5. London: Taylor & Francis.)

In some corn ethanol production scenarios, the use of fossil fuels such as coal to power ethanol distillation plants and diesel to enable the long-distance transportation of corn can actually *increase* total GHG emissions.[65,74,75] If the net energy balance is <1, the total ethanol production cycle would also increase net GHG emissions unless renewable energy sources rapidly supplanted fossil fuels in power generation.[62,75,76]

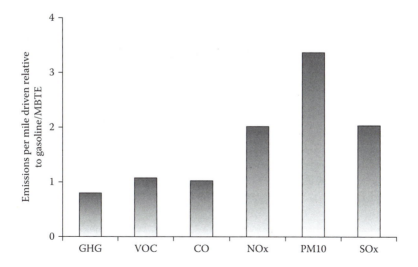

FIGURE 1.25 Total fuel cycle emissions. (Data from Wu, M., Wu, Y., and Wang, M. 2005. Mobility chains analysis of technologies for passenger cars and light-duty vehicles fueled with biofuels: Application of the GREET model to the role of biomass in America's energy future [RBAEF] project. Center for Transportation Research, Argonne National Laboratory.)

For other priority pollutants, ethanol production and the use of E85 as a transportation fuel have ambiguous effects: increased nitric oxides (NO_x), particulate matter (PM_{10}), carbon monoxide (CO), volatile organic carbon (VOC), and sulfur oxides (SO_x) when measured in toto in both urban and rural locations (Figure 1.25).

For production sources other than sugarcane, therefore, the actual reductions in GHG emissions resulting from the adoption of biofuels may be considerably less than anticipated. One estimate for corn ethanol concluded that, by itself, fuel alcohol use in the United States would struggle to reduce transportation-dependent emissions by more than 10%.[77] This study used the following data and arguments:

- For U.S. gasoline consumption of 460×10^9 L per year, corn ethanol replaced 0.8% of this while using 1% of the total cropland.
- To replace 10% of the gasoline consumption, corn ethanol would need to be produced on 12% of the total U.S. cropland.
- Corn ethanol only avoids 25% of the CO_2 emissions of the substituted gasoline emissions when the fossil-fuel-dependent energy consumed to grow and process the corn is accounted for.
- Offsetting 10% of the CO_2 emissions from gasoline consumption would require a fourfold higher production of corn ethanol (i.e., from 48% of U.S. cropland).
- "A challenge for providing transportation fuels is to be able to substitute a joule of energy in harvestable biomass for a joule of primary energy in fossil fuel, and to do this without significant fossil energy consumption. If a joule-for-joule consumption can be achieved, dedicated biomass plantations may be able to displace some 15% of the CO_2 emissions expected from all uses of fossil fuels globally by 2030. For biofuels to replace more than about 15% of fossil-fuel CO_2 emissions in 2030 will require a more rapid improvement in the energy efficiency of the global economy than is apparent from past trends in efficiency."

This pessimistic—or realistic—view of the quantitative impacts of biofuels on GHG scenarios in the twenty-first century was strongly reinforced by a mid-2007 study of *alternatives* to biofuels—specifically, conserving and (if possible) extending natural forests, savannahs, and grasslands. Compared to even the best biomass-to-biofuel case, forestation of an equivalent area of land was calculated to sequester at least twice the amount of CO_2 over a 30-year period than the emissions avoided by biofuel use.[78] As the authors concluded:

If the prime object of policy on biofuels is mitigation of carbon dioxide-driven global warming, policy makers may be better advised in the short term (30 years or so) to focus on increasing the efficiency of fossil fuel use, to conserve the existing forests and savannahs, and to restore natural forest and grassland habitats on cropland that is not needed for food.

1.7 SUMMARY

Both corn- and sugarcane-derived ethanols have—unarguably—emerged as industrial realities: sugar-derived ethanol as a major transportation fuel in Brazil and corn-derived ethanol as a niche but growing market in the United States. As ethanol production processes, both were simply extrapolated from preexisting technologies and, to varying extents, they reflect the limitations of add-on manufacturing strategies.

Together and separately they have been criticized as unsuitable for sustainable alternatives to gasoline in the absence of tax incentives. Highly integrated production processes, with maximum use of coproducts and high degrees of energy efficiency, clearly do (sugarcane ethanol) and may (corn ethanol) result in more chemical energy being present in the output than energy in fossil fuels required for their production.

Similarly, claims for the potential of fuel ethanol in mitigating GHG emissions are highly dependent on the energy requirements for its production.

Although technically proven as a group of production processes, the scope of ethanol production from food crops (primarily sugar and corn) is limited by agricultural and geographical factors. If biofuels are to be globally significant alternatives to petroleum products as transportation fuels, other and additional sources of plant (or biological) material must be mobilized.

REFERENCES

1. McGovern, P. E. 2003. *Ancient wines: The search for the origins of viniculture,* chap. 3. Princeton, NJ: Princeton University Press.
2. McGovern, P. E. 2003. *Ancient wines: The search for the origins of viniculture,* chap. 5. Princeton, NJ: Princeton University Press.
3. McGovern, P. E. 2003. *Ancient wines: The search for the origins of viniculture,* chap. 1. Princeton, NJ: Princeton University Press.
4. McGovern, P. E. et al. 2004. Fermented beverages of pre- and proto-historic China. *Proceedings of the National Academy of Sciences USA* 101:17593.
5. Rose, A. H. 1977. History and scientific basis of alcoholic beverage production. In *Economic microbiology,* vol. 1. *Alcoholic beverages,* ed. Rose, A. H., chap. 1. London: Academic Press.
6. Lyons, T. P., and Rose, A. H. 1977. Whisky. In *Economic microbiology,* vol. 1. *Alcoholic beverages,* ed. Rose, A. H., chap. 10. London: Academic Press.
7. McCarthy, T. 2001. The coming wonder? Foresight and early concerns about the automobile. *Environmental History* 6:46.
8. Thomas, D. E. 1987. *Diesel: Technology and society in industrial Germany,* chap. 5. Tuscaloosa: University of Alabama Press.
9. Tugendhat, C., and Hamilton, A. 1975. *Oil: The biggest business,* chaps. 1–7. London: Eyre Methuen.
10. Cheremisinoff, N. P. 1979. *Gasohol for energy production,* chap. 6. Ann Arbor, MI: Ann Arbor Science Publishers, Inc.
11. International Energy Agency. 1995. *Oil, gas and coal supply outlook,* chap. 2. Paris: International Energy Agency.
12. International Energy Agency. 1982. *World energy outlook,* chaps. 1–2. Paris: International Energy Agency.

13. British Petroleum Co. Ltd. 1974. *BP statistical review of the world oil industry 1973.* London: British Petroleum Co. Ltd.
14. International Energy Agency. 1982. *World energy outlook,* chap. 3. Paris: International Energy Agency.
15. Baer, W. 2001. *The Brazilian economy: Growth and development,* 5th ed., chap. 6. Westport, CT: Praeger Publishers.
16. Faria, J. B., Franco, D. W., and Piggott, J. R. 2004. The quality challenge: Cachaça for export in the 21st century. In *Distilled spirits: Tradition and innovation,* ed. Bryce, J. H. and Stewart, G. G., chap. 30. Nottingham, England: Nottingham University Press.
17. Braunbeck, O., Macedo, I., and Cortez, L. A. B. 2005. Modernizing cane sugar production to enhance the biomass base in Brazil. In *Bioenergy—Realizing the potential,* ed. Silveira, S., chap. 6. Amsterdam: Elsevier.
18. British Petroleum Co. Ltd. 2006. *BP Statistical review of world energy.* London: British Petroleum Co. Ltd.
19. Martines-Filho, J., Burnquist, H. L., and Vian, C. E. F. 2006. Bioenergy and the rise of sugarcane-based ethanol in Brazil. *Choices* 21(2): 91.
20. Melges de Andrade, A., Morata de Andrade, C. A., and Bodinaud, J. A. 1998. Biomass energy use in Latin America: focus on Brazil. In *Biomass energy: Data, analysis and trends,* 87. Paris: International Energy Agency.
21. ANFAVEA (Associação Nacional dos Fabricantes de Veículos Automotores). http://www.anfavea.com.br.
22. UNICA (União da Agroindústria Canvieira do São Paolo). http://www.unica.com.br.
23. Moreira, J. R., Noguiera, L. A. H., and Parente, V. 2005. Biofuels for transport, development, and climate change: lessons from Brazil. In *Growing in the greenhouse: Protecting the climate by putting development first,* ed. Bradley, R. and Baumert, K. A., chap. 3. Washington, D.C.: World Resources Institute.
24. Kojima, M., and Johnson, T. 2005. *Potential for biofuels for transport in developing countries,* chap. 2. Washington, D.C.: International Bank for Reconstruction and Development/World Bank.
25. Pessoa, A. et al. 2005. Perspectives on bioenergy and biotechnology in Brazil. *Applied Biochemistry and Biotechnology* 121–124:59.
26. Amorim, H. V., Basso, L. C., and Lopes, M. L. 2004. Evolution of ethanol fermentation in Brazil. In *Distilled spirits: Tradition and innovation,* ed. Bryce, J. H. and Stewart, G. G., chap. 20. Nottingham, England: Nottingham University Press.
27. Zanin, G. M. et al. 2000. Brazilian bioethanol program. *Applied Biochemistry and Biotechnology* 84–86:1147.
28. Telles, G. P. et al. 2001. Bioinformatics of the sugarcane EST project. *Genetics and Molecular Biology* 24:8.
29. Vincentz, M. et al. 2004. Evaluation of monocot and eudicot divergence using the sugarcane transcriptome. *Plant Physiology* 134:951.
30. MacLeod, A. M. 1979. The physiology of malting. In *Brewing science,* vol. 1, ed. Pollock, J. R. A., chap. 2. London: Academic Press.
31. Bryce, J. H. et al. 2004. Optimizing the fermentability of wort in a distillery—the role of limit dextrinase. In *Distilled spirits: Tradition and innovation,* ed. Bryce, J. H. and Stewart, G. G., chap. 11. Nottingham, England: Nottingham University Press.
32. Waddams, A. L. 1978. *Chemicals from petroleum. An introductory survey,* 4th ed., chap. 8. London: John Murray.
33. Putsche, V., and Sandor, D. 1996. Strategic, economic, and environmental issues for transportation fuels. In *Handbook on bioethanol: Production and utilization,* ed. Wyman, C. E., chap. 2. London: Taylor & Francis.

34. Elander, R. T., and Putsche, V. L. 1996. Ethanol from corn: Technology and economics. In *Handbook on bioethanol: Production and utilization,* ed. Wyman, C. E., chap. 15. London: Taylor & Francis.

35. Energy Information Administration, U.S. Department of Energy, http://www.eia.doe.gov.

36. *Ethanol Industry Outlook.* 2006. Washington, D.C.: Renewable Fuels Association.

37. Shapouri, H., Salassi, M., and Fairbanks, J. N. 2006. *The economic feasibility of ethanol production from sugar in the United States.* Washington, D.C.: U.S. Department of Agriculture.

38. Dale, R. T., and Tyner, W. E. 2006. Economic and technical analysis of ethanol dry milling: Model description. Purdue University staff paper #06-04, West Lafayette, IN.

39. Lyons, T. P. 2004. Fuel ethanol: The current global situation. Lessons for the beverage industry. In *Distilled spirits: Tradition and innovation,* ed. Bryce, J. H. and Stewart, G. G., chap. 19. Nottingham, England: Nottingham University Press.

40. Schubert, C. 2006. Can biofuels finally take center stage? *Nature Biotechnology* 24:777.

41. Renewable Fuels Association, Washington, D.C., http://www.ethanolrfa.org.

42. da Silva, J. G. et al. 1978. Energy balance for ethyl alcohol production from crops. *Science* 201:903.

43. Sama, D. A. 1989. Net energy balance, ethanol from biomass. In *Biomass handbook,* ed. Kitani, O. and Hall, C. W., chap. 3.2.10. New York: Gordon and Breach Science Publishers.

44. International Energy Agency. 2006. *World energy outlook,* annex A. Paris: International Energy Agency.

45. Lewis, C. 1976. Energy relationships of fuel from biomass. *Process Biochemistry* 11 (part 9, November): 29.

46. Hopkinson, C. S., and Day, J. W. 1980. Net energy analysis of alcohol production from sugarcane. *Science* 207:302.

47. Geller, H. S. 1985. Ethanol fuel from sugar cane in Brazil. *Annual Review of Energy* 10:135.

48. von Blottnitz, H., and Curran, M. A. 2007. A review of assessments conducted on bioethanol as a transportation fuel from a net energy, greenhouse gas, and environmental life-cycle perspective. *Journal of Cleaner Production* 15:607.

49. Chambers, R. S. et al. 1979. Gasohol: Does it or doesn't it produce positive net energy? *Science* 206:789.

50. Wu, M., Wu, Y., and Wang, M. 2005. Mobility chains analysis of technologies for passenger cars and light-duty vehicles fueled with biofuels: Application of the GREET model to the role of biomass in America's energy future (RBAEF) project. Center for Transportation Research, Argonne National Laboratory.

51. Ho, S. P. 1989. Global warming impact of ethanol versus gasoline. Presented at 1989 National Conference, "Clean Air Issues and America's Motor Fuel Business." Washington, D.C., October.

52. Marland, G., and Turhollow, A. F. 1990. CO_2 emissions from the production and combustion of fuel ethanol from corn. Oak Ridge National Laboratory, Oak Ridge, TN, Environmental Sciences Division, no. 3301, U.S. Department of Energy, May 1990.

53. Pimentel, D. 1991. Ethanol fuels: Energy security, economics, and the environment. *Journal of Agricultural and Environmental Ethics* 4:1.

54. Keeney, D. R., and DeLuca, T. H. 1992. Biomass as an energy source for the Midwestern U.S. *American Journal of Alternative Agriculture* 7:137.

55. Lorenz, D., and Morris, D. 1995. *How much energy does it take to make a gallon of ethanol?* Washington, D.C.: Institute for Local Self-Reliance.

56. Shapouri, H., Duffield, J. A., and Graboski, M. S. 1995. *Estimating the Net Energy Balance of Corn Ethanol.* Washington, D.C.: U.S. Department of Agriculture, Economic Research Service, AER-721.

57. Henderson, S. 2000. *Assessment of net emissions of greenhouse gases from ethanol–gasoline blends in Southern Ontario.* Agriculture and Agri-Food Canada.

58. Wang, M., Saricks, C., and Santini, D. 1999. *Effects of fuel ethanol use on fuel-cycle energy and greenhouse gas emissions.* U.S. Department of Energy, Argonne National Laboratory, Center for Transportation Research, Argonne, IL.

59. Pimentel, D. 2001. The limits of biomass utilization. In *Encyclopedia of physical science and technology,* 3rd ed., 159–171. New York: Academic Press.

60. Shapouri, H., Duffield, J. A., and Wang, M. 2002. The energy balance of corn ethanol: An update. Washington, D.C.: Agricultural economic report no. 813, U.S. Department of Agriculture, Economic Research Service, Office of the Chief Economist, Office of Energy Policy and New Uses.

61. Berthiaume, R., Bouchard, C., and Rosen, M. A. 2001. Exergetic evaluation of the renewability of a biofuel. *Exergy International Journal* 1:256.

62. Patzek, T. W. 2004. Thermodynamics of the corn–ethanol biofuel cycle. *Critical Reviews in Plant Science* 23:519.

63. Pimentel, D., and Patzek, T. W. 2005. Ethanol production using corn, switchgrass, and wood; biodiesel production using soybean and sunflower. *Natural Resource Research* 14:65.

64. Dias de Oliveira, M. E., Vaughan, B. E., and Rykiel, E. J. 2005. Ethanol as a fuel: Carbon dioxide balances, and ecological footprint. *Bioscience* 55:593.

65. Farrell, A. E. et al. 2006. Ethanol can contribute to energy and environmental goals. *Science* 311:506.

66. Cleveland, C. J., Hall, C. A. S., and Herendeen, R. A. 2006. Letter. *Science* 312:1746.

67. Hill, J. et al. 2006. Environmental, economic, and energetic costs and benefits of biodiesel and ethanol biofuels. *Proceedings of the National Academy of Sciences USA* 103:11206.

68. Kaufmann, R. K. 2006. Letter. *Science* 312:1747.

69. Henke, J. M., Klepper, G., and Schmitz, N. 2005. Tax exemption for biofuels in Germany: is bio-ethanol really an option for climate policy? *Energy* 30:2617.

70. Malça, J., and Friere, F. 2006. Renewability and life-cycle energy efficiency of bioethanol and bio-ethyl tertiary butyl ether (bioETBE): Assessing the implications of allocation. *Energy* 31:3362.

71. Bourne, J. K. 2007. Green dreams. Making fuel from crops could be good for the planet—after a breakthrough or two. *National Geographic* October: 38–59.

72. Perez-Blanco, H. 2009. *The dynamics of energy. Supply, conversion, and utilization,* chap. 6. Boca Raton, FL: CRC Press.

73. Bergeron, P. 1996. Environmental impacts of bioethanol. In *Handbook on bioethanol: Production and utilization,* ed. Wyman, C. E., chap. 5. London: Taylor & Francis.

74. Wang, M., Wu, M., and Huo, H. 2007. Life-cycle energy and greenhouse gas emission impacts of different corn ethanol plant types. *Environmental Research Letter* 2:024001.

75. Patzek, T. W. et al. 2005. Ethanol from corn: Clean renewable fuel for the future, or drain on our resources and pockets? *Environmental Development and Sustainability* 7:319.

76. Pimentel, D., Patzek, T., and Cecil, G. 2007. Ethanol production: Energy, economic, and environmental losses. *Reviews of Environmental Contamination and Toxicology* 189:25.

77. Kheshgi, H. S., Prince, R. C., and Marland, G. 2000. The potential of biomass fuels in the context of global climate change: Focus on transportation fuels. *Annual Reviews of Energy and the Environment* 25:199.

78. Righelato, R., and Spracklen, D. V. 2007. Carbon mitigation by biofuels or by saving and restoring forests? *Science* 317:902.

2 Cellulosic Ethanol as a "Second-Generation" Biofuel

2.1 INTRODUCTION

Ethanol bioproduced from major food crops is insufficient to meet global demand for liquid fuels. Plant biomass is a massive resource, but with this comes the inevitable processing and logistic problems of structural polymers—the "lignocellulose complex" of plant cell wall cellulose, hemicelluloses, and lignin.

How can these materials be converted to soluble sugars to support ethanologenic fermentations? This is where industrial chemistry meets enzymology and molecular biology to devise successful procedures to transform woody trees, shrubs, grasses, and plant material remaining after food crop harvesting into a range of organic molecules that can be taken up and metabolized by microbes.

At the heart of "green" bioprocessing lies a complex group of enzymes—cellulases, hemicellulases, and lignin-degrading peroxidases—that have evolved in countless organisms that feed off living and decaying plant material. How have these biological resources generated a modern biotech industry of enzyme manufacture? What do we know about the molecular control of the production of these enzymes and how do we use this knowledge in assembling a processing industry for lignocellulosic biomass?

2.2 BIOETHANOL AND CELLULOSIC ETHANOL: THE RISE OF BIOMASS-BASED BIOFUELS

There was a definite golden age of biofuels, datable to 1996–2006, when ethanol production from corn and sugarcane was rapidly expanded (Figures 1.9, 1.16, and 1.17 in Chapter 1). In July 2006, the journal *Nature Biotechnology* published in a single issue a cluster of commentaries and articles on biofuels as well as a two-page editorial that directed its scientific readers to consult an article ("Ethanol Frenzy") in *Bloomberg Markets* that emphasized the resurgence of interest from financial institutions in biofuels as a commercial sector for investment.[1,2] Those years also saw massive international, national, and local publicizing of biofuels as a major technical fix to atmospheric CO_2 levels and global warming. Beginning in 2006, however, a trend of media coverage against biofuels commenced. Journalists cited a wide spectrum of problems, but especially the impacts of biofuels production on food prices and the collapse of environmental credibility as bioenergy crop plantations cleared rain forests in Southeast Asia.

Fortunately, there was an immediate answer at hand. Right at the beginning of the decade of intense development of food crop biofuels, a landmark publication for alternative fuels had appeared in 1996, edited by Charles E. Wyman of the National Renewable Energy Laboratory (Golden, Colorado).[3] That single-volume, encyclopedic compilation summarized scientific, technological, and economic data and information on biomass-derived ethanol ("bioethanol"), contrasted to and conceptually separated from ethanol biologically produced from cereal grains and sugarcane juice.

The opportunities for such a potentially massive production base as woody plants were enormous and highly attractive for both commercial and environmental reasons. Lignocellulose quickly became a preferred option as a low-cost, uncontroversial substrate for fuel ethanol production. This has broadly spread around the globe as, beyond Europe and North America, the competing desires for rural biofuels industries and doubts about the green credentials of biofuels have repeatedly surfaced, to be submerged in a rhetoric of "renewables," "recyclables," and "sustainables."

Semantically, however, the restricted use of the "bio" epithet was questionable and has simply not been widely or consistently followed in trade or scientific journals. All biological production routes for ethanol—whether from sugarcane, cornstarch, cellulose, or lignocellulose—share important features and are converging on technical solutions to the efficient utilization of feedstocks. For example, sugarcane-derived ethanol facilities have for many years used cane sugar waste (bagasse, the aboveground portions of sugarcane plants) for on-site combustion and steam generation; if such facilities begin to exploit the sugars in lignocellulosic components present in sugarcane for ethanol production, do we term the product more "bio" or fully "bioethanol"?

In this book, ethanol production routes—chemical or biotechnological—will be clearly differentiated but the product is, in all cases, *ethanol*. "Cellulosic ethanol" will be used to denote fuel ethanol prepared from plant biomass sources to differentiate the product from ethanol biomanufactured from either starch or sucrose.

2.3 STRUCTURAL AND INDUSTRIAL CHEMISTRY OF CELLULOSIC BIOMASS

2.3.1 CELLULOSE, HEMICELLULOSES, AND LIGNIN

Sucrose functions in plants as a highly water-soluble and easily transported product of carbon fixation in leaves (but can also accumulate in storage organs such as sugar beets), and starch is mainly a storage polymer in, for example, cereal grains; however, cellulose is essentially a *structural* polymer in plants (Figure 2.1). This statement masks the sheer quantity of cellulosic material on Earth: Cellulose is the most abundant global polymer; its synthesis rate is estimated at 10^{10}–10^{11} tonnes per year.

Cellulose is highly insoluble, organized into crystalline macroscopic fibers, mixed with other polysaccharides (e.g., hemicelluloses), and protected from enzymic attack in native woods by the physical presence of lignin (Figure 2.2). Lignins are polyphenolic polymers generated by enzyme-catalyzed free radical reactions from phenylpropanoid alcohols. Unlike nucleic acids and proteins, they have no informational content but are neither inert to enzyme-catalyzed degradation nor incapable

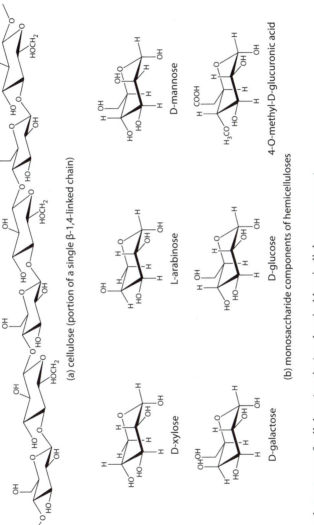

FIGURE 2.1 Chemical structures of cellulose (portion) and typical hemicellulose components.

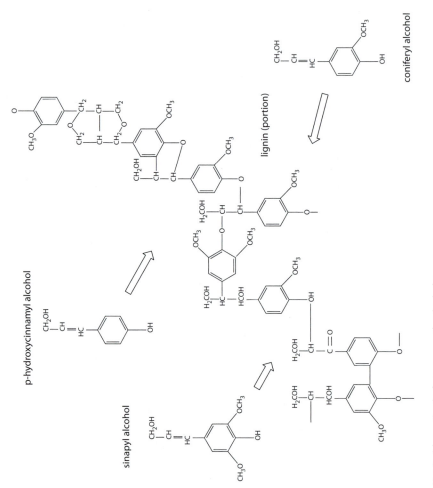

FIGURE 2.2   Outline of lignin biosynthesis from aromatic alcohol monomers.

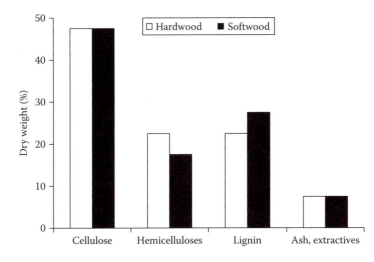

FIGURE 2.3 Chemical composition of wood. (Data from Sudo, S., Takahashi, F., and Takeuchi, M. 1989. In *Biomass Handbook,* ed. Kitani, O. and Hall, C. W., chap. 5.3. New York: Gordon and Breach Science Publishers.)

of being converted (by hydrogenolysis or oxidative breakdown) to useful chemical intermediates—for example, in the manufacture of synthetic resins, perfume, dyes, and pharmaceuticals.[4]

As wood chemicals, extracted celluloses are widely known, especially in the paper industry, but also (as acetylated, nitrated, and other derivatives) find applications as varied as components of explosives, cigarette filters, cosmetics, and medical products such as gauze and bandages. Hardwood and softwood tree species differ in their compositions (Figure 2.3). Detailed data for over 120 tree species list cellulose contents as high as 57% and as low as 38%, with lignin in the range of 17–37% by weight.[4] This plasticity of the chemical composition of biomass suggests that breeding programs and genetic technologies can accelerate the evolution of novel cultivars of plant species dedicated for biofuels.

Hemicelluloses are a diverse group of polysaccharides, with different plant species elaborating structures with two or three types of sugar (sometimes further modified by *O*-methylation or *O*-acetylation) and a sugar acid; the major sugar monomers are the pentoses xylose and arabinose and the hexoses glucose, galactose, and mannose (Figure 2.1). Most plant species contain xylans (1,4-linked polymers of xylose). In addition, hardwood and softwood trees contain copolymers of glucose and mannose (glucomannans), but larchwoods are unusual in having a core polymer of galactose. Table 2.1 summarizes compositional data based on sugar type for a range of tree species, paper products for recycling, cereals, and grasses.[5]

Pectins are the third but minor class of polysaccharides in plant cell walls as a biomass source. They consist of linear galacturonans composed of a 1→4αGalA (galacturonic acid) backbone with varying degrees of methyl esterification, rhamnogalacturonan I (consisting of a backbone of rhamnose and GalA units, the

TABLE 2.1

Compositional Analyses of Tree Species, Paper Recyclates, Cereal Wastes, and Grasses[a]

	Hexose Sugars		Pentose Sugars			Lignin	Water and Alcohol Extractives	Ash
	Glucans	Galactans	Mannans	Xylans	Arabinans			
Tree species[b]	46.2	1.0	3.3	16.0	2.4	24.2	4.0	0.8
Paper recyclates[c]	62.3	0.9	6.7	10.4	—	20.5	8.2	20.7
Cereal wastes[d]	37.4	0.9	0.4	20.5	2.3	20.5	8.2	7.6
Cane sugar bagasse[e]	40.2	0.7	0.3	21.1	1.9	25.2	4.4	4.0
Grasses[f]	31.8	1.1	0.2	18.3	2.8	17.3	18.1	6.3

[a] Percent dry weight basis.

[b] Mean of 18 softwood and hardwood species.

[c] Municipal solid waste and office paper.

[d] Corn stover, wheat straw, and rice straw.

[e] Cane sugar after removal of cane juice.

[f] Four species including switchgrass.

rhamnoses substituted with arabinose and galactose), and highly branched rhamnogalacturonan II. Pectins comprise 10–30% of plant primary walls but only 2–5% of secondary walls. The secondary cell wall (deposited once plant cells have commenced growing) is a much thicker wall and dominates the mechanical properties of woody material.

2.3.2 LIGNOCELLULOSE AS A BIOCHEMICAL RESOURCE

In principle, a bioprocess for producing ethanol from a lignocellulosic substrate could be modeled on those developed for starch from corn grains (Figures 1.20 and 1.21 in Chapter 1). If only cellulosic glucose is considered as a substrate, the essential stages are

1. Milling/grinding of the plant material to reduce particle size
2. Chemical and/or physical pretreatment of the plant material to increase the exposure of the cellulose to enzyme (cellulase) attack
3. Saccharification of cellulose to a mixture of soluble sugars and oligosaccharides
4. Fermentation by a suitable ethanologenic microbe

This is a relatively conservative biotechnological approach and more advanced processes have been much discussed.

STEM TOPIC 2.1: SIMULTANEOUS SACCHARIFICATION AND FERMENTATION (SSF) AND CONSOLIDATED BIOPROCESSING (CBP)

Highly concentrated solutions of simple sugars such as those produced by the hydrolysis of starch, cellulose, and other polysaccharides run the risk of adventitious contamination by unwanted microbes incapable of producing ethanol but adept at metabolizing carbohydrates.

In principle, both SSF and CBP simplify ethanol production by eliminating a separate saccharification step. In SSF, polymeric carbohydrate inputs are hydrolyzed with enzymes when the ethanologen is present, and the resulting monosaccharides and small oligosaccharides are taken up and metabolized to ethanol by the cell population. In CBP as it is normally defined, cellulase production by cellulases secreted by the ethanologenic cells, cellulose hydrolysis, and fermentation occurs in one step, but there is no reason for not extending this to include secreted hemicellulases.

A sophisticated version of SSF was described by Bio-Process Innovation (West Lafayette, Indiana; www.bio-process.com) at the Bioenergy 2000 conference and included two fermentations with yeast cells capable of glucose utilization followed by a different yeast species that could also use pentose sugars (see Figure STEM 2.1).

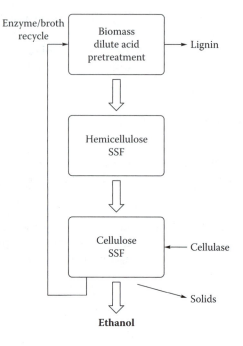

FIGURE STEM 2.1   Schematic outline of simultaneous saccharification and fermentation (SSF).

A variant of SSF occasionally mentioned is SSCF (simultaneous saccharification and cofermentation), which assumes that the ethanologen will coferment hexoses and pentose (usually glucose and xylose).[6] This is useful if SSF is restricted to when only cellulose is the substrate (the sole practical example would be recycled paper) or if an ethanologen is used that can only metabolize glucose. Because recycled paper is a minor biomass substrate and commercial cellulases are mixtures of cellulose and hemicellulases, in practice these restrictions appear unnecessary.

CBP is the simplest fermentation strategy that requires only one fermentation vessel into which the inoculum and the pretreated lignocellulosic material are mixed. CBP was being developed in 2009 by the Mascoma Corporation (Lebanon, New Hampshire; www.mascoma.com).

If the hemicellulosic sugars are also to be utilized, then more complex bioproduction strategies for biochemical engineering must be considered:

- Hemicellulases could be added after cellulase digestion and prior to adding an ethanologen capable of using all the various sugar substrates.
- A mixture of organisms could be used to coferment the sugars or sequential fermentations run for glucose and the remaining sugars.

- Biomass processing could generate separate streams of glucose and hemicellulosic sugars for two parallel fermentations with different ethanologens.

As more of the total potential substrate is included in the fermentation step, the biology inevitably becomes more complex—more so if variable feedstocks are to be used during the year or growing season—because the total available carbohydrate input to the fermentation step will alter significantly (Table 2.1).

Chemical and/or enzymic decomposition of lignins to yield phenolic compounds and other materials has intrigued generations of biological chemists; lignin is not intrinsically and totally refractory because microorganisms have evolved to degrade the polymer.[7] Research into the chemistry of lignin conversion continues, exploring novel means of solubilization of lignin components.[8,9] The majority view inside the ethanol industry, however, is that combustion of recovered and dried lignin is an immediate and successful means of generating energy to offset fossil fuel demands in ethanol production (see Chapter 1, Section 1.6.1).

2.3.3 PRETREATMENT OF LIGNOCELLULOSIC MATERIALS

Cellulose, hemicellulose, and lignin are the polymers that provide the structural rigidity in higher plants that grow vertically from a few centimeters to tens of meters—at the extreme, the giant redwood (*Sequoiadendron giganteum*) reaches up to 300 feet (90 m) in height. At the cellular level, plants derive their remarkable resilience to physical and microbial weathering and attack by having evolved the means to thicken their cell walls greatly, using cellulose in linear polymers of high molecular weight (500,000–1,500,000) that are overlapped and aggregated into macroscopic fibers.[10] Linear strands of cellulose have a close molecular arrangement in fibrillar bundles that are sufficiently regular to have x-ray diffraction patterns characteristic of crystals. This not only augments the structural cohesion but also limits access by water-soluble components and enzymes, and native cellulose is essentially insoluble in water.

Hemicelluloses are diverse in both sugar components and structure, with polymeric molecular weights below 50,000.[10] The heterogeneity of lignins is even greater; any estimate of molecular weight is highly dependent on the method used for extraction and solubilization; average molecular weight distributions may be less than 10,000.[11] Lignin and hemicelluloses may form chemically linked complexes that bind water-soluble hemicelluloses into a three-dimensional array, cemented together by lignin, that sheaths the cellulose microfibrils and protects them from enzymic and chemical degradation.

2.3.3.1 Physical and Chemical Pretreatment of Lignocellulosic Biomass

Given the refractory nature of native lignocelluloses, it is not surprising that chemical processing techniques using acids or alkalis and elevated temperatures have been essential for their use as industrial materials. The starting point is feedstock material such as wood chips, sawdust, and chopped stalks and stems from herbaceous plants.

Mechanical size reduction is unavoidable, but has both economic and energetic costs unless fragmented waste (e.g., sawdust) is used.[12,13]

Diverse techniques have been explored and described for the pretreatment of size-reduced biomass materials with the aim of producing substrates that can be more rapidly and efficiently hydrolyzed to yield mixtures of fermentable sugars. Physical and thermochemical methods taken from extensive R&D work for the pulp, paper, and fiberboard industries have been used to increase greatly the surface area to which aqueous reactants and/or enzymes have access.

Milling has been little favored because the fibrous nature of lignocellulosic materials necessitates lengthy processing times and uneconomically high requirements for energy; only compression milling has been taken to a testing scale beyond the laboratory. Milling can greatly increase the susceptibility to enzymic depolymerization of cellulose and gives a single product stream with only minor degradation of the lignocellulosic polymers.

Thermochemical methods have been more extensively explored and applied in pilot schemes and on near-industrial scales (Table 2.2). The major disadvantage of thermochemistry is the extensive degradation of hemicelluloses.[14] Therefore, a twin-product stream process can be devised by separating solid and liquid phases, the former containing the bulk of the cellulose and the latter the pentose and hexose components of hemicelluloses, although these may be predominantly present in

TABLE 2.2
Chemical Pretreatment Methods for Lignocellulosic Biomass

Method	Principle
Acids	**Hemicellulose solubilization**
Dilute sulfuric	
Dilute hydrochloric	
Dilute nitric	
Dilute phosphoric	
Steaming with sulfuric acid impregnation	
Steam explosion/sulfuric acid impregnation	
Steam explosion/sulfur dioxide	
Steam explosion/carbon dioxide	
Alkalis	**Delignification + hemicellulose removal**
Sodium hydroxide	
Sodium hydroxide + peroxide	
Steam explosion/sodium hydroxide	
Aqueous ammonia	
Calcium hydroxide	
Solvents	**Delignification**
Methanol	
Ethanol	
Acetone	

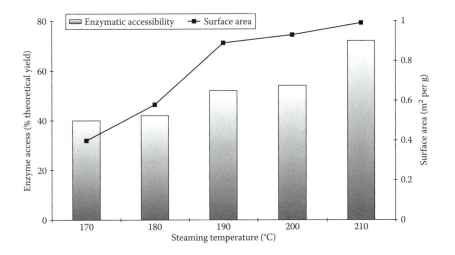

FIGURE 2.4 Efficacy of steaming pretreatments with birch wood. (Data from Puls, J. et al. 1985. *Applied Microbiology and Biotechnology* 22:416.)

oligosaccharides.[15] At temperatures close to 200°C, even short (10 min) pretreatment times have major impacts on surface area and enzyme accessibility (Figure 2.4). Lignin–carbohydrate bonds are disrupted, some of the lignin is depolymerized, and much of morphological coherency of the lignified plant cell wall is destroyed. In addition, aqueous extraction at elevated temperatures removes much of the inorganic salts. This is of particular importance with feedstocks such as wheat straw whose combustion to provide steam and/or energy is impeded by their high salt content and the consequent corrosion problems.[16,17]

STEM TOPIC 2.2: THE SEVERITY OF THERMOCHEMICAL PRETREATMENT OF LIGNOCELLULOSE

Thermochemical pretreatments can differ in several parameters but most notably maximum temperature, residence time at that temperature, and the addition of an acid (or other chemical). The combined effects of temperature and time can be expressed mathematically as the "severity factor," R_o:

$$\log_{10}R_o = \log_{10}[t \times e^{((T-100)/14.75)}]$$

where t is the treatment time (in minutes) at a temperature T (°C); 100°C is taken as the reference temperature and subtracted in the full equation.

In practice, it takes a finite time to reach the desired temperature and an additional time to cool down the reactants; during both times, pretreatment is

TABLE STEM 2.2
Combined Severity Factor Calculations for Methanol Organosolv Pretreatment of Poplar Tree Wood

Acid/Salt Added	t (hr)	T (°C)	$t*\exp(65/14.75)$	Ramping Factor	$\log R_o$	pH	$\log R_o$-pH	Insoluble Pentose (% dry weight)	Xylan Removal (%)
None	2.5	165	12300.7	8317.6	4.31	4.30	0.01	29.2	0
0.04 M NaHSO$_4$	2.5	165	12300.7	8317.6	4.31	2.61	1.70	18.4	37
0.10 M NaHSO$_4$	2.5	165	12300.7	8317.6	4.31	2.01	2.30	12.9	56
0.05 M H$_2$SO$_4$	2.5	165	12300.7	8317.6	4.31	1.63	2.68	0	100

occurring. The exponential can, therefore, be modified to

$$\log_{10}R_o = \log_{10}[(t \times e^{((T-100)/14.75)}) + \text{ramping factor}]$$

where the "ramping factor" measures the contribution to R_o of the heating and cooling stages.

When an acid is used, the effect of acidic pH is to increase the severity of the treatment:

$$\text{Combined severity} = \log_{10}R_o - \text{pH}$$

The calculations in Table STEM 2.2 are taken from the originators of the pH effect equation,[18] although in the original table of calculations the combined factor was misprinted as "$\log(R_o\text{-pH})$"—a very different function!—and the data illustrate the benefit of lowering the pH on solubilizing xylans.

Chemical pretreatment methods have mostly been hydrolysis with aqueous acids and alkalis, although oxidizing agents have also been considered (Table 2.2). The use of such chemical reactants introduces a much higher degree of polysaccharide breakdown and greater opportunities for separately utilizing the various potential substrates in lignocellulosic materials. With wheat straw, for example, sequential treatments with an aqueous methanol, sodium chlorite, and alkali yield four fractions: low molecular mass extractives, lignin, cellulose, and hemicellulose.[19] Similarly, sequential treatments with alkali (lignin removal) and dilute acid (hemicellulose hydrolysis) to leave a highly enriched cellulosic residue have been devised for a variety of feedstocks including switchgrass, corn cob, and aspen woodchip.[20]

Developments for bioindustrial applications have invariably focused on faster, simpler, and more advanced engineering options, including some that have progressed to the pilot plant scale. Pretreatments involving acids (including SO_2-steam explosion) primarily solubilize the hemicellulose component of the feedstock; the use of organic solvents and alkalis tends, on the other hand, to cosolubilize lignin and hemicelluloses. As with thermochemical methods, the product streams can be separated into liquid and solid (cellulosic) phases; if no separation is included in the process, inevitably a complex mixture of hexoses and pentoses will be carried forward to the fermentation step.

Combinations of physical, thermochemical, and chemical pretreatments have often been advocated to maximize cellulose digestibility by subsequent chemical or enzymic treatments. This usually involves higher capital and processing costs, and the potential economic benefits of increased substrate accessibility have seldom been assessed in detail. Different biomass feedstocks may require different technologies for optimized upstream processing; for example, ammonia-based pretreatments (ammonia fiber explosion and ammonia-recycled percolation) are more effective with agricultural residues (including corn stover and corn straw) than with woody materials.[20] Hardwoods yield higher degrees of saccharification after steam explosion than do softwoods.[14] An organic base, *n*-butylamine, has been recommended for pretreatment of rice straw on the evidence of efficient delignification, highly enhanced cellulose hydrolysis by cellulase, the ease of recovery of the amine, and the almost complete re-precipitation of the solubilized lignin when the butylamine is removed by distillation.[21]

A multiauthor review in 2005 of four thermochemical methods, two pretreatments with acid, two with ammonia, and one with lime (calcium hydroxide) as an alkali, concluded that although all nine approaches gave positive outcomes on increasing accessible surface area and solubilizing hemicelluloses and all but one altered lignin structure, only five could reduce the lignin content and only two (the ammonia-based methods) decrystallized cellulose.[22] Exceptions and caveats were noted, however; for example, ammonia fiber/freeze explosion worked well with herbaceous plants and agricultural residues and moderately well with hardwoods, but poorly with softwoods.

A collaboration between the National Renewable Energy Laboratory and six universities in the United States compared ammonia explosion, aqueous ammonia recycle, controlled pH, dilute acid, flow-through with compressed hot water, and lime approaches to prepare corn stover for subsequent biological conversion to sugars. Material balances and energy balances were estimated for the processes, and the digestibilities of the solids were assessed by a standardized cellulase procedure.[23–29] With this feedstock (a major waste product resulting from the corn ethanol industry), all six pretreatment options resulted in high yields of glucose from cellulose by subsequent treatment with cellulase; in addition, the use of high-pH methods offered potential for reducing cellulase amounts required in cellulose hydrolysis.[30]

However, the differences observed in the kinetics of sugar release were sufficient to influence the choice of process, enzymes, and fermentative organisms. A similar conclusion was provided by examination of the steam pretreatment of fast-growing willow (*Salix*) with or without SO_2 impregnation. Although glucose yields of >90% and overall xylose yields >80% could be obtained both with and without SO_2, the most favorable pretreatment conditions for the separate yields of glucose and xylose were closest when using SO_2-impregnated wood chips.[31]

To a large extent, therefore, all pretreatment strategies are likely to include a partial compromise because of the very different susceptibilities to hydrolytic breakdown and solubilization of cellulose and hemicelluloses. Highly efficient industrial solutions will require biotechnological approaches to provide fermenting organisms capable of using both hexoses and pentoses and both monomeric and oligomeric (and possibly polymeric) carbohydrates (this is discussed in detail in Chapters 3 and 4). Even for a single choice of pretreatment method, variation in the biological material (the feedstock) will inevitably occur, for example, in the water content that will necessitate a flexible technology or extra cost outlay to standardize and microman-age the inflow of biomass material.[32]

A recent development that could provide a pure glucose stream is the solubiliza-tion of microcrystalline cellulose with the so-called "ionic liquids"; these are salts that are liquids at room temperature and are also stable up to 300°C. With *n*-butyl-methylimidazolium chloride, cellulose is solubilized by comparatively short (<3 h) treatments at 300°C. The cellulose is then recovered by addition of antisolvents such as water, methanol, and ethanol; the resulting cellulose was 50-fold more susceptible to enzyme-catalyzed hydrolysis as compared to untreated cellulose.[33] Ionic liquids can successfully pretreat materials such as wheat straw.[34,35]

2.3.3.2 Acid Hydrolysis of Pretreated Lignocellulosic Biomass

Acid hydrolysis has been historically the most likely means of liberating ferment-able sugars from pretreated biomass for pilot plant and demonstration facilities for cellulosic ethanol. Two-stage processes employ mild hydrolysis conditions (e.g., 0.7% sulfuric acid at 190°C) to recover pentose sugars efficiently; the more acid-resistant cellulose requires a second stage at higher temperature (215°C). Sugars are recovered from both stages for subsequent fermentation steps.[36] Concentrated (30–70%) sulfuric acid hydrolysis can be performed at moderate temperature (40°C) and result in >90% recovery of glucose, but the procedure is lengthy (2–6 h) and requires efficient recovery of the acid post-treatment for economic feasibility.[37]

The major drawback of acid hydrolysis is the degradation of hexoses and pentoses to growth-inhibitory products: hydroxymethylfurfural (HMF) from glucose, furfural from xylose, together with acetic acid (Figure 2.5). HMF is also known to break down in the presence of water to produce formic acid and other inhibitors of ethanol-producing organisms.[38] To varying extents, *all* ther-mochemical methods of pretreatments suffer from this problem; even total inhi-bition of ethanol production in a fermentation step subsequent to biomass steam treatment has been observed (Figure 2.6). Two contrasting views have become apparent for dealing with this:

- The growth-inhibitory aldehydes are removed by adsorption.[38]
- They can be considered to be an additional coproduct stream capable of purification for resale.[39]

Slowly feeding a cellulosic hydrolysate with high concentrations of furfurals and acetic acid to yeast cells may condition the microorganism to detoxify and/or metabolize inhibitory products of sugar degradation.[40,41] A more proactive strategy is to remove the inhibitors by microbiological means. A U.S. patent describes a fun-gus (*Coniochaeta lignaria*) that can metabolize and detoxify furfural and HMT in agricultural biomass hydrolysates prior to their conversion to substrates for ethanol production.[42]

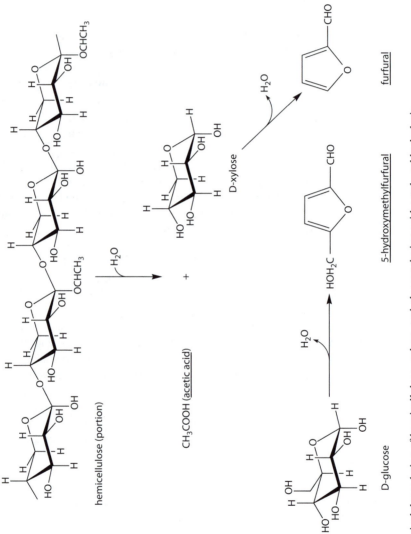

FIGURE 2.5 Chemical degradation of hemicellulose, xylose, and glucose during acid-catalyzed hydrolysis.

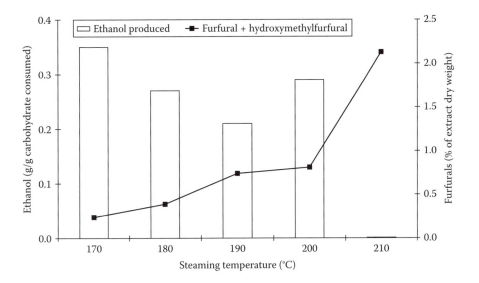

FIGURE 2.6 Effects of the generation by steaming pretreatment of wood of furfural inhibitors on ethanol production by *Fusarium oxysporum*. (Data from Puls, J. et al. 1985. *Applied Microbiology and Biotechnology* 22:416.)

2.4 CELLULASES: BIOCHEMISTRY, MOLECULAR BIOLOGY, AND BIOTECHNOLOGY

2.4.1 ENZYMOLOGY OF CELLULOSE DEGRADATION BY CELLULASES

Cellulase hydrolysis of pretreated biomass has gradually supplanted acid hydrolysis ethanol production from lignocellulosic substrates. The word "cellulase" is deceptively complex—a shorthand term for the four enzyme activities and molecular entities; each has its Enzyme Commission (EC) identifying numbers that are required for the complete hydrolytic breakdown of macromolecular cellulose to monomeric glucose[43–45]:

1. Endoglucanases (1,4-β-D-glucan-4-glucanohydrolases, EC 3.2.1.4) decrease the degree of polymerization of macromolecular cellulose by attacking accessible sites and breaking the linear cellulose chain (Table 2.3).
2. Cellodextrinases (1,4-β-D-glucan glucanohydrolases, EC 3.2.1.74) attack the chain ends of the cellulose polymers, liberating glucose.
3. Cellobiohydrolases (1,4-β-D-glucan cellobiohydrolases, EC 3.2.1.91) attack the chain ends of the cellulose polymers, liberating the disaccharide cellobiose (1,4-β-D-glucosyl glucose).
4. Finally, β-glucosidases (EC 3.2.1.21) hydrolyze soluble cellodextrins (1,4-β-D-glucans) and cellobiose to glucose.

Cellulolytic organisms often possess multiple genes and enzymatically active proteins; for example, the fungus *Hypocrea jecorina* contains two

TABLE 2.3

Substrate Selectivities of Different Cellulase Components

Substrate	Exoglucanase		Endoglucanase	
	Cellobiohydrolase CBHI	Cellobiohydrolase CBHII	EG1	EGII
Macromolecular				
β-Glucan	0[a]	4	5	3
Hydroxyethyl cellulose	0	1	4	2
Carboxymethyl cellulose	1	2	4	5
Crystalline cellulose	4	3	1	1
Amorphous cellulose	1	3	5	3
Small Molecule				
Cellobiose	0	0	0	1
p-Nitrophenyl glucoside	0	0	0	1
Methylumbelliferyl cellotrioside	0	0	0	1

Source: Data from Tolan, J. S., and Foody, B. 1999. *Advances in Biochemical Engineering/ Biotechnology* 65:41.

[a] Relative activity: 0 = inactive, 5 = maximum activity.

cellobiohydrolases, five endoglucanases, and two β-glucosidases.[46–48] This fungus was named until recently *Trichoderma reesei* (and is sometimes still so named in reviews and scientific journal papers); it represented the beginning of biotechnological interest in cellulase because it caused the U.S. Army major equipment and supply problems in the Pacific during World War II by digesting military cotton garments.

Cellulases are widely distributed throughout the global biosphere because cellulose is the single most abundant polymer and many organisms have evolved in widely different habitats to feed on this most abundant of resources. Bacteria and fungi produce cellulases both in natural environments and while contained in the digestive systems of ruminant animals and wood-decomposing insects (e.g., termites); insects themselves may also produce cellulases and other higher life forms—plants and plant pathogenic nematodes certainly do.[44,49] Higher plants need to reversibly "soften" or irreversibly destroy cell wall structures in defined circumstances as part of normal developmental processes, including plant cell growth, leaf and flower abscission, and fruit ripening; these are highly regulated events in cellular morphology.

The enormous taxonomic diversity of cellulase producers has aroused much speculation; it is likely, for example, that once the ability to produce cellulose had evolved with algae and land plants, cellulase producers arose on separate occasions in different ecological niches. Moreover, gene transfer between widely different organisms is thought to occur easily in such densely populated microbial environments as the rumen.[50] From the biotechnological perspective, hover, fungal, and bacterial cellulase producers are the actual and potential industrial sources:

- More than 60 cellulolytic fungi have been reported, including soft-rot, brown-rot, and white-rot species; the last group includes members that can degrade both cellulose and lignin in wood samples.[51] The penetration of fungal hyphae through intact wood results in an enormous surface area of contact between the microbial population and lignocellulosic structures; the release of soluble enzymes results in an efficient hydrolysis of accessible cellulose as different exo- and endoglucanases attack macroscopic cellulose individually at separate sites, a process often known as "synergy."

- Aerobic bacteria have a similar strategy in that physical adherence to cellulose microfibers is not a prerequisite for cellulose degradation, and a multiplicity of cellulases is secreted for maximal cellulose degradation—an excellent example is a marine bacterium whose extraordinary metabolic versatility is embodied in 180 enzymes for polysaccharide hydrolysis, including 13 exo- and endoglucanases, two cellodextrinases, and three cellobiases.[52]

- Anaerobic bacteria, however, contain many examples of a quite different biochemical approach: the construction of multienzyme complexes (cellulosomes) on the outer surface of the bacterial cell wall; anaerobic cellulolytics grow optimally when attached to the cellulose substrate, and for some species this contact is obligatory. The ability of such anaerobic organisms to break down cellulose and to ferment the resulting sugars to a variety of products, including ethanol, has prompted several investigators to promote them as ideal candidates for ethanol production from lignocellulosic biomass.[53]

The drive to commercialize cellulases—in applications as diverse as the stone-washing of denims, household laundry detergent manufacture, animal feed production, textile "biopolishing," paper de-inking, baking, and fruit juice and beverage processing—has ensured that the biochemistry of the exo- and endoglucanases has been extensively researched. The majority of these enzymes share a fundamental molecular architecture comprising two domains or modules[54]: (1) a cellulose-binding region (CBD or CBM), and (2) a catalytic module or core.

As more cellulase enzymes have been sequenced at the levels of either amino acids or DNA, families of conserved polypeptide structures for CBD/CBM have been recognized; they form part of 57 carbohydrate-binding modules collated in a continuously updated database (http://www.cazy.org/). All proteins in three families (CBM1, CBM5, and CBM10) bind to crystalline cellulose, and proteins in the CBM4 and CBM6 families bind to cellulose as well as to xylans and other polysaccharides using different polysaccharide binding sites.[55]

Removal of the portion of the cellulase responsible for binding to cellulose reduces cellulase activity with cellulose as the substrate but not with cellodextrins. Conversely, the isolated binding domains retain their affinity for cellulose but lack catalytic action.[56] The contribution of cellulose binding to overall cellulase activity is important: The endoglucanase II from *H. jecorina* has five amino acid residues,

TABLE 2.4

Engineering-Improved Cellulose Binding and Endoglucanase Activity

	Residue 29	Residue 34	Cellulose Binding (Relative)	Cellulose Hydrolysis (Relative)
Parent	Asparagine	Glutamine	100	100
	Asparagine	Alanine	80	62
	Histidine	Glutamine	120	115
	Valine	Glutamine	90	98
	Alanine	Valine	70	40
	Threonine	Alanine	150	130

Source: Data from Fukuda, T. et al. 2006. *Biotechnology Progress* 22:933.

and selectively altering two of these positions changes cellulose binding affinity and cellulose hydrolysis rate in parallel (Table 2.4).[57]

To enzymologists, the active site of an enzyme is where the important chemistry and catalysis really occur. The active sites in exo- and endoglucanases have different geometries[48,58,59]:

- Cellobiohydrolases have a tunnel-like structure at the active site in which steric hindrance freezes the cellulose polymer chain and restricts the hydrolytic reaction to specifically liberating cellobiose.
- In contrast, endoglucanases have more open active sites and the reaction mechanism is more flexible; significant amounts of glucose, cellobiose, and cellotriose are generated as well as larger oligosaccharides resulting from the random scission of the cellulose polymer chain.

The carboxyl side chains of two acidic amino acids are thought to be directly involved in the chemical mechanism of hydrolysis and cleavage of the β-1,4 glycosidic bond.[59,60]

Joining the separate sites for cellulose binding and cellulose hydrolysis is a linker region of variable length (6–59 amino acid residues) and with O-glycosylated residues in fungal cellulases.[45,60] The linker appears to optimize the geometry between the catalytic and binding regions of the protein, and the degree of glycosylation (in fungal cellulases) determines the flexibility of the linker peptide. With the CBM temporarily anchored, the catalytic site hydrolyzes bonds within a linear range set by the linker and then the CBM translocates along the surface of the cellulose in what has been described as a caterpillar-like motion.[61] This image of cellulase acting as a nanomachine has been supported by evidence that the CBM locates hydrolyzed cellulose chain ends and then exerts a thermodynamic driving force to translate away along the polymeric structure.[62]

Anaerobic cellulolytic bacteria exhibit a very different solution to the multienzyme problem posed by the insoluble cellulose macrosubstrate. The cellulosome multienzyme complex was first described in 1983 for *Clostridium theromocellum* and the same basic assembly occurs in other species[63–65]:

- Noncatalytic cellulosome integrating protein (Cip) contains cohesin domains that act as receptors for dockerin domains on the catalytic proteins.
- Multiple Cip proteins bind cellobiohydrolases, endoglucanases, hemicellulases, mannanases, pectinase, and other enzymes.
- Hydrophilic modules and cellulose binding modules complete the array.
- The Cip units are anchored to the cell wall via other cohesin domains.
- Large and stable complexes are formed with molecular masses in the range of 2–16 MDa, and polycellulosomes occur with molecular masses up to 100 MDa.

The elaboration of such complex structures may aid energy-deficient anaerobes by maximizing the uptake of cellodextrins, cellobiose, and glucose by the spatially adjacent bacteria and ensuring a greatly increased binding affinity to the cellulose. A similar strategic logic drove the evolution of anaerobic fungi, first described in 1975.[66] Resident in herbivorous animals, 17 distinct anaerobic fungi are known and solubilize lignocellulose and produce all enzymes needed to hydrolyze cellulose and hemicelluloses efficiently. Although some of these enzymes are found free in the medium, most of them are associated with cellulosomal and polycellulosomal complexes, in which the enzymes are attached through fungal dockerins to scaffolding proteins. Cellulosomes from anaerobic fungi share many properties with cellulosomes of anaerobic bacteria, but their structures differ at the primary protein level of amino acid sequences.

The cellulases of cellulolytic anaerobes may also be more catalytically efficient than those of typical aerobes, especially the soluble cellulases secreted by fungi. However, this is controversial because enzyme kinetics of the cellulose/cellulase system are problematical: The equations used for soluble enzyme and low molecular weight substrates are inadequate to describe the molecular interactions for cellulases dwarfed by and physically binding to macroscopic and insoluble celluloses.[43] Nevertheless, some results can be interpreted as showing that clostridial cellulases are up to 15-fold more catalytically efficient than are fungal cellulases (based on specific activity measurements, i.e., units of enzyme activity per unit enzyme protein).[44] Similarly, a comparison of fungal and aerobic bacterial cellulases found a 100-fold higher specific activity with the bacterial enzymes.[51]

2.4.2 CELLULASES IN LIGNOCELLULOSIC FEEDSTOCK PROCESSING

Three serious practical drawbacks to the efficient saccharification of cellulose and lignocellulosic materials on the industrial and semi-industrial scales have been noted. First, cellulases have often been described as being catalytically inferior to other glycosidases. This statement is certainly true when crystalline cellulose is the substrate for cellulase action.[67] When more accessible forms of cellulose are hydrolyzed, the catalytic efficiency increases, but comparison with other glycosidases shows how relatively poor are cellulases, even with low molecular mass "model" substrates (Table 2.5).

TABLE 2.5

Catalytic Parameters for Fungal Cellulases

Organism	Substrate	Turnover Number, k_{cat} (s⁻¹)
	Endoglucanases	
Trichoderma reesei	Crystalline cellulose	0.027–0.051
Trichoderma longibrachiatum	Crystalline cellulose	0.05–0.67
Myceliophtora thermophila	Crystalline cellulose	0.013
Trichoderma reesei	Carboxymethyl cellulose	40–60
Trichoderma longibrachiatum	Carboxymethyl cellulose	19–35
Myceliophtora thermophila	Carboxymethyl cellulose	58–140
Glucoamylase	Starch	58
Other glycosidases	Soluble substrates	>100 to >1,000
	Cellobiohydrolase	
Trichoderma reesei	*p*-Nitrophenyl-β-D-lactoside	0.063

Source: Data from Klyosov, A. A. 1988. In *Biochemistry and Genetics of Cellulose Degradation,* ed. J.-P. Aubert, P. Beguin, and J. Millet, 97. London: Academic Press.

STEM TOPIC 2.3: QUANTITATIVE PARAMETERS FOR ENZYMES

Enzymes catalyze (increase the rate of) chemical reactions in which a reactant or reactants are converted into a product or products. Most enzymologists tend to refer to reactants as "substrates."

To provide a quantitative measure of how much enzyme activity is present, it is necessary to define a unit; this is not the same as the mass of enzyme protein present because enzymes are inherently fragile (i.e., they lose catalytic activity even though the protein remains). In other words, enzyme units are defined *functionally*. In the SI nomenclature, a katal (abbreviated as "kat") of an enzyme is that capable of transforming 1 mol of substrate into 1 mol of product within 1 s. In practice, this is a massive quantity of enzymically active protein, so scaled-down units are usually encountered: μkat (1×10^{-6} kat), nkat (1×10^{-9} kat), or pkat (1×10^{-12} kat). The older literature used an enzyme unit (U) equal to the enzymic capacity to transform 1 μmol of substrate into product per minute. A useful conversion factor is

$$1 \text{ enzyme U} = 1,000/60 = 16.7 \text{ nkat}$$

Enzyme units are vital to assess the degree of purity of an enzyme. This is essentially an empirical process of biochemical purification from the cells of the producing organism. The relevant metric is specific activity:

$$\text{Specific activity} = \text{enzymic activity/protein mass (kat/kg or U/mg)}$$

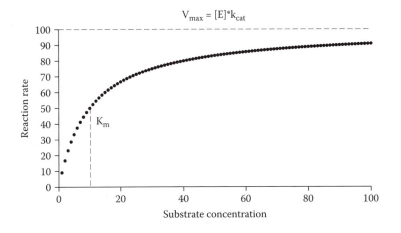

FIGURE STEM 2.3 Graphical representation of parameters of enzyme action.

During purification, the specific activity of an enzyme should increase, but the total quantity of enzyme recovered will decrease as a result of enzyme instability, loss by physical adsorption to surfaces, and the conscious decision to retain only those fractions that represent the highest enzyme concentration (units of activity per unit volume) from chromatographic steps.

Three further quantitative parameters are of importance:

- The V_{max} of an enzyme is the maximum enzyme velocity at infinitely high substrate concentration. Mathematically, V_{max} is the product of the number of enzyme active sites present in the assay and the turnover number of the enzyme.
- k_{cat} is the turnover number—the number of times each enzyme site converts substrate to product per second.
- K_m, the Michaelis–Menten constant, is the substrate concentration at which half the maximum reaction rate is reached.

The interrelationships among these parameters are illustrated in Figure STEM 2.3.

Second, cellulases from organisms not normally grown at elevated temperatures have poor stability at typically used incubation temperatures for cellulose digestion in excess of 50°C. Fungal cellulases show half-life times at 65°C as low as 10 min, whereas thermophilic clostridial enzymes may be stable for 20 times longer.[67]

Third, there is often a rapid decay of hydrolysis rate when cellulase is mixed with cellulose. Kinetic models can be interpreted to show that the β-glucosidic bonds in cellulose that are accessible by cellulases bound onto the macromolecular substrate surface are "used up" as hydrolysis proceeds.[68,69] This intuitive conclusion has also been reached experimentally by fusing a CBM with a fluorescent marker protein; the cellulose accessibility to cellulase could be visualized and declining reactivity of the cellulose substrate attributed to a loss of this accessibility after the easily hydrolyzed cellulose fraction is digested first by the enzyme.[70]

2.4.3 MOLECULAR BIOLOGY AND BIOTECHNOLOGY OF CELLULASE PRODUCTION

The consequence of these limitations of cellulases as catalysts for the degradation of cellulose is that large amounts of cellulase have been considered necessary to process pretreated lignocellulosic substrates (1.5–3% by weight of the cellulose) rapidly; this would impose a high economic cost on cellulose-based ethanol production.[71] Enzyme manufacturers have achieved massive increases in cellulase fermentation productivity: In the 1980s, space–time yields (units of enzyme per unit volume per unit time) for cellulases increased nearly 10-fold; between 1972 and 1984, total cellulase production doubled every 2 years by the selection of strains and the development of fed-batch fermentation systems.[45] Cost reductions in cellulase bioproduction (mostly with *H. jecorina*) have continued because the potential market for cellulosic ethanol is so large that commercial pressures and competition will provide powerful drivers toward process improvement.

2.4.3.1 Physiological Regulation of Cellulase Production

What factors regulate cellulase production when the producing cells are functioning inside the fermentor? For a fungus such as *H. jecorina,* the evidence is that low-level production of cellulase is constitutive but that an inducer of rapid cellulase synthesis is generated when that cellulase encounters cellulose as a substrate. The disaccharide sophorose (β-1,2-glucopyranosyl-D-glucose) is a strong inducer of cellulases in *H. jecorina*.[72–74] Cellulases can also catalyze transglycosylation reactions, in effect shuffling glucose residues in various possible combinations given the availability of multiple hydroxyl groups on the glucose molecule (Figure 2.1).[75] Putting this information together, it can be hypothesized that low activities of cellulase partially degrade cellulose, liberating cellobiose, which is then transglycosyated to sophorose; the inducer stimulates the transcription of cellulase genes.

Cellulase gene transcription is inhibited, however, if glucose accumulates in the environment.[76–78] In addition, sophorose also represses β-glycosidase; because sophorose is hydrolyzed by β-glycosidase, this repression acts to maintain sophorose concentrations and thus maximally stimulate cellulase formation.[74] The overall strategy built up from these regulatory features is an elaborate mechanism typically employed by microbes to prevent the unnecessary and energy-dependent synthesis and secretion of degradative enzymes if readily utilizable carbon or nitrogen is already present. The synthesis of protein is a metabolic burden that is avoided where necessary, and evolution has equipped microbes with efficient mechanisms to avoid this waste of resources.

Can sophorose be used to increase fungal cellulase expression in fermentations to manufacture the enzyme on a large scale? As fine chemical, sophorose is orders of magnitude more expensive than is glucose and its use (even at low concentrations) would be economically unfeasible in the large fermentors mandated for large-scale

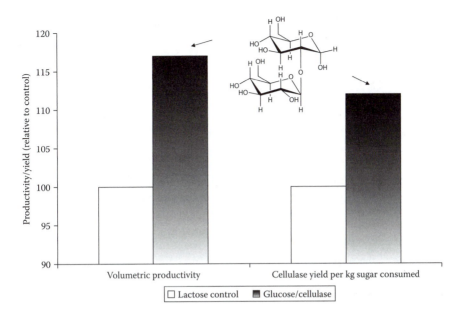

FIGURE 2.7   Productivity of cellulase fermentations in the presence of the inducer sophorose generated from glucose by the transglycosylating action of cellulase. (Data from Mitchinson, C. 2004. Stanford GCEP Biomass Energy Workshop, April 2004, accessed at http:/gcep. stanford.edu/pdfs/energy_workshops_04_04/biomass_mitchinson.pdf.)

commercial enzyme manufacture. Scientists at Genencor discovered, however, that simply treating glucose solutions with *H. jecorina* cellulase could generate sophorose—taking advantage of the transglycosylase activity of β-glucosidase mentioned before—to augment cellulase expression and production in *H. jecorina* cultures.[79] Lactose is used industrially as a carbon source for cellulase fermentations to bypass the catabolite repression imposed by glucose acting to keep cellulase production minimal; adding cellulase-treated glucose increased both cellulase production and the yield of enzyme per unit sugar consumed (Figure 2.7).

The traditional and long established four major components of *H. jecorina* cellulase—two cellobiohydrolases and two endoglucanases—together constitute >50% of the total cellular protein produced by the cells under inducing conditions and can reach 50 g per liter in contemporary industrial strains that are the products of many years of strain development and selection.[80,81] However, the relatively low activities of β-glycosidases in fungal cellulase preparations—possibly an unavoidable consequence of the sophorose induction system—have been considered a barrier to the quantitative saccharification of cellulose. Supplementing *H. jecorina* cellulase preparations with β-glucosidase reduces the inhibitory effects caused by the accumulation of cellobiose.[82]

2.4.3.2   Regulatory Genes Involved in Cellulase Production

H. jecorina remains the focus of much R&D aiming to assemble improved industrial processing of lignocellulosics. The complete genome was published in 2008, and this should lead to advances in the understanding of how cellulase gene expression is regulated in fine detail and, extrapolating from this, how cellulase production in enzyme fermentors might be radically increased.[83] Some of the pieces of the jigsaw puzzle are already in place:

- Deletion of the genes coding for discrete cellulase components prevents the expression of other cellulase genes.[84,85]
- The general carbon catabolite repressor protein CRE1 represses the transcription of cellulase genes and a hyperproducing mutant has a *cre1* mutation rendering cellulase production insensitive to glucose.[86,87]
- Production of cellulases is regulated at the transcriptional level, and two genes encoding transcription factors have been identified.[88–90]

An extensive and detailed study of gene expression in *H. jecorina* revealed 12 previously unrecognized enzymes or proteins involved in polysaccharide degradation. Some of these novel proteins may not function directly in cellulose hydrolysis, but they could be involved in the production and secretion of the cellulase complex or be relevant when other polysaccharides serve as growth substrates.[80]

2.4.3.3 New-Generation Cellulases

The prospects for cellulase usage in lignocellulosic material processing has engendered an intense interest in novel sources of cellulases and in cellulase-degrading enzymes with properties better matched to high-intensity cellulose saccharification processes.[79] Enzyme manufacturers Genencor and Novozymes have both demonstrated tangible improvements in the catalytic properties of cellulases, in particular, thermal stability; such enzyme engineering has involved site-directed mutagenesis and DNA shuffling (Table 2.6).

Evident from Table 2.6 is a firm interest in extending cellulose biotechnology beyond the *Trichoderma* cellulase paradigm. The poor performance of *H. jecorina* as a cellulase producer—sardonically described as the result of nature opting for an organism secreting very large amounts of enzymically incompetent protein rather than choosing an organism elaborating small amounts of highly active enzymes[67]— has engendered many innovative and speculative studies on radical alternatives to this source.

Novel sources of cellulases have a barely explored serendipitous potential to increase the efficiency of saccharification; for example, cellulases from such non-standard (i.e., relatively obscure) fungi as *Chaetomium thermophilum, Thielavia terrestris, Thermoascus aurantiacus, Corynascus thermophilus,* and *Mycellophthora thermophila* (all thermophiles with optimum growth temperatures in the range of 45–60°C) improved the sugar yield from steam-pretreated barley straw incubated with a benchmark cellulase/β-glucosidase mix.[91] Thermostable enzymes from bacterial and fungal sources are rapidly multiplying and there is potential to move cellulase treatment temperatures from 45–50 to 55–60°C.[92] This ability to operate at higher temperatures is important; a well-known factor in chemical kinetics is that an increase of just 10°C results in *doubling* the reaction rate.

TABLE 2.6
A Selection of Patents and Patent Applications in Cellulase Enzymology and Related Areas

Date, Filing Date	Title	Assignee/Applicant	Patent, Application
1/18/2000	Genetic constructs and genetically modified microbes for enhanced production of β-glucosidase	Iogen Corporation, Canada	US 6,015,703
6/12/2001	Carboxymethyl cellulose [sic] from Thermotoga maritima	Diversa, San Diego, CA	US 6,245,547
6/26/2002	Polypeptides having cellobiohydrolase I activity…	Novozymes Biotech, Inc., Davis, CA	US 2004/0197890
12/19/2003	Polypeptides having cellobiohydrolase II activity…	Novozymes Biotech, Inc., Davis, CA	US 2006/0053514
4/30/2004	Variants of β-glucosidases	Novozymes Biotech, Inc., Davis, CA	US 2004/0253702
8/25/2004	Variants of glycoside hydrolases	Novozymes Biotech, Inc., Davis, CA	US 2005/0048619
10/26/2004	Cell-wall degrading enzyme variants	Novozymes A/S, Denmark	US 6,808,915
1/28/2005	Polypeptides having cellulolytic enhancing activity…	Novozymes Biotech, Inc., Davis, CA	US 2006/0005279
2/15/2005	Endoglucanases	Novozymes A/S, Denmark	US 6,855,531
9/1/2005	Polypeptides having cellulolytic enhancing activity…	Novozymes Biotech, Inc., Davis, CA	US 2005/0191736
9/29/2005	Polypeptides having beta-glucosidase activity…	Novozymes Biotech, Inc., Davis, CA	US 2005/0214920
1/6/2006	Polypeptides having cellobiohydrolase activity…	Novozymes Biotech, Inc., Davis, CA	US 2006/0218671
4/6/2006	Polypeptides having cellobiase activity…	Novozymes Biotech, Inc., Davis, CA	US 2006/0075519
2/28/2003	Cellulase-degrading enzymes of Aspergillus	Gielkens et al.	US2004/0001904
3/20/2003	Endoglucanase mutants and mutant hydrolytic depolymerizing enzymes	NERL, Golden, CO	US 2003/0054535
5/22/2003	Thermal tolerant exoglucanase from Acidothermus cellulyticus	NERL, Golden, CO	US 2003/0096342
6/12/2003	Thermal tolerant avicelase from Acidothermus cellulyticus	NERL, Golden, CO	US 2003/0108988
6/13/2006	Thermal tolerant cellulase from Acidothermus cellulyticus	Midwest Research Institute, Kansas City, MO	US 7,059,993
9/20/2005	Method for enhancing cellobiase activity of Termitomyces clypeatus using a glycosylation factor	CSIR, New Delhi (India)	US 6,946,277
12/18/2000	Novel cellulase-producing Actinomycetes…	Genencor International, Inc., Palo Alto, CA	US 2002/0076792

(continued)

TABLE 2.6 (CONTINUED)

A Selection of Patents and Patent Applications in Cellulase Enzymology and Related Areas

Date, Filing Date	Title	Assignee/Applicant	Patent, Application
6/14/2002	Cellulase for use in industrial processes	Genencor International, Inc., Palo Alto, CA	US 2002/0193272
6/26/2003	BGL4 β-glucosidase and nucleic acids encoding the same	Genencor International, Inc., Palo Alto, CA	WO 03/052118
6/26/2003	BGL5 β-glucosidase and nucleic acids encoding the same	Genencor International, Inc., Palo Alto, CA	WO 03/052054
6/26/2003	EGVI endoglucanase and nucleic acids encoding the same	Genencor International, Inc., Palo Alto, CA	WO 03/052057
6/26/2003	EGVII endoglucanase and nucleic acids encoding the same	Genencor International, Inc., Palo Alto, CA	WO 03/052055
6/26/2003	EGVIII endoglucanase and nucleic acids encoding the same	Genencor International, Inc., Palo Alto, CA	WO 03/052056
8/15/2003	Novel variant *Hypocrea jecorina* CBHI cellulases	Genencor International, Inc., Palo Alto, CA	US 2005/0127172
11/5/2003	BGL6 β-glucosidase and nucleic acids encoding the same	Genencor International, Inc., Palo Alto, CA	US 2006/0258554
3/19/2004	Novel CBHI homologs and variant CHBI cellulases	Genencor International, Inc., Palo Alto, CA	US 2005/0054039
1/27/2005	BGL3 β-glucosidase and nucleic acids encoding the same	Genencor International, Inc., Palo Alto, CA	US 2005/0214912
3/23/2005	Exo-endo cellulase fusion protein	Genencor International, Inc., Palo Alto, CA	US 2006/0057672
12/22/2005	Novel variant *Hypocrea jecorina* CBHII cellulases	Genencor International, Inc., Palo Alto, CA	US 2005/0205042
8/22/2006	Variant EGIII-like cellulase compositions	Genencor International, Inc., Palo Alto, CA	US 7,094,588

STEM TOPIC 2.4: THERMOSTABLE CELLULASES (AND OTHER ENZYMES)

Enzymologists often fight shy of the concept of a temperature optimum for an enzyme, stating that the assays are highly dependent on the time over which the enzyme-catalyzed reaction is measured. Enzymes denature at high temperatures; the longer the assay incubation time, the greater the degree of enzyme inactivation. The optimum temperature determined will therefore be influenced by the precise temperature instability of the enzyme. These concerns are illustrated by data quoted in Reference 92 (see STEM Table 2.4).

TABLE STEM 2.4

Parameters of Thermotolerant Cellulases

Species	Cellulase	T Optimum (°C)	Stability (% Activity at 100°C)
Anaerocellum thermophilum	Endoglucanase	95–100	50 after 40 min
Bacillus sp. KSM-S237	Endoglucanase	45	30 after 10 min
Rhodothermus marinus	Endoglucanase	95	50 after 210 min
Streptomyces sp.	Exoglucanase	60	30 after 30 min
Thermotoga neapolitana	Endoglucanase (CelA)	95	50 after 240 min

Source: Viikari, L. et al. 2007. *Advances in Biochemical Engineering/Biotechnology* 108:121.

More detailed analysis, however, can explore thermal inactivation of enzymes. The inactivation process can be expressed mathematically as

$$E/E_o = A_l \exp(k_l t) + A_s \exp(k_s t)$$

where

E_o is the initial enzyme activity

E is the activity at time t

k_l and k_s are the inactivation rate constants of the heat-labile and heat-stable groups

A_l and A_s are the heat-labile and heat-stable fractions of E_o

At long incubation times, the equation simplifies to

$$E/E_o = A_s \exp(k_s t)$$

and a plot of $\ln(E/E_o)$ against t will have a gradient of k_s and intercept of k_s.

By substituting experimental values, a plot of $[E/E_o - A_s \exp(k_s t)]$ against $A_l \exp(k_l t)$ has a gradient of k_l and intercept of A_l.[93]

There is a caveat with thermostable enzymes: Their specific activities are often similar to those of enzymes from mesophiles (with much lower optimum growth temperatures) when the latter are assayed at these lower temperatures. Are thermophile enzymes therefore poor catalysts? One answer is that life at 85°C or higher temperature cannot solve the chemical task of "bending" the polypeptide structure to achieve abnormal acidity functions of the side chains of basic amino acid residues in active sites at elevated temperature.[94]

Widening further the search for cellulases with enhanced features for industrial use, "biogeochemistry" aims to explore the natural diversity of coding sequences available in wild-type DNA; forest floors are an obvious source of novel microbes

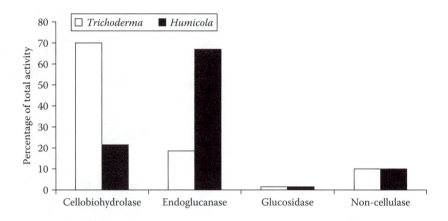

FIGURE 2.8 Differential distribution of cellulase components in cellulases. (Data from Tolan, J. S., and Foody, B. 1999. *Advances in Biochemical Engineering/Biotechnology* 65:41.)

and microbial communities adept at recycling lignocelluloses.[95] Termites are probably the best known wood-degrading organisms but this nutrition is dependent on the bacterial community resident inside each insect. This internal ecosystem has a diverse set of bacterial genes for cellulase and hemicellulose highly specialized for plant lignocellulose degradation, representing an unexploited source of genes for novel cellulases.[96]

Different producing organisms yield cellulases with different profiles of enzyme components (Figure 2.8). Mixtures of cellulases from different cellulolytic organisms have the advantage of maximally exploiting their native traits, both as individual enzymes with differing kinetic properties and as conglomerates of endo- and exoglucanases.[97] The practical value of this flexible approach would be to optimize the saccharification of a lignocellulosic feedstock that exhibited significant seasonal or yearly variation in its chemical composition and to process differing feedstock materials being delivered to a cellulosic ethanol production unit.

2.4.3.4 Novel Molecular Features of Cellulases

A prediction made by some of the pioneers of cellulase biotechnology was that swelling factors would be secreted by the fungus to render cellulose more susceptible to cellulase-catalyzed attack.[98] A family of "swollenin" proteins that bind to macroscopic cellulose and disrupt the structure of the cellulose fibers without any endoglucanase action has subsequently been identified.[99] Fusing cellulose-disrupting protein domains with cellulase catalytic domains could generate more powerful artificial exo- and endoglucanases.

Whether cellulose-binding domains or modules in known cellulases disrupt cellulose structures remains unclear.[100] An unexpected potential resource for laboratory-based evolution of a new generation of cellulases is the strong affinity for cellulose exhibited not by a cellulase but rather by a cellobiose dehydrogenase; combining this binding activity with cellobiohydrolases is an interesting option.[101]

2.4.3.5   Immobilized Cellulases

While enzyme stability is limited (even when thermophiles are used as the sources), extensive experience with immobilized enzymes for biotransformations indicates that cellulases immobilized on inert carriers can offer significant cost savings to commercial use by the repeated use of batches.[102,103] With a commercial β-glucosidase from *Aspergillus niger,* immobilization resulted in two important benefits: greatly improved thermal stability at 65°C and an eightfold increase in maximal enzyme activity at saturating substrate concentration, as well as operational stability during at least six rounds of lignocellulose hydrolysis.[104] This technology has also been used in conjunction with an ionic liquid (Section 2.3.3.1) to accelerate cellulose hydrolysis by immobilized *H. jecorina* cellulase.[105]

2.5   HEMICELLULASES: NEW HORIZONS IN ENERGY BIOTECHNOLOGY

2.5.1   A MULTIPLICITY OF HEMICELLULASES

Mirroring the variety of polysaccharides containing pentoses, hexoses, or both (and with or without sugar hydroxyl group modifications) collectively described as hemicelluloses (Figure 2.1), hemicellulolytic organisms are known across many species and genera, including terrestrial and marine bacteria, yeasts and fungi, rumen bacteria and protozoa, and marine algae. Farther up the evolutionary tree, wood-digesting insects harbor microbial communities with cellulase and hemicellulase activities; however, mollusks, crustaceans, and higher plants all elaborate their own hemicellulases.[9]

STEM TOPIC 2.5: STRUCTURAL CHEMISTRY AND PHYSICS OF HEMICELLULOSES

Polyxylan (β-1,4 linked) backbones are *O*-substituted with hexoses and pentose, glucuronic (GlcA) and phenolic acids, *O*-methylated glucuronic acid, and acetic acid. Xyloglucans are similarly substituted while galactoglucomannans have only hexoses and pentose in the short side chains (see Figure STEM 2.5).

Hemicelluloses do not aggregate with themselves and therefore do not form microfibrils (unlike cellulose). Hemicelluloses form hydrogen bonds with cellulose as part of their main role in cell walls (i.e., providing structural integrity) and three main roles are usually discussed:

- in maintaining and determining the shapes of cell walls and therefore of cells
- acting as a physical barrier to plant pathogens
- cell-to-cell signaling—oligosaccharides are released when, for example, a fungus releases hemicellulases to break down plant wall components that stimulate the defensive systems of the host

```
    xyl                    feruoyl
    2                      2
    xyl                    ara          xyl
    2                      2            2
O-acetyl    ara  glcA  ara  ara            ara  ara  4-O-Me-glcA  O-acetyl
2           2    2    2    2              2    3    2            3
-xyl-xyl-xyl-xyl-xyl-xyl-xyl-xyl-xyl-xyl-xyl-xyl-xyl-xyl-xyl-xyl-xyl-xyl-xyl-xyl-xyl-xyl-xyl-xyl-xyl-xyl-
```

```
                                          ara
                                          3
            5-O-acetyl-ara    5-O-acetyl-ara   gal                      5-O-acetyl-ara
            2                 2                2                        2
O-acetyl    xyl               xyl              xyl        O-acetyl       xyl
6           6                 6                6          6             6
-glc-glc-glc-glc-glc-glc-glc-glc-glc-glc-glc-glc-glc-glc-glc-glc-glc-glc-glc-glc-glc-glc-glc-glc-glc-glc-
```

```
    gal                                    gal
    2                                      2
    gal    gal              ara            gal    gal
    6      6                6              6      6
-glc-man-glc-man-man-man-man-glc-man-glc-man-glc-glc-man-glc-man-glc-man-glc-man-glc-man-glc-man-
```

FIGURE STEM 2.5 Typical hemicellulose polymer structures (numerals refer to carbon atoms on the carbohydrates).

Young's modulus (E) is a measure of how a compressive or stretching force (stress) causes a compression or elongation:

$$E = (F/A)/(\Delta L/L)$$

where F is the force acting on a material with cross-sectional area A and causing a change in length ΔL from an original length L.

For a softwood (fir tree species), the modulus is of the order of 1.0×10^{10} Pa (N m^{-2}), about 5% of the value for structural steel but much closer to bone and concrete. Young's modulus of hemicellulose extracted from *Pinus radiata* wood varied by almost three orders of magnitude, from 8.0×10^9 Pa in nearly dry hemicellulose to 1.0×10^7 Pa in nearly saturated hemicellulose. The mechanical behavior of the wood cell wall in living tissue is therefore very different from that in fully dried material.[106]

Table 2.7 summarizes major classes of hemicellulases, their general sites of action, and the released products. However, microorganisms capable of degrading hemicelluloses have multiple genes encoding many individual hemicellulases. For example, *Bacillus subtilis* (a simple bacterium widely used in academic and commercial biotech laboratories) has in its completely sequenced genome at least 16 separate genes for enzymes involved in hemicellulose degradation.[107]

TABLE 2.7
Major Hemicellulases, Their Enzymic Sites of Action, and Their Products

Hemicellulase	EC Number	Site(s) of Action	Released Products
Xylanases			
Endo-β-1,4-xylanase	3.2.1.8	Internal β-1,4-linkages in Xylans, L-arabino-D-xylans, etc.	Xylose, xylobiose, xylan oligomers, xylan-arabinan oligomers, etc.
Exo-β-1,4-xylosidase	3.2.1.37	External β-1,4-linkages in xylan oligomers, etc.	Xylose
Arabinanases			
Endoarabinanase	3.2.1.99	Internal α-1,5- and/or α-1,3-linkages in arabinans	Arabinose
α-L-Arabinofuranosidase	3.2.1.55	Side chain α-1,2- and/or α-1,3-linkages in xyloarabinans and external α-1,5 linkages in arabinans	Arabinose, xylan oligomers
Mannanases			
Endo-β-1,4-mannanase	3.2.1.78	Internal β-1,4-linkages in mannans, galactomannans, and glucomannans	Mannose, mannan oligomers, etc.
Exo-β-1,4-mannosidase	3.2.1.25	External β-1,4-linkages in mannan oligomers	Mannose
Galactanases			
Endo-β-1,4-galactanase	3.2.1.89	Internal β-1,4-linkages in galactans and arabinogalactans	Galactose, galactan oligomers, etc.
α-Galactosidase	3.2.1.25	Side chain α-1,6-linkages in galactomannan oligomers	Galactose, mannan oligomers
Other			
β-Glucosidase	3.2.1.21	External β-1,4-linkages in glucomannan oligomers	Glucose, mannan oligomers
α-Glucuronidase	3.2.1.139	Side chain 4-O-methyl-α-1,2-linkages in glucuronoxylans	Galactose, mannan oligomers
Esterases			
Acetyl esterase	3.2.1.6	2- or 3-O-acetyl groups on mannan and xylose	Acetic acid, mannose, xylose
Arylesterase	3.2.1.2	3-O-feruoyl/coumaryl-α-L-arabinofuranose side chains	Ferulic, coumaric acids, arabinoxylans

TABLE 2.8

A Selection of Patents and Patent Applications in Hemicellulase Enzymology

Filing Date	Title	Assignee/Applicant	Patent, Application
12/31/2002	Xyloglucanase from *Malbranchea*	Novozymes A/S, Denmark	US 6,500,658 B2
10/7/2003	Family 5 xyloglucanases	Novozymes A/S, Denmark	US 6,630,340 B2
11/9/2004	Family 44 xyloglucanases	Novozymes A/S, Denmark	US 6,815,192 B2
4/25/2006	Polypeptides having xyloglucanase activity…	Novozymes, Inc., Palo Alto (CA)	US 7,033,811 B2
9/17/2002	Novel recombinant xylanases derived from anaerobic fungi…	Hseu and Huang	US 2004/0053238 A1
6/27/2004	Xylanase from *Trichoderma reesei*…	Genencor International, Inc., Palo Alto, CA	US 6,768,001 B2
3/11/2004	Novel xylanases and their use	Georis et al.	EP 1 574 567 A1

Hemicelluloses structures are either linear or branched heteropolysaccharides based on a spine of xylose residues (polyxylans) in angiosperms; in gymnosperms, the predominant hemicellulose basic structures are glucomannans of D-mannosyl and D-glucosyl residues. Endoxylanases fragment xylan backbones and xylosidases cleave the resulting xylan oligosaccharides into xylose; removal of the side chains is catalyzed by glucuronidases, arabinofuranosidases, and acetylesterases. The action of these enzymes can limit the overall rate of hemicellulose saccharification because endo-acting enzymes cannot bind to and cleave xylan polymers close to sites of side chain attachment.[108]

Much of the fine detail of hemicellulase catalytic action is emerging and will be vital for directed molecular evolution of improved hemicellulase biocatalysts.[109,110] For example, a thermostable arabinofuranosidase has been identified and shown to have a unique selectivity in being able to degrade both branched and de-branched arabinans.[111] Synergistic interactions among different microbial arabinofuranosidases have also been demonstrated to result in a more extensive degradation of wheat arabinoxylan than found with individual enzymes.[112] The activity of biotech companies in patenting novel hemicellulase activities is evident in exploring hemicellulases from unconventional microbial sources (Table 2.8). A pertinent example is deep-sea thermophilic bacteria from the Pacific that synthesize thermotolerant xylanases active over a wide pH range and are capable of degrading cereal hemicelluloses.[113]

2.5.2 HEMICELLULASES IN THE PROCESSING OF LIGNOCELLULOSIC BIOMASS

The importance of including hemicellulosic sugars in the conversion of lignocellulosic feedstocks to ethanol to ensure process efficiency and an economic base for biofuel production has been emphasized.[114,115] The numerical basis for this is that hemicellulosic sugars constitute a fermentable resource (weight basis) of approximately 50%

of the cellulosic glucose residues in trees, 60% in sugarcane bagasse, 65% in cereal stalk wastes, and 70% in grass species (Table 2.1).

Commercial cellulase preparations contain variable but often high enzymic activities of hemicellulases; this will adventitiously contribute some hemicellulose sugars from lignocellulosic materials processed to the fermentation stages of ethanol production.[116,117] Thermochemical and acid-catalyzed pretreatments of lignocellulosic biomass materials extensively degrade hemicelluloses (see Section 2.3.3). Depending on the pretreatment method and the feedstock, however, hemicellulose solubilization may approach 100% or be as low as 10%, and the hemicellulose sugars may be present primarily as monomers (xylose, arabinose, etc.) or as oligomers of these pentoses.[118]

2.5.3 MOLECULAR BIOLOGY OF HEMICELLULASES

Both cellulase and hemicellulase genes of the filamentous fungus *H. jecorina* have been shown to be under carbon catabolite repression mediated by the regulatory gene *cre1,* which is therefore a valid target gene in strain engineering for improved enzyme production.[119] The main transcription activator of hemicellulase-encoding genes is *xyr1* (xylanase regulator 1), which mediates the induction derived from various inducing carbon sources and compounds.

Cultivation on glucose as the carbon source causes carbon catabolite repression of *xyr1* transcription mediated by the *cre1* protein product, and *xyr1* transcription is repressed by the specific transcription factor *ACE1*. Constitutive expression of *xyr1* leads to higher xylanolytic enzyme activities.[120] The two major xylanases (*XYN I* and *XYN II*) of *H. jecorina* are simultaneously expressed during growth on xylan as the substrate, but respond differently to low-molecular-weight inducers.[121]

In another filamentous fungi, *Aspergillus niger,* the transcriptional regulator protein *XLNR* controls the transcription of about 20–30 genes encoding hemicellulases and cellulases in a broadly similar fashion to the regulation of gene expression in *H. jecorina.*[122]

2.5.4 MULTIFUNCTIONAL HEMICELLULASES?

Because different hemicellulases presumably act in concert in real-time hemicellulose digestion in nature, enhancing that synergy but engineering multifunctional enzymes would be a logical step. Such a chimera (a biologically active form not previously found in nature) was created by fusing the xylanase domain of the *Clostridium thermocellum* xylanase (*xynZ*) and a dual functional arabinofuranosidase/xylosidase (*DeAFc*) from a compost starter mixture via a flexible peptide linker. The product was found to possess the activities of xylanase, arabinofuranosidase, and xylosidase (plus endoglucanase when the novel gene was expressed in *Escherichia coli* as a laboratory bacterial host) and could hydrolyze natural xylans.[123]

Two highly active trifunctional chimeras were constructed by linking the catalytic portion of a xylanase with an arabinofuranosidase and a xylosidase, using either flexible peptide linkers or linkers containing a cellulose-binding domain. These multifunctional enzymes exhibit synergistic effects in the hydrolysis of xylans and corn stover.[124]

2.6 LIGNIN-DEGRADING ENZYMES
AS AIDS TO SACCHARIFICATION

In contrast to thermochemical pretreatments, the use of microbial degradation of lignin to increase feedstock digestibility has several major advantages:

- Energy inputs are low.
- Hardware demands are modest.
- No environmentally damaging waste products are generated.
- Hazardous chemicals and conditions are avoided.

All of these features have associated economic cost savings. Against this, the need for lengthy pretreatment times and the degradation of polysaccharides (reducing the total fermentable substrate) have acted to keep interest in biological preprocessing of lignocellulosic materials firmly in the laboratory. A careful choice of organism (usually a wood-rotting fungus) or a mixture of suitable organisms, however, can ensure a high degree of specificity of lignin removal.[125,126] Extrapolations of this approach could involve either the preprocessing of in situ agricultural areas for local production facilities or the sequential use of the biomass feedstock first as a substrate for edible mushroom production before further use of the partially depleted material by enzymic hydrolysis to liberate sugars from polysaccharides.

White rot fungi secrete peroxidase enzymes dependent on iron (Fe) or manganese (Mn) ion cofactors but little is known about the regulation of enzyme production by these microbes that would render them predictable for industrial use.[127,128] Fungal lignin-degrading systems include membrane-associated proteins associated with the uptake and oxidation of lignin fragments, the production of ligninolytic secondary metabolites, and defense against ligninolytic oxidants. Catalase, alcohol oxidase, and a transporter protein occur in the outer envelope of the white rot fungi.[129]

A strategic problem with lignin degraders is that they will also be efficient cellulose degraders, but judicious choice of fungi can give isolates capable of high lignin but low cellulose losses during biomass processing.[130] Taking this to the opposite extreme, brown rot fungi completely hydrolyze polysaccharides but without secreting an exoglucanase and without removing lignin.[131] The unraveling of a brown rot fungus genome provides new opportunities for investigating their enzymic mechanisms of cellulose conversion and has shown that an evolutionary shift from white rot to brown rot occurred during which the capacity for depolymerization of lignin was lost.[132]

Therefore, there is now the possibility of being able to mix enzymes from wood-rotting fungi and other species to decompose untreated lignocellulosic materials directly in one stage into mixtures of sugars and lignin degradation products that could be exploited for fermentation to ethanol and as a source of fine chemicals.

2.7 COMMERCIAL CHOICES OF CELLULOSIC FEEDSTOCKS FOR ETHANOL PRODUCTION

The previous discussion has outlined the tool kits with which to convert plant biomass in its broadest sense (rather than just selected portions, such as starch grains or sugar-rich phloem saps). For a start-up lignocellulosic ethanol facility, there are crucial issues of cost and availability. For example, an industrial plant may require close to a million tons of feedstock a year; that feedstock should be (for operational stability) as uniform and free from high levels of toxic impurities and contaminations as possible.

Some materials (most obviously, wood bark) have compositions that are incompatible with the high yields achieved in starch- and sugar-based ethanologenesis, and some softwood materials demand high inputs of cellulase for their saccharification.[133] Any lignocellulosic material is subject to some competitive use, and this may dictate cost considerations (Table 2.9). Some of these direct competitors are long established, mature industries; others have unarguably green credentials for recycling waste materials or in renewable energy generation.[134] In addition, agricultural waste materials have great potential as substrates for the solid-state fermentative

TABLE 2.9
Competing Uses for Lignocellulosic Biomass Materials Considered for Bioethanol Production

Material	Source	Uses
	Agriculture	
Grain straw, cobs, stalks, husks	Grain harvesting	Animal feed, burning as fuel, composting, soil conditioning
Grain bran	Grain processing	Animal feed
Seeds, peels, stones, rejected fruit	Fruit and vegetable harvesting	Animal feed, fish feed, seeds for oil extraction
Bagasse	Sugarcane industries	Burning as fuel
Sheels, husks, fiver, presscake	Oils and oilseed plants	Animal feed, fertilizer, burning as fuel
	Forestry	
Wood residues, bark, leaves	Logging	Soil conditioning and mulching, burning as fuel
Woodchips, shavings, sawdust	Milling	Pulp and paper, chip and fiber board
Fiber waste, sulfite liquor	Pulping	Use in pulp and board industries as fuel
Paper, cardboard, furniture	Municipal solid waste	Recycling, burning as fuel

Source: After Howard, R. L. et al. 2003. *African Journal of Biotechnology* 2:603.

production of a wide spectrum of fine chemicals, including enzymes, biopesticides, bioinsecticides, and plant growth regulators.[135,136]

Much of what is calculable as available biomass may not, therefore, be commercially harvestable on a short-term basis without the large-scale planting of dedicated energy crops. Even then, issues of responsible soil and water management remain unspecified or ignored.

In other words, we have left behind the scientific discussion of what is technologically feasible and entered a more open sector where economics and social concerns play significant roles. Chapters 6, 9, and 10 further explore these issues.

2.8 BIOTECHNOLOGY AND PLATFORM TECHNOLOGIES FOR CELLULOSIC ETHANOL

In 1996, several years' experience with pilot plants worldwide using either enzyme conversion or acid-catalyzed hydrolysis of candidate cellulosic feedstocks inspired the prediction that technologies for the conversion of lignocellulosic biomass to ethanol would be rapidly commercialized.[137] More than a decade later, generic technologies have signally failed to emerge on large-scale production sites; in April 2004, Iogen Corporation (Ottawa, Canada) opened a demonstration facility capable of processing 40 tonnes of feedstock per day and producing 3 million L of ethanol annually from wheat, oat and barley straw, corn cobs, and corn stalk.

Iogen was founded in 1974 and has received research funding from the government of Canada, Petro-Canada, and Shell Global Solutions International B.V. Iogen and its partners are studying the feasibility of producing cellulosic ethanol in Germany, and in 2006 attracted attention from Wall Street investor Goldman Sachs. Significantly, Iogen is also an industrial producer of enzymes used in textiles, pulp and paper, and animal feed.

The Verenium Corporation has a facility in Jennings, Louisiana, for the production of cellulosic ethanol utilizing regionally available feedstocks, including sugarcane bagasse, with a capacity of 1.5 billion L per year. Since 2005, Abengoa Bioenergy has been constructing the world's first industrial-scale cellulosic ethanol plant (to use wheat straw as the feedstock) immediately adjacent to its existing 195 million L per year cereal ethanol plant (Biocarburantes de Castilla y Leon, BcyL) at Babilfuente, Salamanca (Spain). The biomass plant will process over 25,000 tonnes of wheat straw and other materials to produce 5 million L of ethanol annually using enzymatic hydrolysis to effect substrate saccharification.

Other start-up companies are seeking financing for cellulosic ethanol production sites in North America, the United Kingdom, and elsewhere, usually commercializing technologies from university research. Pursuing basic or applied research has, of course, always been much easier than attracting investor funds for establishing even pilot-scale process facilities, let alone major industrial complexes based on radically new technologies such as lignocellulosic biomass. Investors could see the immediate potential and markets for corn starch ethanol—so much so that overcapacity developed and, during 2009, more than 10% of productive capacity was idle.

More research aimed at fine-tuning cellulosic ethanol using cellulases and/or hemicellulases and/or combinations of biomass pretreatments appears regularly in scientific journals. A considerable backlog of (ligno)cellulosic ethanol processes seeking scale-up has developed but, with overall energy demand falling in 2009, the energy landscape remains uncertain.

Before starting a review of the biochemical engineering of cellulosic ethanol and other biofuels (Chapter 5), we will survey the microbial "cell factories" that have been engineered for cellulosic ethanol production by the fermentation of hexose and pentose sugars. At the heart of any successful biomass-based ethanol process will be an intelligently designed microorganism, its genes, and its various biochemical capabilities.

2.9 SUMMARY

Cellulose is the most abundant plant polymer on the surface of the Earth and could, by conversion to ethanol and other biofuels, substitute for at least 50% of contemporary fossil fuel usage.

Plant cell walls—lignified cellulose and hemicelluloses—present challenges to efficient processing and the liberation of the fermentable sugars for ethanol production. Nevertheless, physical size reduction of the materials and subsequent chemical pretreatment combine to yield partially or fully solubilized hemicellulose carbohydrates and cellulose more open to acid or cellulase digestion. All potential candidate plant biomass sources (cereal straw, corn stover, sugarcane bagasse, grasses, and softwood and hardwood trees) can be converted to mixtures or monosaccharides (hexose and pentose sugars) for fermentation to ethanol.

Cellulases and hemicellulases are widely available in nature, and modern biotech research has generated high-yielding microbial producers. With production costs falling, enzymic digestion of cellulosic and hemicellulosic substrates is becoming the standard route for bioprocessing. Novel microbial sources of enzymes are being actively sought—often in microorganisms from extreme environments—to identify enzymes with properties optimally suited for fast, high-temperature digestions of plant polysaccharides.

Developments in gene technology, industrial chemistry, and biochemical engineering have seen the first near-industrial-scale facilities for the production of cellulosic ethanol. However, scientific and technical advances have outpaced the provision of investor capital for the construction of a global biomanufacturing sector based on lignocellulosic, sustainable feedstocks.

REFERENCES

1. Herrera, S. 2006. Bonkers about biofuels. *Nature Biotechnology* 24:755.
2. Bioethanol needs biotech now. 2006. Editorial. *Nature Biotechnology* 24:725.
3. Wyman, C. E., ed. 1996. *Handbook on bioethanol: Production and utilization.* London: Taylor & Francis.
4. Sudo, S., Takahashi, F., and Takeuchi, M. 1989. Chemical properties of biomass. In *Biomass handbook,* ed. Kitani, O. and Hall, C. W., chap. 5.3. New York: Gordon and Breach Science Publishers.

5. Wiselogel, A., Tyson, S., and Johnson, D. 1996. Biomass feedstock resources and composition. In *Handbook on bioethanol: Production and utilization,* ed. Wyman, C. E., chap. 6. London: Taylor & Francis.

6. McMillan, J. D. et al. 1999. Simultaneous saccharification and cofermentation of dilute-acid pretreatment yellow poplar hardwood to ethanol using xylose-fermenting *Zymomonas mobilis. Applied Biochemistry and Biotechnology* 79:649.

7. Ribbons, D. W. 1987. Chemicals from lignin. *Philosophical Transactions of the Royal Society London A* 321:485.

8. Okuda, K. et al. 2004. Disassembly of lignin and chemical recovery—rapid depolymerization of lignin without char formation in water–phenol mixtures. *Fuel Processing Technology* 85:803.

9. Wahyudionoa, S. M., and Goto, M. 2008. Recovery of phenolic compounds through the decomposition of lignin in near and supercritical water. *Chemical Engineering Processes: Process Intensification* 47:1609.

10. Singh, A., and Mishra, P. 1995. Microbial pentose utilization. In *Current Applications in Biotechnology (Progress in Industrial Microbiology, vol. 33)*, chap. 1. Amsterdam: Elsevier.

11. Froment, P., and Pla, F. 1989. Determinations of average molecular weight distributions of lignin. In *Lignin. Properties and materials (ACS Symposium Series 397),* ed. Glasser, W. G. and Sarkanen, S., chap. 10. Washington, D.C.: American Chemical Society.

12. Hsu, T.-A. 1996. Pretreatment of biomass. In *Handbook on bioethanol: Production and utilization,* ed. Wyman, C. E., chap. 10. Washington, D.C.: Taylor & Francis.

13. McMillan, J. D. 1994. Pretreating lignocellulosic biomass. A review. In *enzymatic conversion of biomass for fuels production (ACS Symposium Series 566),* Himmel, M. E., Baker, J. O., and Overend, R. P., chap. 15. Washington, D.C.: American Chemical Society.

14. Higuchi, T. 1989. Steam explosion of wood. In *Biomass handbook,* ed. Kitani, O. and Hall, C. W., chap. 2.5.3. New York: Gordon and Breach Science Publishers.

15. Puls, J. et al. 1985. Biotechnical utilization of wood carbohydrates after steaming pretreatment. *Applied Microbiology and Biotechnology* 22:416.

16. Thygesen, A. et al. 2004. Hydrothermal treatment of wheat straws on pilot plant scale. *Proceedings of the World Conference and Technology Exhibition on Biomass for Energy, Industry and Climate Protection.* Rome, Italy, 10–15 May 2004, ETA-Florence.

17. Thomsen, M. H. et al. 2006. Preliminary results on optimization of pilot scale pretreatment of wheat straw used in coproduction of bioethanol and electricity. *Applied Biochemistry and Biotechnology* 129–132:448.

18. Chum, H. L., Johnson, D. K., and Black, S. K. 1990. Organosolv pretreatment for enzymatic hydrolysis of poplars. 2. Catalyst effects and the combined severity parameter. *Industrial and Engineering Chemistry Research* 29:156.

19. Chahal, D. S., Moo-Young, M., and Dhillon, G. S. 1979. Bioconversion of wheat straw and wheat straw components into single-cell protein. *Canadian Journal of Microbiology* 25:793.

20. Gong, C. S. et al. 1999. Ethanol production from renewable resources. *Advances in Biochemical Engineering/Biotechnology* 65:207.

21. Tanaka, M. et al. 1985. Evaluation of effectiveness of pretreating rice straw with *n*-butylamine for improvement of sugar yield. *Applied Microbiology and Biotechnology* 22:19.

22. Mosier, N. et al. 2005. Features of promising technologies for pretreatment of lignocellulosic biomass. *Bioresource Technology* 96:673.

23. Wyman, C. E. et al. 2005. Coordinated development of leading biomass pretreatment technologies. *Bioresource Technology* 96:1959.

24. Lloyd, T. A., and Wyman, C. E. 2005. Combined sugar yields for dilute sulfuric acid pretreatment of corn stover followed by enzymatic hydrolysis of the remaining solids. *Bioresource Technology* 96, 1967, 2005.

25. Liu, C., and Wyman C. E. 2005. Partial flow of compressed-hot water through corn stover to enhance hemicellulose sugar recovery and enzymatic digestibility of cellulose. *Bioresource Technology* 96:1978.

26. Mosier, N. et al. 2005. Optimization of pH controlled liquid hot water pretreatment of corn stover. *Bioresource Technology* 96:1986.

27. Kim, S., and Holtzapple, M. T. 2005. Lime pretreatment and enzymatic hydrolysis of corn stover. *Bioresource Technology* 96:1994.

28. Kim, T. H., and Lee, Y. Y. 2005. Pretreatment and fractionation of corn stover by ammonia recycle percolation process. *Bioresource Technology* 96:2007.

29. Teymouri, F. et al. 2005. Optimization of the ammonia fiber explosion (AFEX) treatment parameters for enzymatic hydrolysis of corn stover. *Bioresource Technology* 96:2014.

30. Wyman, C. E. et al. 2005. Comparative sugar recovery data from laboratory scale application of leading pretreatment technologies to corn stover. *Bioresource Technology* 96:2026.

31. Sassner, P., Galbe, M., and Zacchi, G. 2005. Steam pretreatment of *Salix* with and without SO_2 impregnation for production of bioethanol. *Applied Biochemistry and Biotechnology* 121–124:1101.

32. Teymouri, F. et al. 2004. Ammonia fiber explosion treatment of corn stover. *Applied Biochemistry and Biotechnology* 113–116:951.

33. Heinze, T., Schwikal, K., and Barthel, S. 2005. Ionic liquids as reaction medium in cellulose functionalization. *Macromolecular Bioscience* 5:520.

34. Dadi, A. P., Varanasi, S., and Schall, C. A. 2006. Enhancement of cellulose saccharification kinetics using an ionic liquid pretreatment step. *Biotechnology and Bioengineering* 95:904.

35. Zhu, S. et al. 2006. Dissolution of cellulose with ionic liquids and its application: A mini-review. *Green Chemistry* 8:325.

36. Lee, Y. Y., Iyer, P., and Torget, R. W. 1999. Dilute-acid hydrolysis of lignocellulosic biomass. *Advances in Biochemical Engineering/ Biotechnology* 65:93.

37. Hamelinck, C., van Hooijdonck, G., and Faaij, A. P. C. 2003. Prospects for ethanol from lignocellulosic biomass: Techno-economic performance as development progresses. Report NWS-E-2003-55, Copernicus Institute, Utrecht University.

38. Weil, J. R. et al. 2002. Removal of fermentation inhibitors formed during pretreatment of biomass by polymeric adsorbents. *Industrial and Engineering Chemistry Research* 41:6132.

39. Brink, D. L., Merriman, M. M., and Gullekson, E. E. 1987. Ethanol fuel, organic chemicals, single-cell proteins: A new forest products industry. Gen. Tech. Rep. PSW-100, Pacific Southwest Forest and Range Experiment Station, Forest Service, U.S. Department of Agriculture.

40. Taherzadeh, M. J., Niklassoen, C., and Lidén, G. 1999. Conversion of dilute-acid hydrolyzates of spruce and birch to ethanol by fed-batch fermentation. *Bioresource Technology* 69:59.

41. Taherzadeh, M. J. et al. 1999. Conversion of furfural in aerobic and anaerobic batch fermentation of glucose by *Saccharomyces cerevisiae*. *Journal of Bioscience and Bioengineering* 87:169.

42. Nichols, N. N. et al. 2006. Culture containing biomass acid hydrolysate and *Coniochaeta lignaria* fungus. US Patent 7,067,303, June 27, 2006.

43. Mosier, N. S. et al. 1999. Reaction kinetics, molecular action, and mechanisms of cellulolytic proteins. *Advances in Biochemical Engineering/Biotechnology* 65:23.

44. Lynd, L. R. et al. 2002. Microbial cellulose utilization: Fundamentals and biotechnology. *Microbiology and Molecular Biology Reviews* 66:506002.

45. Tolan, J. S., and Foody, B. 1999. Cellulase from submerged fermentation. *Advances in Biochemical Engineering/Biotechnology* 65:41.

46. Takashima, S. et al. 1999. Molecular cloning and expression of the novel fungal β-galactosidase genes from *Humicola grisea* and *Trichoderma reesei*. *Journal of Biochemistry* 125:728.

47. Nogawa, M. et al. 2001. L-sorbose induces cellulase gene transcription in the cellulolytic fungus *Trichoderma reesei*. *Current Genetics* 38:329.

48. Kleywegt, G. J. et al. 1997. The crystal structure of the catalytic core domain of the endoglucanase I from *Trichoderma reesei* at 3.6Å resolution, and a comparison with related enzymes. *Journal of Molecular Biology* 272:383.

49. Wilson, D. B., and Irwin, D. C. 1999. Genetics and properties of cellulases. *Advances in Biochemical Engineering/Biotechnology* 65:1.

50. Garcia-Vallvé, S., Romeu, A., and Palau, J. 2000. Horizontal gene transfer of glycosyl transferases of the rumen fungi. *Molecular Biology and Evolution* 17:352.

51. Himmel, M. E. et al. 1996. Cellulases: Structure, function, and applications. In *Handbook on bioethanol: Production and utilization,* ed. Wyman, C. E., chap. 8. London: Taylor & Francis.

52. Taylor, L. E. et al. 2006. Complete cellulase system in the marine bacterium *Saccharophagus degradans* strain 2-40^T. *Journal of Bacteriology* 188:3849.

53. Demain, A. L., Newcomb, M., and Wu, J. H. D. 2005. Cellulase, clostridia, and ethanol. *Microbiology and Molecular Biology Reviews* 69:124.

54. Gilkes, N. R. et al. 1991. Domains in microbial β-1,4-glycanases: Sequence conservation, function, and enzyme families. *Microbiology and Molecular Biology Reviews* 55:303.

55. Pires, V. M. R. et al. 2004. The crystal structure of the family 6 carbohydrate binding module from *Cellvibrio mixtus* endoglucanase 5A in complex with oligosaccharides reveals two distinct binding sites with different ligand specificities. *Journal of Biological Chemistry* 279:21560.

56. van Tilbeurgh, H. et al. 1986. Limited proteolysis of the cellobiohydrolase I from *Trichoderma reesei*: separation of functional domains. *FEBS Letters* 204:223.

57. Fukuda, T. et al. 2006. Enhancement of cellulase activity by clones selected from the combinatorial library of the cellulose-binding domain by cell surface engineering. *Biotechnology Progress* 22:933.

58. Rouvinen, J. et al. 1990. Three-dimensional structure of cellobiohydrolase II from *Trichoderma reesei*. *Science* 249:380.

59. Divne, C. et al. 1994. The three-dimensional crystal structure of the catalytic core of cellobiohydrolase I from *Trichoderma reesei*. *Science* 265:524.

60. Davies, G., and Henrissat, B. 1995. Structures and mechanisms of glycosyl hydrolases. *Structure* 3:853.

61. Srisodsuk, M. et al. 1993. Role of the interdomain linker peptide of *Trichoderma reesei* cellobiohydrolase I in its interaction with crystalline cellulose. *Journal of Biological Chemistry* 268:20756.

62. Bu, L. et al. 2009. The energy landscape for the interaction of the family 1 carbohydrate binding module and the cellulose surface is altered by hydrolyzed glycosidic bonds. *Journal of Physical Chemistry B* 113:10994.

63. Receveur, V. et al. 2002. Dimension, shape, and conformational flexibility of a two domain fungal cellulase in solution probed by small angle x-ray scattering. *Journal of Biological Chemistry* 277:40887.

64. Bayer, E. A. et al. 2004. The cellulosomes: Multienzyme machines for degradation of cell wall polysaccharides. *Annual Review of Microbiology* 58:521.

65. Schwartz, W. H. 2001. The cellulosome and cellulose degradation by anaerobic bacteria. *Applied Microbiology and Biotechnology* 56:634.

66. Ljungdahl, L. G. 2008. The cellulase/hemicellulase system of the anaerobic fungus *Orpinomyces* PC-2 and aspects of its applied use. *Annals of the New York Academy of Sciences* 1125:308.

67. Klyosov, A. A. 1988. Cellulases of the third generation. In *Biochemistry and genetics of cellulose degradation*, ed. Aubert, J.-P., Beguin, P., and Millet, J., 97. London: Academic Press.
68. Eriksson, T., Karlsson, J., and Tjerneld, F. 2002. A model explaining declining rate in hydrolysis of lignocellulose substrates with cellobiohydrolase I (Cel7A) and endoglucanase I (Cel7B) of *Trichoderma reesei*. *Applied Biochemistry and Biotechnology* 101:41.
69. Zhang, Y.-H. P., and Lynd, L. R. 2006. A functionally based model for hydrolysis of cellulose by fungal cellulase. *Biotechnology and Bioengineering* 94:888.
70. Hong, J., Ye, X., and Zhang, Y. H. 2007. Quantitative determination of cellulose accessibility to cellulase based on adsorption of a nonhydrolytic fusion protein containing CBM and GFP with its applications. *Langmuir* 23:12535.
71. Mandels, M. 1985. Applications of cellulases. *Biochemical Society Transactions* 13:414.
72. Mandels, M., Parrish, F. W., and Reese, E. T. 1962. Sophorose as an inducer of cellulase in *Trichoderma reesei*. *Journal of Bacteriology* 83:400.
73. Sternberg, D., and Mandels, G. R. 1979. Induction of cellulolytic enzymes in *Trichoderma reesei* by sophorose. *Journal of Bacteriology* 139:761.
74. Nisizawa, T. et al. 1970. Inductive formation of cellulase by sophorose in *Trichoderma viride*. *Journal of Biochemistry* 70–375.
75. Vaheri, M., Leisola, M., and Kaupinnen, V. 1979. Transglycosylation products of cellulase system of *Trichoderma reesei*. *Biotechnology Letters* 1:41.
76. Carle-Urioste, J. C. et al. 1997. Cellulase induction in *Trichoderma reesei* by cellulose requires its own basal expression. *Journal of Biological Chemistry* 272:10169.
77. Suto, M., and Tomita, F. 2001. Induction and catabolite repression mechanisms of cellulase in fungi. *Journal of Bioscience and Bioengineering* 92:305.
78. El-Gogary, S. et al. 1989. Mechanism by which cellulose triggers cellobiohydrolase I gene expression in *Trichoderma reesei*. *Proceedings of the National Academy of Sciences USA* 86:6138.
79. Mitchinson, C. 2004. Improved cellulases for the biorefinery: A review of Genencor's progress in the DOE subcontract for cellulase cost reduction for bioethanol. Stanford GCEP Biomass Energy Workshop, April 2004, accessed at http:/gcep.stanford.edu/pdfs/energy_workshops_04_04/biomass_mitchinson.pdf.
80. Foreman, P. K. et al. 2003. Transcriptional regulation of biomass-degrading enzymes in the filamentous fungus *Trichoderma reesei*. *Journal of Biological Chemistry* 278:31988.
81. Durand, H., Clanet, M., and Tiraby, G. 1988. Genetic improvement of *Trichoderma reesei* for large scale cellulase production. *Enzyme and Microbial Technology* 10:341.
82. Berlin, A. et al. 2007. Optimization of enzyme complexes for lignocellulose hydrolysis. *Biotechnology and Bioengineering* 97:286.
83. Martinez, D. et al. 2008. Genome sequence analysis of the cellulolytic fungus *Trichoderma reesei* (syn. *Hypocrea jecorina*) reveals a surprisingly limited inventory of carbohydrate active enzymes. *Nature Biotechnology* 26:553.
84. Seiboth, B. et al. 1997. Role of four major cellulases in triggering cellulase gene expression by cellulose in *Trichoderma reesei*. *Journal of Bacteriology* 179:5318.
85. Fowler, T., and Brown, R. D. 1992. The *bgl1* gene encoding extracellular β-glucosidase from *Trichoderma reesei* is required for rapid induction of the cellulase complex. *Molecular Microbiology* 6:3225.
86. Strauss, J. et al. 1995. *Cre1*, the carbon catabolite repressor protein from *Trichoderma reesei*. *FEBS Letters* 376:103.
87. Ilmén, M. et al. 1996. Functional analysis of the cellobiohydrolase I promoter of the filamentous fungus *Trichoderma reesei*. *Molecular and General Genetics* 253:303.
88. Ilmén, M. et al. 1997. Regulation of cellulase gene expression in the filamentous fungus *Trichoderma reesei*. *Applied and Environmental Microbiology* 63:1298.

89. Saloheimo, A. et al. 2000. Isolation of the *ace1* gene encoding a Cys_2-His_2 transcription factor involved in the regulation of activity of the cellulase promoter *cbh1* of *Trichoderma reesei*. *Journal of Biological Chemistry* 275:5817.

90. Aro, N. et al. 2001. ACEII, a novel transcriptional activator involved in regulation of cellulase and xylanase genes of *Trichoderma reesei*. *Journal of Biological Chemistry* 276:24309.

91. Rosgaard, L. et al. 2006. Efficiency of new fungal cellulase systems in boosting enzymatic degradation of barley straw lignocellulose. *Biotechnology Progress* 22:493.

92. Viikari, L. et al. 2007. Thermostable enzymes in lignocellulose hydrolysis. *Advances in Biochemical Engineering/Biotechnology* 108:121.

93. Nath, S. et al. 1997. Evaluation of enzyme thermostability by enzyme assay and differential scanning calorimetry. A study of alcohol dehydrogenase. *Journal of the Chemical Society, Faraday Transactions* 93:3351.

94. Danson, M. J. et al. 1996. Enzyme thermostability and thermoactivity. *Protein Engineering* 9:629.

95. Pace, N. R. 1997. A molecular view of microbial diversity and the biosphere. *Science* 276:734.

96. Warnecke, F. 2007. Metagenomic and functional analysis of hindgut microbiota of a wood-feeding higher termite. *Nature* 450:560.

97. Kim, E. et al. 1998. Factorial optimization of a six-cellulase mixture. *Biotechnology and Bioengineering* 58:494.

98. Reese, E. T., Sui, R. G. H., and Levinson, H. S. 1950. The biological degradation of soluble cellulose derivatives and its relationship to the mechanism of cellulose hydrolysis. *Journal of Bacteriology* 59:485.

99. Saloheimo, M. et al. 2002. Swollenin, a *Trichoderma reesei* protein with sequence similarity to the plant expansins, exhibits disruption activity on cellulosic materials. *European Journal of Biochemistry* 269:4202.

100. Hildén, L., and Johansson, G. 2004. Recent developments on cellulases and carbohydrate-binding modules with cellulose affinity. *Biotechnology Letters* 26:1663.

101. Henriksson, G. et al. 1997. Studies of cellulose binding by a cellobiose dehydrogenase and comparison with cellobiohydrolase I. *Biochemical Journal* 324:833.

102. Saville, B. A. et al. 2004. Characterization and performance of immobilized amylase and cellulase. *Applied Biochemistry and Biotechnology* 113–116:251.

103. Yuan, X. et al. 1999. Immobilization of cellulase using acrylamide grafted acrylonitrile copolymer membranes. *Journal of Membrane Science* 55:101.

104. Tu, M. et al. 2006. Immobilization of β-glucosidase on Eupergit C for lignocellulose hydrolysis. *Biotechnology Letters* 28:151.

105. Jones, P. O., and Vasudevan, P. T. 2010. Cellulose hydrolysis by immobilized *Trichoderma reesei* cellulase. *Biotechnology Letters* 2009 32:103.

106. Cousins, W. J. 1978. Young's modulus of hemicellulose as related to moisture content. *Wood Science and Technology* 12:161.

107. Sonenshein, A. L., Hoch, J. A., and Losick, R., eds. 2002. Bacillus subtilis *and its closest relatives. From genes to cells,* appendix 2. Washington, D.C.: ASM Press.

108. Shallom, D., and Shoham, Y. 2003. Microbial hemicellulases. *Current Opinions in Microbiology* 6:219.

109. Zaide, G. et al. 2001. Biochemical characterization and identification of catalytic residues in α-glucuronidase from *Bacillus stearothermophilus* T-6. *European Journal of Biochemistry* 268:3006.

110. Numan, M. T., and Bhosle, N. B. 2006. α-L-Arabinofuranosidases: The potential applications in biotechnology. *Journal of Industrial Microbiology and Biotechnology* 33:247.

111. Birgisson, H. et al. 2004. A new thermostable α-L-Arabinofuranosidase from a novel thermophilic bacterium. *Biotechnology Letters* 26:1347.

112. Sørensen, H. R. et al. 2006. A novel GH43 α-L-Arabinofuranosidase from *Humicola insolens*: Mode of action and synergy with GH51 α-L-Arabinofuranosidases on wheat arabinoxylan. *Applied Microbiology and Biotechnology* 73:850.

113. Wu, S., Liu, B., and Zhang, X. 2006. Characterization of a recombinant thermostable xylanase from deep-sea thermophilic bacterium *Geobacillus* sp. MT-1 in East Pacific. *Applied Microbiology and Biotechnology* 72:1210.

114. Hinman, N. D. et al. 1989. Xylose fermentation: An economic analysis. *Applied Biochemistry and Biotechnology* 20–21:391.

115. Schell, D. J. et al. 1991. A technical and economic analysis of acid-catalyzed steam explosion and dilute sulfuric acid pretreatments using wheat straw or aspen wood chips. *Applied Biochemistry and Biotechnology* 28–29:87.

116. Brigham, J. S., Adney, W. S., and Himmel, M. E. 1996. Hemicellulases: Diversity and applications. In *Handbook on bioethanol: Production and utilization,* ed. Wyman, C. E., chap. 7. London: Taylor & Francis.

117. Kabel, M. A. et al. 2006. Standard assays do not predict the efficiency of commercial cellulase preparations towards plant materials. *Biotechnology and Bioengineering* 93:56.

118. Galbe, M., and Zacchi, G. 2007. Pretreatment of lignocellulosic materials for efficient bioethanol production. *Advances in Biochemical Engineering/Biotechnology* 108:43.

119. Nakari-Setälä, T. et al. 2009. Genetic modification of carbon catabolite repression in *Trichoderma reesei* for improved protein production. *Applied and Environmental Microbiology* 75:4853.

120. Mach-Aigner, A. R. et al. 2008. Transcriptional regulation of *xyr1*, encoding the main regulator of the xylanolytic and cellulolytic enzyme system in *Hypocrea jecorina*. *Applied and Environmental Microbiology* 74:6554.

121. Rauscher, R. et al. 2006. Transcriptional regulation of *xyn1*, encoding xylanase I, in *Hypocrea jecorina*. *Eukaryotic Cell* 5:447.

122. Stricker, A. R., Mach, R. L., and de Graaff, L. H. 2008. Regulation of transcription of cellulases- and hemicellulases-encoding genes in *Aspergillus niger* and *Hypocrea jecorina* (*Trichoderma reesei*). *Applied Microbiology and Biotechnology* 78:211.

123. Fan, Z., Werkman, J. R., and Yuan, L. 2009. Engineering of a multifunctional hemicellulase. *Biotechnology Letters* 31:751.

124. Fan, Z. et al. 2009. Multimeric hemicellulases facilitate biomass conversion. *Applied and Environmental Microbiology* 75:1754.

125. Akin, D. E. et al. 1995. Alterations in structure, chemistry, and biodegradability of grass lignocellulose treated with the white rot fungi *Ceriporiopsis subvermispora* and *Cyathus stercoreus*. *Applied and Environmental Microbiology* 61:1591.

126. Taniguchi, M. et al. 2005. Evaluation of pretreatment with *Pleurotus ostreatus* for enzymatic hydrolysis of rice straw. *Journal of Bioscience and Bioengineering* 100:637.

127. Singh, D., and Chen, S. 2008. The white-rot fungus *Phanerochaete chrysosporium*: Conditions for the production of lignin-degrading enzymes. *Applied Microbiology and Biotechnology* 81:399.

128. Mendonça, R. T. et al. 2008. Evaluation of the white-rot fungi *Ganoderma australe* and *Ceriporiopsis subvermispora* in biotechnological applications. *Journal of Industrial Microbiology and Biotechnology* 35:1323.

129. Shary S. et al. 2008. Differential expression in *Phanerochaete chrysosporium* of membrane-associated proteins relevant to lignin degradation. *Applied and Environmental Microbiology* 74:7252.

130. Kuhar, S., Nair, L. M., and Kuhad, R. C. 2008. Pretreatment of lignocellulosic material with fungi capable of higher lignin degradation and lower carbohydrate degradation improves substrate acid hydrolysis and the eventual conversion to ethanol. *Canadian Journal of Microbiology* 54:305.

131. Schilling, J. S., Tewalt, J. P., and Duncan, S. M. 2009. Synergy between pretreatment lignocellulose modifications and saccharification efficiency in two brown rot fungal systems. *Applied Microbiology and Biotechnology* 84:465.
132. Martinez, D. et al. 2009. Genome, transcriptome, and secretome analysis of wood decay fungus *Postia placenta* supports unique mechanisms of lignocellulose conversion. *Proceedings of the National Academy of Sciences USA* 106:1954.
133. Foody, B., Tolan, J. S., and Bernstein, J. D. 1997. Pretreatment process for conversion of cellulose to fuel ethanol. US Patent 5,916,780, issued June 29, 1997.
134. Howard, R. L. et al. 2003. Lignocellulose biotechnology: Issues of bioconversion and enzyme production. *African Journal of Biotechnology* 2:603.
135. Rosales, E., Rodriguez Couto, S., and Sanromán, A. 2002. New uses of food waste: Application to laccase production by *Trametes hirsuta*. *Biotechnology Letters* 24:701.
136. Pandey, A., Soccol, C. R., and Mitchell, D. 2000. New developments in solid state fermentation. I. Bioprocesses and products. *Process Biochemistry* 35:1153.
137. Schell, D., and Duff, B. 1996. Review of pilot plant programs for bioethanol conversion. In *Handbook on bioethanol: Production and utilization,* ed. Wyman, C. E., chap. 17. London: Taylor & Francis.

3 Microbiology of Cellulosic Ethanol Production I
Yeasts

3.1 INTRODUCTION

What makes a good ethanol-forming organism (an ethanologen)? Traditionally, yeasts have been the go-to organisms, improved empirically and selected for centuries by brewers. Microbial physiology has defined highly useful traits for ethanologenic yeasts but can these be pursued to biochemical and molecular levels of understanding?

Saccharomyces yeasts are extremely powerful ethanologens—given a strictly limited range of carbohydrate substrates. How can such yeasts be improved by genetic engineering to utilize both hexoses and pentoses or should naturally pentose-fermenting yeasts be chosen and developed for cellulosic ethanol production?

After rounds of genetic manipulation, can the resulting yeasts cope with the rigors of highly concentrated fermentation media containing growth-inhibitory degradation products from lignocellulosic materials and still accumulate ethanol at high conversion efficiencies? Or should the yeast isolates carefully nurtured in breweries be the basis for robust ethanologens to push cellulosic ethanol to its limits of productivity?

3.2 TRADITIONAL ETHANOLOGENIC YEASTS

The fundamental challenge in selecting or tailoring a microorganism to produce ethanol from the mixture of sugars resulting from the hydrolysis of lignocellulosic feedstocks can be descried as follows:

- The best ethanol producers are poor at utilizing pentose sugars (including those that are major components of hemicelluloses, i.e., D-xylose and L-arabinose).
- Species that can efficiently utilize both pentoses and hexoses are less efficient at converting sugars to ethanol, exhibit a low tolerance of high ethanol concentrations, or coproduce high concentrations of metabolites such as acetic, lactic, pyruvic, and succinic acids in amounts to compromise the efficiency of substrate conversion to ethanol.[1–4]

Because bioprospecting microbial species in many natural habitats around the global ecosphere has failed so far to uncover an ideal ethanologen for fuel ethanol or other industrial uses, considerable ingenuity has been exhibited by molecular geneticists following one or the other of two strategies:

1. endowing traditional yeast ethanologens with novel traits, including the ability to utilize pentoses
2. "reforming" other yeasts to be more efficient at converting both pentoses and hexoses to ethanol

A third option (i.e., that of devising conditions for mixed cultures to function synergistically with mixtures of major carbon substrates) is discussed in Chapter 5.

3.2.1 CONVENTIONAL YEASTS

The principal wine/beer yeast *Saccharomyces cerevisiae* has the enormous historic advantage of being generally regarded as safe (GRAS) for industrial use; by extrapolation, it is capable of being sold as an ingredient in animal feed once the fermentation process is completed.[5] In addition, some strains are relatively tolerant of the growth inhibitors found in the acid hydrolysates of lignocellulosic biomass.[6]

Its biotechnological limitations, on the other hand, derive from its relatively narrow range of fermentable substrates[7]:

- Glucose, fructose, and sucrose are rapidly metabolized, as are galactose and mannose (hexose constituents of plant hemicelluloses) and maltose (a disaccharide breakdown product of starch).
- The disaccharides trehalose and isomaltose are slowly utilized, as are the trisaccharides raffinose and maltotriose (another breakdown product of starch), the pentose sugar ribose, and glucuronic acid (a sugar acid in plant hemicelluloses).
- Cellobiose, lactose, xylose, rhamnose, sorbose, and maltotetraose are *not* utilizable.

S. cerevisiae is, from the standpoint of classical microbial physiology, best described as "facultatively fermentative"; that is, it can metabolize sugars such as glucose *either* entirely to CO_2 and water given an adequate O_2 supply *or* (under microaerobic conditions) generate large amounts of ethanol. This ability for dual metabolism is also exhibited by a large number of yeast species.[8] In complete anaerobiosis, however, growth eventually ceases because compounds essential for cell growth (e.g., unsaturated fatty acids and sterols) cannot be synthesized without the involvement of O_2.[9]

At moderate temperature, defined isolates of baker's, brewer's, or wine yeasts can accumulate ethanol as the main fermentation product from glucose, sucrose, galactose, and molasses (Table 3.1).[10–13] Molasses is a particularly relevant sugar source for industrial use; the principal carbohydrate in molasses, from either sugarcane or sugar beet, is sucrose, but there are also variable (and sometimes large) amounts of glucose, fructose, and trisaccharides and tetrasaccharides based on glucose and fructose.[14]

TABLE 3.1

Ethanol Production by Yeasts on Different Carbon Sources

Yeast	Carbon Source	Temperature (°C)	Fermentation Time (hr)	Maximum Ethanol (g/L)	Ref.
Saccharomyces cerevisiae	Glucose, 200 g/L	30	94	91.8	10
Saccharomyces cerevisiae	Sucrose, 220 g/L	28	96	96.7	11
Saccharomyces cerevisiae	Galactose, 20–150 g/L	30	60	40.0	12
Saccharomyces cerevisiae	Molasses, 1.6–5.0 g/L	30	24	18.4	13
Saccharomyces pastorianus	Glucose, 50 g/L	30	30	21.7	10
Saccharomyces bayanus	Glucose, 50 g/L	30	60	23.0	10
Kluyveromyces fragilis	Glucose, 120 g/L	30	192	49.0	10
Kluyveromyces marxianus	Glucose, 50 g/L	30	40	24.2	10
Candida utilis	Glucose, 50 g/L	30	80	22.7	10

Across the many known facultatively fermentative yeasts, the ability to use sugars other than glucose efficiently is highly variable (Table 3.2).[15] A serious limitation for ethanol production is that many potentially suitable yeasts can only *respire* disaccharides (i.e., they can grow on the sugars under aerobic conditions but cannot produce ethanol under any degree of anaerobiosis); this is the so-called Kluyver effect.[16] In a survey of 215 glucose-fermenting yeast species, 96 exhibited the Kluyver effect with at least one disaccharide; and two of these disaccharides are of great importance for ethanol production: maltose and cellobiose (a degradation product of cellulose).[15] The Kluyver effect is only one of four important O_2-related metabolic phenomena in yeasts; the others are

- the Pasteur effect (i.e., the inhibition of sugar consumption by O_2)[17]
- the Crabtree effect (i.e., the occurrence or continuance of ethanol formation in the presence of O_2 at high growth rates or when an excess of sugar is provided)[18]
- the Custers effect (i.e., the inhibition of fermentation by the absence of O_2 found only in a small number of yeast species capable of fermenting glucose to ethanol under *fully* aerobic conditions)[19]

For efficient ethanol producers, the optimum combination of phenotypes is to be

- Pasteur positive (i.e., with efficient use of glucose and other readily utilizable sugars for growth when O_2 levels are relatively high)
- Crabtree positive for high rates of ethanol production when supplied with abundant fermentable sugar from as soon as possible in the fermentation
- Custers negative (i.e., insensitive to fluctuating, sometimes very low, O_2 levels)
- Kluyver negative for the widest possible range of fermentable sugars

TABLE 3.2

Fermentation of Galactose, Five Disaccharides, and Two Trisaccharides by Yeasts

Yeast	Galactose	Maltose	Sucrose	Trehalose	Melibiose	Lactose	Cellobiose	Melezitose	Raffinose
Ambrosiozyma monospora	–	–	–	–	–	–	–	–	–
Candida chilensis	K	K	K	K	–	–	K	K	–
Candida salmanticensis	+	+	+	+	+	K	+	+	+
Candida silvicultrix	+	–	+	–	+	–	+	K	+
Candida shehatae	+	+	–	+	–	–	K	K	–
Kluyveromyces marxianus	+	+	+	+	–	K	K	K	K
Pachysolen tannophilus	K	–	–	–	–	+	K	–	–
Pichia hampshirensis	–	K	K	K	–	–	K	K	–
Pichia stipitis	+	+	+	+	–	K	K	K	–
Pichia subpelliculosa	K	+	+	+	–	–	K	K	+
Saccharomyces bayanus	+	+	+	+	+	–	–	K	+
Saccharomyces cerevisiae	+	+	+	K	+	–	–	+	+
Saccharomyces kluyveri	+	K	+	K	+	–	K	–	+
Saccharomyces pastorianus	+	+	+	K	+	–	–	–	+
Schizosaccharomyces pombe	–	+	+	–	+	–	–	–	+
Zygosaccharomyces fermentati	+	+	+	+	+	–	K	+	+

Source: Data from Barnett, J. A., Payne, R. W., and Yarrow, D. 2000. *Yeasts: Characteristics and Identification*, 3rd ed. Cambridge, England: Cambridge University Press.

Notes: K = exhibits aerobic respiratory growth but no fermentation (Kluyver effect); – = may include delayed use (after 7 days).

Not all of these effects can be demonstrated with common wine yeasts (or only under special environmental or laboratory conditions). However, they are all of relevance when considering the use of other yeasts or when adapting the growth and fermentative capacities of yeast ethanologens to unstable fermentation conditions (e.g., low O_2 supply and intermittent sugar inflow).

Nevertheless, *S. cerevisiae,* with many other yeast species, faces the serious metabolic challenges posed by the use of mixtures of mono-, di-, and oligosaccharides as carbon sources (Figure 3.1). Potential biochemical bottlenecks arise from the conflicting demands of growth, cell division, and the synthesis of cellular constituents in a relatively fixed set of ratios, together with the requirement to balance redox cofactors with an inconsistent supply of both sugar substrates and O_2.

For a yeast species in a particular nutrient medium growing under known physical conditions, specific combinations of these parameters may prove crucial for limiting growth and fermentative ability. For example, over the more than 60 years since its discovery, various factors have been hypothesized to influence the Kluyver effect; however, a straightforward product inhibition by ethanol could be the root cause. In aerobic cultures, ethanol suppresses the utilization of those disaccharides that cannot be fermented, and the rate of their catabolism is adapted to the yeast culture's respiratory capacity.[20] The physiological basis for this preference is that Kluyver-positive yeasts lack high-capacity transporter systems for some sugars to support the high substrate transport into cells necessary for fermentative growth— whereas energy-efficient respiratory growth simply does not require a high rate of sugar uptake.[21,22]

The function of O_2 in limiting fermentative capacity is complex; in excess, it blocks fermentation in many yeasts, but a limited O_2 supply enhances fermentation in other species.[8,18] Detailed metabolic analyses have shown that the basic pathways of carbon metabolism in ethanologenic yeasts are highly flexible for their quantitative expression, with major shifts in how pathways function to direct the traffic flow of glucose-derived metabolites into growth and oxidative or fermentative sugar catabolism.[23,24] The Crabtree effect can be viewed as the existing biochemical networks adapting to consume as much of the readily available sugar (a high-value carbon source for microorganisms) as possible—and always with the capability of being able to reuse the accumulated ethanol as a carbon source when the carbohydrate supply eventually becomes depleted.[20,25,26]

Even when glucose fermentation occurs under anaerobic or microaerobic conditions, the fermentation of xylose (and other sugars) may still require O_2. When xylose metabolism commences by its reduction to xylitol (catalyzed by NADPH-dependent xylose reductase), the subsequent step is carried out under the control of an NAD-dependent xylitol dehydrogenase. This results in a disturbed redox balance of reduced and oxidized cofactors if O_2 is not present; NADH cannot then be reoxidized, and fermentation soon ceases (Figure 3.2).[27,28] This biochemical complexity makes accurate control of an ethanol fermentation difficult and has attracted many researchers to bacterial ethanologens where metabolic regulation is more straightforward.

On the other hand, the remarkably high growth rate attainable by *S. cerevisiae* at very low levels of dissolved O_2 and its efficient transformation of glucose maintain

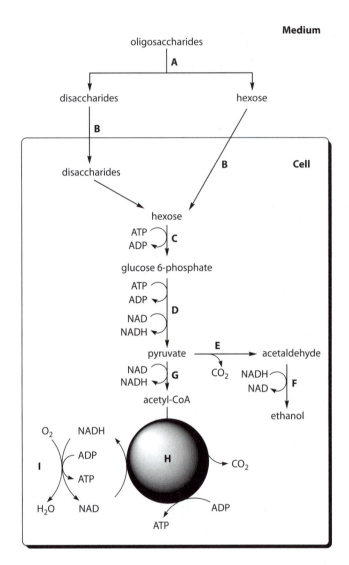

FIGURE 3.1 Biochemical outline of the uptake and metabolism of oligosaccharides and hexoses by yeasts. Indicated steps: A: glycosidases; B: sugar transport and uptake; C: entry into glucose-phosphate pool; D: glycolysis; E: pyruvate decarboxylase; F: alcohol dehydrogenase; G: pyruvate dehydrogenase; H: tricarboxylic acid cycle (mitochondrial); I: electron transport (mitochondrial).

its A-list status in the rankings of biologically useful organisms (Figure 3.3).[29] Whole genome sequencing has shown that the highly desirable evolution of the modern *S. cerevisiae* yeast ethanologen has occurred over 150 million years, resulting in a Crabtree-positive species that can readily generate respiratory-deficient, high alcohol-producing petite cells immune to the Pasteur effect in a readily acquired and efficient fermentative lifestyle.[30,31]

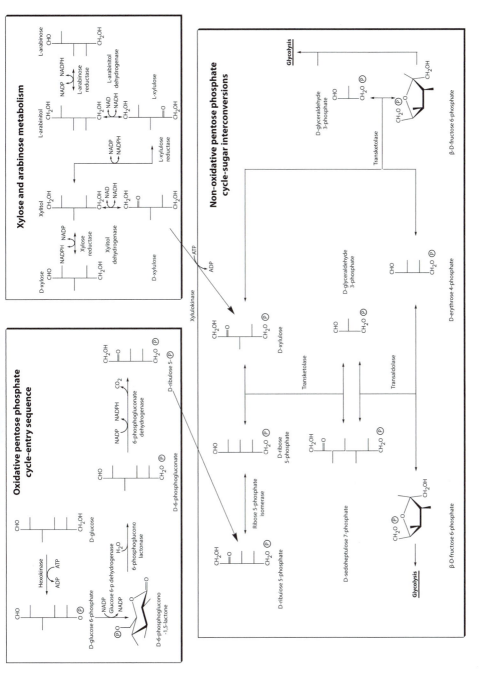

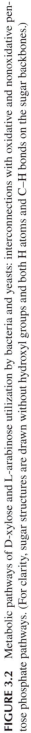

FIGURE 3.2 Metabolic pathways of D-xylose and L-arabinose utilization by bacteria and yeasts: interconnections with oxidative and nonoxidative pentose phosphate pathways. (For clarity, sugar structures are drawn without hydroxyl groups and both H atoms and C–H bonds on the sugar backbones.)

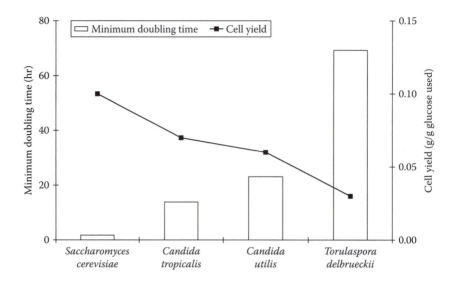

FIGURE 3.3 Growth of yeasts in anaerobic batch cultures after growth previously under O_2 limitation. (Data from Visser, W. et al. 1990. *Applied and Environmental Microbiology* 56:3785.)

STEM TOPIC 3.1: WHAT DOES "FERMENTATION" MEAN?

The word "fermentation" has evolved a long way semantically since its coining by Louis Pasteur (1822–1895) as "life without air" (*la vie sans l'air*). Pasteur was referring to yeast cells and how they altered their respiratory functions when air (or, more precisely, oxygen, O_2) became limiting. Microbes oxidize (or combust) glucose by the following overall chemical reaction:

$$C_6H_{12}O_6 + 6O_2 \rightarrow 6CO_2 + 6H_2O$$

As written, this chemical transformation is of little biological use to a population of yeast, fungal, or bacterial cells because it implies only the generation of metabolic heat by the cells and that all the carbon is evolved as CO_2 and thus cannot be utilized in cell replication. In a real microbiological scenario, glucose represents a valuable organic carbon supply as well as an energy source; part of that carbon is transformed by biochemical reactions inside the cells into the materials for new cells (prior to cell division). Much of the free energy released by the energetically favorable (exothermic) oxidation of the remainder of the glucose is used to drive the energetically unfavorable (endothermic) reactions of biosynthetic pathways; only a small portion of the total energy available is wasted as heat:

$$nC_6H_{12}O_6 + mO_2 \rightarrow cells + pCO_2 + qH_2O + \text{evolved heat (where p,q} \ll 6n)$$

When insufficient O_2 is present to allow the metabolism of glucose to CO_2 and H_2O, yeasts, fungi, and bacteria can produce combinations of alcohols (ethanol, glycerol, n-propanol, n-butanol), acids (formic, acetic, lactic, propionic, butyric), and decarboxylated acids (acetoin, acetone, diacetyl, 2,3-butanediol) instead of CO_2 and H_2O. In the case of ethanol producers such as *S. cerevisiae,* the appropriate chemical equation then becomes

$$nC_6H_{12}O_6 + mO_2 \rightarrow cells + pCO_2 + qH_2O + rC_2H_5OH + \text{evolved heat}$$

Unfortunately, the word fermentation evolved a second and quite distinct meaning in the twentieth century as large-scale industrial processes began to manufacture antibiotics and enzymes and then recombinant proteins, vitamins, and many fine chemicals. All of these processes are routinely described as fermentations even though abundant O_2 is forced at high flow rates into fermentation vessels precisely to avoid microaerobic conditions. The commercial outputs of the process are fermentation products even though the accumulation of ethanol (by *S. cerevisiae* genetically engineered to biomanufacture a recombinant protein) would be catastrophic, representing the failure of the biochemical engineers to maintain adequate airflow properly!

3.2.2 NONCONVENTIONAL YEASTS

For many geneticists and biotechnologists, any yeast other than *Saccharomyces* or *Schizosaccharomyces* is nonconventional. There are good practical reasons for this viewpoint: In particular, the applicability of classical genetic techniques was easy with *S. cerevisiae* and close relatives so that improved traits could be stably inherited from generation to generation. However, this is far from being universally so among yeasts.[32,33]

With the rise of modern biomanufacturing and the production of proteins whose encoding genes are foreign to the host, *S. cerevisiae* was an immediately functional and well-understood option. However, other yeasts had clear advantages in being able to secrete heterologous proteins at higher levels and to have a tighter regulation of their production.[34] This encouraged an active exploration of yeasts such as *Hansulena polymorpha, Pichia pastoris, Kluveromyces lactis,* and *Yarrowia lipolytica*—all of which now are widely accepted for the biotechnological production of proteins.

Traditional alcoholic beverages often have unusual yeast species in their mode of production—over 200 in a South American cereal-based beverage.[35] Such nonconventional yeasts will frequently occur in this and later chapters because their idiosyncratic genetic, physiological, and biochemical properties are highly relevant to the construction of yeasts for the bioproduction of ethanol and fine chemicals from cellulosic biomass.

3.3   METABOLIC ENGINEERING OF YEASTS
FOR CELLULOSIC ETHANOL

3.3.1   INCREASED PENTOSE UTILIZATION BY ETHANOLOGENIC YEASTS BY GENETIC
MANIPULATION WITH YEAST GENES FOR XYLOSE METABOLISM VIA XYLITOL

It has been known for many years that *S. cerevisiae* cells takes up xylose from nutrient media; the transport system is one shared by at least 25 sugars (both natural and synthetic).[36] Moreover, both D-xylose and L-arabinose can be reduced by *S. cerevisiae;* the products are the sugar alcohols, xylitol and L-arabinitol, respectively. Three separate genes encode enzymes with overlapping selectivities for xylose and arabinose as substrates.[37] The wild-type *S. cerevisiae* genome contains genes for both xylose reductase and xylitol dehydrogenase; cells should therefore be able to isomerize xylulose from xylose, and the resulting xylulose (after phosphorylation catalyzed by a specific xylulokinase) can enter the pentose phosphate pathway (Figure 3.2).[38,39]

Overexpressing the endogenous yeast genes for xylose catabolism renders the organism capable of growth on xylose in the presence of glucose as cosubstrate under aerobic conditions, although no ethanol is formed. *S. cerevisiae* may therefore have evolved originally to utilize xylose and other pentoses, but this has been muted, possibly as its natural ecological niche altered or the organism changed its range of favored environments.[38]

Ethanol production from xylose is a rare phenomenon among yeasts; of 200 species tested in the laboratory, only 6 accumulated ethanol to more than 1 g/L (0.1% by volume): *Pichia stipitis, P. segobiensis, Candida shehatae, C. tenuis, Brettanomyces naardenensis,* and *Pachysolen tannophilus.*[39] Even rarer is the ability among yeasts to hydrolyze xylans. Only *P. stipitis* and *C. shehatae* have xylanase activity; *P. stipitis* could, moreover, convert xylan into ethanol at 60% of the theoretical yield as computed from the xylose content of the polymer.[40] Three naturally xylose-fermenting yeasts have been used as donors for genes encoding enzymes of xylose utilization for transfer to *S. cerevisiae: P. stipitis, C. shehatae,* and *C. parapsilosis.*[41–43] These organisms all metabolize xylose by the enzymes of the same low-activity pathway known in *S. cerevisiae* (Figure 3.2), and the relevant enzymes include arabinose as a substrate.

STEM TOPIC 3.2: MAXIMUM THEORETICAL YIELDS
OF ETHANOL FROM SUGAR SUBSTRATES

Once the growth rate in a fermentation of glucose by an ethanol-forming organism becomes very low, if glucose continues to be supplied (or is present in excess), the biochemistry of the process can be written minimally as

$$C_6H_{12}O_6 \rightarrow 2CO_2 + 2C_2H_5OH$$

This represents a fully fermentative state where no glucose is respired and no fermentation side-products are formed. For every molecule of glucose consumed, two molecules of ethanol are formed; for every 180 g of glucose consumed, 92 g of ethanol are formed, or 51.1 g of ethanol are produced per 100 g of glucose utilized (0.51 g ethanol per gram of glucose consumed). Any measurable growth or side-product formation (e.g., glycerol) will reduce the conversion efficiency of glucose to ethanol from this theoretical maximum.

Microbial energetics and biochemistry can, therefore, predict maximum yields of ethanol and other fermentation products from the sugars available in cane juices, starch, and cellulose because these sugars can all be converted biochemically to glucose. How closely this theoretical maximum is reached is a function of fermentation design, control, engineering, and the biochemistry of the microorganism used for ethanol production.

If a lignocellulosic substrate is to be used, however, the range of biochemical pathways required increases greatly both in scope and complexity as more sugars (pentoses as well as hexoses, plus sugar acids such as glucuronic acid) are potentially made available. Hexoses such as galactose and mannose can be converted by yeasts to glucose; correspondingly, their maximal gravimetric conversion to ethanol is the same as that for glucose.

Perhaps surprisingly, this is also true for pentoses. The pathway of xylose metabolism (Figure 3.2) is such that 3 mol of xylose are converted by transaldolase and transketolase actions into 2 mol of hexose (fructose) phosphate and 1 mol of triose phosphate (glyceraldehyde 3-phosphate); each mole of fructose phosphate can yield a maximum of 2 mol of ethanol and each mole of triose phosphate yields 1 mol of ethanol. Therefore:

$$3C_5H_{10}O_5 \rightarrow 5CO_2 + 5C_2H_5OH$$

Each mole of xylose can yield 10/6 or 1.67 mol ethanol; that is, the maximum theoretical yield of ethanol is $(1.67 \times 46)/150$ or 0.51 g per gram of xylose.

P. stipitis has been the most widely used donor of genes for xylose metabolism because it shows relatively little accumulation of xylitol when growing on and fermenting xylose, thus wasting less sugar as the xylitol side product.[44] This advantageous property does not appear to reside in the enzymes for xylose catabolism but rather in the occurrence of an alternative respiration pathway (a cyanide-insensitive route widely distributed among yeasts of industrial importance). Inhibiting this alternative pathway renders *P. stipitis* quite capable of accumulating the sugar alcohols xylitol, arabinitol, and ribitol.[45] Respiration in *P. stipitis* is not repressed by high concentrations of fermentable sugars or by O_2 limitation (i.e., the yeast is Crabtree negative); as an ethanologen, *P. stipitis* suffers from the reduction of fermentative ability by aerobic conditions.[46]

Transferring genetic information from *P. stipitis* in intact nuclei to *S. cerevisiae* produced karyoductants (i.e., diploid cells where nuclei from one species have been introduced into protoplasts of another, the two nuclei subsequently fusing, with the ability to grow on both xylose and arabinose). However, the hybrid organism was inferior to *P. stipitis* in ethanol production and secreted far more xylitol to the medium than did the donor; its ethanol tolerance was, on the other hand, almost exactly midway between the tolerance ranges of *S. cerevisiae* and *P. stipitis*.[47]

For direct genetic manipulation of *S. cerevisiae,* however, the most favored strategy (starting in the early 1990s) has been to insert the two genes (*xyl1* and *xyl2*) from *P. stipitis* coding for xylose reductase (XR) and xylitol dehydrogenase (XDH), respectively.[48] Differing ratios of expression of the two foreign genes resulted in smaller or higher amounts of xylitol, glycerol, and acetic acid, and the optimum XR:XDH ratio of 0.06:1 can yield no xylitol, less glycerol and acetic acid, and more ethanol than with other engineered *S. cerevisiae* strains.[49]

The first patented *Saccharomyces* strain to coferment xylose and glucose to ethanol (not *S. cerevisiae,* but rather a fusion between *S. diastaticus* and *S. ovarum* able to produce ethanol at 40°C) was constructed with four specific traits tailored for industrial use[50–52]:

1. to direct carbon flow effectively from xylose to ethanol production rather than to xylitol and other by-products
2. to coferment mixtures of glucose and xylose effectively
3. to convert industrial strains of *S. cerevisiae* easily to coferment xylose and glucose using plasmids with readily identifiable antibiotic resistance markers controlling gene expression under the direction of promoters of *S. cerevisiae* glycolytic genes
4. to support rapid bioprocesses with growth on nutritionally rich media

In addition to XR and XDH, the yeast's own xylulose-phosphorylating xylulokinase (XK, Figure 3.2) was also overexpressed via high-copy-number yeast–*Escherichia coli* shuttle plasmids.[50] This extra gene manipulation was crucial because both earlier and contemporary attempts to transform *S. cerevisiae* with only genes for XR and XDH produced transformants with slow xylose utilization and poor ethanol production. The synthesis of the xylose-metabolizing enzymes not only did not require the presence of xylose but glucose was incapable of repressing their formation.

It was known that *S. cerevisiae* could consume xylulose anaerobically but only at < 5% of the rate of glucose utilization; XK activity was very low in unengineered cells, and it was reasoned that providing much higher levels of the enzyme was necessary to metabolize xylose via xylulose because the *P. stipitis* XK gene catalyzed a reversible reaction between xylitol and xylulose with the equilibrium heavily on the side of xylitol.[52–54] Such strains were quickly shown to ferment corn fiber sugars to ethanol and were later utilized by the Iogen Corporation in their demonstration process for producing ethanol from wheat straw.[55,56]

The vital importance of increased XK activity in tandem with the XR/XDH pathway for xylose consumption in yeast was demonstrated in *S. cerevisiae*. Not only was xylose consumption increased but xylose as the sole carbon source could be converted to ethanol under both aerobic and anaerobic conditions, although ethanol production was at its most efficient in microaerobiosis (2% O$_2$).[57] Large increases in the intracellular concentrations of the xylose-derived metabolites (xylulose 5-phosphate and ribulose 5-phosphate) were demonstrated in the XK-overexpressing strains, but a major drawback was that xylitol formation greatly exceeded ethanol production when O$_2$ levels in the fermentation decreased.

Subsequently, considerable efforts have been dedicated to achieving higher ethanol productivity with the triple XR/XDH/XK constructs. Apart from continuing attempts to understand the metabolism of xylose more fully by unconventional or little-studied yeast species, two main centers of attention have been evident:

1. strategies for harmonizing the different cofactor requirements in the pathway (i.e., NAPDH-dependent [or -preferring] XR and NAD-requiring XDH), thus reducing xylitol formation
2. overexpressing a wider array of other pentose-metabolizing enzymes to maximize the rate of xylose use or (broadening the metabolic scope) increasing kinetic factors in the central pathways of carbohydrate metabolism

Early attempts to overexpress *P. stipitis* genes for XR and XDH in *S. cerevisiae* often resulted in high rates of xylitol formation; if NADPH formed in the oxidative pentose phosphate pathway (Figure 3.2) equilibrated with intracellular NAD to form NADH, the reduced availability of NAD could restrict the rate of the XDH reaction in the direction of xylitol oxidation to xylulose.[58,59] Adding external oxidants capable of being reduced by NADH (thus regenerating NAD) reduced xylitol formation. Two of these oxidants were furfural and 5-hydroxyfurfural, known as sugar degradation products present in acid hydrolysates of lignocellulose materials (see Chapter 2, Section 2.3.3.2).[60]

The adventitious removal of toxic impurities by these reactions probably explained why xylitol accumulation was very low when lignocellulose acid hydrolysates were used as carbon sources for XR- and XDH-transformed *S. cervisiae*.[61] This line of reasoning does not accord entirely with the high activities of XR measurable in vitro with NADH (63% of the rate with NADPH). However, site-specific mutagenesis on the cloned *P. stipitis* XR gene could increase the activity with NADH to 90% of that with NADPH and greatly reduce xylitol accumulation, although with only a marginally increased xylose utilization rate.[51]

Further optimization of the xylitol pathway for xylose assimilating was therefore entirely possible. Simply coalescing the XR and XDH enzymes into a single fusion protein, with the two active units separated by short peptide linkers, and expressing the chimeric gene in *S. cerevisiae* resulted in the formation of a bifunctional enzyme. The total activities of XR and XDH were similar to the activities when monomeric enzymes were produced. However, the molar yield of xylitol from xylose was reduced, the ethanol yield was higher, and the formation of glycerol was lower, suggesting that the artificially evolved enzyme complex was more selective for NADH in its XR domain as a consequence of the two active sites generating and utilizing NADH being in close proximity.[62]

Direct attempts to alter the preference of XR to use the NADPH cofactor have been made:

- A mutated gene for a *P. stipitis* XR with a lower affinity for NADPH replaced the wild-type XR gene and increased the yield of ethanol on xylose while decreasing the xylitol yield but also increasing the acetate and glycerol yields in batch fermentation.[63]
- The ammonia-assimilating enzyme glutamate dehydrogenase in *S. cerevisiae* (and other yeasts) can be specific to either NADPH or NADH; setting an artificial transhydrogenase cycle by simultaneously expressing genes for both forms of the enzyme improved xylose utilization rates and ethanol productivity.[64]
- Deleting the gene for the NADPH-specific glutamate dehydrogenase aimed to increase the intracellular NADH concentration and the competition between NADH and NADPH for XR but greatly reduced growth rate, ethanol yield, and xylitol yield on a mixture of glucose and xylose. Overexpressing the gene for the NADH-specific enzyme in the absence of the NADPH-requiring form, however, restored much of the loss in specific growth rates and increased both xylose consumption rate when glucose had been exhausted and the ethanol yield, while maintaining a low xylitol yield.[65]
- NADPH regeneration for the XR reaction was approached from a different angle by expressing in a xylose-utilizing *S. cerevisiae* strain the gene for an NADP-dependent D-glyceraldehyde 3-phosphate dehydrogenase, an enzyme providing precursors for ethanol from either glucose or xylose; the resulting strain fermented xylose to ethanol at a faster rate and with a higher yield.[66]
- The selectivity of *P. stipitis* XDH has been changed from NAD to NADP by multiple-site directed mutagenesis of the gene, thereby harmonizing the redox balance with XR.[67]

An NADH-preferring XR has also been demonstrated in the yeast *Candida parapsilopsis* as a source for a new round of genetic and metabolic engineering.[43]

Beyond the initial conversions of xylose and xylitol, pentose metabolism becomes relatively uniform across kingdoms and genera. Most microbial species—and plants, animals, and mammals (including *Homo sapiens*)—can interconvert some pentose structures via the nonoxidative pentose phosphate pathway (Figure 3.2). These reactions are readily reversible; however, extended and reorganized, the pathway can function to oxidize glucose fully via the glucose 6-phosphate dehydrogenase and 6-phosphogluconate dehydrogenase reactions (both forming NADPH), although the pathway is far more important for the provision of essential biosynthetic intermediates for nucleic acids, amino acids, and cell wall polymers.

The reactions can even (when required) run backward to generate pentose sugars from triose intermediates of glycolysis.[68] Increasing the rate of entry of xylulose into the pentose phosphate pathway by overexpressing endogenous XK activity has been shown to be effective for increasing xylose metabolism to ethanol (while reducing xylitol formation) with XR/XDH-transformants of *S. cerevisiae*.[69]

In contrast, disrupting the oxidative pentose phosphate pathway genes for either glucose 6-phosphate or 6-phosphogluconate oxidation increased ethanol production and decreased xylitol accumulation from xylose by greatly reducing (or eliminating) the main supply route for NADPH. This genetic change also increased the formation of the side products acetic acid and glycerol; a further deleterious result was a marked decrease in the xylose consumption rate—again a predictable consequence of low NADPH inside the cells as a coenzyme in the XR reaction.[70] Deleting the gene for glucose 6-phosphate dehydrogenase *in addition to* introducing one for NADP-dependent D-glyceraldehyde 3-phosphate dehydrogenase was an effective means of converting a strain fermenting xylose mostly to xylitol and CO_2 to an ethanologenic phenotype.[66]

The first report of overexpression of selected enzymes of the main nonoxidative pentose phosphate pathway (transketolase and transaldolase) in *S. cerevisiae* harboring the *P. stipitis* genes for XR and XDH concluded that the transaldolase level found naturally in the yeast was insufficient for efficient metabolism of xylose via the pathway. Although xylose could support growth, no ethanol could be produced, and a reduced O_2 supply merely impaired growth and increased xylitol accumulation.[70]

In a more ambitious exercise in metabolic engineering of pentose metabolism by *S. cerevisiae*, high activities of XR and XDH were combined with overexpression of endogenous XK and of four enzymes of the nonoxidative pentose phosphate pathway (transketolase, transaldolase, ribulose-5-phosphate epimerase, and ribose-5-phosphate ketolisomerase) and deletion of the endogenous, nonspecific NADPH-dependent aldose reductase (AR) catalyzing the formation of xylitol from xylose.[71] In comparison with a strain with lower XR and XDH activities and no other genetic modification other than XK overexpression, fermentation performance on a mixture of glucose (20 g/L) and xylose (50 g/L) was improved, with higher ethanol production, much lower xylitol formation, and a faster utilization of xylose. Deleting the nonspecific AR had no effect when XR and XDH activities were high, but glycerol accumulation was higher (Figure 3.4).

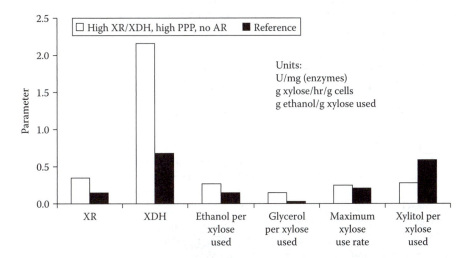

FIGURE 3.4 Effects of increased xylose reductase and xylitol dehydrogenase activities on xylose utilization by *Saccharomyces cerevisiae*. (Data from Karhumaa, K. et al. 2007. *Applied Microbiology and Biotechnology* 73:1039.)

To devise strains more suitable for use with industrially relevant mixtures of carbohydrates, it is essential for strains to be able to use residual oligosaccharides not degraded to free hexose and pentose sugars in hydrolysates of cellulose and hemicelluloses. Research groups have explored combinations of heterologous xylanases and β-xylosidases:

- A fusion protein consisting of the *xynB* β-xylosidase gene from *Bacillus pumilus* and the *S. cerevisiae Mfa1* signal peptide (to ensure the correct posttranslational processing) and the *XYN2* β-xylanase gene from *Hypocrea jecorina* were separately coexpressed in *S. cerevisiae* under the control of the glucose-derepressible *ADH2* alcohol dehydrogenase promoter and terminator. Coproduction of these xylan-degrading enzymes hydrolyzed birch wood xylan; however, no free xylose resulted, probably because of the low affinity of the β-xylosidase for its xylobiose disaccharide substrate.[72]
- A similar fusion strategy with the *xlnD* β-xylosidase gene from *Aspergillus niger* and the *XYN2* β-xylanase gene from *H. jecorina* enabled the yeast to hydrolyze birch wood xylan to free xylose.[73]
- A xylan-utilizing *S. cerevisiae* was constructed using cell surface engineering based on α-agglutinin (a cell surface glycoprotein involved in cell–cell interactions) to display xylanase II from *Hypocrea jecorina* and a β-xylosidase from *Aspergillus oryzae*. With *P. stipitis* XR and XDH and overexpressed endogenous XK, the strain could generate ethanol from birch wood xylan with a conversion efficiency of 0.3 g of ethanol per gram of carbohydrate used.[74]

High XR, XDH, and XK activities combined with the expression of a gene from *Aspergillus acleatus* for displaying β-glucosidase on the cell surface enabled *S. cerevisiae* to utilize xylose and cellulose-derived oligosaccharides from an acid hydrolysate of wood chips.[75]

3.3.2 INCREASED PENTOSE UTILIZATION BY ETHANOLOGENIC YEASTS BY GENETIC MANIPULATION WITH GENES FOR XYLOSE ISOMERIZATION

Historically, the earliest attempts to engineer xylose metabolic capabilities into *S. cerevisiae* involved the single gene for xylose isomerase (XI, catalyzing the interconversion of xylose and xylulose) from bacteria (*E. coli* and *Bacillus subtilis*); however, these failed because the heterologous proteins produced in the yeast cells were enzymically inactive.[4,51] A greater degree of success was achieved using the XI gene (*xylA*) from the bacterium *Thermus thermophilus;* transformants could exhibit ethanol formation in O_2-limited xylose fermentations.[76]

The crucial breakthrough was made when the gene encoding XI in the fungus *Piromyces* sp. strain E2 (capable of anaerobic growth on xylose) was recognized as part of the known bacterial pathway for xylose catabolism revealed for the first time to be functional in a eukaryote.[77] It was then quickly demonstrated that the *xylA* gene expressed in *S. cerevisiae* gave high XI activity but could not by itself induce ethanol production with xylose as the carbon source.[78] The additional genetic manipulations required for the construction of ethanologenic strains on xylose were the overexpression of XK, transketolase, transaldolase, ribulose-5-phosphate epimerase, and ribulose-5-phosphate isomerase and the deletion of nonspecific AR, followed by selection of spontaneous mutants in xylose-limited continuous cultures and anaerobic cultivation in automated sequencing-batch reactors on glucose-xylose media.[79–81]

The outcome was a strain with negligible accumulation of xylitol (or xylulose) and a specific ethanol production three- to fivefold higher than previously publicized strains. Mixtures of glucose and xylose were sequentially but completely consumed by anaerobic cultures of the engineered strain in anaerobic batch culture, with glucose still preferred as the carbon source.[80]

A side-by-side comparison of XR/XDH- and XI-based xylose utilizations in two isogenic strains of *S. cerevisiae* with genetic modifications to improve xylose metabolism (overexpressed XK and nonoxidative pentose phosphate pathway enzymes, and deleted AR) arrived at widely different conclusions for the separately optimal parameters of ethanol production[82]:

- In chemically defined medium, the XI-containing variant showed the highest ethanol yield (i.e., conversion efficiency) from xylose.
- The XR/XDH-transformant had the higher rate of xylose consumption, specific ethanol production, and final ethanol concentration, despite accumulating xylitol.
- In a lignocellulose hydrolysate, neither transformant accumulated xylitol but both were severely affected by toxic impurities in the industrially relevant medium, producing little or no ethanol, xylitol, or glycerol and consuming little or no xylose, glucose, or mannose.

The bacterial XI gene from *T. thermophilus* has been expressed in *S. cerevisiae* along with overexpressed XK and nonoxidative pentose phosphate pathway genes and deleted AR; the engineered strain, despite its low measured XI activity, exhibited for the first time *aerobic* growth on xylose as sole carbon source and *anaerobic* ethanol production at 30°C.[83]

3.3.3 ENGINEERING ARABINOSE UTILIZATION BY ETHANOLOGENIC YEASTS

Xylose reductase is the first step in the pathway of xylose catabolism in most yeast species, but it functions equally well as an enzyme with L-arabinose as with D-xylose, with a slightly higher affinity (lower K_m) and a higher maximal rate (V_{max}) for L-arabinose.[41] In contrast, polyol dehydrogenases that are active with xylitol find L- or D-arabinose to be poor substrates.[84]

An outline of known enzyme-catalyzed metabolic relationships for pentitols and pentoses is given in Figure 3.5; some of these pathways are of increasing contemporary interest because they or their engineered variants could lead to the synthesis of "unnatural" or rare sugars useful for the elaboration of antibiotic or antiviral drugs. This is covered in Chapter 10 when the biorefinery concept for processing agricultural residues into fine chemicals is discussed.

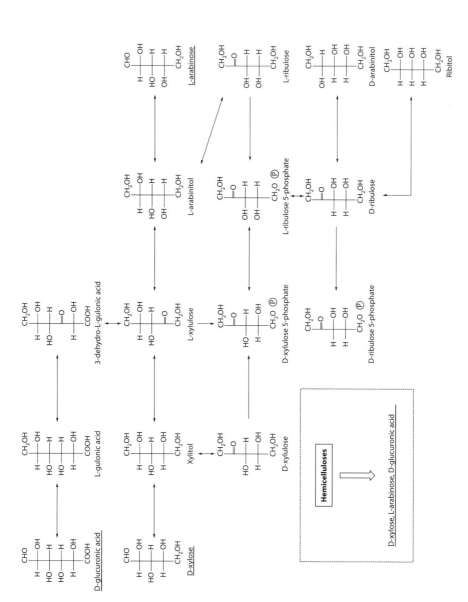

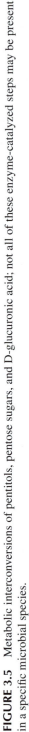

FIGURE 3.5 Metabolic interconversions of pentitols, pentose sugars, and D-glucuronic acid; not all of these enzyme-catalyzed steps may be present in a specific microbial species.

Progress in defining the actual pathways operating in known ethanologenic yeasts was rapid after 2000:

- Analysis of mutations in *P. stipitis* revealed that the catabolism of both D-xylose and L-arabinose proceeded via xylitol.[85]
- An NAD-dependent L-arabinitol 4-dehydrogenase activity was demonstrated in *H. jecorina* induced by growth on L-arabinose; the enzyme forms L-xylulose and will accept ribitol and xylitol as substrates, but not D-arabinitol.[86]
- A gene encoding an L-xylulose reductase (forming xylitol, and NADP dependent) was then demonstrated in *H. jecorina* and overexpressed in *S. cerevisiae*. The L-arabinose pathway uses as its intermediates L-arabinitol, L-xylulose, xylitol, and (by the action of XDH) D-xylulose; the xylulose reductase exhibited the highest affinity to L-xylulose, but some activity was shown toward D-xylulose, D-fructose, and L-sorbose.[87]
- In *H. jecorina*, deletion of the gene for XDH did not abolish growth because *lad1*-encoded L-arabinitol 4-dehydrogenase compensated for this loss. However, doubly deleting the two dehydrogenase genes abolished the ability to grow on either D-xylose or xylitol.[88]

With this knowledge, expressing the five genes for L-arabinose catabolism in *S. cerevisiae* enabled growth on the pentose and, although at a low rate, ethanol production from L-arabinose under anaerobiosis.[89]

The shorter bacterial pathway for L-arabinose catabolism has also been used to adapt yeast cells to utilize arabinose.[90] The bacterial pathway (active in, for example, *Bacillus subtilis* and *E. coli*) proceeds via L-ribulose, L-ribulose 5-phosphate, and D-xylulose 5-phosphate (Figure 3.5), using the enzymes L-arabinose isomerase, L-ribulokinase, and L-ribulose 5-phosphate epimerase.

3.3.4 Comparison of Industrial and Laboratory Yeast Strains for Ethanol Production

Most of the freely available information regarding ethanologenic yeasts has been derived from laboratory strains constructed by research groups in academia; some of these strains (or variants thereof) have certainly been applied to industrial-scale fermentations for cellulosic ethanol production. Published data mostly refer to strains grown in chemically defined media, under laboratory conditions far removed from industrial practice (e.g., continuous culture), or with strains that can accumulate very little ethanol in comparison with modern industrial strains (Figure 3.6).

In addition, strains constructed with plasmids may not have been tested in non-selective media. Plasmid survival in fermentations is generally speculative, even though genetic manipulations are routine for constructing self-selective plasmid-harboring strains where a chromosomal gene in the host is deleted or disrupted and the auxotrophic requirement is supplied as a gene contained on the plasmid.[91]

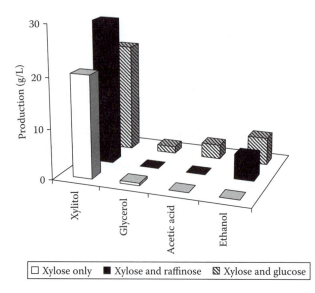

FIGURE 3.6 Product formation by a xylose-utilizing laboratory strain of *Saccharomyces cerevisiae*. (Data from van Zyl, W. H. et al. 1999. *Applied Microbiology and Biotechnology* 52:829.)

STEM TOPIC 3.3: QUANTITATIVE PARAMETERS OF YEAST CELL GROWTH

The kinetics of cell population growth for a yeast (or bacterium) can be represented simply by $dn/dt = \mu n_t$, where n_t is the number of cells at time t and μ is the specific growth rate (with the conventional microbiology unit of per hour). The specific growth rate is important because crucial values for it can be defined and measured in chemostat (continuous) cultures in which μ is controlled by the dilution rate (D) with fresh medium:

- μ_{max} is the maximum specific growth rate for an isolate or species and occurs when increasing D further causes washing out of the cell population.
- μ_{crit} is the specific growth rate at which respiration reaches the critical maximum and beyond which fermentation begins to occur because oxygenation cannot keep pace with sugar substrate inflow.

Biomass yield is denoted by Y; for example, Y_{glc} is the yield of cells (in grams of dry weight) per gram of glucose consumed under the specified conditions and Y_{ATP} is the cell yield from ATP consumption. Y can also signify *product* yield; for example, Y_{eth} is the yield of ethanol in grams per gram of

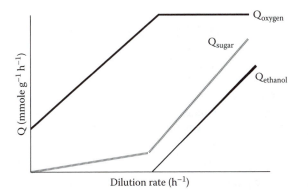

FIGURE STEM 3.3 Chemostat behavior of a *Saccharomyces* yeast under glucose-limiting conditions showing the switch from glucose respiration to ethanologenesis.

substrate consumed. Another useful parameter is the specific metabolic rate, Q; for example, the specific xylose metabolic rate, Q_{xyl}, is xylose used (g) per gram of biomass per hour (see Figure STEM 3.3).[92]

A contentious issue in microbial physiology is that of how to correct measured growth parameters for all the biochemical and molecular events not associated with growth. This is usually referred to as maintenance—a concept that includes all the replacement of macromolecules, the generation of metabolic heat, and all other essential processes in cells in the stationary phase (i.e., not dividing). This forces a change to the overall equation for cell growth:

$$dn/dt = \mu n_t + a n_t$$

where a is the specific maintenance rate (in reciprocal time units). For a modern view of maintenance in microbial physiology, see Reference 93.

Combining data from classic studies of yeast metabolism with computer modeling of the interior workings of the yeast cell, the detailed bioenergetics of yeast cell growth can be estimated (data from Reference 94):

- The energy (as ATP) required for the biosynthesis of all the monomeric precursors (amino acids, nucleic acid bases, etc.) = 9.9 mmol of ATP per gram of dry weight.
- The energy required for the polymerization of monomers into biological macromolecules = 23.9 mmol of ATP per gram of dry weight.
- The energy required to correct polymerization mistakes, manage intracellular and extracellular metabolite traffic, repair oxidative damage to redox proteins, meet osmotic stress, etc. = 35.4 mmol of ATP per gram of dry weight.
- Finally, nongrowth-associated maintenance = 1 mmol of ATP per gram of dry weight per hour.

The final term appears small in rapidly growing cultures: The growth-associated energy requirements at $\mu = 0.4$ per hour total $69.2 \times 0.4 = 27.7$ mmol of ATP per gram of dry weight per hour. In slowly growing cultures ($\mu = 0.01$ per hour), however, the growth-associated energy requirements (0.69 mmol of ATP per gram of dry weight per hour) are *less* than the classically defined maintenance energy. Many industrially important fermentations produce their various products maximally when specific growth rates are very low.

Nevertheless, benchmarking studies comparing laboratory and industrial strains constructed for pentose utilization have appeared. The industrial examples include genetically manipulated polyploid strains typical of *S. cerevisiae* working strains from major brewers or wineries; accounts of engineering such strains for xylose utilization began to be published after 2002.[95,96] A comparison of four laboratory and five industrial strains surveyed both genetically manipulated and genetically undefined but selected xylose consumers (Table 3.3).[97] The industrial strains were

TABLE 3.3

Laboratory and Industrial Strains of *Saccharomyces cerevisiae* for Ethanol Production

Strain	Genetic Description	Xylitol Yield (g/g Xylose Consumed)	Ethanol Yield (g/g Xylose Consumed)	Maximum Hydrolysate (% v/v)[a]
		Laboratory		
TMB3001	XR/XDH/XK overexpressing	0.30	0.33	10
TMB EP	Evolved population from TMB3001	0.31	0.30	
C1	Clone isolated from TMB EP	0.24	0.32	10
C5	Clone isolated from TMB EP	0.28	0.34	10
		Industrial		
F12	XR/XDH/XK overexpressing, polyploid	0.40	0.26	45
A4	XR/XDH/XK overexpressing, polyploid	0.41	0.24	
BH42	Strain selected for improved xylose catabolism	0.36	0.28	10
TMB3399	XR/XDH/XK overexpressing, polypoloid	0.39	0.23	15
TMB3400	Mutagenized and selected from TMB3399	0.41	0.24	15

Source: Data from Sonderegger, M. et al. 2004. *Biotechnology and Bioengineering* 87:90.

[a] Dilute acid hydrolysate of Norway spruce wood.

inferior to the laboratory strains for the yields of both ethanol and xylitol from xylose in minimal media.

Resistance to toxic impurities present in acid hydrolysates of the softwood Norway spruce (*Picea abies*) was higher with genetically transformed industrial strains, but a classically improved industrial strain was no hardier than the laboratory strains (Table 3.3). Similarly, although an industrial strain evolved by genetic manipulation and then random mutagenesis had the fastest rate of xylose use, a laboratory strain could accumulate the highest ethanol concentration on minimal medium.

None of the strains had an ideal set of properties for ethanologenesis in xylose-containing media; long-term chemostat cultivation of one industrial strain in microaerobic conditions on xylose as the sole carbon source definitely improved xylose uptake but neither ethanol nor xylitol yield. Three of the industrial strains could grow in the presence of 10% solutions of undetoxified lignocellulose hydrolysate; the most resistant strain grew best but had a marginally lower ethanol production, perhaps because more of the carbon substrate was used for growth in the absence of any chemical limitation.

The industrial-background strain TMB 3400 (Table 3.3) had no obvious metabolic advantage in anaerobic batch fermentations with xylose-based media when compared with two laboratory strains, one catabolizing xylose by the XR/XDH/XK pathway and the other by the fungal-XI/XK pathway (Figure 3.7).[83] In the presence of undetoxified lignocellulose hydrolysate, however, only the industrial strain could grow adequately and exhibit good ethanol formation (Figure 3.7).

Industrial and laboratory strains engineered for xylose consumption fail to metabolize L-arabinose beyond L-arabitinol.[97] When laboratory and industrial strains endowed with recombinant xylose (fungal) and arabinose (bacterial) pathways were

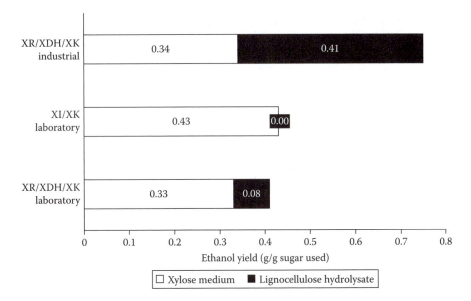

FIGURE 3.7 Ethanol yields from xylose and total sugar of two laboratory and one industrial strain of recombinant *Saccharomyces cerevisiae*. (Data from Karhumaa, K. et al. 2007. *Microbial Cell Factories* 6:5.)

tested in media containing glucose, xylose, and arabinose, the industrial strain accumulated higher concentrations of ethanol. It also had a higher conversion efficiency of ethanol per unit of total pentose utilized and converted less xylose to xylitol and less arabinose to arabinitol—although most of the L-arabinose consumed was still converted to arabinitol.[98]

Interactions between hexose and pentose sugars in the fermentations of lignocellulose-derived substrates have often been considered a serious drawback for ethanol production; this is usually phrased as a type of "carbon catabolite repression" by the more readily utilizable hexose carbon sources, and complex phenotypes can be generated for examination in continuous cultures.[99] In batch cultivation, xylose supports slower growth and much delayed entry into ethanol formation in comparison with glucose.[100] An efficient ethanologen would coutilize multiple carbon sources, funneling them all into the central pathways of carbohydrate metabolism—ultimately to pyruvic acid and thence to acetic acid, acetaldehyde, and ethanol (Figures 3.2 and 3.4).[101,102]

Integration of genes for pentose metabolism is becoming increasingly routine for *S. cerevisiae* strains intended for industrial use; different constructs have quantitatively variable performance indicators (ethanol production rate, xylose consumption rate, etc.), and this suggests that multiple copies of the heterologous genes must be further optimized because gene dosages may differ for the individual genes packaged into the host strain.[103] Such strains can be further improved by a less rigorously defined methodology using evolutionary engineering—that is, selection of strains with incremental advantages for xylose consumption and ethanol productivity, some of whose advantages can be ascribed to increases in measurable enzyme activities for the xylose pathway or the pentose phosphate pathway, and with interesting (but not fully interpretable) changes in the pool sizes of the intracellular pathway intermediates.[104,105]

Efficient utilization of xylose appears to require complex global changes in gene expression. A reexamination of natural *S. cerevisiae* has revealed that classical selection and strain improvement programs can develop yeast cell lines with much shorter doubling times on xylose as the sole carbon source as well as increased XR and XDH activities in a completely nonrecombinant approach.[106] This could easily be applied to improve yeast strains rationally with desirable properties that can be isolated in the heavily selective but artificial environment of an industrial fermentation plant—a practice deliberately pursued for centuries in breweries and wineries but equally applicable to facilities for the fermentation of spent sulfite liquor from the pulp and paper industries.[107]

Defining the capabilities of both industrial and laboratory strains to adapt to the stresses posed by toxic inhibitors in lignocellulose acid hydrolysates is a focus of intense activity.[108–111] Expression of a laccase (from the white rot fungus *Trametes versicolor*) offers some promise as a novel means of polymerizing (and precipitating) reactive phenolic aldehydes derived from the hydrolytic breakdown of lignins. An *S. cerevisiae* expressing the laccase could utilize sugars and accumulate ethanol in a medium containing a spruce wood acid hydrolysate at greatly increased rates in comparison with the parental strain.[112] *S. cerevisiae* also contains the gene for phenylacrylic acid decarboxylase, an enzyme catalyzing the degradation of ferulic acid and other phenolic acids; a transformant overexpressing the gene for the decarboxylase both grew on glucose and accumulated ethanol more rapidly.[113]

3.3.5 Improved Ethanol Production by Naturally Pentose-Utilizing Yeasts

Industrial development of yeasts other than *S. cerevisiae* for cellulosic ethanol has been muted; this is partly due to the ease of genetic transformation of *S. cerevisiae* and closely related strains with bacterial-yeast shuttle vectors. Among nonconventional yeasts, however, *C. shehatae* has some important properties that prove highly desirable with lignocellulosic substrates[114]:

- Ethanol production is more efficient from a mixture of glucose and xylose than from either sugar alone.
- Ethanol production can be demonstrated at elevated temperatures (up to 45°C).
- Ethanol formation from xylose is not affected by wide variation in the xylose concentration in the medium.
- Ethanol can be produced from rice straw hemicellulose hydrolysates.

With a straightforward liquid hot-water pretreatment of alfalfa fibers, *C. shehatae* could produce ethanol in a batch fermentation with a conversion efficiency of 0.47 g per gram of sugar consumed; however, hemicellulose utilization was poor because of the presence of inhibitors.[115] The methylotrophic yeast *Hansenula polymorpha* can, on the other hand, ferment xylose as well as glucose and cellobiose; this species is thermotolerant, actively fermenting sugars at up to 45°C and with a higher ethanol tolerance than *P. stipitis* (although less than *S. cerevisiae*). A vitamin B_2 (riboflavin)-deficient mutant exhibited increased ethanol productivity from both glucose and xylose under suboptimal riboflavin supply and the consequent growth restriction.[116]

P. stipitis, the host organism for genes of a xylose metabolism pathway successfully expressed in *S. cerevisiae,* has been developed as an ethanologen by a research group at the University of Wisconsin (Madison) since the early 1990s.[117] Part of this work was the development of a genetic system for *P. stipitis* that was used to endow the yeast with the ability to grow and produce ethanol anaerobically.

P. stipitis is Crabtree negative and is poorly productive for ethanol. *S. cerevisiae* derives its ability to function anaerobically by the presence of a unique enzyme, dihydroorotate dehydrogenase (DHOdehase), converting dihydroorotic acid to orotic acid in the pyrimidine biosynthetic pathway for nucleic acids. In *S. cerevisiae,* DHOdehase is a cytosolic enzyme catalyzing the reduction of fumaric acid to succinic acid.

This fermentative route is not available to a strictly aerobic yeast such as *Schizosaccharomyces pombe,* which contains a mitochondrial DHOdehase and requires a fully functional mitochondrial electron transport chain for DHOdehase activity. The enzyme may constitute half of a bifunctional protein with a fumarate reductase or be physically associated with the latter enzyme inside the cell.[118] Expression of the *S. cerevisiae* gene for DHOdehase in *P. stipitis* enabled rapid anaerobic growth and ethanol production in a chemically defined medium with glucose as sole carbon source when essential lipids were supplied.[119]

In mixtures of hexoses and pentoses, xylose metabolism by *Pichia stipitis* is repressed while glucose, mannose, and galactose are all used preferentially; this may limit the potential of the yeast for the fermentation of lignocellulosic hydrolysates.

Neither cellobiose nor L-arabinose inhibits induction of the xylose catabolic pathway by D-xylose.[120] Ethanol production from xylose is also inhibited by the $CaSO_4$ formed by the neutralization of sulfuric acid hydrolysates of lignocellulosic materials with $Ca(OH)_2$.

In contrast, Na_2SO_4 (from NaOH) had no effect on either xylose consumption or ethanol production, and $(NH_4)_2SO_4$ (from NH_4OH) reduced growth but enhanced the xylose utilization rate, the rate of ethanol production, and the final ethanol concentration.[121] *P. stipitis* has been shown to produce ethanol on an array of lignocellulosic substrates: sugarcane bagasse, red oak, wheat straw, and hardwood hemicellulose hydrolysates, and corn cob fractions.[122–126]

Strains of the thermotolerant *Kluyveromyces* yeasts are well known to modern biotechnology as vehicles for enzyme and heterologous protein secretion.[127] *K. marxianus* is one of the extraordinarily biodiverse microbial flora known to be present in fermentations for the spirit *cachaça* in Brazil (Chapter 1, Section 1.4); over 700 different yeast species were identified in one distillery over a season, although *S. cerevisiae* was the usual major ethanologen except in a small number of cases where *Rhodotorula glutinis* and *Candida maltosa* predominated.[128] Strains of *K. marxianus* isolated from sugar mills could ferment glucose and cane sugar at temperatures up to 47°C, although long fermentation times and low cell viability were operational drawbacks.[129]

In another study, all eight strains that were screened for D-xylose use were found to be active, and one *K. marxianus* strain was capable of forming ethanol at 55% of the theoretical maximum yield from xylose.[130] A medium based on sugarcane molasses was fermented by *K. marxianus* to ethanol at 45°C, but osmotic stress was evident at high concentrations of molasses or mixtures of sucrose and molasses.[131] Brazilian work has shown that *K. marxianus* is strictly Crabtree negative, requiring (at least in laboratory chemostat experiments) the O_2 supply to be shut down for ethanol to be formed. A high tendency to divert sugars via the oxidative pentose phosphate pathway may be the major obstacle to this yeast as an ethanologen, but metabolic engineering could be applied to redirect carbon flow for fermentative efficiency.[132]

3.4 TOWARD THE PERFECT YEAST ETHANOLOGEN?

Funded research work on *S. cerevisiae* and nonconventional yeasts continues and new variations on two interrelated themes—"*Saccharomyces* + pentose utilization + ???" and genetically engineering natural pentose utilizers to acquire physiological characteristics associated with robust brewer's yeasts—continue to appear, incrementally adjusting and refining metabolism.[133] The likely practical pathway to establishing pilot and industrial processes is the focusing of the "best available local" strain (from a national or regional research or university laboratory) for a particular lignocellulosic biomass source.

For example, Sweden (a relatively small geographical entity) might concentrate on softwood trees while the United States could see biomass ethanol production from softwoods, hardwoods, corn stover, perennial grasses or other feedstocks. Each feedstock would have—at least at any one time—an ideal optimized ethanologen. Convergent evolution might, over time, arrive at production strains that are highly similar; divergent evolution could generate significantly dissimilar microbial physiologies. The biomass ethanol industry might grow to resemble the familiar one of market forces with different rival ethanologens traded and competing or a more fragmented pattern where different technologies do not "travel" well.

With metaphors safely sidelined for the moment, two scientific trends are seeking answers to broader questions than those of micromanaging the catabolic pathways for glucose, xylose, and arabinose:

- If the limits of ethanol productivity are to be reached, should attention be switched to whole-cell biochemistry and gene expression?
- Can any highly engineered ethanologen function in large-scale fermentors with industrially prepared media?

3.4.1 "OMIC" ANALYSES OF YEAST METABOLISM DURING ETHANOL PRODUCTION

For recombinant xylose-utilizing *Saccharomyces cerevisiae,* ethanol yields and productivities can be much lower on xylose than on glucose. Xylose is a novel substrate for this organism and it is not clear how the genome (still almost 100% wild type) is appropriate for xylose metabolism or if all the appropriate genes for xylose utilization are activated and their protein products functioning appropriately.[134] We have already seen the benefits of evolutionary engineering of xylose-utilizing recombinant yeast (Section 3.3.4), where the assumption is that genes other than those manipulated have impacts on growth and ethanol formation, and a rational selection of useful traits is certainly possible in the laboratory.

STEM TOPIC 3.4: GENOMICS AND OMIC BIOTECHNOLOGIES APPLIED TO YEASTS

Four nouns dominate titles in contemporary journals covering molecular biology and bioinformatics:

- *Genomics* is the study of nucleotide sequences encoding structural genes, regulatory genes, and silent (noncoding) segments of the chromosomes of organisms.
- *Proteomics* analyzes the synthesis of proteins encoded by genes; the proteome is the set of proteins synthesized under any one set of environmental and nutritional circumstances.
- *Transcriptomics* is the study of the global analysis of gene expression (genome-wide expression profiling). Genes with similarities in their expression patterns are often functionally related and subject to the same regulatory mechanisms (www.systemsbiology.nl).
- *Metabolomics* is the characterization of the metabolites inside and formed by an organism. This is a very broad discipline and reaches as

far as the Human Metabolome Project (www.metabolmics.ca), but all organisms (including microbes) have metabolite patterns that reflect the interactions between genes, gene expression, enzyme action, and the effects of environment and nutrition.

The adjective "postgenomic" is much used. Its meaning can be limited to human society in that, only a few years ago, the human genome had not been sequenced; now science can exploit the knowledge gained by this monumental effort. The genomes of other organisms are, of course, continuously being sequenced, so any postgenomic era is a definable time only for one organism— and, with new species of plants, animals, insects, and microbes being discovered on a regular basis, this era will truly be reached in the distant future.

What is inside a typical yeast cell, viewed in the light of omic technologies? The molecular makeup of *S. cerevisiae* is reasonably well understood[94]:

- The genome contains over 6,000 genes, but the biochemical functions of only approximately 4,400 are defined.
- A reconstructed metabolic network comprises 1,175 reactions: 702 in the cell cytosol, 124 inside mitochondria, 287 exchanges across the cytoplasmic membrane, and 62 exchanges between cytoplasm and mitochondria.
- A total of 708 open reading frames (ORFs) correspond to proteins responsible for 1,035 of the reactions; the remaining 140 are based on biochemical or other evidence. The reconstructed metabolic network only requires 16% of the characterized ORFs.
- A total of 584 metabolites are distributed as follows: 559 in the cytosol, 164 in the mitochondria, and 121 extracellular.

Such a model can, within appropriate physicochemical constraints from thermodynamics and stoichiometry, account for specific ethanol productivity in anaerobic conditions where glucose is the limiting substrate.[135]

S. cerevisiae, along with the bacteria *Escherichia coli* and *Bacillus subtilis,* formed the initial "Rosetta Stone" for genome research, providing a well-characterized group of genes for comparative sequence analysis; coding sequences of other organisms lacking detailed knowledge of their biochemistry could be assigned protein functionalities based on their degree of homology to known genes. Not only was the genome of *S. cerevisiae* the first eukaryotic genome to be fully sequenced, but the metabolic network of biochemical pathways has also been reconstructed.[94,136]

Genome-scale metabolic networks can now be reconstructed and their properties examined in computer simulations of biotechnological processes. Comparing metabolic profiles in 14 yeast species revealed many important similarities with

S. cerevisiae, but *Pichia augusta* may possess the ability to rebalance the growth demand for NADPH from the pentose phosphate pathway radically (Figure 3.2). This suggests a transhydrogenase activity that could be used to rebalance NADH/NADPH redox cofactors for xylose metabolism (Section 3.3.1).[137]

Although in silico predictions failed to account for continued xylitol accumulation by a recombinant, xylose-utilizing *S. cerevisiae* strain on xylose, this suggested continued problems in balancing cofactor requirements in reductase/dehydrogenase reactions. Computer simulations highlighted glycerol formation as a key target for process improvement: Expressing a nonphosphorylating NADPH-dependent glyceraldehyde 3-phosphate dehydrogenase increased carbon flow to pyruvate (rather than being diverted to glycerol formation), reduced glycerol accumulation, and improved ethanol production by up to 25%.[138,139]

Glycerol is a significant and long recognized coproduct in ethanol fermentations, and acids such as acetate and pyruvate are also known to accumulate during ethanol production by *S. cerevisiae.* Removing the genetic ability to synthesize glycerol can be effective, but broader aspects of yeast biochemistry then come into play:

- Simply deleting the gene-encoding glycerol formation from glycolysis (i.e., glycerol 3-phosphate dehydrogenase [G3PDH]) reduced the growth rate compared to the wild type but reduced acetate and pyruvate accumulation; simultaneously overexpressing the *GLT1* gene for glutamate synthase (catalyzing the reductive formation of glutamate from glutamine and 2-oxo-glutarate) restored the growth characteristics, while reducing glycerol accumulation and increasing ethanol formation.[140]

- Deleting the gene (*FSP1*) encoding the transporter protein for glycerol also improved ethanol production while decreasing the accumulation of glycerol, acetate, and pyruvate. Presumably, increasing the intracellular concentration of glycerol (by blocking its export) accelerates its metabolism by a reversal of the glycerol kinase and G3PDH reactions.[141]

- Improved production of ethanol was also achieved by deleting *FPS1* and overexpressing *GLT1*, accompanied by reduced accumulation of glycerol, acetate, and pyruvate.[142]

Searching through the annotated reactions and enzymes in the yeast metabolic network shows few references to xylose and arabinose (xylulokinase and arabinose 1-dehydrogenase).[94] Although 77 transport proteins are included for movement into the cell across the cytoplasmic membrane, none of these attuned to pentose sugars. Strains of *S. cerevisiae* genetically engineered for xylose utilization depend, unlike naturally xylose-consuming yeasts, on xylose uptake via the endogenous hexose transporters. With a high XR activity, xylose uptake by this ad hoc arrangement limits the pentose utilization rate at low xylose concentrations, suggesting that expression of a yeast xylose/pentose transporter in *S. cerevisiae* would be beneficial for the kinetics and extent of xylose consumption from hemicellulose hydrolysates.[143] High-affinity xylose transporters are present in the cell membranes of the efficient xylose fermenting yeast *Candida succiphila* and in *Kluyveromyces marxianus*[144]:

- In *S. cerevisiae* engineered to express the bacterial arabinose catabolic pathway, the overexpression of arabinose-transporting yeast galactose permease gave high ethanol productivity, at 60% of the theoretical maximum yield from L-arabinose under O_2-limiting conditions.[90]
- When the glucose/xylose transporter *Gxf1* from *Candida intermedia* was expressed in the recombinant xylose-fermenting *S. cerevisiae,* faster rates of xylose uptake and ethanol formation at low xylose concentrations were achieved.[145]
- Most remarkably, two heterologous transporters from the lower plant *Arabidopsis thaliana* expressed in recombinant xylose-utilizing *S. cerevisiae* cells increased xylose consumption that correlated with increased ethanol concentration and productivity.[146]

Given the ever-growing repository of knowledge of the molecular genetics of *S. cerevisiae,* the optimal reformulation of yeast ethanol synthetic ability is often considered to be achievable in the near future. There is a caveat, however: The optimum when glucose is the carbon substrate may not be the best for coutilization of glucose and xylose; furthermore, the exact fermentation strategy used—especially feeding carbon sources rather than including the entire sugar load in the batched medium—must be taken into account.[147]

Industrial biotechnologists appreciate this truth acutely because of the sensitivity of a producing organism to multiple small changes that can occur when different sites attempt to replicate a large-scale fermentation: A microbe is never "isolated," and its gene expression profile depends on the precise (and fluctuating) nutritional environment in which it attempts to survive and grow. We will take up this topic again in Chapter 5 when the biochemical engineering of ethanol fermentations is considered.

3.4.2 Stress Responses in Yeast Ethanologens

However complex the genetic makeup is of a recombinant yeast dedicated to ethanol production from lignocellulosic substrates, the cells must function in the high stresses imposed by the medium and the physical and chemical conditions inside large fermentors. Decades of scientific studies on brewery processes have identified the key set of stresses on yeast physiology[148,149]:

- dissolved oxygen concentration (low but cannot reach zero)
- low pH (generally 3.0–4.0)
- high osmolarity (>500 mg/L Na^+)
- high ethanol concentration (23% v/v can be reached)
- changing nutrient supply (changing continuously as the fermentation proceeds)
- high temperature (must be <35°C)
- high-gravity media (38% w/v sugars)

- preparation and use of dried yeast for use as inocula
- presence of growth inhibitory acids > 0.8% w/v lactic and > 0.05% w/v acetic acid) and of mycotoxins produced by microbial contaminants in grain inputs

Many of these stresses may be imposed simultaneously and synergistic effects occur. Cutting-edge science is now being applied to the problem of multiple stresses in glucose-based ethanol fermentations.[150–152] The extensive literature of studies on *single* stress factors, however, is still increasing.

The most important stress may be ethanol concentration. It has been known for over 20 years that conventional brewer's yeast can accumulate up to 23% v/v ethanol under unexceptional laboratory conditions—far higher concentrations than mentioned in published reports on cellulosic ethanol and recombinant microorganisms.[153] This high ethanol concentration is reached at the end of the fermentation; much lower concentrations will inhibit yeast cell growth earlier in the fermentation but ethanol concentrations are then very much lower.

The crucial point is that yeast cells *can* tolerate up to 23% v/v ethanol. If the large-scale production of ethanol from sugarcane, corn starch, or any cellulosic biomass reached this concentration, industrialists would be more than happy! Potable alcohol manufacturers are starting to employ improved fermentation management strategies to push ethanol productivity to the known limit; eventually, these practices will filter through to cellulosic ethanol production.[154]

3.5 SUMMARY

The *Saccharomyces* yeasts used for brewing potable alcohol are very efficient ethanologens but have a highly restricted substrate range of sugars. Genetic engineering with genes from other yeasts and bacteria has succeeded in endowing conventional yeasts with the ability to utilize xylose and arabinose (from hemicelluloses). Recombinant *Saccharomyces* yeasts can ferment lignocellulosic hydrolysates and were the biological agents used in the first demonstration facility for biomanufacturing ethanol from wheat straw in Canada.

Nonconventional yeasts such as *Candida shehatae* and *Pichia stipitis* have desirable properties as ethanologens but have yet to be fully developed for commercial ethanol production.

Advances in understanding genes and gene expression patterns in *S. cerevisiae* have resulted in the fine-tuning of yeast metabolism to increase ethanol production; computer modeling of yeast metabolism has progressed to the stage where properties of cells can be predicted in detail and compared to experimental results.

Defining one, ideal yeast ethanologen appears unlikely due to the proliferation of experimental strain programs nationally and internationally. Different recombinant yeasts may be optimized for using particular lignocellulosic substrates or mixtures of substrates. The future cellulosic ethanol industry will have a wide selection of biocatalysts available for testing and development with national and regional supplies of plant biomass.

Whatever the choice of yeast ethanologen is, the cells must survive and function in industrial media under conditions where multiple stresses are experienced—low pH, high osmotic concentrations, high ethanol accumulation, fluctuating temperature, etc. Old industrially used strains probably have acquired traits to produce ethanol under prolonged physical and chemical stresses. Recombinant yeasts can also be "black box" selected in continuous cultures to develop similar functionalities.

REFERENCES

1. Wyman, C. E. 1994. Ethanol from lignocellulosic biomass: Technology, economics, and opportunities. *Bioresource Technology* 50:3.
2. Lynd, L. R. 1996. Overview and evaluation of fuel ethanol production from cellulosic biomass: Technology, economics, the environment, and policy. *Annual Review of Energy and the Environment* 21:403.
3. Chandrakant, P., and Bisaria, V. S. 1998. Simultaneous bioconversion of cellulose and hemicellulose to ethanol. *Critical Reviews in Biotechnology* 18:295.
4. Aristidou, A. A. 2007. Application of metabolic engineering to the conversion of renewable resources to fuels and fine chemicals: Current advances and future prospects. In *Fermentation microbiology and biotechnology,* 2nd ed., ed. El-Mansi, E. M. T., Bryce, C. A., Demain, A. L., and Allman, A. R., chap. 9. Boca Raton, FL: CRC Press.
5. Lin, Y., and Tanaka, S. 2006. Ethanol fermentation from biomass resources: current state and prospects. *Applied Microbiology and Biotechnology* 69:627.
6. Hahn-Hägerdal, B. et al. 2001. Metabolic engineering of *Saccharomyces cerevisiae* for xylose utilization. *Advances in Biochemical Engineering/Biotechnology* 73:53.
7. Yoon, S. H., Mukerjee, R., and Robyt, J. F. 2003. Specificity of yeast (*Saccharomyces cerevisiae*) in removing carbohydrates by fermentation. *Carbohydrate Research* 338:1127.
8. van Dijken, J. P. et al. 1986. Alcoholic fermentation by non-fermenting yeasts. *Yeast* 2:123.
9. Lagunas, R. 1986. Misconceptions about the energy metabolism of *Saccharomyces cerevisiae. Yeast* 2:221.
10. Vallet, C. et al. 1996. Natural abundance isotopic fractionation in the fermentation reaction: influence of the nature of the yeast. *Bioorganic Chemistry* 24:319.
11. Çaylak, B., and Vardar Sukan, F. 1998. Comparison of different production processes for bioethanol. *Turkish Journal of Chemistry* 22:351.
12. da Cruz, S. H., Batistote, M., and Ernandes, J. R. 2003. Effect of sugar catabolite repression in correlation with the structural complexity of the nitrogen source on yeast growth and fermentation. *Journal of the Institute of Brewing* 109:349.
13. Ergun, M., and Mutlu, S. F. 2000. Application of a statistical technique to the production of ethanol from sugar beet molasses by *Saccharomyces cerevisiae. Bioresource Technology* 73:251.
14. Cejka, A. 1985. Preparation of media. In *Biotechnology,* vol. 2, ed. Brauer, H., chap. 26. Weinheim, Germany: VCH Verlagsgesellschaft.
15. Barnett, J. A., Payne, R. W., and Yarrow, D. 2000. *Yeasts: Characteristics and identification,* 3rd ed. Cambridge, England: Cambridge University Press.
16. Sims, A. P., and Barnett, J. A. 1976. The requirement of oxygen for the utilization of maltose, cellobiose, and D-galactose by certain anaerobically fermenting yeasts (Kluyver effect). *Journal of General Microbiology* 106:277.
17. Lloyd, D., and James, C. J. 1987. The Pasteur effect in yeasts: mass spectrometric monitoring of oxygen uptake, and carbon dioxide and ethanol production. *FEMS Microbiology Letters* 42:27.
18. van Dijken, J. P., and Scheffers, W. A. 1986. Redox balances in the metabolism of sugars by yeasts. *FEMS Microbiology Reviews* 32:199.

19. Wijsman, M. R. et al. 1984. Inhibition of fermentation and growth in batch cultures of the yeast *Brettanomyces intermedius* upon a shift from aerobic to anaerobic conditions (Custers effect). *Antonie van Leeuwenhoek* 50:183.

20. Weusthuis, R. A. et al. 1994. Is the Kluyver effect caused by product inhibition? *Microbiology* 140:1723.

21. Malluta, E. F., Decker, P., and Stambuk, B. U. 2000. The Kluyver effect for trehalose in *Saccharomyces cerevisiae. Journal of Basic Microbiology* 40:199.

22. Fukuhara, H. 2003. The Kluyver effect revisited. *FEMS Yeast Research* 3:327.

23. Fredlund, E. et al. 2004. Oxygen- and glucose-dependent regulation of central carbon metabolism in *Pichia anomola. Applied and Environmental Microbiology* 70:5905.

24. Frick, O., and Wittman, C. 2005. Characterization of the metabolic shift between oxidative and fermentative growth in *Saccharomyces cerevisiae* by comparative ^{13}C flux analysis. *Microbial Cell Factories* 4:30.

25. Sierkstra, L. N. et al. 1993. Regulation of glycolytic enzymes and the Crabtree effect in galactose-limited continuous cultures of *Saccharomyces cerevisiae. Yeast* 9:787.

26. Kiers, J. et al. 1998. Regulation of alcoholic fermentation in batch and chemostat cultures of *Kluyveromyces lactis* CBS 2359. *Yeast* 14:459.

27. Bruinenberg, P. M. et al. 1983. The role of the redox balance in the anaerobic fermentation of xylose by yeasts. *European Journal of Applied Microbiology and Biotechnology* 18:287.

28. Bruinenberg, P. M. et al. 1984. NADH-linked aldose reductase: the key to alcoholic fermentation of xylose by yeasts. *Applied Microbiology and Biotechnology* 19:256.

29. Visser, W. et al. 1990. Oxygen requirements of yeasts. *Applied and Environmental Microbiology* 56:3785.

30. Hutter, A., and Oliver, S. G. 1998. Ethanol production using nuclear petite yeast mutants. *Applied Microbiology and Biotechnology* 49:511.

31. Merico, A. et al. 2007. Fermentative lifestyle in yeasts belonging to the *Saccharomyces* complex. *FEBS Journal* 274:976.

32. Wolf, K. 1996. Preface to *Nonconventional yeasts in biotechnology. A handbook,* ed. Wolf, K. Berlin: Springer.

33. Spencer, J. F., Spencer, D. M., and Reynolds, N. 1988. Genetic manipulation of nonconventional yeasts by conventional and non-conventional methods. *Journal of Basic Microbiology* 28:321.

34. Domínguez, A. et al. 1998. Non-conventional yeasts as hosts for heterologous protein production. *International Microbiology* 1:131.

35. Osorio-Cadavid, E. et al. 2008. Detection and identification of wild yeasts in *Champús,* a fermented Colombian maize beverage. *Food Microbiology* 25:771.

36. Cirillo, V. P. 1968. Relationship between sugar structure and competition for the sugar transport system in bakers' yeast. *Journal of Bacteriology* 95:603.

37. Träff, K. L., Jönsson, L. J., and Hahn-Hägerdal, B. 2002. Putative xylose and arabinose reductases in *Saccharomyces cerevisiae. Yeast* 19:1233.

38. Toivari, M. H. et al. 2004. Endogenous xylose pathway in *Saccharomyces cerevisiae. Applied and Environmental Microbiology* 70:3681.

39. Toivola, A. et al. 1984. Alcoholic fermentation of D-xylose by yeasts. *Applied and Environmental Microbiology* 47:1221.

40. Lee, H. et al. 1986. Utilization of xylan by yeasts and its conversion to ethanol by *Pichia stipitis* strains. *Applied and Environmental Microbiology* 52:320.

41. Verduyn, C. et al. 1985. Properties of the NAD(P)H-dependent xylose reductase from the xylose-fermenting yeast *Pichia stipitis. Biochemical Journal* 226:669.

42. Ho, N. W. Y. et al. 1990. Purification, characterization, and amino acid terminal sequence of xylose reductase from *Candida shehatae. Enyzme and Microbial Technology* 12:33.

43. Lee, J. K., Bong-Seong, K., and Sang-Yong, K. 2003. Cloning and characterization of the *xyl1* gene, encoding an NADH-preferring xylose reductase from *Candida parapsilosis,* and its functional expression in *Candida tropicalis. Applied and Environmental Microbiology* 69:6179.

44. Skoog, K., and Hahn-Hägerdal, B. 1990. Effect of oxygenation on xylose fermentation by *Pichia stipitis. Applied and Environmental Microbiology* 56:3389.

45. Jeppsson, H., Alexander, N. J., and Hahn-Hägerdal, B. 1995. Existence of cyanide-insensitive respiration in the yeast *Pichia stipitis* and its possible influence on product formation during xylose utilization. *Applied and Environmental Microbiology* 61:2596.

46. Passoth, V., Zimmermann, M., and Klinner, U. 1996. Peculiarities of the regulation of fermentation and respiration in the Crabtree-negative, xylose-fermenting yeast *Pichia stipitis. Applied Biochemistry and Biotechnology* 57–58:201.

47. Kordowska-Wiater, M., and Targonski, Z. 2001. Application of *Saccharomyces cerevisiae* and *Pichia stipitis* karyoductants to the production of ethanol from xylose. *Acta Biochimica Polonica* 50:291.

48. Kötter, P. et al. 1990. Isolation and characterization of the *Pichia stipitis* xylitol dehydrogenase gene, *XYL2,* and construction of a xylose utilizing *Saccharomyces cerevisiae* transformant. *Current Genetics* 18:493.

49. Walfridsson, M. et al. 1997. Expression of different levels of enzymes from the *Pichia stipitis XYL1* and *XYL2* genes in *Saccharomyces cerevisiae* and its effects on product formation during xylose utilization. *Applied Microbiology and Biotechnology* 48:218.

50. Ho, N. W. Y., Chen, Z., and Brainard, A. P. 1998. Genetically engineered *Saccharomyces* yeast capable of effective cofermentation of glucose and xylose. *Applied and Environmental Microbiology* 64:1852.

51. Ho, N. W. Y. et al. 1999. Successful design and development of genetically engineered *Saccharomyces* yeast capable of effective cofermentation of glucose and xylose from cellulosic biomass to fuel ethanol. *Advances in Biochemical Engineering/Biotechnology* 65:163.

52. Ho, N. W. Y., and Tsao, G. T. 1998. Recombinant yeasts for effective fermentation of glucose and xylose. US Patent 5,789,210, August 4, 1998.

53. Yu, S., Jeppsson, H., and Hahn-Hägerdal, B. 1995. Xylulose fermentation by *Saccharomyces cerevisiae* and xylose-fermenting yeast strains. *Applied Microbiology and Biotechnology* 44:314.

54. Deng, X. X., and Ho, N. W. Y. 1990. Xylulokinase activity in various yeasts including *Saccharomyces cerevisiae* containing the cloned xylulokinase gene. *Applied Biochemistry and Biotechnology* 24–25:193.

55. Tolan, J. S. 2006. Iogen's demonstration process for producing ethanol from cellulosic biomass. In *Biorefineries—Industrial processes and products. Volume 1: Status quo and future directions,* ed. Kamm, B., Gruber, P. R., and Kamm, M., chap. 9. Weinheim, Germany: Wiley-VCH Verlag.

56. Moniruzzaman, M. et al. 1997. Fermentation of corn fiber sugars by an engineered xylose utilizing *Saccharomyces cerevisiae* strain. *World Journal of Microbiology and Biotechnology* 13:341.

57. Toivari, M. H. et al. 2001. Conversion of xylose to ethanol by recombinant *Saccharomyces cerevisiae*: importance of xylulokinase (*XKS1*) and oxygen availability. *Metabolic Engineering* 3:236.

58. Kötter, P., and Ciriacy, M. 1993. Xylose fermentation by *Saccharomyces cerevisiae. Applied Microbiology and Biotechnology* 38:776.

59. Tantirungkij, M. et al. 1993. Construction of xylose-assimilating *Saccharomyces cerevisiae. Journal of Fermentation and Bioengineering* 75:83.

60. Wahlborn, C. F., and Hahn-Hägerdal, B. 2002. Furfural, 5-hydroxymethyl-furfural, and acetoin act as external electron acceptors during anaerobic fermentation of xylose in recombinant *Saccharomyces cerevisiae. Biotechnology and Bioengineering* 78:172.

61. Öhgren, K. et al. 2006. Simultaneous saccharification and co-fermentation of glucose and xylose in steam-pretreated corn stover at high fiber content with *Saccharomyces cerevisiae* TMB3400. *Journal of Biotechnology* 126:488.

62. Anderlund, M. et al. 2001. Expression of bifunctional enzymes with xylose reductase and xylitol dehydrogenase activity in *Saccharomyces cerevisiae* alters product formation during xylose fermentation. *Metabolic Engineering* 3:226.

63. Jeppsson, M. et al. 2006. The expression of a *Pichia stipitis* xylose reductase mutant with higher K_m for NADPH increases ethanol production from xylose in recombinant *Saccharomyces cerevisiae*. *Biotechnology and Bioengineering* 93:665.

64. Aristidou, A. et al. 2000. Redox balance in fermenting yeast. *Monograph—European Brewing Convention* 28:161.

65. Roca, C., Nielsen, J., and Olsson, L. 2003. Metabolic engineering of ammonium assimilation in xylose-fermenting *Saccharomyces cerevisiae* improves ethanol production. *Applied and Environmental Microbiology* 69:4732.

66. Verho, R. et al. 2003. Engineering redox cofactor regeneration for improved pentose fermentation in *Saccharomyces cerevisiae*. *Applied and Environmental Microbiology* 69:5892.

67. Watanabe, S., Kodaki, T., and Makino, K. 2005. Complete reversal of coenzyme specificity of xylitol dehydrogenase and increase of thermostability by the introduction of structural zinc. *Journal of Biological Chemistry* 280:10340.

68. Michal, G., ed. 1999. *Biochemical pathways. An atlas of biochemistry and molecular biology,* chap. 3. New York: John Wiley & Sons.

69. Johansson, B. et al. 2001. Xylulokinase overexpression in two strains of *Saccharomyces cerevisiae* also expressing xylose reductase and xylose dehydrogenase and its effect on fermentation of xylose and lignocellulosic hydrolysate. *Applied and Environmental Microbiology* 67:4249.

70. Jeppsson, B. et al. 2002. Reduced oxidative pentose phosphate pathway flux in recombinant xylose-utilizing *Saccharomyces cerevisiae* strains improves the ethanol yield from xylose. *Applied and Environmental Microbiology* 68:1604.

71. Walfridsson, M. et al. 1995. Xylose-metabolizing *Saccharomyces cerevisiae* strains overexpressing the *TKL1* and *TKL2* genes encoding the pentose phosphate pathway enzymes transketolase and transaldolase. *Applied and Environmental Microbiology* 61:4184.

72. Karhumaa, K. et al. 2007. High activity of xylose reductase and xylitol dehydrogenase improves xylose fermentation by recombinant *Saccharomyces cerevisiae*. *Applied Microbiology and Biotechnology* 73:1039.

73. La Grange, D. C. et al. 2000. Coexpression of the *Bacillus pumilus* β-xylosidase (*xynB*) gene with the *Trichoderma reesei* β-xylanase 2 (*xyn2*) gene in the yeast *Saccharomyces cerevisiae*. *Applied Microbiology and Biotechnology* 54:195.

74. La Grange, D. C. et al. 2001. Degradation of xylan to D-xylose by recombinant *Saccharomyces cerevisiae* coexpressing the *Aspergillus niger* β-xylosidase (*xlnD*) and the *Trichoderma reesei* xylanase II (*xyn2*) genes. *Applied and Environmental Microbiology* 67:5512.

75. Katahira, S. et al. 2004. Construction of a xylan-fermenting yeast strain through codisplay of xylanolytic enzymes on the surface of xylose-utilizing *Saccharomyces cerevisiae* cells. *Applied and Environmental Microbiology* 70:5407.

76. Katahira, S. et al. 2006. Ethanol fermentation from lignocellulosic hydrolysate by a recombinant xylose- and cellooligosaccharide-assimilating yeast strain. *Applied Microbiology and Biotechnology* 72:1136.

77. Walfridsson, M. et al. 1996. Ethanolic fermentation of xylose with *Saccharomyces cerevisiae* harboring the *Thermus thermophilus xylA* gene, which expresses an active xylose (glucose) isomerase. *Applied and Environmental Microbiology* 62:4648.

78. Xarhangi, H. R. et al. 2003. Xylose metabolism in the anaerobic fungus *Piromyces* sp. strain E2 follows the bacterial pathway. *Archives of Microbiology* 180:134.

79. Kuyper, M. et al. 2003. High-level functional expression of a fungal xylose isomerase: the key to efficient ethanolic fermentation of xylose by *Saccharomyces cerevisiae?* *FEMS Yeast Research* 4:69.

80. Kuyper, M. et al. 2004. Minimal metabolic engineering of *Saccharomyces cerevisiae* for efficient anaerobic xylose fermentation: A proof of principle. *FEMS Yeast Research* 4:655.

81. Kuyper, M. et al. 2005. Metabolic engineering of a xylose-isomerase-expressing *Saccharomyces cerevisiae* strain for rapid anaerobic xylose fermentation. *FEMS Yeast Research* 5:399.

82. Kuyper, M. et al. 2005. Evolutionary engineering of mixed-sugar utilization by a xylose-fermenting *Saccharomyces cerevisiae* strain. *FEMS Yeast Research* 5:925.

83. Karhumaa, K. et al. 2007. Comparison of the xylose reductase-xylitol dehydrogenase and the xylose isomerase pathways for xylose fermentation by recombinant *Saccharomyces cerevisiae*. *Microbial Cell Factories* 6:5.

84. Karhumaa, K., Hahn-Hägerdal, B., and Gorwa-Grauslund, M.-F. 2005. Investigation of limiting metabolic steps in the utilization of xylose by recombinant *Saccharomyces cerevisiae* using metabolic engineering. *Yeast* 22:359.

85. Shi, N. Q. et al. 2000. Characterization and complementation of a *Pichia stipitis* mutant unable to grow on D-xylose or L-arabinose. *Applied Biochemistry and Biotechnology* 84–86:201.

86. Richard, P. et al. 2001. Cloning and expression of a fungal L-arabinitol 4-dehydrogenase gene. *Journal of Biological Chemistry* 276:40631.

87. Richard, P. et al. 2002. The missing link in the fungal L-arabinose catabolic pathway, identification of the L-xylulose reductase gene. *Biochemistry*, 41:6432.

88. Seiboth, B. et al. 2003. D-Xylose metabolism in *Hypocrea jecorina*: Loss of the xylitol dehydrogenase step can be partially compensated for the *lad1*-encoded L-arabinitol-4-dehydrogenase. *Eukaryotic Cell* 2:867.

89. Richard, P. et al. 2003. Production of ethanol from L-arabinose by *Saccharomyces cerevisiae* containing a fungal L-arabinose pathway. *FEMS Yeast Research* 3:185.

90. Becker, J., and Boles, E. 2003. A modified *Saccharomyces cerevisiae* strain that consumes L-arabinose and produces ethanol. *Applied and Environmental Microbiology* 69:4144.

91. van Zyl, W. H. et al. 1999. Xylose utilization by recombinant strains of *Saccharomyces cerevisiae* on different carbon sources. *Applied Microbiology and Biotechnology* 52:829.

92. Griffin, D. H. 1994. *Fungal physiology*, 2nd ed., chap. 8. New York: Wiley-Liss.

93. van Bodegom, P. 2005. Microbial maintenance: A critical review on its quantification. *Microbial Ecology* 53:513.

94. Förster, J. et al. 2003. Genome-scale reconstruction of the *Saccharomyces cerevisiae* metabolic network. *Genome Research* 13:244.

95. Zaldivar, J. et al. 2002. Fermentation performance and intracellular metabolite patterns in laboratory and industrial xylose-fermenting *Saccharomyces cerevisiae*. *Applied Microbiology and Biotechnology* 59:436.

96. Wahlborn, C. F. et al. 2003. Generation of the improved recombinant xylose-utilizing *Saccharomyces cerevisiae* TMB 3400 by random mutagenesis and physiological comparison with *Pichia stipitis* CBS 6054. *FEMS Yeast Research* 3:319.

97. Sonderegger, M. et al. 2004. Fermentation performance of engineered and evolved xylose-fermenting *Saccharomyces cerevisiae* strains. *Biotechnology and Bioengineering* 87:90.

98. Karhumaa, K. et al. 2006. Co-utilization of L-arabinose and D-xylose by laboratory and industrial *Saccharomyces cerevisiae* strains. *Microbial Cell Factories* 5:18.

99. Roca, C., Haack, M. B., and Olsson, L. 2004. Engineering of carbon catabolite repression in recombinant xylose fermenting *Saccharomyces cerevisiae* strains. *Applied Microbiology and Biotechnology* 63:578.

100. Govindaswamy, S., and Vane, L. M. 2007. Kinetics of growth and ethanol production on different carbon substrates using genetically engineered xylose-fermenting yeast. *Bioresource Technology* 98:677.

101. van Maris, A. J. et al. 2006. Alcoholic fermentation of carbon sources in biomass hydrolysates by *Saccharomyces cerevisiae*: Current status. *Antonie van Leeuwenhoek* 90:391.

102. Hahn-Hägerdal, B. et al. 2007. Towards industrial pentose-fermenting yeast strains. *Applied Microbiology and Biotechnology* 74:937.

103. Wang, Y. et al. 2004. Establishment of a xylose metabolic pathway in an industrial strain of *Saccharomyces cerevisiae*. *Biotechnology Letters* 26:885.

104. Sonderegger, M., and Sauer, U. 2003. Evolutionary engineering of *Saccharomyces cerevisiae* for anaerobic growth on xylose. *Applied and Environmental Microbiology* 69:1990.

105. Pitkanen, J. P. et al. 2005. Xylose chemostat isolates of *Saccharomyces cerevisiae* show altered metabolite and enzyme levels compared with xylose, glucose, and ethanol metabolism of the original strain. *Applied Microbiology and Biotechnology* 67:827.

106. Attfield, P. V., and Bell, P. J. 2006. Use of population genetics to derive nonrecombinant *Saccharomyces cerevisiae* strains that grow using xylose as a sole carbon source. *FEMS Yeast Research* 6:862.

107. Lindén, T., Peetre, J., Hahn-Hägerdal, B., et al. 1992. Isolation and characterization of acetic acid-tolerant galactose-fermenting strains of *Saccharomyces cerevisiae* from a spent sulfite liquor fermentation plant. *Applied and Environmental Microbiology* 58:1661.

108. Martín, C., and Jönsson, L. J. 2003. Comparison of the resistance of industrial and laboratory strains of *Saccharomyces cerevisiae* and *Zygosaccharomyces* to lignocellulose-derived fermentation inhibitors. *Enzyme and Microbial Technology* 32:386.

109. Brandber, T., Franzen, C. J., and Gustafsson, L. 2004. The fermentation performance of nine strains of *Saccharomyces cerevisiae* in batch and fed-batch cultures in dilute-acid wood hydrolysate. *Journal of Bioscience and Bioengineering* 98:122.

110. Garay-Arroypo, A. et al. 2004. Response to different environmental stress conditions of industrial and laboratory *Saccharomyces cerevisiae* strains. *Applied Microbiology and Biotechnology* 63:734.

111. Hahn-Hägerdal, B. et al. 2005. Role of cultivation media in the development of yeast strains for large scale industrial use. *Microbial Cell Factories* 4:31.

112. Larsson, S., Cassland, P., and Jönsson, L. J. 2001. Development of a *Saccharomyces cerevisiae* strain with enhanced resistance to phenolic fermentation inhibitors in lignocellulose hydrolysates by heterologous expression of laccase. *Research* 67:1163.

113. Larsson, S., Nilvebrant, N. O., and Jönsson, L. J. 2001. Effect of overexpression of *Saccharomyces cerevisiae Pad1p* on the resistance to phenylacrylic acids and lignocellulose hydrolysates under aerobic and oxygen-limited conditions. *Applied Microbiology and Biotechnology* 57:167.

114. Abbi, M., Kuhad, R. C., and Singh, A. 1996. Fermentation of xylose and rice straw hydrolysate to ethanol by *Candida shehatae* NCL-3501. *Journal of Industrial Microbiology* 17:20.

115. Sreenath, H. K. et al. 2001. Ethanol production from alfalfa fiber fractions by saccharification and fermentation. *Process Biochemistry* 36:1199.

116. Ryabova, O. B., Chmil, O. M., and Sibirny, A. A. 2003. Xylose and cellobiose fermentation to ethanol by the thermotolerant methylotrophic yeast *Hansenula polymorpha*. *FEMS Yeast Research* 4:157.

117. Jeffries, T. W., and Shi, N.-Q. 1999. Genetic engineering for improved xylose fermentation by yeasts. *Advances in Biochemical Engineering/Biotechnology* 65:117.

118. Nagy, M., Lacroute, F., and Thomas, D. 1992. Divergent evolution of pyrimidine bio-synthesis between anaerobic and aerobic yeasts. *Proceedings of the National Academy of Sciences USA* 89:8966.

119. Shi, N.-Q., and Jeffries, T. W. 1998. Anaerobic growth and improved fermentation of *Pichia stipitis* bearing a *URA1* gene from *Saccharomyces cerevisiae*. *Applied Microbiology and Biotechnology* 50:339.

120. Bicho, P. A. et al. 1988. Induction of D-xylose reductase and xylitol dehydrogenase activities in *Pachysolen tannophilus* and *Pichia stipitis* on mixed sugars. *Applied and Environmental Microbiology* 54:50.

121. Agbogbo, F. K., and Wenger, K. S. 2006. Effect of pretreatment chemicals on xylose fermentation by *Pichia stipitis*. *Biotechnology Letters* 28:2065.

122. van Zyl, C., Prior, B. A., and du Preez, J. C. 1988. Production of ethanol from sugar-cane bagasse hemicellulose hydrolysate by *Pichia stipitis*. *Applied Biochemistry and Biotechnology* 17:357.

123. Tran, A. V., and Chambers, R. P. 1986. Ethanol fermentation of red oak acid prehydro-lysate by the yeast *Pichia stipitis* CBS 5776. *Enzyme and Microbial Technology* 8:439.

124. Nigam, J. N. 2001. Ethanol production of from wheat straw hemicellulose hydrolysate by *Pichia stipitis*. *Journal of Biotechnology* 87:17.

125. Nigam, J. N. 2001. Development of xylose-fermenting yeast *Pichia stipitis* for ethanol production through adaptation on hardwood hemicellulose acid prehydrolysate. *Journal of Applied Microbiology* 90:208.

126. Eken-Saraçoğlu, N., and Arslan, Y. 2000. Comparison of different pretreatments in eth-anol fermentation using corn cob hemicellulosic hydrolysate with *Pichia stipitis* and *Candida shehatae*. *Biotechnology Letters* 22:855.

127. van den Berg, J. A. et al. 1990. *Kluyveromyces* as a host for heterologous gene expres-sion and secretion of prochymosin. *Bio/Technology* 8:135.

128. Schwan, R. F. et al. 2001. Microbiology and physiology of *cachaça* (*aguardente*) fer-mentations. *Antonie van Leeuwenhoek* 79:89.

129. Anderson, P. J., McNeil, K., and Watson, K. 1986. High-efficiency carbohydrate fermen-tation to ethanol at temperatures above 40°C by *Kluyveromyces marxianus* var. *marxi-anus* isolated from sugar mills. *Applied Microbiology and Biotechnology* 51:1314.

130. Margaritis, A., and Bajpal, P. 1982. Direct fermentation of D-xylose to ethanol by *Kluyveromyces* strains. *Applied and Environmental Microbiology* 44:1039.

131. Gough, S. et al. 1996. Fermentation of molasses using a thermotolerant yeast, *Kluyveromyces marxianus* IMB3: Simplex optimization of media supplements. *Applied Microbiology and Biotechnology* 6:187.

132. Bellaver, L. H. et al. 2004. Ethanol formation and enzyme activities around glucose-6-phosphate in *Kluyveromyces marxianus* CBS 6556 exposed to glucose or lactose excess. *FEMS Yeast Research* 4:691.

133 Matsushika, A. et al. 2009. Ethanol production from xylose in engineered *Saccharomyces cerevisiae* strains: Current state and perspectives. *Applied Microbiology and Biotechnology* 84:37.

134. Souto-Maior, A. M., Runquist, D., and Hahn-Hägerdal B. 2009. Crabtree-negative characteristics of recombinant xylose-utilizing *Saccharomyces cerevisiae*. *Journal of Biotechnology* 143:119.

135. Famili, I. et al. 2003. *Saccharomyces cerevisiae* phenotypes can be predicted by using constraint-based analysis of a genome-scale reconstructed metabolic network. *Proceedings of the National Academy of Sciences USA* 100:13134.

136. Goffeau, A. 1997. The yeast genome directory. *Nature* 387:5.

137. Blank, L. M., Lehmbeck, F., and Sauer, U. 2005. Metabolic-flux and network analysis in fourteen hemiascomycetous yeasts. *FEMS Yeast Research* 5:545.

138. Jin, Y. S., and Jeffries, T. W. 2004. Stoichiometric network constraints on xylose metabolism by recombinant *Saccharomyces cerevisiae*. *Metabolic Engineering* 6:229.

139. Bro, C. et al. 2006. In silico aided metabolic engineering of *Saccharomyces cerevisiae* for improved bioethanol production. *Metabolic Engineering* 8:102.

140. Kong, Q. X. et al. 2007. Overexpressing *GLT1* in *gpd1Δ* mutant to improve the production of ethanol in *Saccharomyces cerevisiae*. *Applied Microbiology and Biotechnology* 73:1382.

141. Zhang, A. et al. 2007. Effect of *FPS1* deletion on the fermentative properties of *Saccharomyces cerevisiae*. *Letters in Applied Microbiology* 44:212.

142. Kong, Q. X. et al. 2006. Improved production of ethanol by deleting *FPS1* and overexpressing *GLT1* in *Saccharomyces cerevisiae*. *Biotechnology Letters* 28:2033.

143. Gardonyi, M. et al. 2003. Control of xylose consumption by xylose transport in recombinant *Saccharomyces cerevisiae*. *Biotechnology and Bioengineering* 82:818.

144. Stambuk, B. U. et al. 2003. D-Xylose transport by *Candida succiphila* and *Kluyveromyces marxianus*. *Applied Biochemistry and Biotechnology* 105–108:255.

145. Runquist, D. et al. 2008. Expression of the *Gxf1* transporter from *Candida intermedia* improves fermentation performance in recombinant xylose-utilizing *Saccharomyces cerevisiae*. *Applied Microbiology and Biotechnology* 82:123.

146. Hector, R. E. et al. 2008. Expression of a heterologous xylose transporter in a *Saccharomyces cerevisiae* strain engineered to utilize xylose improves aerobic xylose consumption. *Applied Microbiology and Biotechnology* 80:675.

147. Hjersted, J. L., Henson, M. A., and Mahadevan, R. 2007. Genome-scale analysis of *Saccharomyces cerevisiae* metabolism and ethanol production in fed-batch culture. *Biotechnology and Bioengineering* 97:1190.

148. Gibson, B. R. et al. 2007. Yeast responses to stresses associated with industrial brewery handling. *FEMS Microbiology Reviews* 31:535.

149. Ingledew, W. M. 1999. Alcohol production by *Saccharomyces cerevisiae*: A yeast primer. In *The alcohol textbook,* 3rd ed., ed. Jacques, K. A., Lyons, T. P., and Kelsall, D. R., chap. 5. Nottingham, England: Nottingham University Press.

150. Hou, L. 2010. Improved production of ethanol by novel genome shuffling in *Saccharomyces cerevisiae*. *Applied Biochemistry and Biotechnology* 160:1084.

151. Auesukaree, C. et al. 2009. Genome-wide identification of genes involved in tolerance to various environmental stresses in *Saccharomyces cerevisiae*. *Journal of Applied Genetics* 50(3):301.

152. Li, L. et al. 2009. The induction of trehalose and glycerol in *Saccharomyces cerevisiae* in response to various stresses. *Biochemical and Biophysical Research Communications* 387:778.

153. Ingledew, W. M., and Casey, G. P. 1986. Rapid production of high concentrations of ethanol using unmodified industrial yeast. In *Biotechnology and renewable energy,* ed. Moo-Young, M., Hasnain, S., and Lamptey, J., 246. London: Elsevier Applied Science Publishers.

154. Kelsall, D. R., and Lyons, T. P. 1999. Management of fermentations in the production of alcohol: Moving toward 23% ethanol. In *The alcohol textbook,* 3rd ed., ed. Jacques, K. A., Lyons, T. P., and Kelsall, D. R., chap. 3. Nottingham, England: Nottingham University Press.

4 Microbiology of Cellulosic Ethanol Production II
Bacteria

4.1 INTRODUCTION

Given the long established use of yeasts in ethanol production, why would biotechnology turn to bacteria—microorganisms more associated with problems in the ethanol industry and in human pathogenesis?

Bacteria are more thermophilic than yeasts; some species can utilize pentoses in their native state while others are used in some traditional potable alcohol beverages around the world. Can these features and traits be maximized by genetic and metabolic engineering?

Bacteria are often exquisitely suited to modern biotechnologies such as genomics, proteomics, and transcriptomics. Contemporary research has pushed to the limits of how few genes and gene products could function in minimal bacteria-like organisms. Could these (selected, constructed, and fine-tuned) prove to be ideal future ethanologens?

4.2 ASSEMBLING GENE ARRAYS IN BACTERIA FOR ETHANOL PRODUCTION

Bacteria are traditionally unwelcome to wine producers and merchants because they are spoiling agents. For fuel ethanol production with yeasts, bacteria are frequent contaminants in nonsterile mashes where they produce lactic and acetic acids, which in high concentration inhibit growth and ethanol production by yeasts.[1–3]

Bacteria are much less widely known as ethanol *producers,* but *Escherichia, Klebsiella, Erwinia,* and *Zymomonas* species have all received serious and detailed consideration for industrial use and have been the hosts for recombinant DNA technologies for starch and cellulosic ethanol within the last 25 years (Table 4.1).[4–10] The ability of bacteria to grow at much higher temperatures than is possible with conventional yeast ethanologens led to proposals early in the history of the application of modern technology to fuel ethanol production. Being able to run high-yielding alcohol fermentations at 70°C or above (to accelerate the process and reduce the economic cost of ethanol recovery) could have far reaching industrial implications.[11,12]

TABLE 4.1

Bacterial Species as Candidate Fuel Ethanol Producers

Species	Strain Type	Carbon Source	Ethanol Productivity (g/g Sugar Used)	Ref.
Erwinia chrysanthemi	PDC transconjugant	Xylose	0.45	4
Erwinia chrysanthemi	PDC transconjugant	Arabinose	0.33	4
Klebsiella planticola	PDC transconjugant	Xylose	0.40	5
Zymomonas mobilis	Patented laboratory strain	Amylase-digested starch	0.46	6
Klebsiella oxytoca	*Z. mobilis pdc* and *adhB* genes	Xylose	0.42	7
Klebsiella oxytoca	*Z. mobilis pdc* and *adhB* genes	Arabinose	0.34	7
Klebsiella oxytoca	*Z. mobilis pdc* and *adhB* genes	Glucose	0.37	7
Bacillus stearothermophilus	Lactate dehydrogenase mutant	Sucrose	0.30	8
Escherichia coli	*Z. mobilis pdc* and *adhB* genes	Corn fiber acid hydrolysate	0.41	10

4.2.1 METABOLIC ROUTES IN BACTERIA FOR SUGAR METABOLISM AND ETHANOL FORMATION

Bacteria can mostly accept pentose sugars and a variety of other carbon substrates as inputs for ethanol production (Table 4.1). Unusually, *Z. mobilis* can only use glucose, fructose, and sucrose; however, it can be easily engineered to utilize pentoses by gene transfer from other organisms.[13] This lack of pentose use by the wild type probably restricted its early commercialization because, otherwise, *Z. mobilis* has extremely desirable features as an ethanologen:

- It is a GRAS organism.
- It accumulates ethanol in high concentration as the major fermentation product with a 5–10% higher ethanol yield per unit of glucose used and with a 2.5-fold higher specific productivity than *S. cerevisiae*.[14]
- The major pathway for glucose catabolism is the Entner–Douderoff pathway (Figure 4.1); the inferior bioenergetics of this pathway in comparison with glycolysis means that more glucose is channeled to ethanol production than to growth, and the enzymes required comprise up to 50% of the total cellular protein.[14]
- No Pasteur effect on glucose consumption rate is detectable, although interactions between energy and growth are important.[15]
- *Z. mobilis* also was developed and proposed for ethanol production over 25 years ago, including its pilot-scale use in a high-productivity continuous process using hollow fiber membranes for cell retention and recycling.[16]

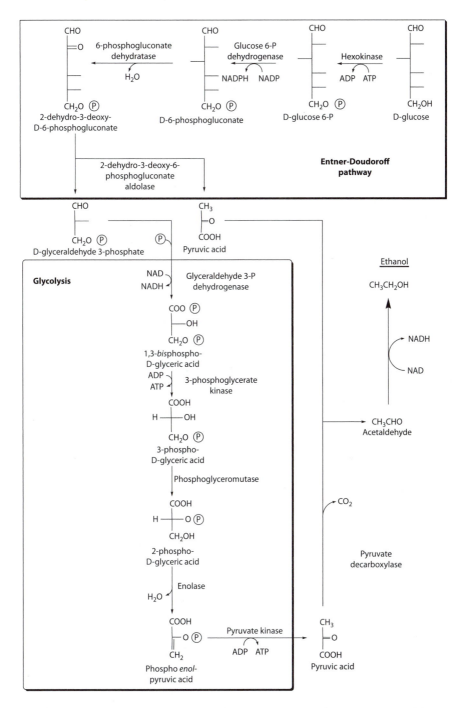

FIGURE 4.1   Entner–Doudoroff pathway of glucose catabolism in *Zymomonas mobilis* and Embden–Meyerhof–Parnas pathway of glycolysis in most bacteria and yeasts in providing pyruvic acid as a substrate for the homoethanol pathway.

Escherichia coli and other bacteria often incompletely metabolize glucose, accumulating large amounts of carboxylic acids, notably acetic acid.[17,18] For *E. coli* as a vehicle for the production of recombinant proteins, acetate accumulation is an acknowledged inhibitory factor; in ethanol production, it is simply a metabolic waste of glucose carbon. Other than this avoidable diversion of resources, enteric bacteria such as *E. coli* are easily genetically manipulated, grow well in both complex and defined media, can use a wide variety of nitrogen sources for growth, and have been the subjects of decades of experience and expertise for industrial-scale fermentations.

STEM TOPIC 4.1: THE ENERGY POVERTY OF BACTERIAL CELL GROWTH ON XYLOSE

Adenosine triphosphate (ATP) is the central "currency" of energy conversion inside bacterial, yeast, fungal, plant, and animal cells. ATP hydrolysis is coupled to endothermic (energy-requiring) reactions; ATP formation conserves part of the chemical energy liberated during exothermic reactions.

The principal means for xylose uptake by *E. coli* is an ATP-hydrolyzing transport system that yields xylose; this is in contrast to glucose uptake by the phosphotransferase systems, which consumes a mole of phosphoenolpyruvate (bioenergetically equivalent to ATP) but generates a mole of glucose 6-phosphate. This apparently small difference results in a radical alteration in energy metabolism when cells are grown on glucose or xylose.

Each mole of glucose taken up requires 2 mol of ATP to transform the hexose to fructose-1,6-diphosphate, but 4 mol of ATP are generated subsequently from the oxidation of 2 mol of C3 intermediates (Figure 3.1 in Chapter 3). Consequently, each mole of glucose in *E. coli* fermented to ethanol is equivalent to 2 mol of ATP.

For xylose, each mole taken up and converted to a pentose phosphate requires 2 mol of ATP. The conversion of pentose phosphate to fructose 6-phosphate and glyceraldehyde 3-phosphate (Figure 3.2 in Chapter 3) can be written as

$$3 \times \text{pentose-P} \rightarrow 2 \times \text{fructose 6-P} + 1 \times \text{glyceraldehyde 3-P}$$

Each fructose phosphate requires a mole of ATP for conversion to fructose-1,6-*bis*phosphate. The total ATP requirement of the fermentation of 3 mol of xylose is therefore $3 + 3 + 2 = 8$; the ATP formed from hexose and triose phosphate oxidation is $(2 \times 4) + 2 = 10$. Three moles of xylose can maximally generate 2 mol of ATP (i.e., 0.67 mol of ATP per mole of xylose fermented).

The growth yield from ATP is the same on either sugar. Therefore, assuming that the maintenance energy demand is similar for glucose and xylose, the growth yield (Y_{sugar}, gram of dry weight cells per gram of sugar consumed) should be proportional to the net energy yield from sugar metabolism. Experimentally, the value of Y_{sugar} is reduced by approximately 50% when xylose replaces glucose as the growth-limiting carbon and energy source.[19]

Two principal routes for glucose catabolism are known in ethanologenic bacteria: the Embden–Meyerhof–Parnas (EMP) pathway of glycolysis (also present in yeasts, fungi, plants, and animals) and, in a restricted range of bacteria (but more widely for the catabolism of gluconic acid), the Entner–Doudoroff (ED) pathway (Figure 4.1).[20–22] The two initial steps of the ED pathway resulting in 6-phosphogluconic acid are those of the oxidative pentose phosphate pathway, but *Z. mobilis* is unique in operating this sequence of reactions under anaerobic conditions.[23] The EMP and ED pathways converge at pyruvic acid; from pyruvate, a range of fermentative products can be produced, including acids such as lactate and decarboxylated acids such as 2,3-butanediol (Figure 4.2).

In *S. cerevisiae* and other yeasts, ethanol formation requires only two reactions from pyruvate (see Figure 3.1 in Chapter 3):

1. Pyruvate decarboxylase (PDC) catalyzes the formation of acetaldehyde.
2. Alcohol dehydrogenase (ADH) catalyzes the NADH-oxidizing reduction of acetaldehyde to ethanol.

In *E. coli,* however, no PDC naturally exists, and pyruvate is catabolized by the pyruvate dehydrogenase (PDH) reaction under aerobic conditions or by the pyruvate formate lyase (PFL) reaction under anaerobic conditions. To maintain the redox balance under fermentative conditions, a spectrum of products is generated from glucose in *E. coli* and many other enteric bacterial species in which ethanol is often a minor component (Table 4.2).[24]

4.2.2 GENETIC AND METABOLIC ENGINEERING OF BACTERIA FOR CELLULOSIC ETHANOL PRODUCTION

Research and development work for bacterial ethanologenesis has focused on *E. coli* and *Z. mobilis,* but there has been significant work on *Klebsiella* strains. Other bacteria have been considered and these are discussed briefly in this section.

4.2.2.1 Recombinant *Escherichia coli:* Lineages and Metabolic Capabilities

Early attempts in the 1980s to improve *E. coli* and other bacteria genetically for efficient ethanol production with recombinant gene technology floundered because they relied on endogenous ADH activities competing with other product pathways.[20] Success required combining the *Z. mobilis* genes for PDC and ADH in a single plasmid: the PET operon; in various guises, the PET system has been used to engineer ethanol production in *E. coli* and other bacteria.[25,26]

A crucial property of the *Z. mobilis* PDC is its relatively high affinity (K_m) for pyruvate as a substrate: 0.4 mM as compared with 2 mM for PFL, 7 mM for lactate dehydrogenase, and 0.4 mM for pyruvate dehydrogenase. In consequence, many reduced amounts of lactic and acetic acids are formed, and the mix of fermentation products is much less toxic to growth and inhibitory to the establishment of dense cell populations.[22] Further mutations have been utilized to delete genes for the biosynthesis of succinic, acetic, and lactic acids to further reduce the waste of sugar carbon to unwanted metabolic acids.[27]

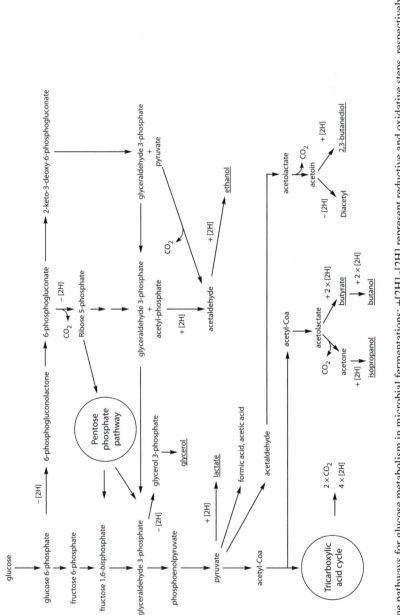

FIGURE 4.2 Enzyme pathways for glucose metabolism in microbial fermentations: +[2H],-[2H] represent reductive and oxidative steps, respectively.

TABLE 4.2

Mixed Acid Fermentation Products Accumulated by
Escherichia coli

Product	Conversion (mol/mol glucose consumed)	Carbon Recovery (%)	Redox Balance[a]
Lactic acid	1.09	54	0.0
Acetic acid	0.32	11	0.0
Formic acid	0.02	0	1.0
Succinic acid	0.18	12	1.0
Ethanol	0.41	14	−2.0
CO_2	0.54	9	2.0
H_2	0.18		−1.0
Σ		100	1.0

Source: Data from Moat, A. G., and Foster, J. W. 1995. *Microbial Physiology,* 3rd ed., chap. 7. New York: John Wiley & Sons.

[a] ΣO atoms − $0.5\Sigma H$ atoms.

When a number of the well-characterized *E. coli* laboratory strains were evaluated for their overall suitability for PET transformation, strain B exhibited the best hardiness to environmental stresses (ethanol tolerance and plasmid stability in nonselective media) and superior ethanol yield on xylose.[28,29] This strain was isolated in the 1940s and was widely used for microbiological research in the 1960s; more importantly, it lacks all known genes for pathogenicity.[30] A strain-B-derived, chromosomally integrated isolate (KO11), also with a disrupted fumarate reductase gene (for succinate formation), emerged as a leading candidate for industrial ethanol production.[31] It shows ethanol production with much improved selectivity against the mixed-acid range of fermentation products (Figure 4.3 and Table 4.2).[32]

An encouragingly wide range of carbon substrates have been shown to support ethanol production with strain KO11:

- pine wood acid hydrolysates[33]
- sugarcane bagasse and corn stover[34]
- corn cobs, hulls, and fibers[35,36]
- dilute acid hydrolysate of rice hulls[37]
- sweet whey and starch[38,39]
- galacturonic acid and other components in orange peel hydrolysates[40]
- the trisaccharide raffinose (a component of corn steep liquors and molasses)[41]

To increase its hardiness to ethanol and inhibitors present in acid hydrolysates of lignocellulosic materials, strain KO11 was adapted to progressively higher ethanol concentrations over a period of months.[42] Increased ethanol tolerance was accompanied (fortuitously) by increased resistance to various growth inhibitors, including aromatic alcohols and acids derived from ligninolysis, and various aromatic and nonaromatic

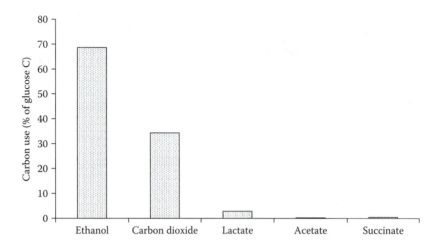

FIGURE 4.3 Fermentative efficiency of recombinant *Escherichia coli strain* KO11 on 10% (v/v) glucose. (Data from Ohta, K. et al. 1991. *Applied and Environmental Microbiology* 57:893; and Dien, B. S., Cotta, M. A., and Jeffries, T. W. 2003. *Applied Microbiology and Biotechnology* 63:258.)

aldehydes (including HMF and furfural).[43–45] Of this multiplicity of inhibitors, the aromatic alcohols proved to be the least toxic to bacterial growth and metabolism, and *E. coli* strains can be at least as refractory to growth inhibitors as are yeast ethanologens.

More wide-ranging genetic manipulation of ethanologenic *E. coli* strains has explored features of the molecular functioning of the recombinant cells as measured by quantitative gene expression and the activities of the heterologous gene products.[46–49] A long recognized problem with high-growth-rate bacterial hosts engineered to contain and express multiple copies of foreign genes is that of metabolic burden; that is, the diversion of nutrients from biosynthesis and cell replication to supporting the expression and copying of the novel gene complement often results in a reduced growth rate in comparison with the host strain.

Chromosomal integration of previously plasmid-borne genes does not avoid this metabolic demand, as became evident when attempts were made to substitute a rich laboratory growth medium with possible cheaper industrial media based on ingredients such as corn steep liquor. Uneconomically high concentrations were required to match the productivities observable in laboratory tests, but this could be only partially improved by lavish additions of vitamins, amino acids, and other putative growth-enhancing medium ingredients.[46]

More immediately influential was increasing PDC activity by inserting plasmids with stronger promoters into the chromosomally integrated KO11 strain. Because the laboratory trials often achieved quite low cell densities (3 g dry weight of cells per liter)—lower by one or two orders of magnitude than those attainable in industrial fermentations—it is highly probable that some ingenuity will be required to develop adequate media for large-scale fermentations while minimizing operating costs.

Genetic manipulation, however, can aid the transition of laboratory strains to commercially relevant media and the physical conditions in high-volume fermentors.

For example, growth and productivity by the KO11 strain in suboptimal media can be greatly increased by the addition of simple additional carbon sources such as pyruvic acid and acetaldehyde. This has no practical significance because ethanol production necessarily cannot be a biotransformation from more expensive precursors; however, together with other physiological data, this implies that the engineered *E. coli* cells struggle to partition carbon flow adequately between the demands for growth (amino acids, etc.) and the requirement to reoxidize NADH and produce ethanol as an end-product.[47,48]

Expressing in KO11 a *Bacillus subtilis* citrate synthase, whose activity is not affected by intracellular NADH concentrations, improves both growth and ethanol yield by >50% in a xylose-containing medium; this novel enzyme in a coliform system may act to achieve a better balance and direct more carbon to 2-oxoglutarate and thence to a family of amino acids required for protein and nucleic acid biosynthesis.[47] Suppressing acetate formation from pyruvate by deleting the endogenous *E. coli* gene (*ackA*) for acetate kinase probably has a similar effect by altering carbon flow around the crucial junction represented by pyruvate (Figure 4.2).[48]

In nutrient-rich media, expressing the *Z. mobilis* homoethanol pathway genes in *E. coli* increases growth rate by up to 50% during the anaerobic fermentation of xylose.[49] Gene array analysis reveals that, of the nearly 4,300 total open reading frames in the genome, only 8% were expressed at a higher level in KO11 in anaerobic xylose fermentations when compared with the B strain parent. However, nearly 50% of the genes involved in xylose catabolism to pyruvate were expressed at higher levels in the recombinant (Figure 4.4).

A further broadening of the substrate range of the KO11 strain was effected by expressing genes (from *Klebsiella oxytoca*) encoding an uptake mechanism for cellobiose, the

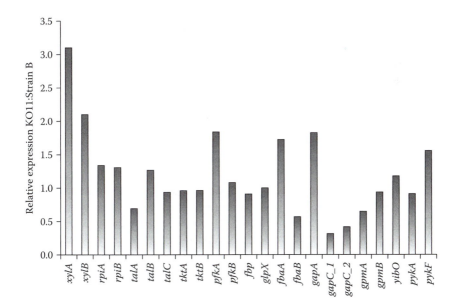

FIGURE 4.4   Gene expression pertinent to xylose metabolism by recombinant *Escherichia coli* strain KO11. (Data from Tao, H. et al. 2001. *Journal of Bacteriology* 183:2979.)

disaccharide product of cellulose digestion. An operon was introduced on a plasmid into KO11 containing the two genes for the phosphoenolpyruvate (PEP)-dependent phosphotransferase (PTS) transporter for cellobiose (generating phosphorylated cellobiose) and phospho-β-glucosidase (for hydrolyzing the cellobiose phosphate intracellularly).[50] The *K. oxytoca* genes proved to be poorly expressed in the *E. coli* host, but spontaneous mutants with elevated specific activities for cellobiose metabolism were isolated and shown to have mutations in the plasmid that eliminated the engineered *casAB* promoter and operator regions. Such mutants rapidly fermented cellobiose to ethanol with an ethanol yield > 90% of the theoretical maximum and (with the addition of a commercial cellulase) fermented mixed-waste office paper to ethanol.

A second major lineage of recombinant *E. coli* started with the incomplete stability of the KO11 strain: Phenotypic instability was reported in repeated batch or continuous cultivation, resulting in declining ethanol but increasing lactic acid production.[51,52] Moreover, the results were different when glucose or xylose provided the carbon supply for continuous culture on glucose alone, KO11 appeared to be stable, but ethanol productivity declined after 5 days and the antibiotic (chloramphenicol) selective marker began to be lost after 30 days.[53,54]

Novel ethanologenic strains were created by expressing the PET operon on a plasmid in *E. coli* FMJ39, a strain with deleted genes for lactate dehydrogenase and PFL and, in consequence, incapable of fermentative growth on glucose.[55] The introduced homoethanol pathway genes complemented the mutations and positively selected for plasmid maintenance to enable active growth by fermentation pathways (i.e., self-selection under the pressure of fermenting a carbon source).[56] The plasmid was accurately maintained by serial culture and transfer under anaerobic conditions with either glucose or xylose as the carbon source and with no selective antibiotic present; it quickly disappeared during growth in the aerobic conditions in which the parental strain grew normally.

One of the resulting strains was further adapted for growth on xylose (FBR3); this construct can ferment a 10% (w/v) concentration of glucose, xylose, arabinose, or a mixture of all three sugars at 35°C over a period of 70–80 h, producing up to 46.6 g/L of ethanol at up to 91% of the theoretical maximum yield.[57] The strains are also able to ferment hydrolysates prepared from corn hulls and germ meal within 60 h and at a yield of 0.51 g of ethanol per gram of sugar consumed.[58,59]

Variants of the strains have been constructed that are relatively deficient in glucose uptake because they carry a mutation in the PEP-glucose PTS system. In organisms with this transport mechanism, the presence of glucose represses the uptake of other sugars. Obviating this induces the cells to utilize xylose and arabinose simultaneously with glucose—rather than sequentially after the glucose supply begins to be exhausted—and to make ethanol production more rapidly, although the overall productivity (carbon conversion efficiency) is minimally affected.[60]

Molecular evolution of an efficient ethanol pathway in *E. coli* without resorting to heterologous gene expression was finally accomplished in 2006.[61] The pyruvate dehydrogenase complex—the route of pyruvate oxidation in *E. coli* under aerobic conditions—is encoded by a four-gene operon, with pyruvate (or a derivative of pyruvate) acting as the inducing agent.[62] A mutant of *E. coli* strain K12 was isolated with a mutation occurring in the pyruvate dehydrogenase operon; the phenotype was a novel pathway endowing the capacity to ferment glucose or xylose to ethanol with a yield of 82% under anaerobic conditions, combining a PDC-type enzyme activity with the endogenous ADH activity.[61]

As with *Saccharomyces* yeasts, work on incrementally adjusting *E. coli* metabolism to improve ethanologenesis continues to produce interesting findings. Two research publications from 2008 took the recombinant *E. coli* story back to its roots in the expression of *Z. mobilis* ethanol pathway enzymes:

- The much studied KO11 strain converts glucose or xylose to ethanol with yields close to 100% of the theoretical maxima when growing in rich medium. In a minimal medium, however, the conversion efficiencies are much lower but can be improved by overexpressing the gene for *Z. mobilis* PDC.[63]
- A different *E. coli* strain, CCE14, has a higher ethanol production rate in mineral medium. This results from the elevated heterologous expression of the chromosomally integrated genes encoding the *Z. mobilis* PDC and ADH genes and their corresponding higher intracellular enzymic activities.[64]

Commercial take-up of recombinant ethanologenic *E. coli* has proved disappointingly slow.[65] In 2009, a demonstration facility using *E. coli* strains produced ethanol from sugarcane bagasse; published details are sketchy, but the process used separate hexose and pentose fermentations (www.verenium.com; Figure 4.5).

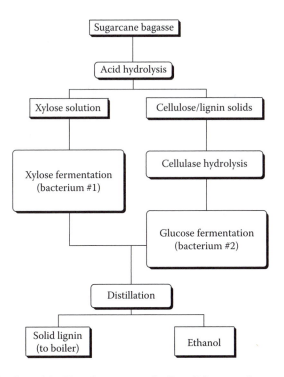

FIGURE 4.5   Outline of the Verenium process for *E. coli* fermentation sugarcane bagasse.

STEM TOPIC 4.2: ENERGY EXCESS: HOW BACTERIA CAN BE BIOENERGETICALLY WASTEFUL

Compared with many natural environments, cultivation of bacteria under laboratory conditions supplies cells with massive quantities of nutrients. Given that set of conditions, bacteria (and probably all microbes) need not be parsimonious with their energy metabolism but can be positively inefficient.

One parameter of energy metabolism is the cell yield from ATP, Y_{ATP} (gram of dry weight per mole of ATP). This is usually quoted as being in the range of 10.0 ± 1.1 g per mole but there is evidence that it can be much more variable in bacteria because of energy spillage.[66] The challenge in accurately measuring energy parameters is the fundamental uncertainty in measuring Q_{ATP}, the yield of ATP from any growth substrate. The substrate level phosphorylation generation of ATP from glycolysis is straightforward, but the majority of cellular ATP in aerobic organisms derives from oxidative phosphorylation, an energy-conserving mechanism requiring ion gradients across biomembranes.

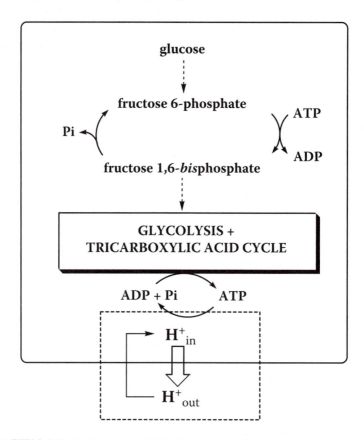

FIGURE STEM 4.2 Futile cycles of ATP formation and hydrolysis in the biochemistry of *strept ococcus bovis.*

The lactic acid bacterium *Streptococcus bovis* is particularly useful because it only has substrate level phosphorylation. Energy auditing shows that it has two major mechanisms that waste ATP, a futile cycle involving fructose 6-phosphate (F6P) and fructose-1,6-*bis*phosphate (FDP)—both intermediates of the standard glycolytic pathway of glucose catabolism—and a membrane leakage of protons from extracellular to intracellular spaces (see Figure STEM 4.2).

Formation of FDP from F6P consumes ATP; hydrolysis of FDP to F6P generates no ATP. The membrane of *S. bovis* can apparently alter significantly, causing another futile cycle. Such cycles are well known to metabolic modeling and are often incorporated into computer simulations of metabolism with yeasts and *E. coli*.

A cellulose-degrading microbe inside rumen stomach contents in herbivores is *Fibrobacter succinogenes,* which has no energy spilling mechanisms.[66] Excess readily utilizable carbohydrate is toxic to this microbe in the laboratory, including cellobiose (the dimeric degradation product of cellulose) but not cellulose itself. This may be because the cellulase of *F. succinogenes* is very sensitive to feedback inhibition by cellobiose (see Chapter 2, Section 2.4.1).

4.2.2.2 Engineering *Zymomonas mobilis* for Xylose and Arabinose Metabolism

Zymomonas mobilis was first known in Europe as a spoiling agent in cider. Its function in the making of beverages such as palm wines is well established in Africa, Central and South America, the Middle East, South Asia, and the Pacific Islands, and it can ferment the sugar sap of the *Agave* cactus to yield pulque.[67] Its unusual biochemistry has already been described (Figure 4.1), but curious metabolic features made *Z. mobilis* a promising target for industrial process development[68]:

- Lacking an oxidative electron transport chain, the species is energetically grossly incompetent (i.e., can capture very little of the potential bioenergy in glucose). In other words, it is nearly ideal from the ethanol fermentation standpoint.
- What little energy production is achieved can be uncoupled from growth by an intracellular wastage (ATPase).
- It shows no Pasteur effect and is seemingly oblivious to O_2 levels regarding glucose metabolism; however, acetate, acetaldehyde, and acetoin are accumulated with increasing oxygenation.

During the 1970s, biotechnological interest in *Z. mobilis* became intense, and a patent for its use in ethanol production from sucrose and fructose was granted in mid-1989. In the same year, Australian researchers demonstrated the ability of *Z. mobilis* to ferment industrial substrates such as potato mash and wheat starch to ethanol with 95–98% conversion efficiencies at ethanol concentrations up to 13.5% (v/v).[69] An Australian process for producing ethanol from starch was scaled up to a 13,000-L volume.[6]

The capability to utilize pentose sugars for ethanol production—with lignocel-lulosic substrates as the goal—was engineered into a strain recognized in 1981 as a superior ethanologen. Strain CP4 (originally isolated from fermenting sugarcane juice) exhibited that a rapid rate of ethanol formation from glucose achieved a high concentration, could ferment both glucose and sucrose at temperatures up to 42°C, and formed less polymeric fructose (levan) from sucrose than the other good ethanol producers. On transfer to high-glucose medium, CP4 had the shortest lag time before growth commenced and a short doubling time.[70] Pilot-scale fermentations evaluated mutant strains selected for increased ethanol tolerance; improving ethanol produc-tion from sucrose and molasses became a target for strain developement.[71] Direct genetic manipulation was explored to broaden the substrate range.[72]

Strains of *Z. mobilis* were first engineered to catabolize xylose when four genes for xylose utilization by *E. coli* were introduced into *Z. mobilis* strain CP4 and expressed: xylose isomerase (*xylA*), xylulokinase (*xylB*), transketolase (*tktA*), and transaldolase (*talB*) on a plasmid under the control of strong constitutive promoters from *Z. mobilis*.[73,74] The chosen transformant, CP4(pZB5), could grow on xylose as the carbon source with an ethanol yield of 86% of the theoretical maximum; cru-cially, xylose and glucose could be taken up by the cells simultaneously using a per-mease because no active (energy-expending), selective transport system for glucose exists in *Z. mobilis*. The transport facilitator for glucose is highly specific and only mannose and (weakly) galactose, xylose, sucrose, and fructose appear to be taken up by this mechanism.[75]

Using a plasmid containing five genes from *E. coli*, *araA* (encoding L-arabinose isomerase), *araB* (L-ribulose kinase), and *araD* (L-ribulose 5-phosphate-4-epimerase), plus *tkta* and *talB*, a strain (ATCC39767[pZB206]) was engineered to ferment L-arabinose and produce ethanol with a very high yield (96%) but at a slow rate, ascribed to the low affinity of the permease uptake mechanism for L-arabinose.[76] A third strain was transformed with a plasmid introducing genes for xylose metabo-lism and subsequently adapted for improved growth in the presence of hydrolysate inhibitors by serial subculture in progressively higher concentrations of the wood hydrolyate.[74,77]

A strain cofermenting glucose, xylose, and arabinose was constructed by cro-mosomal integration of the genes. This strain (AX101) was genetically stable and fermented glucose and xylose much more rapidly than it did arabinose. However, it produced ethanol at a high efficiency (0.46 g/g sugar consumed) and with only minor accumulations of xylitol, lactic acid, and acetic acid.[78–80]

The major practical drawback for the AX101 strain is its sensitivity to acetic acid (formed in lignocellulosic hydrolysates by the breakdown of acetylated sugars). This sensitivity was demonstrated in trials of the strain with an agricultural waste (oat hulls) substrate pretreated by the two-stage acid process developed by the Iogen Corporation in Canada—although the bacterial ethanologen outperformed a yeast in both volumetric productivity and glucose-to-ethanol conversion.[81]

Further strains of *Z. mobilis* have been constructed in attempts to increase etha-nol production from xylose or tolerance to acetic acid.[82–87] A major step forward in understanding the metabolic physiology of *Z. mobilis* was the complete sequencing of its genome in 2005[88]:

- The 2 Mbp circular chromosome encodes for 1,998 predicted functional genes; of these, nearly 20% showed no similarities to known genes, suggesting a high probability of coding sequences of industrial significance for ethanologens.
- The idiosyncratic carbohydrate metabolism of the species can be explained by the absence of three key genes for the common EMP/tricarboxylic acid cycle paradigm: 6-phosphofructokinase, 2-oxoglutarate dehydrogenase, and malate dehydrogenase.
- The nonoxidative pentose phosphate pathway is mostly missing, although the genes are present for the biosynthesis of phosphorylated ribose and thence histidine, as well as nucleotides for both DNA and RNA (Figure 4.6).
- In comparison to a strain of *Z. mobilis* with lower tolerance to ethanol and rates of ethanol production, glucose uptake, and specific growth rate, strain

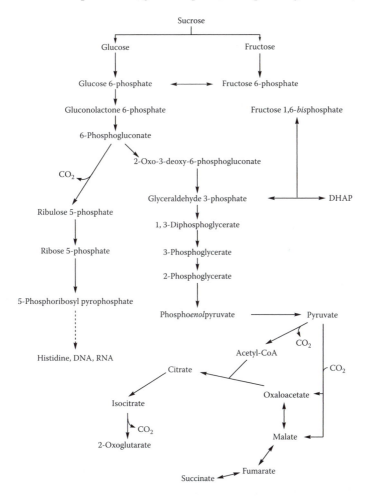

FIGURE 4.6 Fragmentary carbohydrate metabolism and interconversions in *Zymomonas mobilis* as confirmed by whole-genome sequencing. (Seo, J.-S. et al., 2005. *Nature Biotechnology* 23:63.)

ZM4 contains 54 *additional* genes, including four transport proteins and two oxidoreductases. All potentially mediate the higher ethanol productivity of the strain; two genes coding for capsular carbohydrate synthesis may be involved in generating an altered morphology more resistant to osmotic stress.

- Intriguingly, 25 of the new genes showed similarities to bacteriophage genes, indicating a horizontal transfer of genetic material via phages.

In *Z. mobilis,* there is evidence for a glucose-sensing system increasing tolerance to the osmotic stress imposed by high sugar concentrations.[89] Studies of redox biochemistry have revealed new features of cofactor usage, possible means of reducing xylitol formation from xylose, and understanding of cellular metabolism under both aerobic and anaerobic conditions.[90–93]

This continued effort by academic research laboratories was matched in 2007 when DuPont announced a major initiative to develop *Z. mobilis* for lignocellulosic ethanol, linking with an established corn ethanol producer, Broin Companies (now POET Biorefineries). Whether *Z. mobilis* will achieve full industrial status for ethanol or for higher value chemicals (succinic acid, sorbitol, and gluconic acid have been suggested) remains uncertain.[94] However, continued interest in this microbe with such an unusual fundamental biochemistry is ensured.

4.2.2.3 Development of *Klebsiella* Strains for Ethanol Production

The bacterium *Klebsiella oxytoca* has been isolated from paper and pulp mills and grows around other sources of wood; as well as growing on hexoses and pentoses, it can utilize cellobiose and cellotriose but does not secrete endoglucanase activity.[31] A strain transformed with the xylose-directing PET operon could produce ethanol at up to 98% of the theoretical yield and was highly suitable for lignocellulose substrates because it utilized xylose twice as fast as glucose and twice as fast as did *E. coli* strain KO11.[95]

Stabilizing the PET operon was accomplished by chromosomal integration at the site of the pyruvate formate lyase (*pfl* gene); screening for mutants hyper-resistant to the selectable chloramphenicol marker resulted in the P2 strain with improved fermentation kinetics and capability of producing ethanol from glucose or cellobiose.[96] Strain P2 has been demonstrated to generate ethanol from the cellulosic and lignocellulosic materials sugarcane bagasse, corn fiber, and sugar beet pulp.[36,97,98]

As a candidate industrial strain for cellulosic ethanol production, P2 can utilize a wide range of substrates, including the disaccharides cellobiose, cellulose, and xylan.[41,96,99] This relatively nonspecific diet has led to the cloning and expression of a two-gene *K. oxytoca* operon for xylodextrin utilization in *E. coli* strain KO11. The gene product of *xynB* is a xylosidase (which also has weak arabinosidase activity) and that of the adjacent gene in the *K. oxytoca* genome (*xynT*) is a membrane protein probably involved in transport.[100] The enhanced-recombinant *E. coli* could metabolize xylodextrins containing up to six xylose residues; unexpectedly, xylodextrin utilization was more rapid than that by the donor *K. oxytoca*.

Definitive comparisons of recombinant ethanologenic bacteria in tightly controlled, side-by-side comparisons are rare. From data compiled in 2003 from various sources with *E. coli, K. oxytoca,* and *Z. mobilis* strains fermenting mixtures of glucose, xylose, and arabinose, conflicting trends are evident[31]:

- For maximum ethanol concentration, the ranking order was
  Z. mobilis AX101 >> *K. oxytoca* P2 = *E. coli* FBR5.
- For ethanol yield (percentage of maximum possible conversion), *E. coli* was superior:
  E. coli FBR5 > *K. oxytoca* P2 = *Z. mobilis* AX101.
- The rankings of ethanol production rate (g/L/h) were again different:
  E. coli FBR5 >> *Z. mobilis* AX101 >> *K. oxytoca* P2.

All three strains utilize arabinose, xylose, and glucose, but *Z. mobilis* AX101 cannot utilize the hemicellulose component hexose sugars galactose or mannose.

STEM TOPIC 4.3: COMPARISON OF YEASTS AND BACTERIA AS ETHANOLOGENS FROM HEMICELLULOSE SUGARS

When *E. coli* and two yeasts were compared for ethanologenesis from a dilute acid hydrolysis of corn cobs (a hemicellulose sugar fraction), the recombinant bacterium outperformed yeasts in some key parameters.[101]

The cells were compared under various experimental conditions, but the most relevant condition for practical use used the raw hydrolysate (see Table STEM 4.3).

The comparisons were incomplete because different laboratories conducted the tests and not all analytical tests were made. Nevertheless, the ability of recombinant *E. coli* to rival fermentation performance with natural pentose-utilizing and recombinant yeasts was (in 1994) obvious. The highest ethanol concentration was also obtained with *E. coli* on detoxified (overlimed) hemicellulose hydrolysate.

For commercial ethanol production, the highest volumetric productivity obtained with one of the recombinant *S. cerevisiae* was probably conclusive because the fastest production of ethanol by a known and accepted organism outweighs most other considerations.

TABLE STEM 4.3

Comparison of *E. coli* and Yeast Ethanologens Using Hemicellulose Hydrolysates

Organism	Temperature (°C)	Total Sugar (g/L)	Ethanol	Y_{eth} (g/g total sugar)	Maximum Volumetric Productivity (g/L/h)	Maximum Specific Productivity (g/g DW/h)
E. coli KO11	30	39.3	9.4	0.24	0.88	0.18
S. cerevisiae (XI)	30	42.4	2.7	0.06	0.20	0.31
S. cerevisiae (XR + XDH)	30	42.2	2.9	0.07	1.04	0.15
P. stipitis CBS 5773 γ	30	47.6	8.4	0.18	0.76	0.08

4.2.2.4 Other Bacterial Species

Interest in *Erwinia* bacteria for ethanol production dates back at least to the late 1950s. In 1971, the explanation for the unusually high ethanol production by *Erwinia* species was identified as a PDC/ADH pathway, decarboxylating pyruvate to acetaldehyde followed by reduction to ethanol, akin to that in *Z. mobilis;* ethanol is the major fermentative product, accompanied by smaller amounts of lactic acid.[102] Soft-rot bacteria secrete hydrolases and lyases to solubilize lignocellulosic polymers, and the PET operon was used to transform *E. carotovora* and *E. chrysanthemi* to produce ethanol from cellobiose, glucose, and xylose.

Both strains fermented cellobiose at twice the rate shown by cellobiose-utilizing yeasts.[9] The genetically engineered *E. chrysanthemi* could ferment sugars present in beet pulp, but was inferior to *E. coli* strain KO11 in ethanol production, generating more acetate and succinate in mixed-acid patterns of metabolism.[98]

Zymobacter palmae was isolated on the Japanese island of Okinawa from palm sap; a facultative anaerobe, the bacterium can ferment glucose, fructose, galactose, mannose, sucrose, maltose, melibiose, raffinose, mannitol, and sorbitol, converting maltose efficiently to ethanol with only a trace of fermentative acids.[103] Its metabolic characteristics indicate potential as an ethanologen; broadening its substrate range to include xylose followed previous work with *Z. mobilis* expressing *E. coli* genes for xylose isomerase, xylulokinase, transaldolase, and transketolase.[104] The recombinant *Zb. palmae* completely cofermented a mixture of 40 g/L each of glucose and xylose simultaneously within 8 h at 95% of the theoretical yield.

Introducing a *Ruminococcus albus* gene for β-glucosidase transformed *Zb. palmae* to cellobiose utilization; the heterologous enzyme was >50% present on the cell surface or inside the periplasm, and the recombinant could transform 2% cellobiose to ethanol at 95% of the theoretical yield.[105] The PDC enzyme of the organism is an interesting target for heterologous expression in ethanologenic bacteria; it has the highest specific activity and lowest affinity for its substrate pyruvate of any bacterial PDC, and it has been expressed in *E. coli* to approximately 33% of the soluble protein.[106]

Enterobacter species would not, at first glance, be an immediate choice for the production of ethanol because they are often opportunistic pathogens in people with precarious immune systems, especially in hospitals where infection control has broken down (infections are becoming more common in intensive care). *Enterobacter cloacae,* however, has been used in biological control of plant diseases, and interest in *Enterobacter* as biosynthetic organisms for fine chemicals dates back to the late 1990s.

Enterobacter species are in the bacterial community present during the fermentation of pulque, a traditional Mexican alcoholic beverage.[107] A strain of *Enterobacter asburiae* is the first microorganism described that ferments methylglucuronoxylose generated along with xylose during the acid-mediated saccharification of hemicellulose; genetic definition of the methylglucuronoxylose utilization pathway could lead to the metabolic engineering of established bacterial biocatalysts for complete bioconversion of acid hydrolysates of methylglucuronoxylan.[108] The strain has also been successfully engineered for the efficient conversion of acid hydrolysates of hemicellulose to biofuels and chemical feedstocks.[109]

Cyanobacteria (blue-green algae) have generally lost their fermentative capabilities, now colonizing marine, brackish, and freshwater habitats where photosynthetic metabolism predominates; out of 37 strains in a German culture collection, only five accumulated fermentation products in darkness and under anaerobic conditions, and acids (glycolic, lactic, formate, and oxalate) were the major products.[110] Nevertheless, expression of Z. *mobilis pdc* and *adh* genes under the control of the promoter from the operon for the CO_2-fixing ribulose1,5-*bis*phosphate carboxylase in a *Synechococcus* strain synthesized ethanol phototrophically from CO_2 with an ethanol/acetaldehyde molar ratio in excess of 75:1.[111]

Cyanobacteria have simple growth nutrient requirements and use light, CO_2, and inorganic elements efficiently. Thus, they represent a system for longer-term development for the bioconversion of solar energy (and CO_2) by genetic transformation, strain and process evolution, and metabolic modeling to redirect intermediary metabolism by channeling intermediates of light-driven carbon fixation into fermentative metabolic pathways.[112,113]

4.3 THERMOPHILIC SPECIES AND CELLULOSOME BIOPRODUCTION TECHNOLOGIES

By 1983, experimental laboratory programs with *Bacillus* strains had achieved ethanologenesis at 60°C; ethanol was the major fermentation product, but acetic and formic acids remained serious by-products, and evidence from laboratory studies suggested that ethanol accumulation followed (and depended on) the formation of these growth-inhibiting acids.[12] The ability to run ethanol fermentations at even higher temperatures (>70°C) with thermophilic microbes remains both a scientific fascination and a conscious attempt to accelerate bioprocesses, despite the low ethanol tolerance and poor hexose-converting abilities of some anaerobic thermophilic bacteria. Isolates from various sources (hot springs, paper pulp mills, and brewery wastewater) have been examined using three main criteria for suitable organisms[114]:

1. the ability to ferment D-xylose to ethanol
2. high viability and ethanol productivity with pretreated wheat straw
3. tolerance to high sugar concentrations

Five good (but unidentified) strains were identified by this screening program, all from hot springs in Iceland; the best could grow in xylose solutions of up to 60 g/L. Also, isolated from geothermal springs in Iceland, a strain similar to *Thermoanaerobacterium aciditolerans* produces ethanol from glucose at 80% of the theoretical maximum and from xylose at 66% efficiency.[115] *Thermoanaerobacterium saccharolyticum* has been intensively studied by the Mascoma Corporation (Lebanon, New Hampshire) for consolidated bioprocessing (CBP; see Chapter 2, STEM Topic 2.1), using gene deletion to eliminate wastage to organic acids and H_2 formation.[116,117]

Thermophilic and mesophilic clostridia also have their advocates, especially with reference to the direct fermentation of cellulosic polymers by the cellulosome multienzyme complexes, as discussed in Chapter 2 (Section 2.4.1). Bypassing the

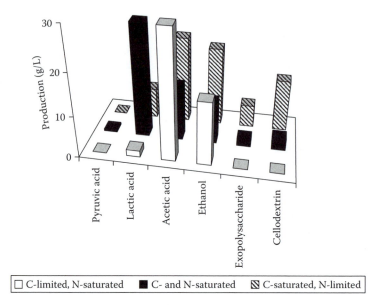

FIGURE 4.7   Cellobiose utilization by *Clostridium cellulolyticum* in continuous culture at low dilution rate, 0.023–0.035 per hour. (Data from Desvaux, M. 2006. *Biotechnology Progress* 22:1229.)

cellulosome is possible if cellulose degradation products (rather than polymeric celluloses) are used as carbon sources; this equates to using bacteria with cellulase-treated materials, including agricultural residues and paper recyclates.

Laboratory studies with *C. cellulolyticum* tested cellobiose in this fashion but with chemostat culture so as to control growth rates and metabolism more closely.[118] The results demonstrated that a more efficient partitioning of carbon flow to ethanol was possible than with cellulose as the substrate, but that the fermentation remained complex; acids were the major products (Figure 4.7). Nevertheless, clostridia are open to metabolic engineering to reduce the waste of carbohydrates as acids and polymeric products, or as vehicles for CBP; this is discussed in Chapter 5 (Section 5.6).

4.4   "DESIGNER" CELLS AND SYNTHETIC ORGANISMS

How few macromolecules are necessary for a self-replicating bacterial cell? Such a question might appear purely academic were it not for the patent application filed in 2007 by J. Craig Venter Institute (Rockville, Maryland).[119] A minimal organism into which are inserted a small group of genes encoding the enzymes required for ethanol synthesis from a single sugar substrate would be the first logical step; obviously, more genes would be required for multiple sugars in lignocellulosic biomass sources but would avoid all the complications of "real" organisms whose genomes need multiple deletions to focus carbon flow toward ethanol.[116,117]

The answer to the question posed above may be: 151 genes encoding 38 RNA molecules and 113 proteins.[120] Such an organism would be entirely dependent on the supply of small molecule precursors for macromolecule assembly because of the complete lack of any biosynthetic capabilities for amino acids, nucleic acid bases, and triglycerides ("architectural" molecules) or ATP. The time line required for practical application and demonstration of such a synthetic organism is presently unclear.[121]

STEM TOPIC 4.4: MACROMOLECULES FOR A PUTATIVE MINIMAL BACTERIUM

DNA-acting enzymes	3
RNA-acting enzymes	6
Ribosomal RNA molecules	3
Ribosomal RNA-acting enzymes	6
Ribosomal proteins	54
Ribosome functional protein factors	11
Transfer RNA molecules	29
Transfer RNA-acting enzymes	35
Other RNA for DNA/RNA synthesis	2
Chaperonins (for protein folding)	2
Total	151

These functionally active macromolecules (plus the DNA genome encoding them) are contained within a lipid bilayer vesicle (a minimal cell membrane) across which all the protein and nucleic acid precursors and ATP (and any other requirements) are assumed to diffuse.[120] The minimal cell would also be very fragile and the membrane vesicle easily lysing and thus incapable of replicating outside the laboratory.

Work undertaken at the J. Craig Venter Institute (Rockville, Maryland) has defined a minimal set of 381 protein-encoding genes from *Mycoplasma genitalium,* including pathways for carbohydrate metabolism, nucleotide biosynthesis, phospholipid biosynthesis, and a cellular set of uptake mechanisms for nutrients; this would suffice to generate a free living organism in a nutritionally rich culture medium. Adding in genes for pathways of ethanol from glucose and other sugars would result in a cellulosic ethanol producer with maximum biochemical and biotechnological simplicity.[119]

4.5 SUMMARY

The popular laboratory bacterium *Escherichia coli* can utilize pentoses but is a poor ethanologen. When genes for ethanol biosynthesis from *Zymomonas mobilis* are added and expressed, powerful ethanol producers are generated and incorporated into a novel cellulosic ethanol process.

With its unusual and fragmentary central metabolism, *Z. mobilis* has been the focus of development for ethanol production for 25 years and is now likely to be upscaled for either ethanol and/or fine chemical production from lignocellulosic biomass sources.

Other bacterial species, in particular *Klebsiella oxytoca,* have also received attention. New candidate bacterial ethanologens appear regularly, and hyperthermophiles are now increasingly attractive for consolidated bioprocessing of cellulosic inputs transformed directly to ethanol. Cyanobacteria (blue-green algae) could be engineered to utilize light energy and CO_2 for ethanol biosynthesis.

Contemporary synthetic biology pinpoints theoretical and practical approaches to reaching minimal bacteria-like microbes with highly defined metabolic capabilities that could be genetically assembled for cellulosic ethanol production at high efficiency and with minimal waste of glucose to side metabolites.

REFERENCES

1. Narendranath, N. V., Thomas, K. C., and Ingledew, W. M. 2001. Acetic acid and lactic acid inhibition of growth of *Saccharomyces cerevisiae* by different mechanisms. *Journal of the American Society of Brewing Chemists* 59:187.
2. Graves, T. et al. 2007. Interaction effects of lactic acid and acetic acid at different temperatures on ethanol production by *Saccharomyces cerevisiae* in corn mash. *Applied Microbiology and Biotechnology* 73:1190.
3. Schell, D. J. et al. 2007. Contaminant occurrence, identification and control in a pilot-scale corn fiber to ethanol conversion process. *Bioresource Technology* 98:2942.
4. Tolan, J. S., and Finn, R. K. 1987. Fermentation of D-xylose and L-arabinose to ethanol by *Erwinia chrysanthemi*. *Applied and Environmental Microbiology* 53:2033.
5. Tolan, J. S., and Finn, R. K. 1987. Fermentation of D-xylose to ethanol by genetically modified *Klebsiella planticola*. *Applied and Environmental Microbiology* 53:2039.
6. Millichip, R. J., and Doelle, H. W. 1989. Large-scale ethanol production from milo (*sorghum*) using *Zymomonas mobilis*. *Process Biochemistry* 24:141.
7. Bothast, R. J. et al. 1994. Fermentation of L-arabinose, D-xylose and D-glucose by ethanologenic recombinant *Klebsiella oxytoca* strain P2. *Biotechnology Letters* 16:401.
8. San Martin, R. et al. 1992. Development of a synthetic medium for continuous anaerobic growth and ethanol production with a lactate dehydrogenase mutant of *Bacillus stearothermophilus*. *Journal of General Microbiology* 138:987.
9. Beall, D. S., and Ingram, L. O. 1993. Genetic engineering of soft-rot bacteria for ethanol production from lignocellulose. *Journal of Industrial Microbiology* 11:151.
10. Dien, B. S. et al. 1997. Conversion of corn milling fibrous coproducts into ethanol by recombinant *Escherichia coli* strains KO11 and SL40. *World Journal of Microbiology and Biotechnology* 13:619.
11. Zeikus, J. G. et al. 1981. Thermophilic ethanol fermentation. *Basic Life Science* 18:441.
12. Hartley, B. S., and Payton, M. A. 1983. Industrial prospects for thermophiles and thermophilic enzymes. *Biochemical Society Symposia* 48:133.
13. Feldmann, S. D., Sahm, H., and Sprenger, G. A. 1992. Pentose metabolism in *Zymomonas mobilis* wild-type and recombinant strains. *Applied Microbiology and Biotechnology* 38:354.
14. Sprenger, G. A. 1996. Carbohydrate metabolism in *Zymomonas mobilis*: A catabolic pathway with some scenic routes. *FEMS Microbiology Letters* 145:301.
15. Zikmanis, P., Kruce, R., and Auzina, L. 1997. An elevation of the molar growth yield of *Zymomonas mobilis* during aerobic exponential growth. *Archives of Microbiology* 167:167.
16. Skotnicki, M. L. et al. 1983. High-productivity alcohol fermentations using *Zymomonas mobilis*. *Biochemical Society Symposia* 48:53.
17. Gschaedler, A. et al. 1999. Effects of pulse addition of carbon sources on continuous cultivation of *Escherichia coli* containing recombinant *E. coli gapA* gene. *Biotechnology and Bioengineering* 63:712.

18. Aristidou, A. A., San, K. Y., and Bennett, G. N. 1999. Improvement of biomass yield and recombinant gene expression in *Escherichia coli* by using fructose as the primary carbon source. *Biotechnology Progress* 15:140.

19. Lawford, H. G., and Rousseau, J. D. 1995. Comparative energetics of glucose and xylose metabolism in ethanologenic recombinant *Escherichia coli* B. *Applied Biochemistry and Biotechnology* 51–52:179.

20. Aristidou, A. A. 2007. Application of metabolic engineering to the conversion of renewable resources to fuels and fine chemicals: Current advances and future prospects. In *Fermentation microbiology and biotechnology*, 2nd ed., ed. El-Mansi, E. M. T., Bryce, C. A., Demain, A. L., and Allman, A. R., chap. 9. Boca Raton, FL: CRC Press.

21. Michal, G., ed. 1999. *Biochemical pathways. An atlas of biochemistry and molecular biology,* chap. 3. New York: John Wiley & Sons.

22. Ingram, L. O. et al. 1999. Enteric bacteria catalysts for fuel ethanol production. *Biotechnology Letters* 15:855.

23. Buchholz, S. E., Dooley, M. M., and Eveleigh, D. E. 1987. *Zymomonas*—An alcoholic enigma. *Trends in Biotechnology* 5:199.

24. Moat, A. G., and Foster, J. W. 1995. *Microbial physiology,* 3rd ed., chap. 7. New York: John Wiley & Sons.

25. Ingram, L. O. et al. 1987. Genetic engineering of ethanol production in *Escherichia coli*. *Applied and Environmental Microbiology* 53:2420.

26. Ingram, L. O., and Conway, T. 1988. Expression of different levels of ethanologenic enzymes of *Zymomonas mobilis* in recombinant strains of *Escherichia coli*. *Applied and Environmental Microbiology* 54:397.

27. Ingram, L. O. et al. 1991. Ethanol production in *Escherichia coli* strains coexpressing *Zymomonas* PDC and *ADH* genes. US Patent 5,000,000, March 19, 1991.

28. Alterhum, F., and Ingram, L. O. 1989. Efficient ethanol production from glucose, lactose, and xylose by recombinant *Escherichia coli*. *Applied and Environmental Microbiology* 55:1943.

29. Ohta, K., Alterhum, F., and Ingram, L. O. 1990. Effects of environmental conditions on xylose fermentation by recombinant *Escherichia coli*. *Applied and Environmental Microbiology* 56:463.

30. Kuhnert, P. et al. 1997. Detection system for *Escherichia coli*-specific virulence genes: Absence of virulence determinants in B and C strains. *Applied and Environmental Microbiology* 63:703.

31. Dien, B. S., Cotta, M. A., and Jeffries, T. W. 2003. Bacteria engineered for fuel ethanol production: current status. *Applied Microbiology and Biotechnology* 63:258.

32. Ohta, K. et al. 1991. Genetic improvement of *Escherichia coli* for ethanol production: Chromosomal integration of *Zymomonas mobilis* genes encoding pyruvate decarboxylase and alcohol dehydrogenase II. *Applied and Environmental Microbiology* 57:893.

33. Barbosa, M. F. et al. 1992. Efficient fermentation of *Pinus* sp. acid hydrolysates by an ethanologenic strain of *Escherichia coli*. *Applied and Environmental Microbiology* 58:1382.

34. Ashgari, A. et al. 1996. Ethanol production from hemicellulose hydrolysates of agricultural residues using genetically engineered *Escherichia coli* strain KO11. *Journal of Industrial Microbiology* 16:42.

35. Beall, D. S. et al. 1992. Conversion of hydrolysates of corn cobs and hulls into ethanol by recombinant *Escherichia coli* strain B containing genes for ethanol production. *Biotechnology Letters* 14:857.

36. Moniruzzaman, M. et al. 1996. Ethanol production from AFEX pretreated corn fiber by recombinant bacteria. *Biotechnology Letters* 18:955.

37. Moniruzzaman, M., and Ingram, L. O. 1998. Ethanol production from dilute acid hydrolysate of rice hulls using genetically engineered *Escherichia coli*. *Biotechnology Letters* 20:943.

38. Guimaraes, W. V., Dudley, G. L., and Ingram, L. O. 1992. Fermentation of sweet whey by ethanologenic *Escherichia coli*. *Biotechnology and Bioengineering* 40:41.

39. Guimaraes, W. Y. et al. 1992. Ethanol production from starch by recombinant *Escherichia coli* containing integrated genes for ethanol production and plasmid genes for saccharification. *Biotechnology Letters* 14:415.

40. Grohmann, K. et al. 1994. Fermentation of galacturonic acid and other sugars in orange peel hydrolysates by an ethanologenic strain of *Escherichia coli*. *Biotechnology Letters* 16:281.

41. Moniruzzaman, M. et al. 1997. Extracellular melibiose and fructose are intermediates in raffinose catabolism during fermentation to ethanol by engineered enteric bacteria. *Journal of Bacteriology* 179:1880.

42. Yomano, L. P., York, S. W., and Ingram, L. O. 1998. Isolation and characterization of ethanol-tolerant mutants of *Escherichia coli* KO11 for fuel ethanol production. *Journal of Industrial Microbiology and Biotechnology* 20:132.

43. Zaldivar, J., Martinez, A., and Ingram, L. O. 1999. Effect of selected aldehydes on the growth and fermentation of ethanologenic *Escherichia coli*. *Biotechnology and Bioengineering* 65:24.

44. Zaldivar, J., Martinez, A., and Ingram, L. O. 2000. Effect of alcohol compounds found in hemicellulose hydrolysate on the growth and fermentation of ethanologenic *Escherichia coli*. *Biotechnology and Bioengineering* 68:524.

45. Zaldivar. J., and Ingram, L. O., Effect of organic acids on the growth and fermentation of ethanologenic *Escherichia coli* LY01. *Biotechnology and Bioengineering* 66:203.

46. Martinez, A. et al. 1999. Biosynthetic burden and plasmid burden limit expression of chromosomally integrated heterologous genes (*pdc, adhB*) in *Escherichia coli*. *Biotechnology Progress* 15:891.

47. Underwood, S. A. et al. 2002. Flux through citrate synthase limits the growth of ethanologenic *Escherichia coli* KO11 during xylose fermentation. *Applied and Environmental Microbiology* 68:1071.

48. Underwood, S. A. et al. 2002. Genetic changes to optimize carbon partitioning between ethanol and biosynthesis in ethanologenic *Escherichia coli*. *Applied and Environmental Microbiology* 68:6263.

49. Tao, H. et al. 2001. Engineering a homo-ethanol pathway in *Escherichia coli:* Increased glycolytic flux and levels of expression of glycolytic genes during xylose fermentation. *Journal of Bacteriology* 183:2979.

50. Moniruzzaman, M. et al. 1997. Isolation and molecular characterization of high-performance cellobiose-fermenting spontaneous mutants of ethanologenic *Escherichia coli* KO11 containing the *Klebsiella oxytoca casAB* operon. *Applied and Environmental Microbiology* 63:4633.

51. Lawford, H. G., and Rousseau, J. D. 1995. Loss of ethanologenicity in *Escherichia coli* B recombinant pLO1297 and KO11 during growth in the absence of antibiotics. *Biotechnology Letters* 17:751.

52. Lawford, H. G., and Rousseau, J. D. 1996. Factors contributing to the loss of ethanologenicity of *Escherichia coli* B recombinant pLO1297 and KO11. *Applied Biochemistry and Biotechnology* 57–58:293.

53. Dumsday, G. J. et al. 1997. Continuous ethanol production by *Escherichia coli* B recombinant KO11 in continuous stirred tank and fluidized bed fermenters. *Australian Biotechnology* 7:300.

54. Dumsday, G. J. et al. 1999. Comparative stability of ethanol production by *Escherichia coli* KO11 in batch and chemostat culture. *Journal of Industrial Microbiology and Biotechnology* 23:701.

55. Mat-Jan, F., Alam, K. Y., and Clark, D. P. 1989. Mutants of *Escherichia coli* deficient in the fermentative lactate dehydrogenase. *Journal of Bacteriology* 171:342.

56. Hespell, R. B. et al. 1996. Stabilization of *pet* operon plasmids and ethanol production in *Escherichia coli* strains lacking lactate dehydrogenase and pyruvate—formate-lyase activities. *Applied and Environmental Microbiology* 62:4594.
57. Dien, B. S. et al. 1998. Fermentation of hexose and pentose sugars using a novel ethanologenic *Escherichia coli* strain. *Enzyme and Microbial Technology* 23:366.
58. Dien, B. S., Iten, L. B., and Bothast, R. J. 1999. Conversion of corn fiber to ethanol by recombinant *Escherichia coli* strain FBR3. *Journal of Industrial Microbiology and Biotechnology* 22:575.
59. Dien, B. S. et al. 2000. Development of new ethanologenic *Escherichia coli* strains for fermentations of lignocellulosic biomass. *Applied Biochemistry and Biotechnology* 84–86:181.
60. Nichols, N. N., Dien, B. S., and Bothast, R. J. 2001. Use of carbon catabolite repression mutants for fermentation of sugar mixtures to ethanol. *Applied Microbiology and Biotechnology* 56:120.
61. Kim, Y., Ingram, L.O., and Shanmugam, K.T. 2007. Construction of an *Escherichia coli* K-12 mutant for homo-ethanol fermentation of glucose or xylose without foreign genes. *Applied and Environmental Microbiology* 73:1766.
62. Quail, M. A., Haydon, D. J., and Guest, J. R. 1994. The *pdhR-aceEF-lpd* operon of *Escherichia coli* expresses the pyruvate dehydrogenase complex. *Molecular Microbiology* 12:95.
63. Huerta-Beristain, G. et al. 2008. Specific ethanol production rate in ethanologenic *Escherichia coli* strain KO11 is limited by pyruvate decarboxylase. *Journal of Molecular Microbiology and Biotechnology* 15:55.
64. Orencio-Trejo, M. et al. 2008. Metabolic regulation analysis of an ethanologenic *Escherichia coli* strain based on RT-PCR and enzymatic activities. *Biotechnology for Biofuels* 1:8.
65. Dien, B. S., Cotta, M. A., and Jeffries, T. W. 2003. Bacteria engineered for fuel ethanol production: Current status. *Applied Microbiology and Biotechnology* 63:258.
66. Russell, J. B. 2007. The energy spilling reactions of bacteria and other organisms. *Journal of Molecular Microbiology and Biotechnology* 13:1.
67. Swings, J., and De Ley, J. 1977. The biology of *Zymomonas*. *Bacteriology Review* 41:1.
68. Doelle, H. W. et al. 1993. *Zymomonas mobilis*—science and industrial application. *Critical Reviews in Biotechnology* 13:57.
69. Richards, L., and Doelle, H. W. 1989. Fermentation of potato mash, potato mash/maltrin mixtures and wheat starch using *Zymomonas mobilis*. *World Journal of Microbiology and Biotechnology* 5:307.
70. Skotnicki, M. L. et al. 1981. Comparison of ethanol production by different *Zymomonas* strains. *Applied and Environmental Microbiology* 41:889.
71. Skotnicki, M. L. et al. 1983. High-productivity alcohol fermentations using *Zymomonas mobilis*. *Biochemical Society Symposia* 48:53.
72. Rogers, P. L., Goodman, A. E., and Heyes, R. H. 1984. *Zymomonas* ethanol fermentations. *Microbiology Science* 1:133.
73. Zhang, M. et al. 1995. Metabolic engineering of a pentose metabolism pathway in ethanologenic *Zymomonas mobilis*. *Science* 267:240.
74. Zhang, M. et al. 1995. Promising ethanologens for xylose fermentation. Scientific note. *Applied Biochemistry and Biotechnology* 51–52:527.
75. Parker, C. et al. 1995. Characterization of the *Zymomonas mobilis* glucose facilitator gene-product (*glf*) in recombinant *Escherichia coli:* Examination of transport mechanism, kinetics and the role of glucokinase in glucose transport. *Molecular Microbiology* 15:795.

76. Deanda, K. et al. 1996. Development of an arabinose-fermenting *Zymomonas mobilis* strain by metabolic pathway engineering. *Applied and Environmental Microbiology* 62:4465.
77. Lawford, H. G. et al. 1999. Fermentation performance characteristics of a prehydrolysate-adapted xylose-fermenting recombinant *Zymomonas* in batch and continuous fermentations. *Applied Biochemistry and Biotechnology* 77–79:191.
78. Gao, Q. et al. 2002. Characterization of heterologous and native enzyme activity profiles in metabolically engineered *Zymomonas mobilis* strains during batch fermentation of glucose and xylose mixtures. *Applied Biochemistry and Biotechnology* 98–100:341.
79. Lawford, H. G., and Rousseau, J. D. 2002. Performance testing of *Zymomonas mobilis* metabolically engineered for cofermentation of glucose, xylose, and arabinose. *Applied Biochemistry and Biotechnology* 98–100:429.
80. Mohagheghi, A. et al. 2002. Cofermentation of glucose, xylose, and arabinose by genomic DNA-integrated xylose/arabinose fermenting strain of *Zymomonas mobilis* AX101. *Applied Biochemistry and Biotechnology* 98–100:885.
81. Lawford, H. G., and Rousseau, J. D. 2003. Cellulosic fuel ethanol: Alternative fermentation process designs with wild-type and recombinant *Zymomonas mobilis*. *Applied Biochemistry and Biotechnology* 105–108:457.
82. Joachimsthal, E. L., and Rogers, P. L. 2000. Characterization of a high-productivity recombinant strain of *Zymomonas mobilis* for ethanol production from glucose/xylose mixtures. *Applied Biochemistry and Biotechnology* 84–86:343.
83. Kim, I. S., Barrow, K. D., and Rogers, P. L. 2000. Kinetic and nuclear magnetic resonance studies of xylose metabolism by recombinant *Zymomonas mobilis* ZM4(pZB5). *Applied and Environmental Microbiology* 66:186.
84. Kim, I. S., Barrow, K. D., and Rogers, P. L. 2000. Nuclear magnetic resonance studies of acetic acid inhibition of rec *Zymomonas mobilis* ZM4(pZB5). *Applied Biochemistry and Biotechnology* 84–86:357.
85. Jeon, Y. J. et al. 2002. Kinetic analysis of ethanol production by an acetate-resistant strain of *Zymomonas mobilis*. *Biotechnology Letters* 24:819.
86. Jeon, Y. J., Svenson, C. J., and Rogers, P. L. 2005. Over-expression of xylulokinase in a xylose-metabolizing recombinant strain of *Zymomonas mobilis*. *FEMS Microbiology Letters* 244:85.
87. De Graaf, A. A. et al. 1999. Metabolic state of *Zymomonas mobilis* in glucose-, fructose-, and xylose-fed continuous cultures as analyzed by ^{13}C- and ^{31}P-NMR spectroscopy. *Archives of Microbiology* 171:371.
88. Seo, J.-S. et al. 2005. The genome sequence of the ethanologenic bacterium *Zymomonas mobilis* ZM4. *Nature Biotechnology* 23:63.
89. Christogianni, A. et al. 2005. Transcriptional analysis of a gene cluster involved in glucose tolerance in *Zymomonas mobilis:* Evidence for an osmoregulated promoter. *Journal of Bacteriology* 187:5179.
90. Jeon, B. Y., Hwang, T. S., and Park, D. H. 2009. Electrochemical and biochemical analysis of ethanol fermentation of *Zymomonas mobilis* KCCM11336. *Journal of Microbiology and Biotechnology* 19:666.
91. Fűhrer, T., and Sauer, U. 2009. Different biochemical mechanisms ensure network-wide balancing of reducing equivalents in microbial metabolism. *Journal of Bacteriology* 19:2112.
92. Zhang, X., Chen, G., and Liu, W. 2009. Reduction of xylose to xylitol catalyzed by glucose-fructose oxidoreductase from *Zymomonas mobilis*. *FEMS Microbiology Letters* 293:214.
93. Yang, S. et al. 2009. Transcriptomic and metabolomic profiling of *Zymomonas mobilis* during aerobic and anaerobic fermentations. *BMC Genomics* 10:34.
94. Rogers, P. L. et al. 2007. *Zymomonas mobilis* for fuel ethanol and higher value products. *Advances in Biochemical Enginineering/Biotechnology* 108:264.

95. Ohta, K. et al. 1991. Metabolic engineering of *Klebsiella oxytoca* M5A1 for ethanol production from xylose and glucose. *Applied and Environmental Microbiology* 57:2810.

96. Wood, B. E., and Ingram, L. O. 1992. Ethanol production from cellobiose, amorphous cellulose, and crystalline cellulose by recombinant *Klebsiella oxytoca* containing chromosomally integrated *Zymomonas mobilis* genes for ethanol production and thermostable cellulase genes from *Clostridium thermocellum. Applied and Environmental Microbiology* 58:2130.

97. Doran, J. B., Aldrich, H. C., and Ingram, L. O. 1994. Saccharification and fermentation of sugar cane bagasse by *Klebsiella oxytoca* P2 containing chromosomally integrated genes encoding the *Zymomonas mobilis* ethanol pathway. *Biotechnology and Bioengineering* 44:240.

98. Doran, J. B. et al. 2000. Fermentations of pectin-rich biomass with recombinant bacteria to produce fuel ethanol. *Applied Biochemistry and Biotechnology* 84–86:141.

99. Burchhardt, G., and Ingram, L. O. 1992. Conversion of xylan to ethanol by ethanologenic strains of *Escherichia coli* and *Klebsiella oxytoca. Applied and Environmental Microbiology* 58:1128.

100. Qian, Y. et al. 2003. Cloning, characterization, and functional expression of the *Klebsiella oxytoca* xylodextrin utilization operon (*xynTB*) in *Escherichia coli. Applied and Environmental Microbiology* 69:5957.

101. Hahn-Hägerdal, B. et al. 1994. An interlaboratory comparison of the performance of ethanol-producing microorganisms in a xylose-rich acid hydrolysate. *Applied Microbiology and Biotechnology* 41:62.

102. Haq, A., and Dawes, E. A. 1971. Pyruvic acid metabolism and ethanol formation in *Erwinia amylovora. Journal of General Microbiology* 68:295.

103. Okamoto, T. et al. 1993. *Zymobacter palmae* gen. nov. sp. *nov.,* a new ethanol-fermenting peritrichous bacterium isolated from palm sap. *Archives of Microbiology* 160:333.

104. Yanase, H. et al. 2007. Genetic engineering of *Zymobacter palmae* for ethanol production from xylose. *Applied and Environmental Microbiology* 73:1766.

105. Yanase, H. et al. 2005. Ethanol production from cellobiose by *Zymobacter palmae* carrying the *Ruminococcus albus* β-glucosidase gene. *Journal of Biotechnology* 118:35.

106. Raj, K. C. et al. 2002. Cloning and characterization of the *Zymobacter palmae* pyruvate decarboxylase gene (*pdc*) and comparison to bacterial homologues. *Applied and Environmental Microbiology* 68:2869.

107. Escalante, A. et al. 2008. Analysis of bacterial community during the fermentation of *pulque,* a traditional Mexican alcoholic beverage, using a polyphasic approach. *International Journal of Food Microbiology* 124:126.

108. Bi, C., Rice, J. D., and Preston, J. F. 2009. Complete fermentation of xylose and methylglucuronoxylose derived from methylglucuronoxylan by *Enterobacter asburiae* strain JDR-1. *Applied and Environmental Microbiology* 75:395.

109. Bi, C. et al. 2009. Genetic engineering of *Enterobacter asburiae* strain JDR-1 for efficient production of ethanol from hemicellulose hydrolysates. *Applied and Environmental Microbiology* 75:5743.

110. Heyer, H., and Krumbein, W. E. 1991. Excretion of fermentation products in dark and anaerobically incubated cyanobacteria. *Archives of Microbiology* 155:284.

111. Deng, M.-D., and Coleman, J. R. 1999. Ethanol synthesis by genetic engineering in cyanobacteria. *Applied and Environmental Microbiology* 65:523.

112. Angermayr, S. A. et al. 2009. Energy biotechnology with cyanobacteria. *Current Opinion in Biotechnology* 20:257.

113. Hellingwerf, K. J., and Teixeira de Mattos, M. J. 2009. Alternative routes to biofuels: Light-driven biofuel formation from CO_2 and water based on the "photanol" approach. *Journal of Biotechnology* 142:87.

114. Sommer, P., Georgieva, T., and Ahring, B. K. 2004. Potential for using thermophilic anaerobic bacteria for bioethanol production from hemicellulose. *Biochemical Society Transactions* 32:283.

115. Koskinen, P. E. et al. 2008. Ethanol and hydrogen production by two thermophilic, anaerobic bacteria isolated from Icelandic geothermal areas. *Biotechnology and Bioengineering* 101:679.

116. Shaw, A. J. et al. 2008. Metabolic engineering of a thermophilic bacterium to produce ethanol at high yield. *Proceedings of the National Academy of Sciences USA* 105:13769.

117. Shaw, A. J., Hogsett, D. A., and Lynd, L. R. 2009. Identification of the [FeFe]-hydrogenase responsible for hydrogen generation in *Thermoanaerobacterium saccharolyticum* and demonstration of increased ethanol yield via hydrogenase knockout. *Journal of Bacteriology* 191:6457.

118. Desvaux, M. 2006. Unraveling carbon metabolism in anaerobic cellulolytic bacteria. *Biotechnology Progress* 22:1229.

119. Glass, J. I. et al. 2006. Minimal bacterial genome. US Patent application 2007/0122826, October 12, 2006.

120. Forster, A. C., and Church, G. M. 2006. Towards synthesis of a minimal cell. *Molecular Systems Biology* 2:45.

121. Forster, A. C., and Church, G. M. 2006. Synthetic biology projects in vitro. *Genome Research* 17:1.

5 Biochemical Engineering of Cellulosic Ethanol

5.1 INTRODUCTION

Starch ethanol relies mostly on traditional process engineering, but both the potable and fuel alcohol industries have developed more ambitious engineering concepts, including continuous fermentation systems operated on industrial scales. Could these be adapted for cellulosic production or even be essential for large-scale cellulosic ethanol production to be successful?

Compared with starch ethanol, cellulosic ethanol diverges markedly at the earliest stage when the plant biomass is pretreated by any one of a group of physicochemical methods before the cellulose is hydrolyzed by cellulase. These pretreatment options have been incorporated into the functioning of demonstration plants for cellulosic ethanol production. But is this set of methodologies now complete and can they be adapted for an ever widening set of plant biomass resources?

Should cellulosic ethanol evolve to be as far as possible a pure biotechnology as exemplified by consolidated bioprocessing? Or should cellulosic ethanol remain close to the fermentation designs long used for corn and sugarcane ethanol?

5.2 CASE STUDY: THE IOGEN CORPORATION PROCESS WITH WHEAT STRAW

The Iogen Corporation has operated a demonstration process outside Toronto, Canada, since 2004; the basic operations are outlined in Figure 5.1. In many of its features, the Iogen process is relatively conservative:

- Wheat straw is a substrate—a high-availability feedstock with a low lignin content in comparison with tree wood materials (Figure 5.2).[1,2]
- A dilute acid and heat pretreatment of the biomass—the levels of acid are sufficiently low that recovery of the acid is not needed and corrosion problems are avoided.
- Separate cellulose hydrolysis and fermentation with a single sugar substrate product stream (hexoses plus pentoses) for fermentation in a batch process.
- Cellulase breakdown of cellulose—Iogen is an enzyme producer.
- A *Saccharomyces* yeast ethanologen—engineered for xylose consumption as well as offering a low incidence of contamination, the ability to recycle the cells, and the option for selling on the spent cells for agricultural use.[1]

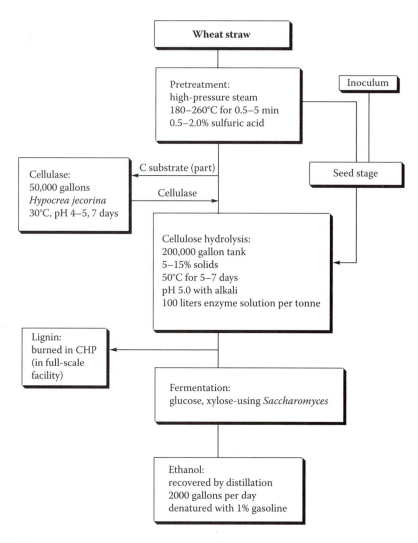

FIGURE 5.1 Outline of the Iogen Corporation demonstration process for bioethanol production. (Information from Tolan, J. S. 2006. In *Biorefineries—Industrial Processes and Products. Volume 1: Status Quo and Future Directions,* ed. Kamm, B., Gruber, P. R., and Kamm, M., chap. 9. Weinheim, Germany: Wiley-VCH Verlag.)

In the first description of the process (1999), agricultural residues such as wheat straw, grasses, and energy crops (aspen, etc.) were equally possible or a possibility.[3] By the next appearance of the article in 2006, cereal straws had become the substrates of choice. Lignin does not form a seriously refractory barrier to cellulase access with wheat straw; this renders organic solvent pretreatment unnecessary. Over 95% of the cellulosic glucose is released by the end of the enzyme digestion step; the remainder is included in the lignin cake that is spray-dried prior to combustion (Figure 5.1).

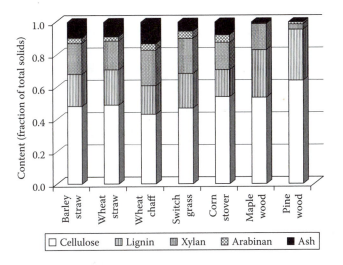

FIGURE 5.2 Fuel ethanol feedstock compositions. (Data from Foody, B., Tolan, J. S., and Bernstein, J. D. 1999. US Patent 5,916,780, June 29, 1999.)

The Iogen process is no more complex than wet mill and dry mill options for corn ethanol production (Figures 1.20 and 1.21 in Chapter 1)—substituting acid pretreatment for corn grinding steps and adding on-site cellulase generation, the latter mostly as a strategy to avoid the costs of preservatives and stabilizer but possibly also to use a small proportion of the hydrolyzed cellulose as a feedstock for the enzyme fermentation itself. The earlier Bio-Hol process had opted for *Zymomonas mobilis* as the ethanologen and established acid hydrolysis pretreatments for wheat straw, soy stalks, corn stover, canola stalks, pine wood, and poplar wood.[4]

For both *Z. mobilis* and *Saccharomyces cerevisiae,* pretreated wheat straw had the distinct advantage of presenting far less of a toxic mixture to the producer organism (Figure 5.3); methods for removing growth inhibitors from the biomass acid hydrolysates could reduce the effect more than 20-fold. The Stake Company Ltd. was founded in 1973 to develop and market a process for biomass conversion to sugar streams for both biofuels and animal feeds as well as chemicals derived from lignin and hemicellulose.[5] A continuous feedstock processing system was constructed to handle 4–10 tons of wood chips per hour and licensed to end-users in the United States and France.

Prior to 2004, more radical processes were examined in detail—including being upscaled to pilot plant operations—for lignocellulosic ethanol. These proposals included those to avoid the need for cellulase fermentations independent of the main ethanolic fermentation as well as the use of thermophilic bacteria in processes that more closely resembled industrial chemistry than they did the traditional potable alcohol manufacture. Indeed, it is clear that Iogen considered sourcing thermophilic bacteria and nonconventional yeasts during the 1990s.[3] The achieved reality of the Iogen process will therefore be used as a guide to how innovations have successfully translated into practical use—or have failed to do so—to review progress over the

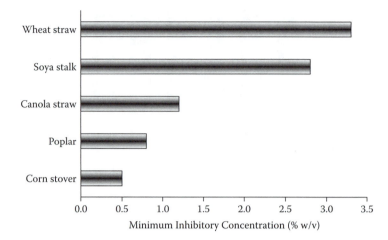

FIGURE 5.3 Growth of *Zymomonas mobilis* on biomass hydrolysates. (Data from Lawford, G. R. et al. 1986. In *Biotechnology and Renewable Energy,* ed. Moo-Young, M., Hasnain, S., and Lamptey, J., 276. New York: Elsevier Applied Science Publishers.)

last three decades and offering predictions for new solutions to well-known problems as the cellulosic ethanol industry expands.

5.3 BIOMASS SUBSTRATE PRETREATMENT STRATEGIES

In the calculations for Figure 5.2 some interesting conclusions are reached. Although wheat straw has undoubted advantages, other feedstocks outperform wheat straw for some key parameters:

- Wheat straw has a lower gravimetric ratio of total carbohydrate (cellulose, starch, xylan, arabinan) to lignin than barley straw, corn stover, switchgrass, or even wheat chaff.
- Wheat straw has a lower cellulose-to-lignin ratio than most of these biomass sources (with the exception of switchgrass).
- Of the seven quoted examples of lignocellulosic feedstocks, wheat chaff and switchgrass have the highest total pentose (xylan and arabinan) contents—quoted as an important quantitative predictor for ethanol yield from cellulose because less cellulase is required.[1,2]

An important consideration included in Iogen's deliberations on feedstock suitability was the reproducible and predictable supply of wheat straw. The USDA's Agricultural Service also itemized sustainable supply as one of its two key factors for biomass feedstocks; the other was cost effectiveness.[6] Financial models indicate that feedstock costs are crucial, and any managed reduction of the costs of biomass crop production, harvesting, and the sequential logistics of collection, transportation, and storage prior to substrate pretreatment will have an impact on the viability of biofuel facilities. (Economic aspects of feedstock supply chains are discussed in the next chapter.)

The Energy Information Administration has constructed the National Energy Modeling System to forecast U.S. energy production, use, and price trends in 25-year predictive segments. The biomass supply schedule includes agricultural resides, dedicated energy crops, forestry sources, and wood waste and mill residues; wheat straw (together with corn stover, barley straw, rice straw, and sugarcane bagasse) is a specified component of the agricultural residue supply.[7]

A brief survey will now be made of wheat straw and other leading candidate lignocellulosics. Special emphasis will be put on how different national priorities place emphasis on different biomass sources and on what evolving agricultural practices and processing technologies may diversify cellulosic ethanol facilities on scales equal to and larger than the Iogen demonstration facility.

STEM TOPIC 5.1: NET ENERGY BALANCE COMPUTATIONS FOR CELLULOSIC ETHANOL

Revisiting the ERG biofuel analysis meta-model (EBAMM) developed at the University of California, Berkeley (http://rael.berkeley.edu/ebamm/), the 2006 version of cellulosic ethanol production contains an agricultural energy total of 1.4 MJ/L ethanol and an ethanol conversion energy at 28 MJ/L. The output energy in ethanol is 21 MJ/L and coproducts amount to 4.8 MJ/L. In contrast to corn ethanol, a large energy output is estimated as the recycled biomass energy of 26.3 MJ/L. Therefore:

$$NEV = (21 + 4.8) - (1.4 + 28 - 26.3) = 22.7 \text{ MJ/L}$$

The single largest energy input for the agricultural phase is direct fossil fuel use (diesel) at 50.5% of the total. For the ethanol conversion phase, the use of biomass to supply input energy (heat for ethanol distillation, etc.) is dominant, almost totally replacing requirements for diesel, coal, etc. The residual energy requirements include transportation of the biomass to the biorefinery, a small amount of diesel, and energy use in water recycling and effluent treatment.

Therefore, the assessment and reputation of cellulosic ethanol as a far superior net energy source depend entirely on the use of agricultural materials to generate energy in the same way as the burning of sugarcane bagasse to provide on-site energy and surplus electricity can make sugarcane ethanol far more ecologically and environmentally acceptable than corn ethanol production.

5.3.1 WHEAT STRAW

While the Iogen process relies on acid pretreatment and cellulase digestion, Danish investigators rank other pretreatment methods (with short residence times of 5 or

6 min) as superior for subsequent wheat straw cellulose digestion by cellulase (24 h at 50°C)[8]:

$$\text{Steam explosion (215°C)} > H_2O_2 \text{ (190°C)} > \text{water (190°C)}$$
$$> \text{ammonia (195°C)} > \text{acid (190°C)}$$

The degradation of cellulose to soluble sugars was enhanced by adding nonionic surfactants and polyethylene glycol during enzymatic hydrolysis; the best results were obtained with a long-chain alcohol ethoxylate in conjunction with steam explosion pretreated wheat straw. The additives may have occupied cellulase binding sites on residual lignin or have helped to stabilize the enzyme during the lengthy digestion.[8]

In addition, the attention of Novozymes (one of the world's major enzyme producers) in Denmark was attracted after 2004 by wheat straw and the problems of its complete conversion to fermentable sugars:

- Arabinoxylans form an undigested fraction in the "vinasse" (the insoluble fermentation residue) after the end of a wheat-based ethanol process. A mixture of depolymerizing enzymes from *Hypocrea jecorina* and *Humicola insolens* could solubilize the insoluble material and release arabinose and xylose—although at different rates with different optimal pH values and temperature ranges for the digestion.[9]
- A subsequent study mixed three novel α-L-arabinofuranosidases with an endoxylanase and a β-xylosidase to liberate pentoses from water-soluble and water-insoluble arabinoxylans and vinasse. Much lower enzyme activities were required than previously, and this may be a technology for pentose release *prior* to wheat-straw substrate fermentations.[10]
- Mixtures of α-L-arabinofuranosidases from *H. insolens,* the white-rot basidomycete *Meripilus giganteus,* and a *Bifidobacterium* species were highly effective in digesting wheat arabinoxylan. The different enzymes acted synergistically on different carbohydrate bonds in the hemicellulose structures.[11]

Arabinans constitute only 3.8% (by weight) of the total carbohydrate (cellulose, starch, xylose, and arabinose) in wheat straw, and a lack of utilization of all the pentose sugars represents a minor inefficiency. Releasing all the xylose as a substrate for the engineered xylose-utilizing yeast (xylose constitutes 24% of the total sugars) and completing the depolymerization of cellulose to (insofar as is possible) free glucose are more significant targets for process improvement.

As in Canada, wheat straw has been assessed to be a major lignocellulosic feedstock in Denmark. Unlike at Iogen, substrate pretreatment studies from Denmark have concentrated on wet oxidation (i.e., heat, water, and high-pressure O_2) to hydrolyze hemicellulose while leaving much of the lignin and cellulose insoluble; various conditions have been explored, including the combining of thermal hydrolysis, wet oxidation, and steam explosion.[12-15] The major concern with this type of pretreatment method—with so many biomass substrates—is the formation of inhibitors that are toxic to ethanologens and reduce ethanol yield.[16-18]

These degradation products of lignocellulosic and hemicellulosic polymers include aromatic acids and aldehydes as well as aliphatic carboxylic acids and sugar-derived components. Laboratory strains of *Saccharomyces cerevisiae* exhibit differential responses to the growth inhibitors, and cell-free enzyme preparations with cellulase and xylanase activities were severely inhibited by chemically defined mixtures of the known wheat straw inhibitors; formic acid was by far the most potent inhibitor.[19] The positives to be drawn were that even a laboratory strain could grow in 60% (w/v) of the wheat straw substrate and that a focused removal of one (or a few) inhibitors (in particular, formic acid) may suffice to render the material entirely digestible by engineered yeast strains.

5.3.2 Switchgrass

Switchgrass, a perennial, warm-season prairie grass and the leading candidate grass energy crop, could be grown in all rural areas in the continental United States east of the Rocky Mountains, from North Dakota to Alabama, with the exception of southern Texas, southern Florida, and Maine.[20] Until the mid-1990s, switchgrass was primarily known in scientific agricultural publications as a forage crop for livestock, but it has been tested as a direct energy source, cocombusted with coal at 7–10% of the energy production levels.[21] Simply burning switchgrass operates at 32% energy efficiency but using pellet grass in space-heating stoves can achieve 85% conversion efficiency. The environmental and agronomic advantages of switchgrass as a direct energy crop are severalfold[21]:

- Like all biomass crops, emissions are low in sulfur and mercury (especially when compared with coal).
- Switchgrass requires modest amounts of fertilizer for optimum growth— much lower application rates than with corn.
- Switchgrass stands are perennial, needing no recurrent soil preparation and thus greatly reducing soil erosion and runoff caused by annual tillage.
- An acre of switchgrass could be the energy equivalent of 2–6 tons of coal; the high variability is associated with fertilizer application, climate variation, etc.

In hard economic terms, however, recycling alternative fuels such as municipal solid waste and used tires has been calculated to be preferable to either switchgrass or *any* form of biomass, independent of the scale of use in mass burn boilers.[22] This analysis is clearly restricted to what can be acquired for recycling and would have very different likely outcomes if lower wastage, non-Western economies and societies were to be similarly analyzed. Using the criterion of bulk burnable material resulting from biomass drying, herbaceous plants have been advocated as the best choice for flexibly harvestable materials intended for power production via steam boilers. This choice is over that of corn stover, tree seedlings such as fast-growing willow, tree trimmings, by-products of lumber production, or switchgrass.[23]

Upland and lowland cultivars of switchgrass differ appreciably in their biomass yield, tolerance to drought, and response to nitrogen fertilizer application;

even between upland and lowland variants the differences were found to be sufficiently great to merit recommendations for specific types of growth habitat if energy cropping were to be practiced.[24] In the northern prairies, nitrogen fertilizer use results in only variable and inconsistent increases in biomass production; a single annual harvest after the first frost is optimal for polymeric material but with reduced total nitrogen and ash as well as coinciding with low infestation by grass weed species. A mixture of switchgrass and big blue stem grass (*Andropogon gerardii* Vitman) has been recommended over dependence on a monoculture approach.[25]

Like all grasses, switchgrass suffers as a substrate for ethanol from its low polymeric sugar content but elevated contributions of low molecular weight material to its dry mass; the lignin component of the insoluble material is reduced when compared with other major lignocellulosic materials (Table 1.5, Chapter 1). The potential bioconversion of carbohydrates is limited by the associated aromatic constituents within the grass fiber, and three strategies have been or are being considered[26]: (1) using lignin-degrading white rot fungi, (2) pretreatment with phenolic acid esterases, or (3) plant breeding to modify cell wall aromatics.

This suboptimal chemistry has spurred attempts to discover means to produce industrially or commercially important biomaterials from switchgrass—in particular, high-value and nutritional antioxidants.[26–28] Soluble phenolics are a potential industrial resource for fine chemicals and are present in the highest concentrations in the top internodes of the grass; lower internodes contain greater amounts of cell-wall-linked phenolics such as coumaric and ferulic acids.[28] Steroidal sapogenins, starting points for the synthesis of pharmacologically active compounds, are possible hepatoxins for grazing animals.[29] Grass fibers can also be used as raw material for biocomposites, packagings, and thermoplastics; switchgrass could be a large-scale substrate for fermentations to biomanufacture biodegradable polyhydroxyalkanoate polymers.[30] Pulps prepared from switchgrass may show promise as reinforcement components in newsprint.[31]

As with the Iogen process, dilute acid hydrolysis has been explored as a pretreatment methodology for switchgrass; in a batch reactor, the optimum conditions were 1.2% sulfuric acid at 180°C for 0.5 min; subsequent cellulase digestion released 91.4% of the cellulose as glucose and cellobiose.[32] Cellulose and lignin in switchgrass pretreated with dilute acid appeared not to interact when cellulase was added to degrade the insoluble polyglucan; acid-extracted lignin had little or no effect on the rate or extent of cellulose reactivity and saccharification.[33] Mixing switchgrass with aqueous ammonia and heating under pressure at 120°C for 20 min aided the subsequent digestion with cellulase and xylanase.[34] However, milder conditions were developed for the ammonia fiber explosion technology—100°C for 5 min—and resulted in a 93% solubilization of the polyglucan content of the grass.[35]

5.3.3 CORN STOVER

Corn stover is the aboveground plant from which the corn grain has been removed; the constituent parts are leaves, stalk, tassel, corn cob, and shuck (the husk around the grains when in the intact cob). Up to 30% by dry weight of the harvested plant is represented by the collected grain. In one of the earliest technological and economic reviews of corn-based fuel alcohol production, corn stover was included for consideration—but solely as an alternative to coal as a boiler fuel for distillation. In late 1978, a report for the U.S Department of Energy estimated that corn stover would *increase* the final cost of fuel ethanol by $0.04 per gallon because the use of corn stover as a fuel entailed costs roughly double those of local Illinois coal.[36]

The use of corn stover was therefore considered to be "justified only if the plant is located in an area where transportation cost would cause a doubling of the coal cost, or environmental considerations would rule against the use of coal; neither of which is very likely." Such arguments left corn stover in the field as an aid against soil erosion for over a decade until the option of lignocellulosic ethanol began to be considered seriously. By 2003 the National Renewable Energy Laboratory (Golden, Colorado) estimated the annual sustainable production of corn stover as 80–100 million dry tonnes per year, of which 20% might be utilized in the manufacture of fiber products and fine chemicals (e.g., furfural); 60–80 million dry tonnes would remain as a substrate for ethanol production.[37]

An estimate of total corn stover availability 5 years earlier had been over 250 million tonnes; 30 million tonnes were left on the fields for erosion control, leaving 100 million available for biofuels production.[38] With expanding corn acreage and a definite future for corn-based ethanol, a supply of corn stover is ensured—and commercial drivers may direct that starch ethanol and cellulosic ethanol facilities might be best sited adjacent to one another.

With corn stover rising up the rankings of biomass substrates for ethanol production in the United States and elsewhere, experimental investigations of pretreatment technologies have proliferated since 2002.[39-50] From this impressive body of practical knowledge, some conclusions can be made:

- A variety of pretreatment methodologies, using acid or alkali, in batch or percolated modes, can yield material with a high digestibility when cellulase is applied to the insoluble residues (Table 5.1).
- With some practical technologies, treatment times can be reduced to a few minutes (Table 5.1).
- Particle size is an important parameter for pretreatment kinetics and effectiveness—diffusion of acid within a particle becomes rate influencing above a critical biomass size and small particle sizes are more readily hydrolyzed with cellulase because of their higher surface area-to-mass ratios—although liquid water at high temperature minimizes this difference and causes ultrastructural changes with the appearance of micrometer-size pores in the material.[40,51]
- Pretreated corn stover is less toxic to ethanologens than other agricultural substrates.[44,46,47]
- Moreover, removal of acetic acid (a degradation product of acetylated hemicellulose sugars) has been demonstrated at 25–35°C using activated carbon powder. A natural fungus has been identified to metabolize furans and actively grow in dilute-acid hydrolysates from corn stover.[52,53]
- Predictive mathematical models have been developed for the rheology and delignification of corn stover during and after pretreatment.[54,55]

TABLE 5.1

Pretreatment Methodologies for Corn Stover: Physical Conditions and Cellulose Digestibility

Hydrolysis	Hydrolysis Conditions	Hemicellulose Solubilization (%)	Enzyme Digestion	Glucan Conversion (%)	Ref.
5% H_2SO_4 or HCl	120°C, 60 min	85	Cellulase	94.7	38
0.2% Na_2CO_3	195°C, 15 min, 12 bar O_2	60	Cellulase	85	40
2% H_2SO_4	190°C, 5 min	—	Cellulase	73[a]	43
Aqueous ammonia	Room temperature, 10–12 min	—	Cellulase	88.5	44
0.5 g $Ca(OH)_2$ per gram of biomass	55°C, 4 weeks	—	Cellulase	93.2[b]	45
Hot water (pH controlled)	190°C, 15 min	—	Cellulase	90	46
Hot water	Intermittent flowthrough, 200°C	—	Cellulase	90	47
SO_2, steam	200°C, 10 min	—	Cellulase	89	48
0.22–0.98% H_2SO_4	140–200°C	—	Cellulase	92.5[c]	49

[a] Total sugars.
[b] Combined posthydrolysis and pretreatment liquor contents.
[c] Combined glucose and xylose recoveries.

- Liquid flow-through enhances hemicellulose sugar yields, increases cellulose digestibility to enzyme treatment, and reduces unwanted chemical reactions—but with the associated penalties of high water and energy use. Some of the benefits of flowthrough can be achieved by limited fluid movement and exchange early in the acid digestion process.[42,48,56,57]

Detailed studies of individual approaches to pretreatment have resulted in important insights into the fundamental sciences as well as guidelines for their large-scale use with corn stover. With long-duration treatment with lime (calcium hydroxide) at moderate temperatures (25–55°C), the enzymic digestibility of the resulting cellulose was highly influenced by both the removal of acetylated hemicellulose residues and delignification; however, deacetylation was not seriously influenced by the levels of O_2 or the temperature.[58] Adding a water washing to ammonia-pretreated material removed lignophenolic extractives and enhanced cellulose digestibility.[59]

Grinding into smaller particles increased the cellulose digestibility after ammonia fiber explosion, but the chemical compositions of the different particle-size classes showed major changes in the contents of xylans and low molecular weight compounds (Figure 5.4). This could be explained by the various fractions of corn stover being differentially degraded in smaller or larger particles on grinding. For example, the cobs are relatively refractive to size reduction; the smaller particle sizes after

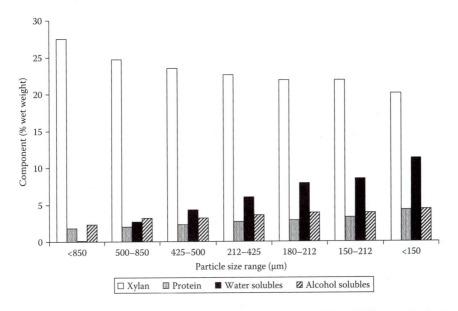

FIGURE 5.4 Size reduction of corn stover and chemical composition of differentially sized particles. (Data from Chundawal, S. P., Venkatesh, B., and Dale, B. E. 2007. *Biotechnology and Bioengineering* 96:219.)

AFEX treatment were more cellulase degradable than were larger particles. Electron microscopic chemical analysis of the surface of the pretreated material provided evidence that lignin–carbohydrate complexes had been disrupted.[59]

The high hemicellulose content of corn cobs has been exploited in a development where aqueous ethanol-pretreated material is washed and then hydrolyzed with an endoxylanase; food-grade xylooligosaccharides can easily be purified, and the cellulosic material is readily digestible with cellulase.[60] An additional advantage of corn cobs is that they can be packed at high density, thus reducing the required water inputs and giving a high concentration in the xylan product stream.

The dominance of inorganic acids for acid pretreatment of biomass substrates has only recently been challenged by the use of maleic acid, one the strongest organic dicarboxylic acids and a potential mimic of the active sites of hydrolase enzymes with two adjacent carboxylic acid residues at their active sites.[61] In comparison with dilute sulfuric acid, maleic acid use resulted in a greatly reduced loss of xylose at high solids loadings (150–200 g of dry stover per liter), resulting in 95% xylose yields, only traces of furfural, and unconditioned hydrolysates that could be used by recombinant yeast for ethanol production. Of the maximum glucose release, 90% could be achieved by cellulase digestion of the pretreated stover within 160 h.

Examination of cellulase digestion of pretreated corn stover has also been undertaken:

- Studies of the binding of cellobiohydrolase to pretreated corn stover identified access to the cellulose in cell wall fragments and the crystallinity of the cellulose microfibrils after pretreatment to be crucial.[62]

- Adding small amounts of surfactant emulsifiers during cellulase digestion of pretreated corn stover also increased the conversion of cellulose, xylan, and total polysaccharide to sugars by acting to disrupt lignocellulose, stabilize the enzyme, and improve the absorption of the enzyme to the macroscopic substrates.[63]
- With steam-pretreated corn stover, near-theoretical glucose yields could be achieved by combining xylanases with cellulase to degrade residual hemicellulose bound to lignocellulosic components.[64]
- The initial rate of cellulase-catalyzed hydrolysis is influenced strongly by the cellulose crystallinity, whereas the extent of cellulose digestion is most influenced by the residual lignin.[65] Modern methods of polymer analysis (e.g., diffusive reflectance infrared and fluorescence techniques) used in this work may be adaptable to on-site monitoring of pretreated biomass substrates.
- The formation of glucose from pretreated corn stover catalyzed by cellulase is subject to product inhibition, and the effects of substrate concentration and the amount ("loading") of the enzyme are important in determining kinetic parameters.[66]
- Cellulase and cellobiohydrolase can both be effectively recovered from pretreated and digested corn stover and recycled with consequent cost savings of approximately 15% (50% if a 90% enzyme recovery could be achieved).[67]
- The solid material used for cellulase-catalyzed hydrolysis itself is a source of potential toxic compounds produced during pretreatment but trapped in the bulk solids; activated carbon is (as discussed previously) effective in removing acidic inhibitors from the liquid phase resulting from digestion of the reintroduced substrate.[68]

A comparative study of several methods for corn stover pretreatment concluded that alkaline methodologies had the potential to reduce the quantities of cellulase necessary in cellulose digestion but that hemicellulase activities may require supplementation.[69]

5.3.4 SOFTWOODS

The widely quoted assessment of softwoods is that, as a biomass substrate, the lignocellulose is too highly lignified and difficult to process to yield cellulose easily digested by cellulase; in practical terms, excess enzyme is required and imposes unrealistically high costs and protracted digestion times.[1,70] Nevertheless, the massive resources of softwood trees in the Pacific Northwest of the United States, Canada, Scandinavia, Northern Europe, and large tracts of Russia maintain softwoods as an attractive potential biomass for fuel alcohol production. Sweden has a particular stake in maximizing the efficiency of ethanol production from softwoods on account of the planned diversion of large amounts of woody biomass to direct heat and power facilities with the phasing out of nuclear generating capacity.[70] Not surprisingly, much of the published work on softwood utilization for ethanol derives from Canadian and Scandinavian universities and research centers.

Although softwoods are low in xylans in comparison with other biomass crops (Table 2.1, Chapter 2), their content of glucan polymers is high; the requirement for xylose-utilizing ethanologens remains a distinct priority, while mannose levels are high and should contribute to the pool of easily utilized hexoses. To make the potential supply of fermentable sugars fully accessible to yeasts and bacteria for fermentation, attention has been focused on steam explosive pretreatments with or without acid catalysts (SO_2 or sulfuric acid) since the 1980s. Pretreatment yields a mixed pentose and hexose stream with 50–80% of the total hemicellulose sugars and 10–35% of the total glucose; a subsequent cellulase digestion liberates a further 30–60% of the theoretical total glucose.[70]

Because only a fraction of the total glucose may be recoverable by such technologies, a more elaborate design has been explored in which a first stage is run at lower temperature for hemicellulose hydrolysis and a second stage is operated at a higher temperature (with a shorter or longer heating time and with the same, higher, or lower concentration of acid catalyst) to liberate glucose from cellulose.[71] Such a two-stage process results in a sugar stream (before enzyme digestion) higher in glucose and hemicellulose pentoses and hexoses but with much reduced degradation of the hemicellulose sugars and no higher levels of potential growth inhibitors such as acetic acid (Figure 5.5).

Two-stage pretreatment suffers from requiring more elaborate hardware and a more complex process management; in addition, attempting to dewater sulfuric acid-impregnated wood chips before steaming decreases the hemicellulose sugar yield from the first step and the glucose yield after the second, higher temperature stage.[72] Such pressing alters the wood structure and porosity, causing uneven heat and mass transfer during steaming. Partial air drying appears to be a more suitable substrate for the dual-stage acid-catalyzed steam pretreatment.

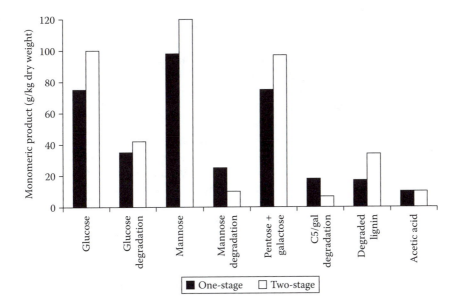

FIGURE 5.5 Sugar stream from single- and dual-stage steam/acid pretreatments of spruce wood chips. (Data from Wingren, A. et al. 2004. *Biotechnology Progress* 20:1421.)

Not only is the monomeric sugar yield higher with the two-stage hydrolysis process, but the cellulosic material remaining also requires only 50% of the cellulase for subsequent digestion.[73] This is an important consideration because steam-pretreated softwood exhibits no evident saturation with added cellulase even at extremely high enzyme loadings that still cannot ensure quantitative conversion of cellulose to soluble sugars. With steam-pretreated softwood material, even lavish amounts of cellulase can liberate only 85% of the glucan polymer at low ratios of solid material to total digestion volume.

Although high temperatures (up to 52°C) and high agitation rates are helpful, enzyme inactivation is accelerated by faster mixing speeds.[74] The residual lignin left after steam pretreatment probably restricts cellulase attack and degradation by forming a physical barrier restricting access and by binding the enzyme nonproductively. Extraction with cold dilute NaOH reduces the lignin content and greatly increases the cellulose-to-glucose conversion; the alkali possibly removes a fraction of the lignin with a high affinity for the cellulase protein.[75]

In addition to the engineering issues, the two-stage technology has two serious economic limitations:

- Hemicellulose sugar recovery is aided after the first step by washing the slurry with water, but the amount of water used is a significant cost factor for ethanol production. To balance high sugar recovery and low water usage, a continuous countercurrent screw extractor that could accept low liquid-to-insoluble solids ratios was developed by the National Renewable Energy Laboratory.[76]
- The lignin recovered after steam explosion has a low product value on account of its unsuitable physicochemical properties; organic solvent extraction of lignin produces a higher value coproduct.[77]

As with other biomass substrates, heat pretreatment at extremes of pH generates inhibitors of microbial ethanol production and, prior to this, cellulase enzyme action.[78,79] A number of detoxification methods have been proposed but simply adjusting the pH to 10 to precipitate low molecular weight sugar and lignin degradation products is effective.[80]

Bioprocessing with enzymes remains an attractive option, especially if different species of wood-rotting fungi are mixed to achieve synergies with their different spectra of enzyme activies.[81] Aqueous ethanol pretreatment of softwoods (the Lignol process) has been strongly advocated on account of its ability to yield highly digestible cellulose as well as lignin, hemicellulose, and furfural product streams. The extraction is operated at acid pH at 185–198°C, and some sugar degradation does inevitably occur.[82] After process optimization, a set of conditions (180°C, 60 min of treatment time, 1.25% v/v sulfuric acid, and 60% v/v ethanol) yielded[83]:

- a solids fraction containing 88% of the cellulose present in the untreated wood chips
- glucose and oligosaccharides equivalent to 85% of the cellulosic glucose released by cellulase treatment (48 h) of the treated lignocellulose

- approximately 50% of the total xylose recovered from the solubilized fraction
- over 70% of the lignin solubilized in a form potentially suitable for industrial use in the manufacture of adhesives and biodegradable polymers

5.3.5 SUGARCANE BAGASSE

As with corn stover, bagasse is the residue from sugarcane juice extraction and, as such, is an obligatory waste product. Development programs for cellulosic ethanol production from bagasse started in Brazil in the 1990s. The commercial Brazilian process uses organic solvent-treated bagasse. Other reports from Cuba, Denmark, Sweden, Japan, Austria, Brazil, and the United States have detailed alternative processes:

- Steam explosion—impregnation with SO_2 prior to steam explosion gives high yields of pentose sugars with no additional formation of toxic inhibitors when compared to the absence of any acid catlyst.[84–86]
- Liquid hot-water pretreatment—this is probably the cheapest method (requiring no catalyst or chemical) and, when operated at <230°C, is effective at solubilizing hemicellulose and lignin while leaving cellulose as an insoluble residue for further processing.[87,88]
- Peracetic acid—alkaline pre-pretreatment followed by the use of peracetic acid—gives synergistic enhancements of cellulose digestibility.[89,90]
- Ammonia and aqueous ammonia of ground bagasse are followed by cellulase and hemicellulase digestions.[91]
- Dilute acid—this has not yet been fully tested for ethanol production but, as a method for preparing xylose as a substrate for xylitol production, it is capable of yielding hydrolysates with high concentrations of free xylose.[92]
- Wet oxidation—alkaline wet oxidation at 195°C for 15 min produces solid material that is 70% cellulase. Approximately 93% of the hemicelluloses and 50% of the lignin are solubilized, and the cellulose can be enzymically processed to glucose with 75% efficiency.[93]

Three different natural species of yeasts have been demonstrated to ferment pentoses and/or hexoses from chemical hydrolysates of sugarcane bagasse: *C. shehatae, Pichia stipitis,* and *Pachysolen tannophilus*.[94–97] Acid hydrolysates are best detoxified by ion exchange materials or activated charcoal; laccase and high-pH precipitation methods are less effective.[95] The novel technique of electrodialysis (migration of ions through membranes under a direct electric field) removed the sulfuric acid and also the acetic acid generated during acid hydrolysis of hemicelluloses so effectively that the reutilization of the sulfuric acid could be contemplated.[97]

Recombinant xylose-utilizing yeast has been desensitized to hydrolysates containing increasing concentrations of phenolic compounds, furfuraldehydes, and carboxylic acids without loss of the xylose-consuming capacity and while retaining the ability to form ethanol rather than xylitol.[96] Acetic acid and furfural at concentrations similar to those measured in sugarcane bagasse hydrolysates adversely affect both laboratory and industrial strains of *S. cerevisiae*.[98] A practical solution is that the predominantly pentose-containing hydrolysates from bagasse pretreatments can also be used to dilute the sugarcane-juice-based medium for sugar ethanol fermentation while maintaining an equivalent sugar concentration and utilizing a pentose-consuming *P. stipitis* to coferment the sugar mixture.[99]

5.3.6 OTHER LARGE-SCALE AGRICULTURAL AND FORESTRY BIOMASS FEEDSTOCKS

Brewer's spent grain is the residue remaining after extraction of cereal grain wort; as a by-product of the brewing process, the material is attracting interest as a low-cost feedstock for cellulosic ethanol production. Fermentations with *Pichia stipitis* and *Kluyveromyces marxianus* have been explored.[100] The spent grains can also be utilized as a solid substrate for fungi producing cellulases and hemicellulases.[101]

In addition to isolating grains for processing, cereal-milling plants also generate fiber-rich fractions as a coproduct stream. In the wet milling of corn, the fiber fraction has traditionally been added into a feed product (Figure 1.20, Chapter 1). A modified dry milling procedure has been developed to recover fiber fractions prior to fermentation; this "quick fiber" contained 65% by weight of total carbohydrate and 32% by weight of glucans, and dilute acid pretreatment was used before fermentation of the substrate to ethanol by either *Escherichia coli* or *S. cervisiae*.[102] Destarched, cellulose-rich, and arabinoxylan-rich fractions of the corn fiber support the growth of strains of *Hypocrea jecorina* and their secretion of hydrolases for plant polysaccharides. These enzymes act synergistically with commercial cellulases on corn fiber hydrolysate and represent a valuable source of on-site enzymes for corn fiber product utilization.[103]

Similarly, wheat bran is produced worldwide in large quantities as a coproduct of wheat milling; residual starch in the bran material can be hydrolyzed to glucose and oligoglucans by amylolytic enzymes. Acid hydrolysis pretreatment followed by cellulase treatment gives a sugar (pentose and hexose) yield of 80% of the theoretical: 135, 228, and 167 g/kg of starch-free bran for arabinose, xylose, and glucose, respectively.[104]

Rice husks are approximately 36% by weight cellulose and 12% hemicellulose. As such, this agricultural by-product could be a major low-cost feedstock for ethanol production; 60% of the total sugars could be released by acid hydrolysis and treatment with a mixture of enzymes (β-glucosidase, xylanase, and esterase) with no formation of furfuraldehyde sugar degradation products.[105] Recombinant *E. coli* could ferment the released sugars to ethanol; high-pH treatment of the hydrolysate reduced the time required for maximal production of ethanol substantially, from 64 to 39 h. In a study from India, rice straw was pretreated with and without exogenous acid and the released hemicellulose sugars fermented by a strain of *C. shehatae*. Ethanol production was also demonstrated by yeast cells immobilized in calcium alginate beads—an example of an advanced fermentation technology discussed in more detail in Section 5.5.3.[106]

Fast-growing willow trees are a major focus of research interest as a bioenergy crop in Scandinavia. High sugar recoveries were achieved from lignocellulosic material by steaming sulfuric-acid-impregnated material for a brief period (4–8 min) at 200°C and then digesting the cellulose enzymatically, liberating glucose with 92% efficiency and xylose with 86% efficiency. The pretreated substrate could also be used for simultaneous saccharification and fermentation with a *S. cerevisiae* strain.[107]

Many exotic plant materials have been included in surveys of and bioenergy sources. As a lignocellulosic, straw from the *brava,* a Bolivian high-plains resident species, can be consid acid-impregnated material gave hemicellulose fractions at 19(mented by three pentose-utilizing yeasts: *P. stipitis, C. shehatae,* *tannophilus.* A higher temperature (230°C) was necessary for cellulose hydroly. *Prosopis juliflora* (a mesquite) is a raw material for long-term sustainable production of cellulosic ethanol, especially on account of the invasive weed characteristics (and reputations) of mesquite species. Pretreated with dilute acid and chemically deligni-fied, cellulose hydrolysis produced sugar mixtures fermentable by both *Pichia stipitis* and *Saccharomyces cerevisiae.*[109]

STEM TOPIC 5.2: CO_2 EMISSIONS FROM CELLULOSIC ETHANOL PRODUCTION

Returning to EBAMM model, the CO_2 emissions are the following:

- agricultural phase = 189 g CO_2e/L
- ethanol conversion = 124 g CO_2e/L
- coproduct credits = 106 g CO_2e/L
- net emissions = (189 + 124) − 106 = 207 g CO_2e/L

This is equivalent to an 88% reduction in comparison with conventional gasoline. Cellulosic ethanol is assessed as using less fertilizer (33% of corn use), no lime, and no gasoline but more (38%) diesel than corn in the agricultural phase. The calculations are dominated by the reduced fertilizer requirement, which accounts for 63% of the total reduction in CO_2 emissions in the field.

Similarly, the replacement of fossil fuels in the provision of energy for the conversion of plant materials to ethanol explains the large reduction (91%) in CO_2 emissions in this phase in comparison with corn ethanol.

Clearly, the extent to which agricultural materials are used for energy generation in cellulosic ethanol production defines how "green" the actual process is!

5.4 FERMENTATION MEDIA AND THE VERY HIGH GRAVITY CONCEPT

After acid hydrolysis (or some other pretreatment) and cellulase digestion, the prod-uct of the process is a mixed carbon source for fermentation by an ethanologenic microbe. Few details have been made public by Iogen about its development of nutritional balances, nitrogen sources, or media recipes for the production stage fer-mentation.[1] This is not surprising because, for most industrial processes, medium

.imization is a category of trade secret unless patenting and publication priorities .eem otherwise.

Few industrial fermentations (for products such as enzymes, antibiotics, acids, and vitamins) rival the conversion efficiency obtained in ethanol production; one notable exception—about which a vast literature is available—is that of citric acid manufacture using yeasts and fungi.[110] Nevertheless, interest in media development for ethanol production has been intense for many years in the potable alcohol industry, and some innovations and developments in this industrial field have been successfully translated to that of fuel ethanol production.

5.4.1 FERMENTATION MEDIA FOR ETHANOL PRODUCTION

Formulating cost-effective media for the recombinant microorganisms developed for broad-spectrum pentose and hexose utilization commenced in the 1990s. For pentose-utilizing *E. coli,* for example, the benchmark was a nutrient-rich laboratory medium suitable for the generation of high cell density cultures.[111] Media were then assessed using the criteria that the final ethanol concentration should be at least 25 g/L, the xylose to ethanol conversion efficiency would be high (90%), and a volumetric productivity of 0.52 g/L/h was to be attained.

In a defined minimal-salts medium, growth was poor—only 15% of that observed in the laboratory medium; supplementation with vitamins and amino acids improved growth but could match only approximately half of the volumetric productivity. As predicted from its wide industrial use in fermentations, the use of corn steep liquor as a complex nitrogen source was the best compromise between the provision of a complete nutritional package with plausible cost implications for a large-scale process.

As an example of the different class of compromise inherent in the use of lignocellulosic substrates, the requirement to have a carbon source with a high content of monomeric xylose and low hemicellulose polymers implied the formation of high concentrations of acetic acid as a breakdown product of acetylated sugar residues. To minimize the associated growth inhibition, one straightforward strategy was that of operating the fermentation at a relatively high pH (7.0) to reduce the uptake of the weak acid inhibitor.

In a study performed with the yeast *P. stipitis,* some surprising interactions were discovered between nitrogen nutrition and ethanol production by the yeast.[112] When the cells had ceased active growth in a chemically defined medium, they were unable to ferment either xylose or glucose to ethanol unless a nitrogen source was also provided. Ethanol production was increased by the amino acids alanine, arginine, aspartic acid, glutamic acid, glycine, histidine, leucine, and tyrosine (although isoleucine was inhibitory); a more practical nitrogen supply for industrial fermentations consisted of a mixture of urea (up to 80% of the nitrogen) and hydrolyzed milk protein supplemented with tryptophan and cysteine (up to 60%); the use of either urea or the protein hydrolysate was less effective than the combination of both. Adding small amounts of minerals—in particular, iron, manganese, magnesium, calcium, and zinc salts—as well as amino acids could more than double the final ethanol concentration to 54 g/L.

Returning to recombinant *E. coli,* attempts to define the minimum salts concentration (to avoid stress imposed by osmotically active solutes) resulted in the formulation of a medium with low levels of sodium and other alkali metal ions and total salts.[113] Although this medium was devised during optimization of lactic acid production, it proved equally effective for ethanol production from xylose. Because many bacteria biosynthesize and accumulate internally high concentrations of osmoprotective solutes when challenged with high exogenous levels of salts, sugars, etc., modulating known osmoprotectants was tested and shown to improve the growth of *E. coli* in the presence of high concentrations of glucose, lactate, sodium lactate, and sodium chloride.[114]

The minimum inhibitory concentrations of these solutes were increased by adding the well-known osmoprotectant betaine, increasing the synthesis of the disaccharide trehalose (a dimer of glucose), or both; the combination of the two was more effective than either alone. Although the cells' tolerance to ethanol was not enhanced, the use of the combination strategy would be expected to improve growth in the presence of the high sugar concentrations ever more frequently encountered in media for ethanol fermentations.

Accurately measuring the potential for ethanol formation represented by a cellulosic biomass substrate for fermentation (or fraction derived from such a material) is complex because any individual fermentable sugar (glucose, xylose, arabinose, galactose, mannose, etc.) may be present in a large array of different chemical forms: monomers, disaccharides, oligosaccharides, and even residual polysaccharides. Precise chemical assays may require considerable time and analytical effort. Bioassay of the material using ethanologens in a set medium and under defined, reproducible conditions is preferable and more cost effective— broadly analogous to the use of shake flask tests to assess potency of new strains and isolates and the suitability of batches of protein and other complex nutrients in conventional fermentation laboratories.[115]

5.4.2 HIGH-CONCENTRATION MEDIA DEVELOPED FOR ALCOHOL FERMENTATIONS

Until the 1980s, the general brewing industry view of yeasts for alcohol production was that most could tolerate only low concentrations (7 or 8% by volume) of ethanol. Consequently, fermentation media (worts) could be formulated to a maximum of 15 or 16° (Plato, Brix, or Balling, depending on the industry subsector), which is equivalent to 15 or 16% by weight of a sugar solution. The events that radically changed this assessment of yeasts and their physiology were precisely and cogently described by one of the key players[116]:

- When brewer's yeasts were grown and measured in the same way as the more ethanol-tolerant distiller's and saké yeasts, differences in ethanol tolerance were smaller than previously thought.
- "Stuck" fermentations (i.e., ones with little or no active growth in supraoptimal sugar concentrations) could easily be rescued by avoiding complete anaerobiosis and supplying additional readily utilizable nitrogen for yeast growth.

- By removing insoluble grain residues (to reduce viscosity), recycling clear mashes to prepare more concentrated media from fresh grain, optimizing yeast nutrition in the wort, and increasing cooling capacity, yeast strains with no prior conditioning and genetic manipulation could produce ethanol up to 23.8% by volume.

Very high gravity (VHG) technologies have four major technical and economic advantages:

- Water use is greatly reduced.
- Plant capacity is increased and fermentor tank volume is more efficiently utilized.
- Labor productivity is improved.
- The energy requirements of distillation are reduced because the fermented broth is more concentrated (>16% v/v ethanol).

With an increased volume of the yeast starter culture added to the wort (higher "pitching rate") and a prolonged growth phase fueled by adequate O_2 and free amino nitrogen (amino acids and peptides), high-gravity worts can be fermented to ethanol concentrations > 16% v/v even at low temperatures (14°C) within a week and with no evidence for any ethanol toxicity.[117–120]

From the work on VHG fermentations, the realization was gained that typical media were seriously suboptimally supplied with free amino acids and peptides for the crucial early growth phase in the fermentation; increasing the free amino nitrogen content by over fourfold still resulted in the exhaustion of the extra nitrogen within 48 h (Figure 5.6). With the correct supplements, brewer's yeast could consume all the

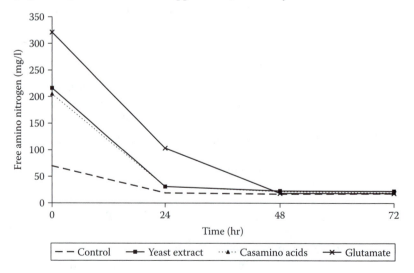

FIGURE 5.6 Utilization of free amino acids and peptides in high-gravity wort for alcohol production. (Data from Thomas, K. C., and Ingledew, W. M. 1990. *Applied and Environmental Microbiology* 56:2046.)

fermentable sugars in a concentrated medium within 8 days at 20°C or accumulate 17% (v/v) ethanol within 3 days.[121]

Fresh yeast autolysate was another convenient (and cost-effective) means of nitrogen supplementation with an industrial distillery yeast from central Europe. However, with such a strain, although nitrogen additions improved final ethanol concentration and glucose utilization, none of them increased cell viability in the late stages of the fermentation, ethanol yield from sugar, or the maximum rate of ethanol formation.[122] Commercial proteases can liberate free amino acids and peptides from wheat mash; the low molecular weight nitrogen sources increase the maximal growth (cell density) of the yeast cells and reduce the fermentation time in VHG worts from 9 to 3 days—without (this being an absolute priority) proteolytic degradation of the glucoamylase added previously to the mash to saccharify the wheat starch. Rather than adding an extra nitrogen ingredient, a onetime protease digestion could replace medium supplementation.[123]

Not all amino acids are beneficial. Lysine is severely inhibitory to yeast growth if the mash is deficient in freely assimilated nitrogen. However, adding extra nitrogen sources such as yeast extract, urea, or ammonium sulfate abolishes this effect, promotes uptake of lysine, increases cell viability, and accelerates the fermentation.[124]

Partial removal of bran from cereal grains (wheat and wheat–rye hybrids) is an effective means of improving the mash in combination with VHG technology with or without nitrogen supplementation (Figure 5.7); in a fuel alcohol plant, this would increase plant efficiency and reduce the energy required for heating the fermentation medium and distilling the ethanol produced from the VHG process.[125] Conversely, adding particulate materials (wheat bran, wheat mash insolubles, soy or horse gram flour, even alumina) improves sugar utilization in VHG media. The mechanism may be to offer some (undefined) degree of osmoprotection.[126,127]

A highly practical goal was in defining optimum conditions for temperature and mash substrate concentration with available yeast strains and fermentation hardware. With a wheat grain-based fermentation, a temperature of 30°C and an initial mash specific gravity of 26% (w/v) gave the best balance of high ethanol productivity, final ethanol concentration, and shortest operating time.[128] The conclusions from such investigations, however, are highly dependent on the yeast strain employed and on the type of beer fermentation being optimized. Brazilian investigators working with a lager yeast strain found that a lower sugar concentration (20% w/v) and temperature (15°C) were optimal together with a triple supplementation of the wort with yeast extract (as a peptidic nitrogen source), ergosterol (to aid growth), and the surfactant Tween-80 (possibly to aid O_2 transfer in the highly concentration medium).[129]

For VHG fermentations, nitrogen nutrition is crucial (i.e., the supply of readily utilizable nitrogen-containing nutrients to support growth) and other medium components also require optimization; adding 50 mM of a magnesium salt in tandem with a peptone (to supply preformed nitrogen sources) increased ethanol concentrations from 14.2 to 17% within a 48-h fermentation.[130] These results were achieved with a medium based on corn flour; the process resembled that in corn ethanol with starch digestion to glucose with amylase and glucoamylase enzyme treatments.

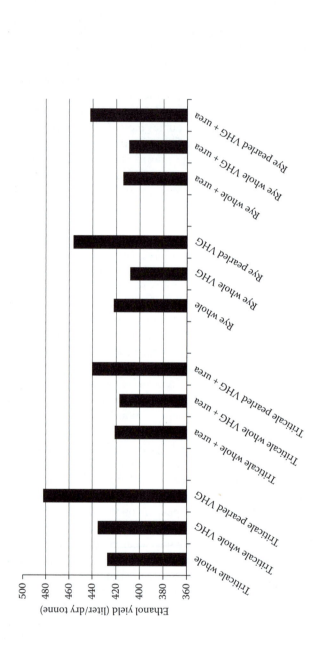

FIGURE 5.7 Grain pearling and very high gravity grain mash fermentations for fuel ethanol production. (Data from Wang, S. et al. 1999. *Process Biochemistry* 34:421.)

With a range of nutrients tested (glycine, magnesium, yeast extract, peptone, biotin, and acetaldehyde), cell densities could dramatically differ: The measured ranges were $74–246 \times 10^6$ cells per gram of mash after 24 h and $62–392 \times 10^6$ cells per gram of mash after 48 h. A cocktail of vitamins added at intervals in the first 28–37 h of the fermentation was another facile strategy for improving final ethanol concentration, average ethanol production rate, specific growth rate, cell yield, ethanol yield, and a reduced glycerol accumulation.[131]

Small amounts of acetaldehyde have been claimed to reduce the time required to consume high concentrations of glucose (25% w/v) in VHG fermentations; the mechanism is speculative but could involve increasing the intracellular NAD:NADH ratio and accelerating general sugar catabolism by glycolysis (Figure 3.1, Chapter 3).[132] Side effects of acetaldehyde addition included increased accumulation of the higher alcohols 2,3-butanediol and 2-methylpropanol, exemplifying again how immune fuel ethanol processes are to the unwanted contaminants and flavor agents so strictly controlled in potable beverage production. Mutants of brewer's yeast capable of faster fermentations, more complete utilization of wort carbohydrates, and higher viability under VHG conditions are easily selected after UV treatment. Some of these variants could also exhibit improved fermentation characteristics at low operating temperature (11°C).[133]

VHG technology has been demonstrated on a test scale with barley, wheat, and corn grains, achieving single-fermentation ethanol yields of up to 13.3–14.5% by volume.[134] No definitive study has appeared for cellulosic ethanol, and the likely problem area is that of toxic products of lignocellulose acid hydrolysis being too concentrated in VHG-type media. Other pretreatment technologies may be mandatory for VHG fermentations for cellulosic substrates.

With the rise of the "omics" technologies, the molecular biology of yeast cells in VHG media has begun to reveal interesting features[135]:

- Amino acid supplementation increases the expression of glucose-utilizing enzymes.
- Up-regulation of trehalose and glycogen biosynthetic enzymes also occurs.
- Aminoacyl–tRNA synthetases and heat-shock proteins are boosted at the end of the lag phase.

VHG media have also been shown to be successful in continuous ethanol-producing and immobilized cell systems, which are discussed in Section 5.5.[136]

STEM TOPIC 5.3: THE DYNAMICS OF ETHANOL EFFLUX FROM CELLS

With highly concentrated media and yeast cells pushed to their limits for ethanol production rates, can ethanol escape from cells rapidly enough to avoid distorting cell volumes or (even worse) denaturing cellular proteins? Studies in the 1980s agreed that ethanol diffused freely across cell membranes of ethanologenic microorganisms so that intra- and extracellular concentrations were equilibrated. Most of the evidence, however, was indirect and interpretative because ethanol transport processes were too rapid for conventional analytical techniques (e.g., preloading cells with radioactively labeled ethanol and following the kinetics of label movement).

The use of in vivo nuclear magnetic resonance (NMR) methodologies changed this picture dramatically.[137] For the first time, influx and efflux rates of ethanol could be determined. For a culture of *Z. mobilis* at 20°C, producing ethanol from glucose at a maximal rate:

- maximal glucose uptake rate = 0.22 μmol per minute per milligram of cells
- maximal ethanol production rate = 0.44 μmol per minute per milligram of cells

For a concentration difference, Δ_c, between intra- and extracellular phases of only 1 mmol/L, the specific ethanol efflux was 0.42 μmol per minute per milligram of cells; at Δ_c = 100 mmol/L, this rate increased to 0.99 mol per minute per milligram of cells. Careful scrutiny of the insert in Figure 3 from the journal paper cited in Reference 137 shows that the Δ_c was much greater than 100 mmol/L. Ethanol efflux would not have been rate limiting and cytoplasmic ethanol accumulation would not have occurred.

At 29°C (a more likely temperature for ethanol production), the rate of ethanol efflux across the cytoplasmic membrane more than doubles. Interestingly, ethanol diffusion *inside* the bacterium was faster than transport across the lipid bilayer, but the differential rates do not result in ethanol accumulation.

5.5 FERMENTOR DESIGN AND NOVEL FERMENTOR TECHNOLOGIES

5.5.1 CONTINUOUS FERMENTATIONS FOR ETHANOL PRODUCTION

With enormous strides made in the development of high-alcohol fermentations prepared using high- and very-high-gravity media compatible with low operating temperatures, fast turnaround times, and high conversion efficiency from carbohydrate feedstocks, what has the traditional alcohol industry achieved with fermentor hardware and process design and control?

Beginning in the late 1950s in New Zealand, the Dominion Breweries introduced and patented a novel continuous brewing process in which part of the fermented beer wort was recycled back to the wort of the start of the fermentation step (Figure 5.8).[138] Within 2 years, a rival continuous process had been announced and patented in Canada by Labatt Breweries, and independently conceived research was being published by the Brewing Industry Research Foundation in the United Kingdom.[139] By 1966, the concept had been simplified and reduced to a single tower

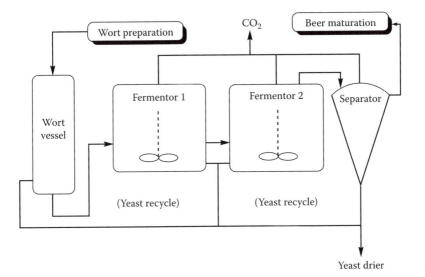

FIGURE 5.8 Continuous ethanol fermentation with yeast recycling for beer manufacture.

structure in which the bulk of the yeast cells was retained (for up to 400 h), while wort might only reside for only 4 h in the highly anaerobic conditions in most of the tower before a "beer" product emerged.[140]

However, despite much initial enthusiasm and the evolution of technologies into a family tree of "open" (where yeast cells emerge at rapid rates from the process), "closed" (where yeast cells are mostly retained), "homogeneous" (approximating a standard stirred-tank fermentor), and "heterogeneous" (with gradients of cells, substrates, and products across several vessels) systems devised for the continuous brewing of beer, very few were developed to the production scale and most became defunct. The innovations failed the test of four basic parameters of traditional brewing practice[139]:

- Commercial brewers offer multiple product lines and flexible production schedules to a marketplace that has become more sophisticated, discerning, and influenced by advertisements.
- Continuous systems offer a fixed rate of beer production (and increasing flow rate tends to lead to washout).
- Only highly flocculent strains of yeast can be used.
- Desired flavors and aromas cannot always be generated; undesirable levels of malodorous compounds are easily formed.

Moreover, contemporaneous developments in brewing technology led, for example, to the marked reduction in malting and fermentation time in batch processes, the losses during malting and wort boiling, increased α-acid content and hops (and, consequently, reductions in hop usage), and the elimination of the prolonged "lagering" storage necessary for lager beer production (an increasingly popular product line worldwide). Such incremental improvements were easy to retrofit in established (often venerably old) brewing facilities, which demanded mostly tried and trusted hardware for any newly opened sites.

Other than in New Zealand, continuous brewing systems have proved generally unpopular. The rise of globally recognized brands has resulted in the backlash of movements for traditional, "real" beers locally produced in microbreweries, thus continuing the trend of consumer choice against a perceived blandness in the output of the market leaders. Large multinationals offering a branded product for global markets remain the most likely to invest in high tech brewing.

In contrast, industrial and fuel ethanol production is immune to such drivers. By the late 1970s the cold-shouldering of new technologies by potable alcohol producers had persuaded academic and research groupings of chemical engineers to focus on ethanol as an alternative fuel, with a renewed interest in highly engineered solutions to the demands for process intensification, process control, and cost reduction—for example:

- Utilizing the cell recycling principle developed in New Zealand but adding a vacuum-enhanced continuous recovery of ethanol from the fermentation vessel, a high-productivity process was devised that incorporated sparging with O_2 to support active cell growth over a prolonged period of time and bleed out the fermented broth to withdraw nonvolatile compounds of potential toxicity to the producing cells. For a 95% ethanol product, such technology offered a 50% reduction in production costs over batch fermentation.[141]
- A single-stage gas-lift tower fermentor with a highly flocculent yeast was designed to run with nearly total retention of the cell population and generate a clear liquid effluent. Analysis of the vapor-phase ethanol concentration in the headspace gas gave, via computer control, an accurate control of input and output flow rates.[142]
- A more straightforward use of laboratory-type continuous fermentors operating on the fluid overflow principle and not requiring a flocculent yeast achieved a maximal ethanol yield (89% of the theoretical from carbohydrate) at a low dilution rate (0.05 per hour), showed 95% utilization of the inflowing sugars up to a dilution rate of 0.15 per hour, and only suffered from washout at 0.41 per hour. The fermentation was operated with a solubilized mixed substrate of sugars from Jerusalem artichoke, a plant species capable of a very high carbohydrate yield on poor soils with little fertilizer application.[143]

For large-scale fuel ethanol manufacture, however, another feature of the continuous process has proved of widest application—that is, the use of multiple fermentors linked in series with (or without) the option of recycling the fermenting broth, sometimes described as "cascading" (Figure 5.9).[122] Several evolutionary variants of this process paradigm have been developed from design concepts for corn ethanol dating from the late 1970s.[144] In the USSR, batteries of 6–12 fermentors were linked in multistage systems for the production of ethanol from miscellaneous raw materials in 70 industrial centers. Unlike in the West, however, such advanced engineering was applied to other types of fermentation including beer, champagne, and fruit wines, as

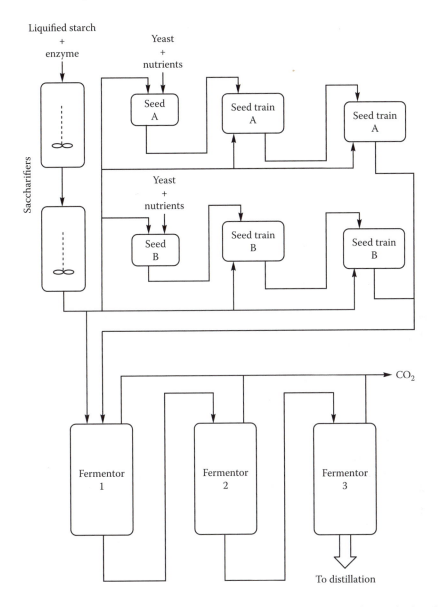

FIGURE 5.9 Cascaded saccharification, yeast propagation, prefermentation, and ethanol fermentation.

well as fodder yeast (as a form of single-cell protein). These were supported by scientists at the All-Union Research Institute of Fermentation Products, who developed sophisticated mathematical analyses—unfortunately, in a literature almost entirely in Russian.[145]

Multistage fermentations have been merged with VHG media in laboratory systems, with 99% of consumption of media containing up to 32% w/v glucose.[146]

In such a continuously flowing system, ethanol yield from glucose increased from the first to the last fermentor in the sequence at all glucose concentrations tested (15–32% w/w); ethanol was therefore produced more efficiently in the later stages of the inhomogeneous set of fermentations that are set up in quasi-equilibrium within the sequence of the linked fermentation vessels. This implies mathematical modeling to control productivity will be difficult in multistage processes because parameters such as ethanol yield from sugars will be variable and possibly difficult to predict with complex feedstocks.

In a similar vein, a Chinese prototype with a working volume of 3.3 L achieved a 95% conversion of glucose to ethanol; although oscillations were observed in residual glucose, ethanol concentrations, and cell densities, some success was found in devising models to predict yeast cell lysis and viability loss.[147] In Brazil, attention was focused on the problems of running ethanol fermentations at high ambient (tropical) temperatures, and a five-fermentor system was devised in which a temperature gradient of 8°C was established. High biomass cultures were generated at up to 43°C, and a high-viability process with continuous ethanol production was demonstrable at temperatures normally considered supraoptimal for brewer's yeast.[148]

A much simpler design, one transposed from brewery work in the United Kingdom, fermented sugarcane juice at sugar concentrations up to 200 g/L in a tower with continuous recycling of a highly flocculent yeast strain; a constant dilution rate and pH (3.3) gave outflow ethanol concentrations of up to 90 g/L with conversion rates as high as 90% of the theoretical maximum.[149] In France, the concept of separate compartments for growth and ethanologenesis was pursued in a two-stage fermentor with efficient recycling of the yeast cells; a steady state could be reached where the residual glucose concentration in the second stage was close to zero.[150]

However described—continuous, cascaded, multistage, etc.—sequential fermentations offer a step change in fermentation flexibility that may be of great value for lignocellulosic feedstocks. For example, different vessels could be operated at higher or lower pH, temperature, and degree of aeration to ferment different sugars and could even accommodate different ethanologenic organisms. Figure 5.10 is a schematic of a process with split pentose and hexose sugar streams with five fermentors in the continuous cascaded sequence; the liquid stream that moves from fermentor to fermentor is in contact with a stripping gas to remove the ethanol (thus avoiding the buildup of inhibitory concentrations). The process is a composite one based on patents granted between 1987 and 2006.[151–153]

A low-energy solvent absorption and extractive distillation process recovers the ethanol from the stripping gas and the gas is then reused; however, the same arrangement of fermentation vessels can be used with conventional removal of ethanol by distillation from the outflow of the final fermentor. Moreover, the pentose and hexose fermentations could employ the same recombinant organism or separate ethanologens. An example of this used *Pichia stipitis* as the organism to ferment pentoses in an airlift loop tower while choosing *S. cerevisiae* for glucose utilization in an overflow tower fermentor for the residual glucose; with an appropriate flow rate, a utilization of 92.7% was claimed from the sugars prepared from sugarcane bagasse.[95]

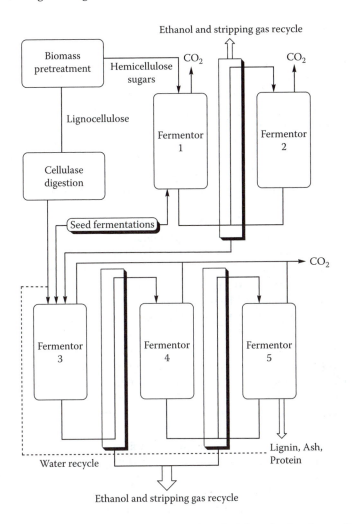

FIGURE 5.10 A cascaded fermentation system for ethanol production from a lignocellulosic biomass with continuous ethanol removal.

5.5.2 FED-BATCH FERMENTATIONS

Beyond ethanol production, the modern mainstream fermentation industry manufacturing primary metabolites including vitamins and amino acids, secondary metabolites (including antibiotics), and recombinant proteins almost invariably opts for fed-batch fermentation technologies; it has invested much time and expertise in devising feeding strategies for carbon sources (including sugars, oligosaccharides, and amylase-digested starches) to operate at a minimum concentration of free sugars and avoid carbon catabolite repressions. Some large-scale processes are run to vanishingly small concentrations of free glucose and the feed rate is regulated not by direct measurement but, rather, indirectly by effects of transient glucose accumulation on physical parameters such as pH (responding to acid accumulation during glucose overfeeding) or the trends in dissolved O_2 concentration.[154]

Yeast (*S. cerevisiae* in its baker's yeast guise) cells are one of the three principal production platforms for recombinant proteins for the biopharmaceuticals market; the others are *E. coli* and mammalian cell cultures. Of the ten biopharmaceutical protein products achieving regulatory approval in the United States or the European Union during 2004, seven were produced in mammalian cell lines, two in *E. coli,* and the tenth in *S. cerevisiae*.[155] In such cell systems, ethanol formation is either to be avoided (as a waste of glucose) or carefully regulated—possibly as a means of feedback control to a complex and variable sugar feed to high cell densities where O_2 supply is critical to maintain biosynthetic reactions rather than a loosely regulated simple fermentation.[156,157]

For fuel ethanol, very high rates of sugar consumption and ethanol production are mandatory for competitive commercial production processes. Temperature is an important parameter: In Europe, 30°C is considered optimal for growth and 33°C for ethanol production.[158] Rather than aiming at microaerobic conditions, a high-aeration strategy is beneficial for stabilizing a highly viable cell mass capable of high ethanol productivity. Glycerol accumulates as a major coproduct, but this waste of sugar metabolism can be minimized by several options for fermentation management:

- Avoiding high-temperature "excursions" in the stage of ethanol production, glycerol formation becomes uncoupled to growth and glycerol accumulation may (in addition to a role in osmotolerance) offer some degree of temperature protection to ethanol-forming yeast cells.[158]
- High-aeration regimes greatly reduce glycerol accumulation.[159]
- Maintaining a high respiratory quotient (the ratio of CO_2 produced to O_2 consumed) results in high ethanol-to-glycerol discrimination ratio when the online data are used to feedback-control the inlet sugar feed rate.[160]

In addition to cane sugar as a substrate, fed-batch technology has been used with conventional yeasts to produce ethanol from sugarcane molasses (a by-product of sugar refining) with an exponentially decreasing feed rate.[161] The bacterial ethanologen *Z. mobilis* is most productive for biomass formation and for ethanol production from glucose when feeding avoids the accumulation of high concentrations of glucose. An important finding from this work was that attempts to regulate a constant glucose concentration do not optimize the process because of the complex relationship between specific growth rate and glucose supply.[162]

In general, experience in the fermentation routes to producing fine chemicals highlights the importance of accurately monitoring analyte levels inside fermentors to avoid excessive accumulations (or depletions) of key substrates and nutrients, triggering repression mechanism, and the appearance of metabolic imbalances—all of which have a negative impact on process economics.[163] Increasingly sophisticated methodologies are being applied to the real-time monitoring of sugars, ethanol, and cell population dynamics inside ethanol fermentors.[164] Because fed-batch processes are widely considered to be the favored route to the contemporary yeast cell limit of 23% v/v ethanol production, bioprocess management using more sophisticated tools to ensure a steady and slow release of glucose and other monomeric sugars is a major future milestone for industrial-scale ethanol production.[165]

5.5.3   Immobilized Yeast and Bacterial Cell Production Designs

Many of the problems encountered in regulating both growth and metabolism in actively growing, high-cell-density systems could, in principle, be avoided by immobilizing the producer cells in or on inert supports and reusing them (if their viability can be maintained) for repetitive cycles of production. This is akin to transforming whole-cell biocatalysis to a more chemically defined form known for more than a century in the fine chemicals manufacturing industry. Cost savings in the power inputs required to agitate large volumes (100,000–500,000 L) in their cooling and their power-consuming aeration spurred the fermentation industry at large to consider such technologies, and serious attempts to immobilize productive ethanologens began in the 1970s.

Alginate beads, prepared from the hydrophilic carbohydrate polymer alginic acid (extracted from kelp seaweed), are porous, are compatible with an aqueous environment, and could be loosely packed; in such a packed-bed bioreactor, perfused with a nutrient and substrate solution, a high volumetric productivity of ethanol production could be maintained for up to 12 days with a less than 10% loss of productive capacity.[166] With cells of the yeast *Kluyveromyces marxianus* supplied with a Jerusalem artichoke tuber extract, the maximum volumetric productivity was 15 times that for a conventional stirred tank fermentor. With *S. cerevisiae* in a fluidized bed contained in a closed circuit, ethanol production was possible with glucose solutions of up to 40% w/v at a relatively low temperature (18°C).[167]

As with continuous fermentation technologies, potable alcohol manufacturers expressed interest in and actively developed demonstration facilities for immobilized yeast cells but failed to exploit the potential of the innovations fully—again because the product failed to meet organoleptic and other exacting specifications for the industry.[168] Many different matrices were investigated beyond alginate and other seaweed polymers, including ceramic materials, synthetic polymers, and stainless steel fibers, to adsorb or entrap cells in up to five different reactor geometries:

1. the packed bed, the simplest arrangement and capable of upward or downward flow of the liquid phase
2. the fluidized bed, where mixing of gas, liquid, and solid phases occurs continuously
3. airlift and bubble column bioreactors
4. conventional stirred tank reactors
5. membrane bioreactors, where the cells are free but retained by a semipermeable membrane

Beer producers have adopted immobilized cells to remove unwanted aroma compounds selectively from the primary fermentation product, "green" beer, and to produce and remove aldehydes from alcohol-free and low-alcohol beers, using strains and mutants of *S. cerevisiae*.[169] For the main fermentation, however, immobilized cells remain a promising option for further development; however, the added cost of immobilization techniques is impossible to justify for commercial processes without much stronger increases in volumetric productivity.

Cost-effective options have been explored for fuel ethanol production. In Brazil, improved fermentation practices and efficiencies have taken up the slack offered by low-intensity production systems, and further progress requires technological innovations. Immobilizing yeast ethanologens on sugarcane stalks is an eminently practical solution to providing an easily sourced and import-substituting support matrix.[170]

A summary of data from other studies of immobilized cells relevant to fuel ethanol production is given in Table 5.2.[171–175] The liquid phase can be recirculated in an immobilized system to maximize the utilization of the fermentable sugars. This is particularly useful to harmonize attaining maximal productivity (grams of ethanol produced per liter per hour) and maximum ethanol yield (as a percentage of the maximum theoretically convertible); these optima occur at different rates of flow with bacterial and yeast ethanologens.[171]

Recombinant xylose-utilizing *Z. mobilis* cells were immobilized in photocross-linked resins prepared from polyethylene or polypropylene glycols and shown to utilize both glucose and xylose efficiently from acid hydrolysates of cedar tree wood, rice straw, newspaper, and bagasse.[176] Alginate-immobilized *S. cerevisiae* cells have also been used in conjunction with a five-vessel cascade reactor in a continuous alcohol fermentation design.[177] Sweet sorghum is a major potential source of fuel ethanol production in China, where laboratory studies have used immobilized yeast cells to ferment stalk juice (equivalent to pressed cane sugar juice).[178]

More advanced engineering has resulted in mixed cultures of immobilized cells of the yeasts *K. marxianus* and *S. cerevisiae* fermenting cheese whey (a solution of the disaccharide lactose) with higher efficiency than either yeast immobilized alone.[179] The most sophisticated system is probably ethanol fermentation by immobilized *S. cerevisiae* cells in magnetic particles in a magnetically stabilized, fluidized bed reactor.[180]

Immobilized cells represent stationary phase cells, which may exhibit enhanced resistance to growth inhibitors and other impurities in substrate solutions that are more injurious to actively dividing cells. Immobilized cells of various microbial species, for example, are considered to be more resistant to aromatic compounds, antibiotics, and low pH. This opens the novel possibility that immobilized yeast could be used to ferment batches of microbial toxin-containing feedstocks. A trichothecene mycotoxin inhibits protein synthesis and mitochondrial function in *S. cerevisiae* and, at 200 mg/L, can cause cell death. Conversely, fermentation is not affected and glucose metabolism may be positively redirected toward ethanol production, suggesting that trichothecene-contaminated grain could be salvaged for fuel ethanol manufacture.[181]

Alginate-encapsulated yeast cells are certainly able to withstand toxic sugar degradation products in acid hydrolysates prepared from softwoods (mainly spruce wood). Such immobilized cells could fully ferment the glucose and mannose sugars within 10 h; free cells could not accomplish this within 24 h, although the encapsulated cells lost their activity in subsequent batch fermentations.[182]

TABLE 5.2
Immobilized and Free Cell Systems for Fuel Ethanol Production: Critical Parameters for Process Efficiency

Ethanologen	Support	Carbon Source	Immobilized Cells			Free Cells			Ref.
			Maximum Ethanol Productivity (g/L/h)	Maximum Ethanol Yield (%)	Maximum Ethanol Outflow (g/L)	Maximum Ethanol Productivity (g/L/h)	Maximum Ethanol Yield (%)	Maximum Ethanol Outflow (g/L)	
S. cerevisiae	75% Polypropylene, 20% soybean hulls, 5% soybean flour	Glucose	499	45		5	24		180
Z. mobilis	75% Polypropylene, 20% soybean hulls, 5% zein	Glucose	536	50		5	26		180
S. cerevisiae	Calcium alginate	Sugarbeet molasses	10.2	83	46.2	8.7	71	39.4	171
S. cerevisiae	Seed waste		51						172
E. coli	Membrane cell recycle bioreactor	Xylose	25.2		31.5	1.8		35.2	173
E. coli	Clay brick	Xylose	4.5		21.9	1.8		35.2	173
E. coli	Calcium alginate	Xylose	2.0		37.1	1.8		35.2	173
S. cerevisiae	Calcium alginate	Glucose	2.8	38	13.1	0.3	31.2	9.8	174

5.5.4 Contamination Events and Buildup in Fuel Ethanol Plants

Contamination was a consideration in the poor take-up of continuous fermentation technologies by potable alcohol producers, especially because holding-prepared wort for sometimes lengthy periods of time without yeast inoculation provided an excellent growth medium for adventitious microbes in the brewery.[168] With the accumulation of operating experience in fuel alcohol facilities, bacterial populations have been identified that not only reduce yield but also can prove difficult to eradicate; some bacteria (including lactic acid producers) form biofilms under laboratory conditions and can colonize many (perhaps every) available surface in complex sequences of linked fermentors and the associated pipe work.[183]

When bacterial contaminants reach 10^6 or 10^7 cells per milliliter, the economic losses for ethanol production can reach 3% of volumetric capacity; if profitability is marginal, this is a serious impact. Thus, antibiotic regimes have been devised to pulse controlling agents through continuous processes.[184] This prophylactic approach has been applied to continuous ethanol facilities where the total losses will be greater because continuous operations have begun to dominate the larger (>40 million gallons a year) production plants.

An antibiotic such as penicillin G is not metabolized and degraded by *S. cerevisiae* and its addition rate can be poised against its expected chemical degradation at the low pH of the fermentation broth. Outside the spectrum of known antibiotics, a useful alternative is the curious (and little known) chemical adjunct between urea and hydrogen peroxide. This bacteriocidal agent can effectively control lactobacilli in wheat mash and provides useful levels of readily assimilable nitrogen and O_2 (by enzyme-catalyzed decomposition of the peroxide) to enhance yeast growth and fermentative capacity.[185]

5.6 SIMULTANEOUS SACCHARIFICATION AND FERMENTATION AND CONSOLIDATED BIOPROCESSING

Extrapolating back up the process stream and considering a totally enzyme-based hydrolysis of polysaccharides, an ideal ethanol process has been defined to include[186]:

- lignin removal during pretreatment to minimize unwanted solids in the substrate
- simultaneous conversion of cellulose and hemicellulose to soluble sugars
- ethanol recovery during the fermentation to high concentrations
- immobilized cells with enhanced fermentation productivity

An even closer approach to the ideal would use enzymes to degrade lignin sufficiently without resorting to extremes of pH to expose cellulose and hemicellulose fully before their degradation to sugars by a battery of cellulases, hemicellulases, and ancillary enzymes (esterases, etc.) in a totally enzymic process with only a minimal biomass pretreatment (i.e., size reduction). Because pretreatment methods could solubilize much of the hemicellulose, two different approaches were suggested with either cellulolytic microbes (whole cell catalysis) or the addition of fungal cellulase

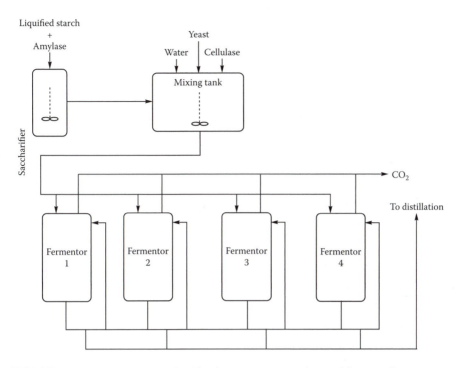

Liquified starch
+
Amylase

Saccharifier

Water | Cellulase

Yeast

Mixing tank

CO_2

To distillation

Fermentor 1

Fermentor 2

Fermentor 3

Fermentor 4

FIGURE 5.11   Simultaneous saccharification, yeast propagation, and fermentation.

and hemicellulase to the fermentation medium.[187,188] These options have become known as consolidated bioprocessing (CBP), simultaneous saccharification and fermentation (SSF; Figure 5.11), and simultaneous saccharification and cofermentation (SSCF; see STEM Topic 2.1, Chapter 2).

Consolidated bioprocessing suffers from the biological problems of low ethanol tolerance by the clostridial ethanologens and poor ethanol selectivity of the fermentation.[189] Commercialization of clostridial species has been correspondingly slow; few studies have progressed beyond the initial laboratory stage. The phytopathogenic fungus *Fusarium oxysporium* is the sole nonbacterial wild-type microbe actively considered for CBP. The ability of the organism to ferment xylose as well as hexose sugars to ethanol was recognized in the early 1980s, and several strains can secrete cellulose-degrading enzymes.[190,191]

Hemicellulose sugars can also be utilized in acid hydrolysates, although with low conversion efficiencies (0.22 g of ethanol per gram of sugar consumed).[192] Brewers' spent grains could be converted directly to ethanol at 60% of the theoretical yield based on total glucose and xylose content.[193] Extensive metabolic engineering of *F. oxysporium* is likely to be required for an efficient ethanologen, and detailed analysis of the intracellular biochemical networks has begun to reveal potential sites for intervention.[194–196]

Metabolic engineering of *S. cerevisiae* to degrade macromolecular cellulose has been actively pursued by research groups in South Africa, the United States, Canada, Sweden, and Japan. Fungal genes encoding various components of the cellulase complex have successfully been expressed in ethanologenic *S. cerevisiae,* yielding strains capable of utilizing and fermenting either cellobiose or cellulose.[197–200] Calculations show that, based on the growth kinetics of and enzyme secretion by cellulose degraders such as *H. jecorina,* approximately 1% of the total cell protein of a recombinant cellulase-secreting *S. cerevisiae* would be required.[201,202]

The key to developing CBP remains that of isolating an efficient natural cellulolytic and hemicellulolytic microbe that can also compete with yeasts as an ethanologen. Attention is focusing on thermophilic microorganisms—in particular, *Thermoanaerobacterium.*[203] Mixtures of isolated strains may mutually complement each other's enzymic capabilities and reduce side product formation.[204]

The SSF technologies suffer from a similar drawback: how to enable efficient ethanologens to coexist with hydrolytic enzymes from thermophiles at elevated temperatures that can accelerate carbohydrate breakdown to readily fermentable sugars:

- If yeasts are to act as the ethanologens, thermotolerant strains would perform more in harmony with the elevated temperatures in which cellulases work efficiently.[205,206]
- Bacteria are more readily operated in high-temperature bioprocesses, and recombinant *Klebsiella oxytoca* produced ethanol more rapidly under SSF conditions than did cellobiose-utilizing yeasts. Coculturing *K. oxytoca* and *S. pastorianus, Kluyveromyces marxianus,* or *Z. mobilis* resulted in increased ethanol production in both isothermal and temperature-profiled SSF to increase the cellulase activity.[207]
- Both *K. oxytoca* and *Erwinia* species have the innate abilities to transport and metabolize cellobiose, thus reducing the need to add exogenous β-glucosidase to the cellulase complex. Moreover, chromosomally integrating the *E. chrysanthemi* gene for endoglucanase and expressing the gene at a high level result in the secretion of high enzyme activities sufficient to hydrolyze cellulose.[208,209]

Reducing the quantity of cellulase added to ensure efficient cellulose digestion would also be beneficial for the economics of the SSF concept. Adding nonionic surfactants, polyethylene glycol, and a "sacrificial" protein to decease nonproductive absorption of cellulase to lignin binding sites has also been demonstrated to increase cellulase action. Thus, cellulose digestion efficiency can be maintained at lower enzyme-to-substrate ratios.[8,210]

The importance of the quantity of cellulase added was underlined by a Swedish study that showed that reducing the enzyme loading by 50% actually increased the production cost of ethanol in SSF by 5% because a less efficient cellulose hydrolysis reduced the ethanol yield.[211] At low enzyme loading, growing the yeast inoculum on the pretreated biomass material (barley straw) has considerable advantages; the conditioned cells can be used at a reduced concentration (2 g/L, down from 5 g/L) and with an increased solids content in the SSF stage.[212]

The cost of commercially used fungal cellulase has decreased by over an order of magnitude due to the efforts of enzyme manufacturers after 1995.[213] Multiple efforts have been made to increase the catalytic efficiencies of cellulases from established and promising novel microbial sources.[214–216] Inserting genes for components of the cellulase complex into efficient recombinant ethanol producers has also continued as part of a strategy to reduce the need to add exogenous enzymes; such cellulases can be secreted at levels that represent significant fractions of the total cell protein and increase ethanol production capabilities.[217–221] This is of particular importance for the accumulation of high concentrations of ethanol because ethanol at >65 g/L inhibits the fungal (*H. jecorina*) cellulase commonly used in SSF studies.[222]

The SSF technology has been claimed to be superior to independent stages of enzymic hydrolysis and fermentation with sugarcane bagasse, utilizing more of both the cellulose and hemicelluloses.[223] A continued commitment to develop SSF is evident in the publications on SSF technologies continuing to appear with a wide variety of lignocellulosic feedstocks (Table 5.3).[32,224–237] Prominent in the list of lignocellulosic feedstocks is corn stover, a material with the unique distinction of having a specific biocatalyst designed for its utilization.[238] This fusion of the biochemical abilities of *Geotrichum candidium* and *Phanerochaete chrysosporium* points toward a long-term option for both SSF and CBP (i.e., that of harnessing the proven hypercapabilities of some known microbes to degrade lignocellulose and converting them to ethanologens by retroengineering into them the ethanol biochemistry of *Z. mobilis*).

TABLE 5.3

Simultaneous Saccharification and Fermentation Applied to Fuel Ethanol Production from Lignocellulosic Feedstocks

Ethanologen	Lignocellulosic Material	Country of Origin	Year of Publication	Ref.
S. cerevisiae + *Pachysolen tannophilus*	Rice straw	India	1995	223
S. cerevisiae	Hybrid poplar, switchgrass, corn stover	United States	1997	224
P. tannophilus	Timothy grass, alfalfa, reed canary grass, corn stalks, barley straw	Canada	1998	225
Kluyveromyces marxianus	Sugarcane leaves, *Antigonum leptopus* leaves	India	2001	226
S. cerevisiae	Corn stover	United States	2003	227
S. cerevisiae	Corn stover	Hungary	2004	228
S. cerevisiae	Switchgrass, poplar	Taiwan	2005	229
E. coli (KO11)	Corn stover	United States	2005	32
S. cerevisiae	Corn stover	Sweden	2006	230
S. cerevisiae	Corn stover	United States	2006	231
S. cerevisiae	Starch	Slovakia	2008	232
S. cerevisiae	Sweet sorghum stem	China	2008	233
S. cerevisiae	Softwood	Sweden	2009	234
S. cerevisiae	Bamboo shoot	Japan	2009	235
Candida tropicalis	Cassava pulp	Thailand	2009	236

TABLE 5.4

Cocultivations of Ethanologenic and Ethanologenic plus Enzyme-Secreting Microbes

Ethanologen	Immobilized?	Enzyme Secretor	Immobilized?	Ref.
S. cerevisiae + Candida shehatae	–	*Sclerotum rolfsii*	–	242
S. cerevisiae + Candida shehatae	+	None (glucose and xylose mix)	–	243
S. cerevisiae + Pichia stipitis	–	None (glucose and xylose mix)	–	244
S. cerevisiae	+	*Aspergillus awamori*	–	245
S. cerevisiae + Pachysolen tannophilus + E. coli	–	None (softwood hydrolysate)	–	246
S. cerevisiae + Candida shehatae	+	None (glucose and xylose mix)	–	247
S. cerevisiae or *Zymomonas mobilis*	–	*Saccharomycopsis fibuligera*	–	248

On parallel tracks, attempts have been made to introduce fungal genes for starch degradative enzymes into candidate industrial ethanologens and explore the possible advantages from combining genetic backgrounds from two microbes into a single hybrid designed for high amylase secretion.[239-241] The commercial use of food wastes such as cheese whey has prompted the construction of strains with β-galactosidase to hydrolyze lactose extracellularly and use the released glucose and galactose simultaneously for ethanol production under anaerobic conditions.[242]

As a final option (one that mimics the evolution of natural microbial communities in soils, forest leaf litter, water-logged areas, and stagnant pools), cocultivation of a good ethanologen together with an efficient secretor of enzymes to degrade polymeric carbohydrates and/or lignocelluloses is a route avoiding introducing genetically manipulated organisms and could be adapted to continuous technologies if a close control of relative growth rates and cell viabilities can be achieved. One or more of the microbial partners can be immobilized; Table 5.4 includes two examples of this approach together with the cocultivation of different ethanologens to ferment glucose/xylose mixtures and pretreated lignocellulosics.[243-249]

5.7 DOWNSTREAM PROCESSING AND BY-PRODUCTS

5.7.1 ETHANOL RECOVERY FROM FERMENTED BROTHS

The distillation of ethanol from fermented broths remains the dominant practice in ethanol recovery in large and small ethanol production facilities.[250] Other physical techniques have been designated as having lower energy requirements than simple distillation, and some (vacuum dehydration [distillation], liquid extraction, supercritical fluid extraction) can yield anhydrous ethanol for fuel purposes from a dilute aqueous alcohol feed (Figure 5.12).[251] Only water removal by molecular sieving has

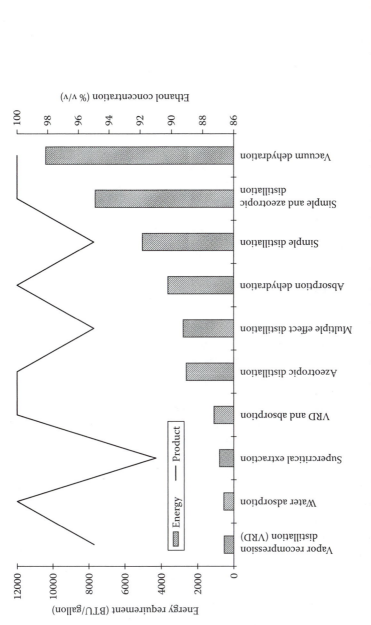

FIGURE 5.12 Energy requirement and ethanol product concentration from technologies developed for separation of ethanol–water mixtures. (Data from Sikyta, B. 1995. *Techniques in Applied Microbiology.* [*Progress in Industrial Microbiology*, vol. 31], chap. 12. New York: Elsevier Science & Technology.)

been successful on an industrial scale, however, and all new ethanol plants are built with molecular sieve dehydrators in place.[252]

Nevertheless, the economic costs of dehydration are high, especially when anhydrous ethanol is to be the commercial product (Figure 5.12). In the early 1980s, the energy requirements were so high that the practical basis for fuel ethanol production was questioned because the energy required for distillation approximated the total combustion energy of the alcohol product.[253,254] However, the investment costs of rivals to distillation were so high (up to 8.5 times that of conventional distillation) that little headway was made. Attention was focused on improving process efficiencies and energy cycling with the development of low-energy hydrous ethanol distillation plants with 50% lower steam-generating requirements.[252]

The economics of downstream processing are markedly affected by the concentration of ethanol in the fermented broth; for example, the steam required to produce an ethanol from a 10% v/v solution of ethanol is only 58% of that required for a more dilute (5% v/v) starting point. Pushing the ethanol concentration in the fermentation to 15% v/v reduces the required steam to approximately half that required for low conversion broth feeds.[253]

5.7.2 CONTINUOUS ETHANOL RECOVERY FROM FERMENTORS

Partly as another route for process cost reduction but also as a means to avoid the accumulation of ethanol concentrations inhibitory to cell growth, technologies to remove ethanol in situ (i.e., during the course of the fermentation) have proved intermittently popular.[254] Seven different modes of separation have been demonstrated in small-scale fermentors:

1. A volatile product such as ethanol can be separated from a fermentation broth under vacuum even at a normal operating temperature; a system with partial medium removal and cell recycling was devised to minimize the accumulation of nonvolatile products inhibitory to yeast growth and productivity.[141]
2. If the fermentor is operated normally but the culture liquid is circulated through a vacuum chamber, the ethanol formed can be removed on a continuous basis. This arrangement avoids the need to supply O_2 to vessels maintained under vacuum.[255]
3. Solvent extraction with a long chain alcohol (n-decanol) with immobilized cells of *S. cerevisiae* and up to 409 g/L of glucose (from glucose syrup) could be metabolized at 35°C.[256]
4. As with water removal from concentrated ethanol, ethanol can be selectively adsorbed by different types of resins with hydrophobic surfaces, including cross-linked divinylbenzene polystyrene resins widely used in modern chromatographic separations of alcohols, sugars, and carboxylic acids. Such resins work efficiently with ethanol at low ethanol concentrations, and the ethanol can be desorbed with warm, dry N_2 gas at 60–80°C.[257,258]
5. Hollow-fiber microfiltration is effective for ethanol and other small molecule products (such as lactic acid), but is slow and difficult to sterilize.[259]

6. In membrane pervaporation, the cells are retained by a semipermeable membrane while a partial vacuum is applied to the permeate side; ethanol concentrations could be maintained below 25 g/L for 5 days while a concentrated ethanol efflux stream of 17% w/v was achieved.[260] Polyvinyl alcohol membranes operate better at elevated temperature, and this suggests that thermophilic ethanologens would be very suitable in a membrane pervaporative process.[261]

7. Gas stripping of ethanol can be effected in an air-lift fermentor—a type of vessel originally developed for viscous microbial fermentation broths but also used for some of the more fragile and shear-sensitive mammalian cells in culture. This is another example of a technology that would inevitably work better with a thermophilic ethanologen and an elevated fermentation temperature.[262] Alternatively (and more economically, with reduced power consumption for gas volume flow), the fermentation broth is circulated through an inert packed column and continuously sparged with a stripping gas (see Figure 5.10). Such arrangements can result in highly stable continuous fermentations (for >100 days), with near theoretical yields of ethanol from concentrated glucose solutions with corn steep water providing nutrients.[263,264]

How many (if any) of these advanced downstream technologies become adopted for industrial use will depend heavily on their economics; for example, ethanol stripping is assessed at providing a significant cost savings for fuel ethanol production from corn starch.[265] With lignocellulosic substrates being used more widely, especially in developing economies, a much simplified technology can provide surprisingly elegant solutions.

Solid-state fermentations have long been used for fermented foods and saké but can easily be adapted to manufacture (under more stringent conditions and with a reduced labor intensity) many fine chemicals and enzymes.[266] A continuous process has been engineered to process and ferment feedstocks such as fodder beet and sweet sorghum in a horizontal tubular bioreactor; the fermenting material (with a low moisture content) is moved along with the aid of a spiral screw.[267,268] Some ethanol volatilization will occur at any temperature above ambient (caused by the fermentation process), but the bulk of the product could be recovered by gas or air flowing through the container before the ethanol is condensed and transferred to a final dehydration step (as in the gas stripping technology). Although originally devised for farm-scale use, this solid-phase bioprocess was sufficiently productive to allow distillation from 8% v/v outputs.

5.7.3 SOLID BY-PRODUCTS FROM ETHANOL FERMENTATIONS

The solids remaining at the end of the fermentation (distiller's dry grains with solids [DDGS]; see Figure 1.21, Chapter 1) are a high-protein animal feed. This is a saleable by-product that has been suggested to be so commercially desirable that reduced ethanol yields could be tolerated to support the increased production, although in practice high-sugar residues pose severe practical difficulties to DDGS drying and

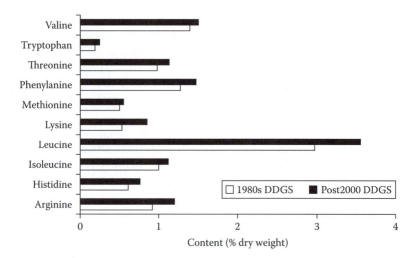

FIGURE 5.13   Essential amino acid content of distiller's dried grain with solids: changes in U.S. compositions from the 1980s. (Data from Jacques, K. A. 2003. In *The Alcohol Textbook*, 4th ed., ed. Jacques, K. A., Lyons, T. P., and Kelsall, D. R., chap. 27. Nottingham, England: Nottingham University Press.)

processing. The rapid rise of ethanol production from corn starch, however, has demanded some remarketing of this coproduct[269]:

- The product is less dark because sugars are more efficiently fermented and less available to react chemically and caramelize in the dried product.
- The essential amino acid contents are higher (Figure 5.13).
- Although ruminant animals can certainly benefit from feeding with DDGS, pigs are geographically much closer to ethanol plants in the U.S. Midwest.

Reducing phosphate content would widen the use of DDGS by addressing animal-waste disposal issues, and the development of more efficient methods for removing water in the preparation of the DDGS could greatly reduce processing costs.[270] Adding a second fermentation (or enzymic biotransformation), a dry-grind processing to generate plant oils, and a higher value animal feed from the DDGS, and separating more useful and saleable fine chemicals from the primary fermentation, would increase the total mass of recovered bioproducts to the maximum achievable (Figure 5.14).[271]

Pricing is crucial because an increased supply of DDGS is likely to reduce its market price significantly, and its alternate use as the feedstock for further ethanol production itself has been worthy of investigation. Steam and acid pretreatments can convert the residual starch and fiber into a substrate for yeast-based ethanol production with a yield 73% of the theoretical maximum from the glucans in the initial solids.[272]

A simpler option is to realize the potential in the fermented solids to provide nutrients and substrates for a new round of yeast growth and ethanol production. Such spent media ("spent wash," stillage, or vinasse) can be recycled in the process

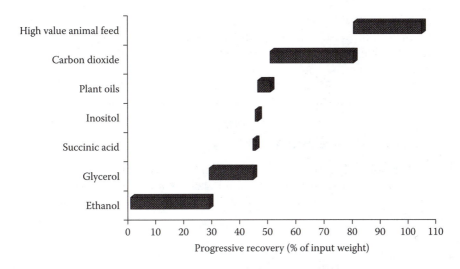

FIGURE 5.14 Projected recovery of product and coproducts from the ethanol fermentation of corn starch. (Data from Dawson, K. A. 2003. In *The Alcohol Textbook,* 4th ed., ed. Jacques, K. A., Lyons, T. P., and Kelsall, D. R., chap. 28. Nottingham, England: Nottingham University Press.)

known as "backsetting," which has been found to be beneficial for yeast growth and a practical means of reducing water usage in a fuel alcohol facility.[273]

Backsetting is not without its accepted potential drawbacks, including the accumulation of toxic nonvolatiles in the fermentor, increased mash viscosity, and dead cells causing problems with viability measurements. However, as a crude means of adapting the fermentation to a semicontinuous basis, it has its advocates on both environmental and economic grounds. Furthermore, mixtures of fungal enzymes decrease vinasse viscosity and liberate pentose sugars from soluble and insoluble arabinoxylans that would be suitable for fermentation by a suitable pentose-utilizing ethanologen.[274–277]

Brazil has by far the longest continuous history of devising methods for economically viable disposal for vinasse and the solid-waste product (bagasse)[278]:

- Bagasse combustion in steam turbines generates electricity at 1 MWh per cubic meter of produced alcohol.
- Anaerobic digestion of vinasse can produce enough biogas for 0.5 MWh per cubic meter of produced alcohol, and both processes have been applied at full scale at distilleries.
- Laboratory studies show that anaerobic digestion would also be beneficial for bagasse, increasing the power output to 2.25 MWh per cubic meter of produced alcohol if the nonbiodegradable residue is burned.
- The total potential power generation from biogas and combustion routes would be equivalent of 4% of national power demand.
- Important for minimizing fertilizer use, material dissolved in the digestion wastewater represents 70% of the nutrient demand of sugarcane fields.

This success is even more remarkable because neither vinasse nor bagasse had consistently been considered as saleable and both might even represent negative value as incurred disposal costs unless used immediately locally as fertilizers or to prevent soil erosion.

STEM TOPIC 5.4: SWITCHGRASS ETHANOL—A FINAL VIEW ON ENERGY AND CO_2

The ERG Biofuel Analysis Meta-Model (EBAMM) is mostly theoretical, but switchgrass grown on 10 farms in North Dakota, South Dakota, and Nebraska have supplied data with which to refine the model for the agricultural phase of cellulosic ethanol production[279]:

- agricultural energy inputs = 4,700 MJ/ha
- switchgrass yield = 7,100 kg/ha

These new data resulted in net energy values of megajoules per liter of ethanol; computed total CO_2 emissions were 6 g CO_2e per megajoule of ethanol produced (i.e., 94% lower than from conventional gasoline).

The largest measured energy input for the agricultural phase was fertilizer use, which accounted for 61.7% of the total; diesel was another 18.8% and herbicides 10.1%. Across the 10 farms, total agricultural energy requirements varied from 3,673 to 6,451 MJ/ha. There was a similar large spread in switchgrass yield over three consecutive seasons from 5,200 to 11,100 kg/ha (means for each farm).

5.8   SUMMARY

The protracted development of cellulosic ethanol processes toward commercial-scale production has centered on biochemical engineering choices very similar to those long used in the cereal grain ethanol industry.

Nevertheless, experimental studies have investigated more radical process options, including some evolved in the potable alcohol industry: continuous fermentations, very high gravity media, fed-batch fermentations, immobilized ethanologenic cells, and continuous ethanol removal from the fermentor space. No clear evidence indicates that any of these advanced technology options is essential for the near-term development of cellulosic ethanol production; however, all of them could aid the achievement of technically and economically viable large production units.

The most likely challenge to the separate hydrolysis and fermentation model remains consolidated bioprocessing with engineered or natural cellulolytic ethanologens or simultaneous saccharification and fermentation with thermotolerant enzymes and thermophilic yeast (or bacterial) cells.

REFERENCES

1. Tolan, J. S. 2006. Iogen's demonstration process for producing ethanol from cellulosic biomass. In *Biorefineries—Industrial processes and products. Volume 1: Status quo and future directions,* ed. Kamm, B., Gruber, P. R., and Kamm, M., chap. 9. Weinheim, Germany: Wiley-VCH Verlag.
2. Foody, B., Tolan, J. S., and Bernstein, J. D. 1999. Pretreatment process for conversion of cellulose to fuel ethanol. US Patent 5,916,780, June 29, 1999.
3. Tolan, J. S. 1999. Alcohol production from cellulosic biomass: the Iogen process, a model system in operation. In *The alcohol textbook,* 3rd ed., ed. Jacques, K. A., Lyons, T. P., and Kelsall, D. R., chap. 9. Nottingham, England: Nottingham University Press.
4. Lawford, G. R. et al. 1986. Scale-up of the Bio-Hol process for the conversion of biomass to ethanol. In *Biotechnology and renewable energy,* ed. Moo-Young, M., Hasnain, S., and Lamptey, J., 276. New York: Elsevier Applied Science Publishers.
5. Taylor, J. D. 1986. Commercial update on the Stake process. In *Biotechnology and renewable energy,* ed. Moo-Young, M., Hasnain, S., and Lamptey, J., 286. New York: Elsevier Applied Science Publishers.
6. Wright, C. T. et al. 2005. Biomechanics of wheat/barley straw and corn stover. *Applied Biochemistry and Biotechnology* 121–124:5.
7. Haq, Z., and Easterley, J. L. 2006. Agricultural residue availability in the United States. *Applied Biochemistry and Biotechnology* 129–132:3.
8. Kristensen, J. B. et al. 2007. Use of surface active additives in enzymatic hydrolysis of wheat straw lignocellulose. *Enzyme and Microbial Technology* 40:888.
9. Sørensen, H. R., Pedersen, S., and Meyer, A. S. 2006. Optimization of reaction conditions for enzymatic viscosity reduction and hydrolysis of wheat arabinoxylan in an industrial ethanol fermentation residue. *Biotechnology Progress* 22:505.
10. Sørensen, H. R. et al. 2007. Enzymatic hydrolysis of wheat arabinoxylan by a recombinant "minimal" enzyme cocktail containing β-xylosidase and novel endo-1,4-β-xylanase and α-L-arabinofuranosidase activities. *Biotechnology Progress* 23:100.
11. Sørensen, H. R. et al. 2006. A novel GH43 α-L-arabinofuranosidase from *Humicola insolens*: Mode of action and synergy with GH51 α-L-arabinofuranosidases on wheat arabinoxylan. *Applied Microbiology and Biotechnology* 73:850.
12. Ahring, B. K. et al. 1996. Pretreatment of wheat straw and conversion of xylose and xylan to ethanol by thermophilic anaerobic bacteria. *Bioresource Technology* 58:107.
13. Bjerre, A. B. et al. 1996. Pretreatment of wheat straw using combined alkaline wet oxidation and alkaline hydrolysis resulting in convertible cellulose and hemicellulose. *Biotechnology and Bioengineering* 49:568.
14. Schmidt, A. S., and Thomsen, A. B. 1998. Optimization of wet oxidation pretreatment of wheat straw. *Bioresource Technology* 64:139.
15. Ahring, B. K. et al. 1999. Production of ethanol from wet oxidized wheat straw by *Thermoanaerobacter mathranii*. *Bioresource Technology* 68:3.
16. Klinke, H. B., Thomsen, A. B., and Ahring, B. K. 2001. Potential inhibitors from wet oxidation of wheat straw and their effect on growth and ethanol production by *Thermoanaerobacter mathranii*. *Applied Microbiology and Biotechnology* 57:631.
17. Klinke, H. B. et al. 2002. Characterization of degradation products from alkaline wet oxidation of wheat straw. *Bioresource Technology* 82:15.
18. Klinke, H. B. et al. 2003. Potential inhibitors from wet oxidation of wheat straw and their effect on ethanol production of *Saccharomyces cerevisiae:* Wet oxidation and fermentation by yeast. *Biotechnology and Bioengineering* 81:738.
19. Panagiotou, G., and Olsson, L. 2007. Effect of compounds released during pretreatment of wheat straw on microbial growth and enzymatic hydrolysis rates. *Biotechnology and Bioengineering* 96:250.

20. McLoughlin, S. B. et al. 2002. High-value renewable energy from prairie grasses. *Environmental Science & Technology* 36:2122.
21. Tenenbaum, D. A. 2002. Switching to switchgrass. *Environmental Health Perspectives* 110:A18–A19.
22. Kaylen, K. S. 2005. An economic analysis of using alternative fuels in a mass burn boiler. *Bioresource Technology* 96:1943.
23. Kamm, J. 2004. A new class of plants for a biofuel feedstock energy crop. *Applied Biochemistry and Biotechnology* 113–116:55.
24. Stroup, J. A. et al. 2004. Comparison of growth and performance in upland and lowland switchgrass types to water and nitrogen stress. *Bioresource Technology* 86:65.
25. Mulkey, V. R., Owens, V. N., and Lee, D. K. 2008. Management of warm-season grass mixtures for biomass production in South Dakota, USA. *Bioresource Technology* 99:609.
26. Anderson, W. F., and Akin, D. E. 2008. Structural and chemical properties of grass lignocelluloses related to conversion for biofuels. *Journal of Industrial Microbiology and Biotechnology* 35:355.
27. Lau, C. S. et al. 2004. Extraction of antioxidant compounds from energy crops. *Applied Biochemistry and Biotechnology* 113–116:569.
28. Sarath, G. et al. 2007. Internode structure and cell wall composition in maturing tillers of switchgrass (*Panicum virgatum* L.). *Bioresource Technology* 98:2985.
29. Lee, S. T. et al. 2001. The isolation and identification of steroidal sapogenins in switchgrass. *Journal of Natural Toxins,* 10:273.
30. Anderson, W. F. et al. 2005. Enzyme pretreatment of grass lignocellulose for potential high-value co-products and an improved fermentable substrate. *Applied Biochemistry and Biotechnology* 121–124:303.
31. Law, K. N., Kokta, B. V., and Mao, C. B. 2001. Fiber morphology and soda-sulphite pulping of switchgrass. *Bioresource Technology* 77:1.
32. Chung, Y. C., Bakalinsky, A., and Penner, M. H. 2005. Enzymatic saccharification and fermentation of xylose-optimized dilute acid-treated lignocellulose. *Applied Biochemistry and Biotechnology* 121–124:947.
33. Meunier-Goddik, L., and Penner, M. H. 1999. Enzyme-catalyzed saccharification of model celluloses in the presence of lignacious residues. *Journal of Agricultural and Food Chemistry* 47:346.
34. Kurakake, M. et al. 2001. Pretreatment with ammonia water for enzymatic hydrolysis of corn husk, bagasse, and switchgrass. *Applied Biochemistry and Biotechnology* 90:251.
35. Alizadeh, H. et al. 2005. Pretreatment of switchgrass by ammonia fiber explosion (AFEX). *Applied Biochemistry and Biotechnology* 121–124:1133.
36. Paul, J. K., ed. 1980. Alcohol from corn (50 million gallons/year). In *Large and small scale ethyl alcohol processes from agricultural raw materials. Part 1.* Park Ridge, NJ: Noyes Data Corporation.
37. Kadam, K. L., and McMillan, J. D. 2003. Availability of corn stover as a sustainable feedstock for bioethanol production. *Bioresource Technology* 88:17.
38. Rooney, T. 1998. Lignocellulosic feedstock resource assessment. Report SR-580-24189, National Renewable Energy Laboratory, Golden, CO, p. 123.
39. Varga, E., Szengyel, Z., and Reczey, K. 2002. Chemical pretreatments of corn stover for enhancing enzymatic digestibility. *Applied Biochemistry and Biotechnology* 98–100:73.
40. Kim, S. B., and Lee, Y. Y. 2002. Diffusion of sulfuric acid within lignocellulosic biomass particles and its impact on dilute-acid treatment. *Bioresource Technology* 83:165.
41. Varga, E. et al. 2003. Pretreatment of corn stover using wet oxidation to enhance enzymatic digestibility. *Applied Biochemistry and Biotechnology* 104:37.
42. Yang, B., and Wyman, C. E. 2004. Effect of xylan and lignin removal by batch and flowthrough pretreatment on the enzymatic digestibility of corn stover cellulose. *Biotechnology and Bioengineering* 86:88.

43. Pimenova, N. V., and Hanley, T. R. 2003. Measurement of rheological properties of corn stover suspensions. *Applied Biochemistry and Biotechnology* 105–108:383.

44. Varga, E., Reczey, K., and Zacchi, G. 2004. Optimization of steam pretreatment of corn stover to enhance enzymatic digestibility. *Applied Biochemistry and Biotechnology* 113–116:509.

45. Kim, S. B., and Lee, Y. Y. 2005. Pretreatment and fractionation of corn stover by ammonia recycle percolation process. *Bioresource Technology* 96:2007.

46. Kim, S., and Holtzapple, M. T. 2005. Lime pretreatment and enzymatic hydrolysis of corn stover. *Bioresource Technology* 96:1994.

47. Mosier, N. et al. 2005. Optimization of pH controlled liquid hot water pretreatment of corn stover. *Bioresource Technology* 96:1986.

48. Liu, C., and Wyman, C. E. 2005. Partial flow of compressed-hot water through corn stover to enhance hemicellulose sugar recovery and enzymatic digestibility of cellulose. *Bioresource Technology* 96:1978.

49. Öhgren, K., Galbe, M., and Zacchi, G. 2005. Optimization of steam pretreatment of SO$_2$-impregnated corn stover for fuel ethanol production. *Applied Biochemistry and Biotechnology* 121–124:1055.

50. Lloyd, T. A., and Wyman, C. E. 2005. Combined sugar yields for dilute sulfuric acid pretreatment of corn stover followed by enzymatic hydrolysis of the remaining solids. *Bioresource Technology* 96:1967.

51. Zeng, M. et al. 2007. Microscopic examination of changes of plant cell structure in corn stover due to hot water pretreatment and enzymatic hydrolysis. *Biotechnology and Bioengineering* 97:265.

52. Berson, R. E. et al. 2005. Detoxification of actual pretreated corn stover hydrolysate using activated carbon powder. *Applied Biochemistry and Biotechnology* 121–124:923.

53. Nichols, N. N. et al. 2005. Bioabatement to remove inhibitors from biomass-derived sugar hydrolysates. *Applied Biochemistry and Biotechnology* 121–124:379.

54. Pimenova, N. V., and Hanley, T. R. 2004. Effect of corn stover concentration on rheological characteristics. *Applied Biochemistry and Biotechnology* 113–116:347.

55. Kim, S., and Holtzapple, M. T. 2006. Delignification kinetics of corn stover in lime pretreatment. *Bioresource Technology* 97:778.

56. Zhu, Y., Lee, Y. Y., and Elander, R. T. 2004. Dilute-acid pretreatment of corn stover using a high-solids percolation reactor. *Applied Biochemistry and Biotechnology* 117:103.

57. Zhu, Y., Lee, Y. Y., and Elander, R. T. 2005. Optimization of dilute-acid pretreatment of corn stover using a high-solids percolation reactor. *Applied Biochemistry and Biotechnology* 121–124:1045.

58. Kim, S., and Holtzapple, M. T. 2006. Effect of structural features on enzyme digestibility of corn stover. *Bioresource Technology* 97:583.

59. Chundawal, S. P., Venkatesh, B., and Dale, B. E. 2007. Effect of size based separation of milled corn stover on AFEX pretreatment and enzymatic digestibility. *Biotechnology and Bioengineering* 96:219.

60. Zhu, Y. et al. 2005. Enzymatic production of xylooligosaccharides from corn stover and corn cobs treated with aqueous ammonia. *Applied Biochemistry and Biotechnology* 129–132:586.

61. Lu, Y., and Mosier, N. S. 2007. Biomimetic catalysis for hemicellulose hydrolysis in corn stover. *Biotechnology Progress* 23:116.

62. Jeoh, T. et al. 2007. Cellulase digestibility of pretreated biomass is limited by cellulose accessibility. *Biotechnology and Bioengineering* 98:112.

63. Kaar, W. E., and Holtzapple, M. T. 1998. Benefits from Tween during enzymic hydrolysis of corn stover. *Biotechnology and Bioengineering* 59:419.

64. Öhgren, K. et al. 2007. Effect of hemicellulose and lignin removal on enzymatic hydrolysis of steam pretreated corn stover. *Bioresource Technology* 98:2503.

65. Laureano-Perez, L. et al. 2005. Understanding factors that limit enzymatic hydrolysis of biomass: characterization of pretreated corn stover. *Applied Biochemistry and Biotechnology* 121–124:1081.

66. O'Dwyer, J. P. et al. 2007. Enzymatic hydrolysis of lime-pretreated corn stover and investigation of the HCH-1 model: Inhibition pattern, degree of inhibition, validity of simplified HCH-1 model. *Bioresource Technology* 98:2969.

67. Steele, B. et al. 2005. Enzyme recovery and recycling following hydrolysis of ammonia fiber explosion-treated corn stover. *Applied Biochemistry and Biotechnology* 121–124:901.

68. Berson, R. E., Young, J. S., and Hanley, T. R. 2006. Reintroduced solids increase inhibitor levels in a pretreated corn stover hydrolysate. *Applied Biochemistry and Biotechnology* 129–132:612.

69. Wyman, C. E. et al. 2005. Comparative sugar recovery data from laboratory scale application of leading pretreatment technologies to corn stover. *Bioresource Technology* 96:2026.

70. Galbe, M., and Zacchi, G. 2002. A review of the production of ethanol from softwood. *Applied Microbiology and Biotechnology* 59:618.

71. Wingren, A. et al. 2004. Process considerations and economic evaluation of two-step steam pretreatment for production of fuel ethanol from softwood. *Biotechnology Progress* 20:1421.

72. Kim, K. H., Tucker, M. P., and Nguyen, Q. A. 2002. Effects of pressing lignocellulosic biomass on sugar yield in two-stage dilute-acid hydrolysate process. *Biotechnology Progress* 18:489.

73. Nguyen, Q. A. et al. 2002. Two-stage dilute-acid pretreatment of softwoods. *Applied Biochemistry and Biotechnology* 84–86:561.

74. Tengborg, C., Galbe, M., and Zacchi, G. 2001. Influence of enzyme loading and physical parameters on the enzymatic hydrolysis of steam-pretreated softwood. *Biotechnology Progress* 17:110.

75. Pan, X. et al. 2005. Strategies to enhance the enzymatic hydrolysis of pretreated softwood with high residual lignin content. *Applied Biochemistry and Biotechnology* 121–124:1069.

76. Kim, K. H. et al. 2001. Continuous countercurrent extraction of hemicellulose from pretreated wood residues. *Applied Biochemistry and Biotechnology* 91–93:253.

77. Mabee, W. E. et al. 2006. Updates on softwood-to-ethanol process development. *Applied Biochemistry and Biotechnology* 129–132:55.

78. Larsson, S. et al. 1999. The generation of fermentation inhibitors during dilute acid hydrolysis of softwood. *Enzyme and Microbial Technology* 24:151.

79. Tengborg, C., Galbe, M., and Zacchi, G. 2001. Reduced inhibition of enzymatic hydrolysis of steam-pretreated softwood. *Enzyme and Microbial Technology* 28:835.

80. Larsson, S. et al. 1999. Comparison of different methods for detoxification of lignocellulose hydrolyzates of spruce. *Applied Biochemistry and Biotechnology* 77–79:91.

81. Lee, J. W. et al. 2008. Enzymatic saccharification of biologically pretreated *Pinus densiflora* using enzymes from brown rot fungi. *Journal of Bioscience and Bioengineering* 106:162.

82. Pan, X. et al. 2005. Biorefining of softwoods using ethanol organosolv pulping: Preliminary evaluation of process streams for manufacture of fuel-grade ethanol and co-products. *Biotechnology and Bioengineering* 90:473.

83. Pan, X. et al. 2006. Bioconversion of hybrid poplar to ethanol and co-products using an organosolv fractionation process: Optimization of process yields. *Biotechnology and Bioengineering* 94:851.

84. Morjanoff, P. J., and Gray, P. P. 1987. Optimization of steam explosion as a method for increasing susceptibility of sugarcane bagasse to enzymatic saccharification. *Biotechnology and Bioengineering* 29:733.

85. Kaar, W. E., Gutierrez, C. V., and Kinoshita, C. M. 1998. Steam explosion of sugarcane bagasse as a pretreatment for conversion to ethanol. *Biomass & Bioenergy* 14:277.

86. Martín, C. et al. 2002. Comparison of the fermentability of enzymatic hydrolysates of sugarcane bagasse pretreated by steam explosion using different impregnating agents. *Applied Biochemistry and Biotechnology* 98–100:699.
87. Laser, M. et al. 2002. A comparison of liquid hot water and steam pretreatments of sugar cane bagasse for bioconversion to ethanol. *Bioresource Technology* 81:33.
88. Sasaki, M., Adschiri, T., and Arai, K. 2003. Fractionation of sugarcane bagasse by hydrothermal treatment. *Bioresource Technology* 86:301.
89. Teixera, L. C., Linden, J. C., and Schroeder, H. A. 1999. Optimizing peracetic acid pretreatment conditions for improved simultaneous saccharification and cofermentation (SSCF) of sugar cane bagasse to ethanol fuel. *Renewable Energy* 16:1070.
90. Teixera, L. C., Linden, J. C., and Schroeder, H. A. 2000. Simultaneous saccharification and cofermentation of peracetic acid-pretreated biomass. *Applied Biochemistry and Biotechnology* 84–86:111.
91. Prior, B. A., and Day, D. F. 2008. Hydrolysis of ammonia-pretreated sugar cane bagasse with cellulase, β-glucosidase, and hemicellulase preparations. *Applied Biochemistry and Biotechnology* 146:151.
92. Silva, S. S., Matos, Z. R., and Carvalho, W. 2005. Effects of sulfuric acid loading and residence time on the composition of sugarcane bagasse hydrolysate and its use as source of xylose for xylitol production. *Biotechnology Progress* 21:1449.
93. Martín, C., Klinke, H. B., and Thomsen, A. B. 2007. Wet oxidation as a pretreatment method for enhancing the enzymatic convertibility of sugarcane bagasse. *Enzyme and Microbial Technology* 40:426.
94. Yang, B. et al. 1997. Study on the hydrolyzate of sugarcane bagasse to ethanol by fermentation. *Chinese Journal of Biotechnology* 13:253.
95. Chandel, A. K. et al. 2007. Detoxification of sugarcane bagasse hydrolysate improves ethanol production by *Candida shehatae* NCIM 3501. *Bioresource Technology* 98:1947.
96. Martín, C. et al. 2007. Adaptation of a recombinant xylose-utilizing *Saccharomyces cerevisiae* to a sugarcane bagasse hydrolysate with high content of fermentation inhibitors. *Bioresource Technology* 98:1767.
97. Cheng, K.-K. et al. 2007. Fermentation of pretreated sugarcane bagasse hemicellulose hydrolysate to ethanol by *Pachysolen tannophilus*. *Biotechnology Letters* 29:1051.
98. Garay-Arroyo, A. et al. 2004. Response to different environmental stress conditions of industrial and laboratory *Saccharomyces cerevisiae* strains. *Applied Microbiology and Biotechnology* 63:734.
99. de Castro, H. F., Oliveira, S. C., and Furlan, S. A. 2003. Alternative approach for utilization of pentose stream from sugarcane bagasse by an induced flocculent *Pichia stipitis*. *Applied Biochemistry and Biotechnology* 105–108:547.
100. White, J. S. et al. 2008. Bioconversion of brewers' spent grains to bioethanol. *FEMS Yeast Research* 8:1175.
101. Xiros, C. et al. 2008. Hydrolysis and fermentation of brewers' spent grain by *Neurospora crassa*. *Bioresource Technology* 99:5427.
102. Dien, B. S. et al. 2004. Fermentation of "quick fiber" produced from a modified corn-milling process into ethanol and recovery of corn fiber oil. *Applied Biochemistry and Biotechnology* 113–116:937.
103. Li, X. L. et al. 2005. Profile of enzyme production by *Trichoderma reesei* grown on corn fiber fractions. *Applied Biochemistry and Biotechnology* 121–124:321.
104. Palmarola-Adrados, B. et al. 2005. Ethanol production from nonstarch carbohydrates of wheat bran. *Bioresource Technology* 96:843.
105. Saha, B. C. et al. 2005. Dilute acid pretreatment, enzymatic saccharification, and fermentation of rice hulls to ethanol. *Biotechnology Progress* 21:816.
106. Abbi, M., Kuhad, R. C., and Singh, A. 1996. Fermentation of xylose and rice straw hydrolysate to ethanol by *Candida shehatae* NCL-3501. *Journal of Industrial Microbiology* 17:20.

107. Sassner, P. et al. 2008. Steam pretreatment of H$_2$SO$_4$-impregnated *Salix* for the production of bioethanol. *Bioresource Technology* 99:137.

108. Sanchez, G. et al. 2004. Dilute-acid hydrolysis for fermentation of the Bolivian straw material *Paja brava*. *Bioresource Technology* 93:249.

109. Gupta, R., Sharma, K. K., and Kuhad, R. C. 2009. Separate hydrolysis and fermentation (SHF) of *Prosopis juliflora,* a woody substrate, for the production of cellulosic ethanol by *Saccharomyces cerevisiae* and *Pichia stipitis*-NCIM 3498. *Bioresource Technology* 100:1214.

110. Karaffa, L., and Kubicek, C. P. 2003. *Aspergillus niger* citric acid accumulation: Do we understand this well working black box? *Applied Microbiology and Biotechnology* 61:189.

111. Lawford, H. G., and Rousseau, J. D. 1996. Studies on nutrient requirements and cost-effective supplements for ethanol production by recombinant *E. coli. Applied Biochemistry and Biotechnology* 57–58:307.

112. Slininger, P. J. et al. 2006. Nitrogen source and mineral optimization enhance D-xylose conversion to ethanol by the yeast *Pichia stipitis* NRRL Y-7124. *Applied Microbiology and Biotechnology* 72:1285.

113. Martinez, A. et al. 2007. Low salt medium for lactate and ethanol production by recombinant *E. coli* B. *Biotechnology Letters* 29:397.

114. Miller, E. N., and Ingram, L. O. 2007. Combined effect of betaine and trehalose on osmotic tolerance of *Escherichia coli* in mineral salts medium. *Biotechnology Letters* 29:213.

115. Weimer, P. J. et al. 2005. In vitro gas production as a surrogate measure of the fermentability of cellulosic biomass to ethanol. *Applied Microbiology and Biotechnology* 67:52.

116. Ingledew, W. M. 1999. Alcohol production by *Saccharomyces cerevisiae:* A yeast primer. In *The alcohol textbook,* 3rd ed, ed. Jacques, K. A., Lyons, T. P., and Kelsall, D. R., chap. 5. Nottingham, England: Nottingham University Press.

117. Ingledew, W. M., and Casey, G. P. 1986. Rapid production of high concentrations of ethanol using unmodified industrial yeast. In *Biotechnology and renewable energy,* ed. Moo-Young, M., Hasnain, S., and Lamptey, J., 246. New York: Elsevier Applied Science Publishers.

118. Casey, G. P., and Ingledew, W. M. 1983. High gravity brewing: Influence of pitching rate and wort gravity on early yeast viability. *Journal of the American Society of Brewing Chemists* 41:148.

119. Casey, G. P., Magnus, C. A., and Ingledew, W. M. 1983. High gravity brewing: Nutrient enhanced production of high concentrations of ethanol by brewing yeasts. *Biotechnology Letters* 5:429.

120. Casey, G. P., Magnus, C. A., and Ingledew, W. M. 1984. High gravity brewing: Effects of nutrition on yeast composition, fermentative ability, and alcohol production. *Applied Microbiology and Biotechnology* 48:639.

121. Thomas, K. C., and Ingledew, W. M. 1990. Fuel alcohol production: Effects of free amino nitrogen on fermentation of very-high-gravity wheat mashes. *Applied and Environmental Microbiology* 56:2046.

122. Bafrncová, P. et al. 1999. Improvement of very high gravity ethanol fermentation by medium supplementation using *Saccharomyces cerevisiae. Biotechnology Letters* 21:337.

123. Jones, A. M., and Ingledew, W. M. 1994. Fuel alcohol production: Assessment of selected commercial proteases for very high gravity wheat mash fermentations. *Enzyme and Microbial Technology* 16:683.

124. Thomas, K. C., and Ingledew, W. M. 1992. Relationship of low lysine and high arginine concentrations to efficient ethanolic fermentation of wheat mash. *Canadian Journal of Microbiology* 38:626.

125. Wang, S. et al. 1999. Grain pearling and the very high gravity (VHG) fermentation technologies from rye and triticale. *Process Biochemistry* 34:421.

126. Thomas, K. C., Hynes, S. H., and Ingledew, W. M. 1994. Effects of particulate materials and osmoprotectants on very-high-gravity ethanolic fermentation by *Saccharomyces cerevisiae. Applied Environmental Biotechnology* 60:1519, 1994.

127. Reddy, L. V., and Reddy, O. V. 2005. Improvement of ethanol production in very high gravity fermentation by horse gram (*Dolichos biflorus*) flour supplementation. *Letters in Applied Microbiology* 41:440.

128. Wang, S. et al. 1999. Optimization of fermentation temperature and mash specific gravity for fuel ethanol production. *Cereal Chemistry* 76:82.

129. Dragone, G. et al. 2003. Improvement of the ethanol productivity in a high gravity brewing at pilot plant scale. *Biotechnology Letters* 25:1171.

130. Wang, F. Q. et al. 2007. Optimization of an ethanol production medium in very high gravity fermentation. *Biotechnology Letters* 29:233.

131. Alfenore, S. et al. 2002. Improving ethanol production and viability of *Saccharomyces cerevisiae* by a vitamin-feeding strategy during fed-batch process. *Applied Microbiology and Biotechnology* 60:67.

132. Barber, A. R., Henningsson, M., and Pamment, N. B. 2002. Acceleration of high gravity yeast fermentations by acetaldehyde addition. *Biotechnology Letters* 24:891.

133. Blieck, L. et al. 2007. Isolation and characterization of brewer's yeast variants with improved fermentation performance under high-gravity conditions. *Applied Microbiology and Biotechnology* 73:815.

134. Gibreel, A. et al. 2009. Fermentation of barley by using *Saccharomyces cerevisiae:* Examination of barley as a feedstock for bioethanol production and value-added products. *Applied and Environmental Microbiology* 75:1363.

135. Pham, T. K., and Wright, P. C. 2008. The proteomic response of *Saccharomyces cerevisiae* in very high glucose conditions with amino acid supplementation. *Journal of Proteome Research* 7:4766.

136. Bai, F. W. et al. 2008. Parameter oscillation attenuation and mechanism exploration for continuous VHG ethanol fermentation. *Biotechnology and Bioengineering* 102:113.

137. Schoberth, S. M. et al. 1996. Ethanol transport in *Zymomonas mobilis* measured by using in vivo nuclear magnetic resonance spin transfer. *Journal of Bacteriology* 178:1756.

138. Campbell, S. L. *The continuous brewing of beer.* Available as a download from the New Zealand Institute of Chemistry, www.nzic.org.nz/ChemProcesses/food.html.

139. Hornsey, I. S. 2003. *A history of beer and brewing,* chap. 9. Cambridge, England: The Royal Society of Chemistry.

140. Royston, M. G. 1966. Tower fermentation of yeast. *Process Biochemistry* 1:215.

141. Cysewski, G. R., and Wilke, C. R. 1977. Rapid ethanol fermentations using vacuum and cell recycle. *Biotechnology and Bioengineering* 19:1125.

142. Comberbach, D. M., Ghommidh, C., and Bu'Lock, J. D. 1986. Kinetic modeling and computer control of continuous alcohol fermentation in the gas-lift tower. In *Biotechnology and renewable energy,* ed. Moo-Young, M., Hasnain, S., and Lamptey, J., 208. New York: Elsevier Applied Science Publishers.

143. Margaritis, A., and Bajpai, P. 1982. Continuous ethanol production from Jerusalem artichoke tubers. I. Use of free cells of *Kluyveromyces marxianus. Biotechnology and Bioengineering* 24:1473.

144. Madson, P. W., and Monceaux, D. A. 1999. Fuel ethanol production. In *The alcohol textbook,* 3rd ed., ed. Jacques, K. A., Lyons, T. P., and Kelsall, D. R., chap. 17. Nottingham, England: Nottingham University Press.

145. Yarovenko, V. L. 1978. Theory and practice of continuous cultivation of microorganisms in industrial alcoholic processes. *Advances in Biochemical Engineering/Biotechnology* 9:1.

146. Bayrock, D., and Ingledew, W. M. 2001. Application of multistage continuous fermentation for production of fuel alcohol by very-high-gravity fermentation technology. *Journal of Industrial Microbiology and Biotechnology* 27:87.

147. Bai, F. W. et al. 2004. Continuous ethanol production and evaluation of yeast cell lysis and viability loss under very high gravity medium conditions. *Journal of Biotechnology* 10:287.
148. Laluce, C. et al. 2002. Continuous ethanol production in a nonconventional five-stage system operating with yeast cell recycling at elevated temperatures. *Journal of Industrial Microbiology and Biotechnology* 29:140.
149. Paiva, T. C. et al. 1996. Continuous alcoholic fermentation process in a tower reactor with recycling of flocculating yeast. *Applied Biochemistry and Biotechnology* 57–58:535.
150. Ben Chaabane, F. et al. 2006. Very high ethanol productivity in an innovative continuous two-stage bioreactor with cell recycle. *Bioprocess and Biosystems Engineering* 29:49.
151. Dale, M. C., Wankat, P. C., and Okos, M. R. 1987. Immobilized cell reactor-separator with simultaneous product separation and methods for design and use thereof. US Patent 4,665,027, May 12, 1987.
152. Dale, M. C. 1992. Method of use of a multistage reactor-separator with simultaneous product separation. US Patent 5,141,861, August 25, 1992.
153. Dale, M. C., and Okos, M. R. 2006. High speed, consecutive batch or continuous, low effluent process for the production of ethanol from molasses, starches, or sugars. US Patent 7,070,967, July 4, 2006.
154. Åkesson, M. et al. 1999. Online detection of acetate formation in *Escherichia coli* cultures using dissolved oxygen responses to feed transients. *Biotechnology and Bioengineering* 64:590.
155. Walsh, G. 2005. Biopharmaceuticals: Approvals and approval trends in 2004. *BioPharm International* 18(5):58.
156. Belo, I., Pinheiro, R., and Mota, M. 2003. Fed-batch cultivation of *Saccharomyces cerevisiae* in a hyperbaric bioreactor. *Biotechnology Progress* 19:665.
157. Cannizzaro, C., Valentinotti, S., and van Stockar, U. 2004. Control of yeast fed-batch process through regulation of extracellular ethanol concentration. *Bioprocess and Biosystems Engineering* 26:377.
158. Aldiguier, A. S. et al. 2004. Synergistic temperature and ethanol effect on *Saccharomyces cerevisiae* dynamic behaviour in ethanol bio-fuel production. *Bioprocess and Biosystems Engineering* 26:217.
159. Alfenore, S. et al. 2004. Aeration strategy: A need for very high ethanol performance in *Saccharomyces cerevisiae* fed-batch process. *Applied Microbiology and Biotechnology* 63:537.
160. Bideaux, C. et al. 2006. Minimization of glycerol production during the high-performance fed-batch ethanolic fermentation process in *Saccharomyces cerevisiae*, using a metabolic model as a prediction tool. *Applied and Environmental Microbiology* 72:2134.
161. Carvalho, J. C. M. et al. 2003. Ethanol production by *Saccharomyces cerevisiae* grown in sugarcane blackstrap molasses through a fed-batch process: Optimization by response surface methodology. *Applied Biochemistry and Biotechnology* 110:151.
162. Bravo, S., Mahn, A., and Shene, C. 2000. Effect of feeding strategy on *Zymomonas mobilis* CP4 fed-batch fermentations and mathematical modeling of the system. *Applied Microbiology and Biotechnology* 54:487.
163. Petersson, A., and Lidén, G. 2006. Fed-batch cultivation of *Saccharomyces cerevisiae*, using a metabolic model as a prediction tool. *Applied and Environmental Microbiology* 72:2134.
164. Odman, P. et al. 2009. On-line estimation of biomass, glucose and ethanol in *Saccharomyces cerevisiae* cultivations using in-situ multi-wavelength fluorescence and software sensors. *Journal of Biotechnology* 144:102.
165. Kelsall, D. R., and Lyons, T. P. 1999. Management of fermentations in the production of alcohol: moving toward 23% ethanol. In *The alcohol textbook,* 3rd ed., ed. Jacques, K. A., Lyons, T. P., and Kelsall, D. R., chap. 3. Nottingham, England: Nottingham University Press.

166. Margaritis, A., and Bajpai, P. 1982. Continuous ethanol production from Jerusalem artichoke tubers. II. Use of immobilized cells of *Kluyveromyces marxianus. Biotechnology and Bioengineering* 24:1483.
167. Moebus, O., and Teuber, M. 1982. Production of ethanol by solid particles of *Saccharomyces cerevisiae* in a fluidized bed. *Applied Microbiology and Biotechnology* 15:194.
168. Verbelen, P. J. et al. 2006. Immobilized yeast cell systems for continuous fermentation applications. *Biotechnology Letters* 28:1515.
169. Willaert, R. 2007. Cell immobilization and its applications in biotechnology: Current trends and future prospects. In *Fermentation microbiology and biotechnology,* ed. El-Mansi, E. M. T., Bryce, C. A., Demain, A. L., and Allman, A. R., chap. 10. Boca Raton, FL: CRC Press.
170. de Vasconcelos, J. N., Lopes, C. E., and de França, F. P. 2004. Continuous ethanol production using yeast immobilized on sugar-cane stalks. *Brazillian Journal of Chemical Engineering* 21:357.
171. Kunduru, M. R., and Pometto, A. L. 1996. Continuous ethanol production by *Zymomonas mobilis* and *Saccharomyces cerevisiae* in biofilm reactors. *Journal of Industrial Microbiology* 16:249.
172. Göksungur, Y., and Zorlu, N. 2001. Production of ethanol from beet molasses by Ca-alginate immobilized yeast cells in a packed-bed bioreactor. *Turkish Journal of Biology* 25:265.
173. Dey, S. 2002. Semicontinuous production of ethanol from agricultural wastes by immobilized coculture in a two-stage bioreactor. *Journal of Environmental Biology* 23:399.
174. Qureshi, N. et al. 2005. Continuous production of ethanol in high-productivity bioreactors using genetically engineered *Escherichia coli* Fbr5: Membrane and fixed cell reactors. *American Institute of Chemical Engineers* paper no. 589g (abstract only).
175. Najafpour, G., Younesi, H., and Syahidah Ku Ismail, K. 2004. Ethanol fermentation in an immobilized cell reactor using *Saccharomyces cerevisiae. Bioresource Technology* 92:251.
176. Yamada, T., Fatigati, M. A., and Zhang, M. 2002. Performance of immobilized *Zymomonas mobilis* 31821 (pZB5) on actual hydrolysates produced by Arkenol technology. *Applied Biochemistry and Biotechnology* 98–100:899.
177. Smogroicová, D., Dömény, Z., and Svitel, J. 2001. Modeling of saccharide utilization in primary beer fermentations with yeasts immobilized in calcium alginate. *Applied Biochemistry and Biotechnology* 94:147.
178. Liu, R., and Shen, F. 2008. Impacts of main factors on bioethanol fermentation from stalk juice of sweet sorghum by immobilized *Saccharomyces cerevisiae* (CICC 1308). *Bioresource Technology* 99:847.
179. Guo, X., Zhou, J., and Xiao, D. 2010. Improved ethanol production by mixed immobilized cells of *Kluyveromyces marxianus* and *Saccharomyces cerevisiae* from cheese whey powder solution fermentation. *Applied Biochemistry and Biotechnology* 160:532.
180. Liu, C. Z., Wang, F., and Ou-Yang, F. 2009. Ethanol fermentation in a magnetically fluidized bed reactor with immobilized *Saccharomyces cerevisiae* in magnetic particles. *Bioresource Technology* 100:878.
181. Koshinsky, H. A., Cosby, R. H., and Khachatourians, G. G. 1992. Effects of T-2 toxin on ethanol production by *Saccharomyces cerevisiae. Biotechnology and Applied Biochemistry* 16:275.
182. Talebnia, F., Niklasson, C., and Taherzadeh, M. J. 2005. Ethanol production from glucose and dilute-acid hydrolyzates by encapsulated *S. cerevisiae. Biotechnology and Bioengineering* 90:345.
183. Skinner-Nemec, K. A., Nichols, N. N., and Leathers, T. D. 2007. Biofilm formation by bacterial contaminants of fuel ethanol production. *Biotechnology Letters* 29:379.

184. Ingledew, W. M. 2003. Continuous fermentation in the fuel alcohol industry: How does the technology affect yeast? In *The alcohol textbook,* 4th ed., ed. Jacques, K. A., Lyons, T. P., and Kelsall, D. R., chap. 11. Nottingham, England: Nottingham University Press.

185. Narandranath, N. V., Thomas, K. C., and Ingledew, W. M. 2000. Urea hydrogen peroxide reduces the numbers of lactobacilli, nourishes yeast, and leaves no residues in the ethanol fermentation. *Applied and Environmental Microbiology* 66:4187.

186. Padukone, N. 1996. Advanced process options for bioethanol production. In *Handbook on bioethanol: Production and utilization,* ed. Wyman, C. E., chap. 14. London: Taylor & Francis.

187. Cooney, C. L. et al. 1978. Simultaneous cellulose hydrolysis and ethanol production by a cellulolytic anaerobic bacterium. *Biotechnology and Bioengineering Symposium Series* 8:103.

188. Wright, J. D. 1998. Ethanol from biomass by enzymatic hydrolysis. *Chemical Engineering Progress* August:62.

189. Philippidis, G. P. 1996. Cellulose bioconversion technology. In *Handbook on bioethanol: Production and utilization,* ed. Wyman, C. E., chap. 12. London: Taylor & Francis.

190. Linko, M., Viikari, L., and Suihko, M. L. 1984. Hydrolysis of xylan and fermentation of xylose to ethanol. *Biotechnology Advances* 2:233.

191. Singh, A., and Kumar, P. K. 1991. *Fusarium oxysporium:* Status in bioethanol production. *Critical Reviews in Biotechnology* 11:129.

192. Targonski, Z., Bujak, S., and Baraniak, A. 1985. Acid hydrolysis of beech sawdust hemicellulose and ethanol fermentation of hydrolysates by *Fusarium* sp. 27. *Acta Microbiologica Polonica* 34:261.

193. Panagiotou, G., and Christakopoulos, P. 2004. NADPH-dependent D-aldose reductases and xylose fermentation in *Fusarium oxysporium. Journal of Bioscience and Bioengineering* 97:299.

194. Xiros, C., and Christakopoulos, P. 2009. Enhanced ethanol production from brewer's spent grain by a *Fusarium oxysporum* consolidated system. *Biotechnology and Biofuels* 2:4.

195. Panagiotou, G. et al. 2006. Engineering of the redox imbalance of *Fusarium oxysporium* converting glucose to ethanol. *Metabolic Engineering* 8:474.

196. Panagiotou, G. et al. 2005. Intracellular metabolite profiling of *Fusarium oxysporium* enables anaerobic growth on xylose. *Journal of Biotechnology* 115:425.

197. Fujita, Y. et al. 2004. Synergistic saccharification, and direct fermentation to ethanol, of amorphous cellulose by use of an engineered yeast strain codisplaying three types of cellulolytic enzyme. *Applied and Environmental Microbiology* 70:1207.

198. van Rooyen, R. et al. 2005. Construction of cellobiose-growing and fermenting *Saccharomyces cerevisiae. Journal of Biotechnology* 120:284.

199. McBride, J. E. et al. 2005. Utilization of cellobiose by recombinant β-glucosidase-expressing strains of *Saccharomyces cerevisiae:* Characterization and evaluation of the sufficiency of expression. *Enzyme and Microbial Technology* 37:93.

200. Den Haan, R. et al. 2007. Hydrolysis and fermentation of amorphous cellulose by recombinant *Saccharomyces cerevisiae. Metabolic Engineering* 9:87.

201. Lynd, L. R. et al. 2005. Consolidated bioprocessing of cellulosic biomass: An update. *Current Opinion in Biotechnology* 16:577.

202. Den Haan, R. et al. 2007. Functional expression of cellobiohydrolases in *Saccharomyces cerevisiae* towards one-step conversion of cellulose to ethanol. *Enzyme and Microbial Technology* 40:1291.

203. Shaw, A. J., Hogsett, D. A., and Lynd, L. R. 2009. Identification of the [FeFe]-hydrogenase responsible for hydrogen generation in *Thermoanaerobacterium saccharolyticum* and demonstration of increased ethanol yield via hydrogenase knockout. *Journal of Bacteriology* 191:6457.

204. Miyazaki, K. et al. 2008. An ability of isolated strains to efficiently cooperate in ethanolic fermentation of agricultural plant refuse under initially aerobic thermophilic conditions: Oxygen deletion process appended to consolidated bioprocessing (CBP). *Bioresource Technology* 99:1768.

205. Ballesteros, I. et al. 1991. Selection of thermotolerant yeasts for simultaneous saccharification and fermentation (SSF) of cellulose to ethanol. *Applied Biochemistry and Biotechnology* 28–29:307.

206. Ballesteros, I. et al. 1993. Optimization of the simultaneous saccharification and fermentation process using thermotolerant yeasts. *Applied Biochemistry and Biotechnology* 39–40:201.

207. Golias, H. et al. 2002. Evaluation of a recombinant *Klebsiella oxytoca* strain for ethanol production from cellulose by simultaneous saccharification and fermentation: Comparison with native cellobiose-utilizing yeast strains and performance in co-culture with thermotolerant yeasts and *Zymomonas mobilis*. *Journal of Biotechnology* 96:155.

208. Ingram, L. O., and Doran, J. B. 1995. Conversion of cellulosic materials to ethanol. *FEMS Microbiology Reviews* 16:235.

209. Zhou, S., and Ingram, L. 1999. Engineering endoglucanase-secreting strains of ethanologenic *Klebsiella oxytoca* P2. *Journal of Industrial Microbiology and Biotechnology* 22:600.

210. Yang, B., and Wyman, C. E. 2006. BSA treatment to enhance enzymatic hydrolysis of cellulose in lignin containing substrates. *Biotechnology and Bioengineering* 94:611.

211. Wingren, A. et al. 2005. Effect of reduction in yeast and enzyme concentrations in a simultaneous-saccharification-and-fermentation-based bioethanol process: Technical and economic evaluation. *Applied Biochemistry and Biotechnology* 121–124:485.

212. Linde, M., Galbe, M., and Zacchi, G. 2007. Simultaneous saccharification and fermentation of steam-pretreated barley straw at low enzyme loadings and low yeast concentration. *Enzyme and Microbial Technology* 40:1100.

213. Schubert, C. 2006. Can biofuels finally take center stage? *Nature Biotechnology* 24:777.

214. Li, X. L. et al. 2004. Properties of a recombinant β-glucosidase from polycentric anaerobic fungus *Orpinomyces* PC-2 and its application for cellulose hydrolysis. *Applied Biochemistry and Biotechnology* 113–116:233.

215. Hughes, S. R. et al. 2006. High-throughput screening of cellulase F mutants from multiplexed plasmid sets using an automated plate assay on a functional proteomic robotic workcell. *Proteome Science* 4:10.

216. Li, X. L. et al. 2007. Expression of an AT-rich xylanase gene from the anaerobic fungus *Orpinomyces* sp. strain PC-2 in and secretion of the heterologous enzyme by *Hypocrea jecorina*. *Applied Microbiology and Biotechnology* 74:1264.

217. Murai, T. et al. 1998. Assimilation of cellooligosaccharides by a cell-surface engineered yeast expressing β-glucosidase and carboxymethylcellulase from *Aspergillus aculeatus*. *Applied and Environmental Microbiology* 64:4857.

218. Zhou, S. et al. 1999. Enhancement of expression and apparent secretion of *Erwinia chrysanthemi* endoglucanase (encoded by *celZ*) in *Escherichia coli* B. *Applied and Environmental Microbiology* 65:2439.

219. Cho, K. M., and Yoo, Y. J. 1999. Novel SSF process for ethanol production from microcrystalline cellulose using δ-integrated recombinant yeast, *Saccharomyces cerevisiae*. *Journal of Microbiology and Biotechnology* 9:340.

220. Zhou, S. et al. 2001. Gene integration and expression and extracellular secretion of *Erwinia chrysanthemi* endoglucanase CelY (*celY*) and CelZ (*celZ*) in ethanologenic *Klebsiella oxytoca* P2. *Applied and Environmental Microbiology* 67:6.

221. Fujita, Y. et al. 2002. Direct and efficient production of ethanol from cellulosic material with a yeast strain displaying cellulolytic enzymes. *Applied and Environmental Microbiology* 68:5136.

222. Wu, Z., and Lee, Y. Y. 1997. Inhibition of the enzymatic hydrolysis of cellulose by ethanol. *Biotechnology Letters* 19:977.

223. Martín, C. et al. 2006. Investigation of cellulose convertibility and ethanolic fermentation of sugarcane bagasse pretreated by wet oxidation and steam explosion. *Journal of Chemical Technology and Biotechnology* 81:1669.

224. Chadha, B. S. et al. 1995. Hybrid process for ethanol production from rice straw. *Acta Microbiologica Immunologica Hungarica* 42:53.

225. Chung, Y. C., Bakalinsky, A., and Penner, M. H. 1997. Analysis of biomass cellulose in simultaneous saccharification and fermentation processes. *Applied Biochemistry and Biotechnology* 66:249.

226. Belkacemi, K. et al. 1998. Ethanol production from AFEX-treated forages and agricultural residues. *Applied Biochemistry and Biotechnology* 70–72:441.

227. Hari Krishna, S., Janardhan Reddy, T., and Chowdary, G. V. 2001. Simultaneous saccharification and fermentation of lignocellulosic wastes to ethanol using a thermotolerant yeast. *Bioresource Technology* 77:193.

228. Schell, D. J. et al. 2003. Dilute-sulfuric acid pretreatment of corn stover in pilot-scale reactor: Investigation of yields, kinetics, and enzymatic digestibilities of solids. *Applied Biochemistry and Biotechnology* 105–108:69.

229. Varga, E. et al. 2004. High solid simultaneous saccharification and fermentation of wet oxidized corn stover to ethanol. *Biotechnology and Bioengineering* 88:567.

230. Kim, T. H., and Lee, Y. Y. 2005. Pretreatment of corn stover by soaking in aqueous ammonia. *Applied Biochemistry and Biotechnology* 121–124:1119.

231. Öhgren, K. et al. 2006. Simultaneous saccharification and co-fermentation of glucose and xylose in steam-pretreated corn stover at high fiber content with *Saccharomyces cerevisiae* TMB3400. *Journal of Biotechnology* 126:488.

232. Kim, T. H. et al. 2006. Pretreatment of corn stover by low-liquid ammonia percolation process. *Applied Biochemistry and Biotechnology* 133:41.

233. Rebroš, M. et al. 2008. Ethanol production from starch hydrolyzates using *Zymomonas mobilis* and glucoamylase entrapped in polyvinylalcohol hydrogel. *Applied Biochemistry and Biotechnology* 158:561.

234. Yu. J. et al. 2008. Ethanol production from H_2SO_3-steam-pretreated fresh sweet sorghum stem by simultaneous saccharification and fermentation. *Applied Biochemistry and Biotechnology* 160:401.

235. Bertilsson, M., Olofsson, K., and Lidén, G. 2009. Prefermentation improves xylose utilization in simultaneous saccharification and co-fermentation of pretreated spruce. *Biotechnology and Biofuels* 2:8.

236. Shimokawa, T. et al. 2009. Effects of growth stage on enzymatic saccharification and simultaneous saccharification and fermentation of bamboo shoots for bioethanol production. *Bioresource Technology* 100:6651.

237. Rattanachomsri, U. et al. 2009. Simultaneous nonthermal saccharification of cassava pulp by multi-enzyme activity and ethanol fermentation by *Candida tropicalis*. *Journal of Bioscience and Bioengineering* 107:488.

238. Zhang, Y. et al. 2006. Protoplast fusion between *Geotrichum candidium* and *Phanerochaete chrysosporium* to produce fusants for corn stover fermentation. *Biotechnology Letters* 28:1351.

239. Skotnicki, M. L. et al. 1983. High-productivity alcohol fermentations using *Zymomonas mobilis*. *Biochemical Society Symposia* 48:53.

240. Gorkhale, D., and Deobagkar, D. 1994. Isolation of intergeneric hybrids between *Bacillus subtilis* and *Zymomonas mobilis* and the production of thermostable amylase by hybrids. *Biotechnology and Applied Biochemistry* 20:109.

241. Ma, Y.-J. et al. 2000. Efficient utilization of starch by a recombinant strain of *Saccharomyces cerevisiae* producing glucoamylase and isoamylase. *Biotechnology and Applied Biochemistry* 31:55.

242. Ramakrishnan, S., and Hartley, B. S. 1993. Fermentation of lactose by yeast cells secreting recombinant fungal lactase. *Applied and Environmental Microbiology* 59:4230.

243. Palnitkar, S. S., and Lachke, A. H. 1990. Efficient simultaneous saccharification and fermentation of agricultural residues by *Saccharomyces cerevisiae* and *Candida shehatae,* the D-xylose fermenting yeast. *Applied Biochemistry and Biotechnology* 26:151.

244. Lebeau, T., Jouenne, T., and Junter, G. A. 1998. Continuous alcoholic fermentation of glucose/xylose mixtures by co-immobilized *Saccharomyces* and *Candida shehatae.* *Applied Microbiology and Biotechnology* 50:309.

245. Kordowska-Wister, M., and Targonski, Z. 2002. Ethanol fermentation on glucose/ xylose mixture by co-cultivation of restricted glucose catabolite repressed mutants of *Pichia stipitis* with respiratory deficient mutants of *Saccharomyces cerevisiae.* *Acta Microbiologica Polonica* 51:345.

246. Farid, M. A., El-Enshasy, H. A., and El-Deen, A. M. 2002. Alcohol production from starch by mixed cultures of *Aspergillus awamori* and immobilized *Saccharomyces cerevisiae* at different agitation speeds. *Journal of Basic Microbiology* 42:162.

247. Qian, M. et al. 2006. Ethanol production from dilute-acid softwood hydrolysate by co-culture. *Applied Microbiology and Biotechnology* 134:273.

248. Lebeau, T., Jouenne, T., and Junter, G. A. 2007. Long-term incomplete xylose fermentation, after glucose exhaustion, with *Candida shehatae* co-immobilized with *Saccharomyces cerevisiae.* *Microbiological Research* 162:211.

249. Chi, Z. et al. 2009. *Saccharomycopsis fibuligera* and its applications in biotechnology. *Biotechnology Advances* 27:423.

250. Madson, P. W. 2003. Ethanol distillation: The fundamentals. In *The alcohol textbook,* 4th ed., ed. Jacques, K. A., Lyons, T. P., and Kelsall, D. R., chap. 22. Nottingham, England: Nottingham University Press.

251. Singh, A., and Mishra, P. 1995. Microbial pentose utilization. *Current applications in biotechnology* (*Progress in Industrial Microbiology,* vol. 33), chap. 14.

252. Swain, R. L. B. 2003. Development and operation of the molecular sieve: An industry standard. In *The alcohol textbook,* 4th ed., ed. Jacques, K. A., Lyons, T. P., and Kelsall, D. R., chap. 23. Nottingham, England: Nottingham University Press.

253. Essien, D., and Pyle, D. L. 1983. Energy conservation in ethanol production by fermentation. *Process Biochemistry* 18:31.

254. Sikyta, B. 1995. *Techniques in applied microbiology.* (*Progress in industrial microbiology,* vol. 31), chap. 12. New York: Elsevier Science & Technology.

255. Ghose, T. K., Roychoudhury, P. K., and Ghosh, P. 1984. Simultaneous saccharification and fermentation (SSF) of lignocellulosics to ethanol under vacuum cycling and step feeding. *Biotechnology and Bioengineering* 26:377.

256. Minier, M., and Goma, G. 1982. Ethanol production by extractive fermentation. *Biotechnology and Bioengineering* 24:1565.

257. Pitt, W. W., Haag, G. L., and Lee, D. D. 1983. Recovery of ethanol from fermentation broths using selective sorption-desorption. *Biotechnology and Bioengineering* 25:123.

258. Einicke, W. D., Gläser, B., and Schöoullner, R. 1991. In-situ recovery of ethanol from fermentation broth by hydrophobic adsorbents. *Acta Biotechnologica* 11:353.

259. Nishizawa, Y. et al. 1983. Ethanol production by cell recycling with hollow fibers. *Journal of Fermentation Technology* 61:599.

260. O'Brien, D. J. et al. 2004. Ethanol recovery from corn fiber hydrolysate fermentations by pervaporation. *Bioresource Technology* 92:15.

261. Li, G. et al. 2007. Time-dependence of pervaporation performance for the separation of ethanol/water mixtures through poly(vinyl alcohol) membrane. *Journal of Colloid Interface Science* 306:337.
262. Dominguez, J. M. et al. 2000. Ethanol production from xylose with the yeast *Pichia stipitis* and simultaneous product recovery by gas stripping using a gas-lift fermentor with attached side-arm (GLSA). *Biotechnology and Bioengineering* 67:336.
263. Taylor, F. et al. 1995. Continuous fermentation and stripping of ethanol. *Biotechnology Progress* 11:693.
264. Taylor, F. et al. 1997. Effects of ethanol concentration and stripping temperature on continuous fermentation rate. *Applied Microbiology and Biotechnology* 48:311.
265. Taylor, F. et al. 2000. Dry-grind process for fuel ethanol by continuous fermentation and stripping. *Biotechnology Progress* 16:541.
266. Mazumdar-Shaw, K., and Suryanarayan, S. 2003. Commercialization of a novel fermentation concept. *Advances in Biochemical Engineering/Biotechnology* 85:29.
267. Gibbons, W. R., Westby, C. A., and Dobbs, T. L. 1984. A continuous, farm-scale, solid-phase fermentation process for fuel ethanol and protein feed production from fodder beets. *Biotechnology and Bioengineering* 26:1098.
268. Gibbons, W. R., Westby, C. A., and Dobbs, T. L. 1986. Intermediate-scale, semicontinuous solid-phase fermentation process for fuel ethanol from sweet sorghum. *Applied and Environmental Microbiology* 51:115.
269. Jacques, K. A. 2003. Ethanol production and the modern livestock feed industry: A relationship continuing to grow. In *The alcohol textbook,* 4th ed., ed. Jacques, K. A., Lyons, T. P., and Kelsall, D. R., chap. 27. Nottingham, England: Nottingham University Press.
270. Rausch, K. D., and Belyea, R. L. 2006. The future of coproducts from corn processing. *Applied Biochemistry and Biotechnology* 128:47.
271. Dawson, K. A. 2003. Biorefineries: The versatile fermentation plants of the future. In *The alcohol textbook,* 4th ed., ed. Jacques, K. A., Lyons, T. P., and Kelsall, D. R., chap. 28. Nottingham, England: Nottingham University Press.
272. Tucker, M. P. et al. 2004. Conversion of distiller's grain into fuel alcohol and a higher value animal feed by dilute-acid pretreatment. *Applied Biochemistry and Biotechnology* 113–116:1139.
273. Leiper, K. A. et al. 2006. The fermentation of beet sugar syrup to produce bioethanol. *Journal of the Institute of Brewing* 112:122.
274. Sørensen, H. R., Meyer, A. S., and Pederson, S. 2003. Enzymatic hydrolysis of water-soluble wheat arabinoxylan. I. Synergy between α-L-arabinofuranosidases, endo-1, 4-β-xylanases, and β-xylosidase activities. *Biotechnology and Bioengineering* 81:726.
275. Sørensen, H. R. et al. 2005. Efficiencies of designed enzyme combinations in releasing arabinose and xylose from wheat arabinoxylan in an industrial ethanol fermentation residue. *Enzyme and Microbial Technology* 36:773.
276. Sørensen, H. R., Pederson, S., and Meyer, A. S. 2006. Optimization reaction conditions for enzymatic viscosity reduction and hydrolysis of wheat arabinoxylan in an industrial ethanol fermentation residue. *Biotechnology Progress* 22:505.
277. Sørensen, H. R., Pederson, S., and Meyer, A. S. 2007. Synergistic enzyme mechanisms and effects of sequential enzyme additions on degradation of water insoluble wheat arabinoxylan. *Enzyme and Microbial Technology* 40:908.
278. van Haandel, A. C. 2005. Integrated energy production and reduction of the environmental impact at alcohol distillery plants. *Water Science Technology* 52:49.
279. Schmer, M. R. et al. 2008. Net energy of cellulosic ethanol from switchgrass. *Proceedings of the National Academy of Sciences USA* 105:464.

6 The Economics of Fuel Ethanol

6.1 INTRODUCTION

The twentieth century witnessed cycles of waxing and waning interest in and funding for biofuels programs, most of which were wrecked by economic arguments about the high estimated price of any biofuel relative to the actual prices of gasoline and diesel. When the projected costs of biofuels were 10 times the current costs of conventional fuels, R&D programs were essentially blue sky investigations. Toward the end of the first decade of the twenty-first century, that differential was greatly eroded by technical developments and by global trends in world oil prices. Can the economic status of biofuels production be accurately defined—in particular, with ethanol biomanufactured from sugarcane, corn, and (eventually) plant biomass? Is the case for fuel ethanol still dependent on tax incentives and special pleading by some environmentalists and—most importantly—by politicians concerned by global warming or energy security?

Can fermentation-based production routes for liquid fuels ever compete in the market place with the established and global infrastructure of crude oil, gasoline, and diesel? Are test flights for aviation biofuels and heavily publicized events with racing cars fueled with ethanol anything more than stunts or propaganda?

Massive price inflation in oil and motor fuels after 2000 certainly eroded much of the traditional price-based arguments against biofuels but, after the market turmoil of 2007 and 2008, has the situation really changed? If science is replaced by speculation and all forecasts are more or less educated guesses, can the case for continued R&D in biofuels be justified?

6.2 MARKET FORCES AND INCENTIVES

6.2.1 The Impact of Oil Prices on the Future of Biofuels after 1980

Economists attract ridicule and resentment in equal measures.[1]

The most telling aspect of this quote is not that it derives from a collection of essays originally published in *The Economist,* one of the leading opinion formers in Western liberal economic thought, but rather that it is the *first* sentence in the introduction to that volume. Graphical representations from many economic sources share one obvious common factor: a short time axis. In the world of practical economics, hours, days, weeks, and months dominate the art of telling the near future—for price movements in stock markets, in the profitability of major corporations and their mergers, in the collapse of currencies, or in surges in commodity prices. Projections are usually linear extrapolations from small historical databases; predictions may very soon be completely forgotten. Economic models may, with hindsight, appear optimistic or wildly inaccurate but, by the time hindsight is possible, the original set of parameters may have become irrelevant.

The history of biofuels since the early 1970s exhibits such cycles of optimism and pessimism, of exaggerated claims or dire prognostications; a series of funding programs have blossomed but—sometimes equally rapidly—faded.[2] The prime mover in that sequence has invariably been the market price of oil, and however undesirable a driver this is in the ongoing discussions on the development of biofuels from the viewpoint of the scientific research community, it should never be ignored.[3] A high cost of any biofuel relative to that of gasoline, diesel fuel, and heating oil is the main plank in the logic used by skeptics: that however worthy are the goal and vision of biofuels for the future, they simply cannot be afforded and—in a sophisticated twist of the argument—may themselves contribute to the continuing deprivation of energy-poor nations and societies while the energy-rich developed economies impose rationing of fossil fuel use and access to maintain their privileged position.

Lobbyists for the global oil industry clearly have a vested interest in continuously challenging the economic costs of biofuel production. However, an underlying fear is that—whether significant climate change could be lessened by the adoption of biofuels for private transportation and whether "energy security" is simply a novel means of subsidizing inefficient farmers to grow increasingly larger harvests of monoculture crops to maintain agricultural incomes and/or employment for a few decades more—only a clear understanding of the financial implications of biofuels can help fix the agenda for rational choices to be made about investing in new technologies across the wide spectrum of rival possible biofuel options in the twenty-first century.

Moreover, it is undeniable that oil price volatility can shake the confidence of any investor in bioenergy. In the two decades after 1983, the average retail price for gasoline (averaged over all available grades) was $0.83 per gallon, but transient peaks and troughs reached $1.29 and $0.55, respectively (Figure 6.1). Slumping oil prices almost wiped out the young sugar-ethanol-fueled car fleet in Brazil in the 1980s (Chapter 1, Section 1.4). After 2000, the conclusion that the era of cheap oil was irreversibly over was accepted but has been partially refuted after the price collapse of 2008; a more persistent concern is that the demands of the burgeoning economies of India and China will place unavoidable stresses on oil availability whatever the price of crude oil may be.[4]

6.2.2 Production Price, Taxation, and Incentives in the Market Economy

It is vital at this point to differentiate commercial realities from strategic (or geopolitical) and all other considerations. Although historical, environmental, and political arguments have all been adduced in support of bioenergy programs (as discussed in Chapter 1), fiscal considerations now play an important role in encouraging the take-up of novel alternative fuels and in partitioning the market for first-generation

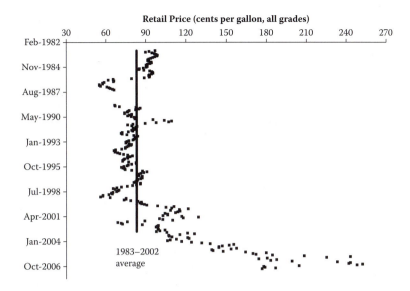

FIGURE 6.1   U.S. gasoline retail prices, 1983–2007: total sales by all sellers, incorporating prices for all available grades. (U.S. Department of Energy data.)

ethanol and biodiesel and subsequent generations of rival (but not all equally readily commercialized) biofuels.

Indeed, taxation issues were quickly recognized and seized on by proponents of cellulosic ethanol, particularly because they were useful to counter the gasoline versus gasohol price differential. For example, in the United States, the indirect costs of regulating air pollution and of military protection for oil supplies from the Middle East are calculable; they greatly inflate the nominal price of crude oil but are not (explicitly) passed on to the consumer.[5] This distortion of the transportation fuel market by hidden subsidies has also led to economics models in which other indirect factors are included in the cost-benefit analysis:

- technological developments that improve the national scientific base for employment, patents, and overseas licensing, and engineering advances that spill over into related fields
- reduced foreign currency payments and associated banking costs highly important for a developing economy such as Brazil's
- higher income and sales tax returns from greater rural employment
- reduced longer-term economic impacts of climate change and air pollution

To varying degrees, all these arguments are contentious; skeptics can be found from opposite ends of the economic spectrum, including both oil industry analysts and academics who foresee only accelerated land degradation from the industrial agronomy of bioenergy crop cultivation.[6,7]

Moreover, taxation as an instrument of social and economic policy has obvious limitations if wasteful subsidies or punitive levels of taxation on standard gasoline and diesel products are to be avoided. Consider the following three scenarios:

1. Fuel ethanol production can generate a commercial fuel with pump prices no greater than those of standard gasoline grades at equivalent tax rates. The comparison is valid when average prices over a period of 1–5 years are calculated, thus avoiding false comparisons at peaks and troughs caused by fluctuations in both agricultural feedstock prices (as an important cost input to biofuel production) and oil price movements if they continue to move inside the wide limits evident since the early 1980s (Figure 6.1).
2. Fuel ethanol can be produced commercially at a total (production, distribution, and resale) cost averaged over a 5- to 10-year period that is 10–50% higher than that of gasoline.
3. Fuel ethanol production can only generate an unsubsidized product with a total cost more than twice that of the refinery gate price of standard gasoline (a price differential quoted for the United States in the late 1990s[8])—or perhaps even up to 10 times higher than conventional fuels where, for example, local conditions of climate and biomass availability are consistently much less favorable than for sugarcane production in Brazil or corn in the United States or where only refractory lignocellulosic feedstocks can be accessed with poorly developed bioprocessing technology.

The first (optimistic) case approximates that of Brazilian consumers with flexibly fueled cars after 2000.[9] The second case is the conclusion most often widely reached in techno-economic studies. The third scenario is parallel to the emergency or wartime case discussed when the energy yields of conventional and alternative fuels were considered in Chapter 1: Even if biofuels are prohibitively expensive *now*, technical developments may erode that differential or be obviated if fossil fuel shortages become acute. In all three cases, taxation policy can be (and has been) an influence on consumer choice and purchasing patterns, whether for short-term (tactical) or longer term (strategic) reasons and when legislation enforces alternative or reconstituted fuels to achieve environmental targets.

A snapshot of data from October 2002 in Brazil, however, revealed the complexity of the interaction between production/distribution costs and imposed taxation on the final at-pump selling price.[10] Although gasohol mixtures, hydrous ethanol, and diesel all had very similar production costs—equivalent to approximately $0.15 per liter ($0.57 per gallon) at that time, the final cost to the user was determined by the much higher taxes applied to gasohol (Figure 6.2). Brazil exemplified the extensive use of taxation to determine and direct the perceived prices of gasoline and alternative fuels as a deliberate instrument of national policy.

Such deliberate management of the fuel economy is likely to be instigated in societies where not only economics but also social and environmental considerations are taken into account. However, it runs the risk of experiencing budgetary shortfalls if the total tax raised is severely reduced when the policy is too successful in achieving

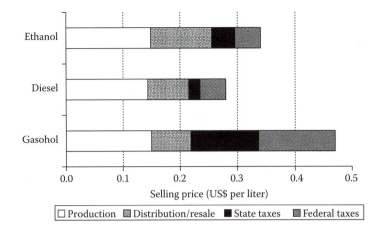

FIGURE 6.2 Production and distribution costs and taxation for motor fuels in Brazil in 2002. (Data from Moreira, J. R., Noguiera, L. A. H., and Parente, V. 2005. In *Growing in the Greenhouse: Protecting the Climate by Putting Development First,* ed. Bradley, R. and Baumert, K. A., Chap. 3. Washington, D.C.: World Resources Institute.)

its aims. This becomes even worse if private transportation is perceived as being subsidized by other taxation sources—for example, sales and income tax. Therefore, for all the various interest groups in biofuels development, the priority is to establish viable production processes with the minimum requirement for tax incentives.

6.3 COST MODELS FOR FUEL ETHANOL PRODUCTION

Economic considerations have been featured in both primary analyses and reviews of the biotechnology of fuel ethanol production published in the last 25 years.[11] Because a cellulosic ethanol industry has yet to mature fully, most of those studies have been derived from laboratory or (at best) small pilot plant data. Estimates for feedstock and capital investment costs have varied greatly, as have assumptions on the scale of commercial production required to achieve any intended price/cost target for the product. As with estimates of net energy yield and greenhouse gas reductions, the conclusions reached are heavily influenced by the extent to which costs can be offset by coproduct generation (as a source of income) and the complexity of the total production process—not only as a primary cause for increased setup costs but also as a potential source of process efficiencies and additional, saleable coproducts.

Few of the influential studies are full business models for fuel ethanol—in particular avoiding any computations for profitability—often because the main driver has been to establish and substantiate grounds for initial or continued investment by national and/or international funding agencies. The implicit assumption has been that any production process for fuel ethanol outside Brazil suffers by that very comparison because of the lack of such favorable climatic and economic features (in particular, land use, labor cost, and the dovetailing of ethanol production with a fully mature sugarcane industry). Nevertheless, a historical survey of key points in the development of the economic case for sugarcane, corn, and cellulosic ethanol reveals the convergence toward a set of key parameters that will be crucial for any biofuel candidate in the next 10–50 years.

6.3.1 EARLY BENCHMARKING STUDIES OF CORN AND LIGNOCELLULOSIC ETHANOL IN THE UNITED STATES

During the 1970s, the U.S. Department of Energy (DOE) commissioned four detailed technical and economic reports on possible production routes for ethanol as a fuel supplement[12]:

1. corn-based manufacturing facilities at scales from 10 million up to 100 million gallons per year
2. wheat-straw conversion via enzymic hydrolysis at a 25 million-gallon-per-year scale of production
3. another intermediate scale process for molasses fermentation to produce 14 million gallons per year
4. a farm-based model (25 gallons per hour)

6.3.1.1 Corn-Derived Ethanol in 1978

The assessment of corn-derived ethanol was the most extensive of the reports (60% of the total printed pages in the final collection of papers) and formed a notional blueprint for a facility sited in Illinois with a projected working life of 20 years and operating costs of approximately $0.95 per gallon of hydrous ethanol (Table 6.1). The final factory gate selling price was computed to be $1.05 per gallon (1978 prices, equivalent to $3.45 in 2009) in the base case of the 50 million-gallon-per-year capacity, including the results of a 15% discounted cash flow/interest rate of return analysis.

The selling price was a little lower ($0.98 per gallon), with twice the annual capacity, but considerably higher ($1.55 per gallon) at only 10 million gallons per year. The quoted comparative price for refinery gasoline was $0.40 per gallon; after allowing for the lower energy content of ethanol (70% of that of gasoline), the real cost of corn-derived ethanol would have been $1.50 per gallon for the 50 million-gallon facility (i.e., 3.75-fold higher than gasoline at that time).

Various options were explored in the study to define the sensitivity of the required selling price for ethanol:

- The DOE required the analyses to define a selling price that would not only cover the annual operating expenses but also yield a return on equity; the base case is a 15% discounted cash flow/interest rate of return. Increasing this factor to 20% resulted in a higher selling price ($1.16 per gallon for the base case scenario).
- Lengthening the depreciation schedule from 10 years to 20 years increased the selling price by $0.02 per gallon.
- Increasing the working capital to 20% of the total production cost increased the selling price by $0.03 per gallon.

TABLE 6.1
Cost Estimates for Ethanol Production from Corn Grain

Manufacturing Input	Annual Costs ($ Million)	Production Cost (Cents/Gallon)
Raw Materials		
Corn	44.77	89.5
Yeast	0.32	0.6
Ammonia	0.37	0.7
Coal	2.41	4.8
Other chemicals	0.18	0.4
Utilities		
Cooling water (from plant)	0.00	0.0
Steam (from plant)	0.00	0.0
Electricity	1.65	3.3
Diesel fuel	0.01	0.0
Labor		
Management	0.24	0.5
Supervisors/operators	2.19	4.4
Office and laborers	1.20	2.4
Fixed Charges		
Depreciation	5.80	11.6
License fees	0.03	0.1
Maintenance	1.83	3.7
Tax and insurance	0.91	1.8
Miscellaneous		
Freight	2.50	5.0
Sales	1.93	3.9
General/administrative overheads	0.64	1.3
Coproduct Credits		
Dark grains	19.18	38.4
Ammonium sulfate	0.41	0.8
Total	**47.41**	**94.8**

Source: Data from Paul, J. K., ed. 1980. *Large and Small Scale Ethyl Alcohol Processes from Agricultural Raw Materials.* Park Ridge, NJ: Noyes Data Corporation.

- A higher investment tax credit (50%) would reduce the selling price by $0.02 per gallon.
- Financing only 80% of the plant investment could reduce the selling price by $0.10 per gallon.
- For every 10% rise in the price of corn, the selling price would increase by $0.08 per gallon (after allowing for a triggered rise in the selling price of the solid animal feed coproduct).
- For every 10% rise in the price of the animal feed coproduct, the selling price would decrease by $0.04 per gallon.
- Replacing local coal by corn stover as the fuel for steam generation would increase the selling price by $0.04 per gallon—although a lower total investment (by approximately $1 million) would have been an advantage resulting from the removal of the need for flue gas desulfurization.

All of these changes are comparatively minor, and other quantified changes to the overall process were likely to have been equally small. Ammonium sulfate (a coproduct arising from flue gas desulfurization) was only generated in small amounts (approximately 3 tons per day), and no allowance was made for capturing and selling the CO_2 generated in the fermentation step. No denaturant was included in the final cost breakdown.

Alternative feedstocks were also explored. Milo (grain sorghum) offered a slight reduction in the selling price of ethanol (down to $1.02 per gallon) but was considered a small-acreage crop at that time. Both wheat and sweet sorghum were likely to increase the final factory gate selling price to $1.31 and $1.40 per gallon, respectively. Although wheat and milo grain could be processed in essentially the same equipment used for corn, sweet sorghum required a higher investment in plant facilities.

6.3.1.2 Wheat-Straw-Derived Ethanol in 1978

The first detailed economic costing of a lignocellulosic ethanol process envisaged a process similar in outline to that eventually adopted by the Iogen Corporation in Canada (see Chapter 5), where wheat straw was pretreated with acid before fungal cellulase was used to digest the cellulose for a *Saccharomyces* fermentation of the liberated glucose (no pentose sugars were included as substrates at that date). Results from laboratory studies were extrapolated to a 25 million-gallon-per-year facility that was designed to be stand-alone and capable of generating the required fungal cellulase on site (Table 6.2).

The calculated factory gate plant was much higher than that estimated for corn-derived ethanol: $3.34 per gallon (equivalent to $11.32 in 2009). This cost included raw materials, utilities, maintenance materials and labor, operating labor and supplies, the facility laboratory, plant overhead, taxes, insurance, and depreciation. However, it made no estimates for general and administrative, sales, or research costs; profit; and any by-product credits or disposal charges. Of this total, 61% was attributed to materials, of which the wheat straw feedstock accounted for 11 or 12%.

TABLE 6.2
Cost Estimates for Ethanol Production from Wheat Straw

Manufacturing Input	Production Cost (Cents/Gallon)	Cost (% of Total)
Raw Materials		
Wheat straw	38.8	11.6
Cellulose, newsprint	55.9	16.7
Peptone	79.2	23.7
Other chemicals	20.3	6.1
Utilities		
Cooling water	1.0	0.3
Process water	2.8	0.8
Steam	30.2	9.0
Electricity	14.2	4.2
Variable Cost		
Operating labor	9.4	2.8
Maintenance labor	8.6	2.6
Control laboratory	1.9	0.6
Maintenance material	8.6	2.6
Operating supplies	0.9	0.3
Direct Cost		
Plant overhead	15.9	4.8
Taxes and insurance	7.7	2.3
Depreciation	38.7	11.6
Total	**334.2**	**100**

Source: Data from Paul, J. K., ed. 1980. *Large and Small Scale Ethyl Alcohol Processes from Agricultural Raw Materials.* Park Ridge, NJ: Noyes Data Corporation.

The other components of the price estimate and various options for reducing the total were assessed:

- Utilities and capital costs were each 14% of the total.
- Labor costs were 11% of the total.
- A high-cost peptone (proteased protein) nitrogen source was included in the fermentation media; adopting a lower price product could reduce the cost of the product by 9%.
- The conversion of cellulose to soluble sugars was 45% (w/w basis, calculated on the straw weight); increasing this to 60% could effect a 6% reduction in product cost.
- Reducing the enzyme loading (ratio of enzyme to cellulose) by fourfold could reduce the product price to approximately $2 per gallon.

No allowance is made in Table 6.2 for capital costs or profit. On the basis of 15% per year of fixed capital for general and administrative, sales, and research costs; a 15-year life expectancy of the plant; and a 48% tax rate, the selling price for the 95% ethanol product would have to have been $4.90 or $4.50 per gallon for a discounted cash flow return on investment (after taxes) or 15 or 10%, respectively.[12]

An independent estimate for ethanol production from corn stover was also funded by the DOE. The conclusion was that, considering scaling up an advanced process option with cell recycling and converting the postdistillation stillage into methane by anaerobic fermentation to generate a combustible source of steam generation, a total production cost of $3.38 per gallon could be estimated—again with no allowance for plant profitability at a 14 million-gallon-per-year scale of production.[13]

The main conclusion was therefore that the costs of generating fermentable sugars by enzymic hydrolysis dominated production economics; consequently, any major reduction in the cost of generating the cellulase or increase in its specific activity (or stability or ease of recovery for recycling) would be highly effective in lowering the production costs. Although federal and state initiatives in operation by 1979 offered some mitigation of the high costs of biomass-derived ethanol, the sums were small. For example, under the National Energy Act of 1978, alcohol fuels were eligible for DOE entitlements worth $0.05 per gallon, and 16 states had reduced or eliminated entirely state gasoline taxes on gasohol mixtures; the largest amount (worth $0.095 per gallon) was in Arkansas.

6.3.1.3   Fuel Ethanol from Sugarcane Molasses

A 14.3 million-gallon-per-year facility for 95% ethanol production from sugarcane molasses was technically the easiest production process to design and cost in the 1970s (Table 6.3). To simplify even further, although coproducts (CO_2 and fusel oils) were considered, no economic calculations were made for these; however, sale of the yeasts grown in the fermentations was included to effect a cost reduction of approximately 6.5%. The final calculated manufacturing price of 95% ethanol was $0.995 per gallon (equivalent to $3.37 in 2009)—"a cost which is today very comparable to producing alcohol from ethylene."[12]

Of the production costs, the raw material molasses was the dominating factor, accounting for 63% of the total. No exact geographical location for the hypothetical facility was given, but the quoted molasses price ($50 per ton) was the 1978 summer average of molasses delivered to the East Coast and Midwest of the United States.

6.3.1.4   Farm-Scale Ethanol Production

How small can a viable rural ethanol production site be? The final section of the 1978–1980 collection of case studies projected fuel ethanol production from corn on a family-run farm in the Midwest. The cost calculations were very different, however, because there were already a working market for fuel alcohol and a known price ($1.74 per gallon in Iowa, November 1979). Using this figure, a net operating profit of total revenues could be projected for the first year of operation, ignoring factors such as finished goods and work-in-process inventories that would reduce the actual production cost of ethanol (Table 6.4).

TABLE 6.3

Cost Estimates for Ethanol Production from Molasses

Manufacturing Input	Annual Cost ($ × 10³)	Cost (% of Total)
Molasses[a]	9,100	63.2
Other materials	100	0.7
Power	44	0.3
Steam	315	2.2
Water	30	0.2
Labor	770	5.3
Administration	400	2.8
Interest	1,178	8.2
Depreciation	1,405	9.8
Maintenance	662	4.6
Taxes and insurance	397	2.8
Total	**14,401**	**100**
Ethanol sales (gallons)	13,538	
Yeast sales	942	
Ethanol production cost[b] ($/gallon)	0.994	

Source: Data from Paul, J. K., ed. 1980. *Large and Small Scale Ethyl Alcohol Processes from Agricultural Raw Materials.* Park Ridge, NJ: Noyes Data Corporation.

[a] 182,000 tons per year.

[b] After allowing for sales of the yeast coproduct.

The operating profit, however, was entirely represented by sales of the fermentation stillage delivered (by truck) to neighbors within 5 miles of the farm. The selling price of such stillage would be much depressed if a large brewery or distiller were located nearby; if no net income could be generated by these means, then the facility would run at a loss. The capture of CO_2 from the fermentation was not considered because the capital cost of the equipment was too high to give a good return on the investment. A localized, small-scale production of fuel ethanol could therefore provide all the fuel requirements for running a farm's gasoline-consuming operations and provide a reasonable financial return as a commercial venture—but only if the agrobusiness were run as an early example of a biorefinery (see Chapter 10) and produced not only ethanol but also a saleable fermentation- and corn-derived coproduct.

Published in 1982, a second survey of technology and economics for farm-scale ethanol production at >100 gallons per hour (or up to 1 million gallons per year) estimated a total annual cost of $1.97 per gallon as a breakeven figure. [14] The technical aspects of the process had been investigated in a facility with fermentation vessels of up to 5,750 L capacity. The projected price included a $0.41-per-gallon sales income from the wet grain coproduct, annual amortized capital cost, operating costs, and fixed costs (including insurance, maintenance, and property taxes). Also designed for farm-scale use, a fermentation process using sweet sorghum as the feedstock and

TABLE 6.4

Cost Estimates for Farm-Scale Ethanol Production from Corn

Input	Annual Cost ($ × 10³)	Cost (% of Total)
Corn[a]	138	51.1
Enzymes	24	8.8
Electricity	3	1.1
Straw	11	4.0
Miscellaneous	18	6.7
Labor	21	7.7
Interest	24	8.8
Depreciation	7	2.5
Sales and marketing	26	9.5
Total	**270**	**100**
Ethanol sales (gallons)	132	
Stillage sales	56	
Ethanol production cost[b] ($/gallon)	1.63	
Ethanol sales at $1.74/gallon ($)	230	
Total income ($)	285	
Gross profit ($)	15	

Source: Data from Paul, J. K., ed. 1980. *Large and Small Scale Ethyl Alcohol Processes from Agricultural Raw Materials.* Park Ridge, NJ: Noyes Data Corporation.

[a] 60,000 bushels per year (3.4 million lb or 1.5 million kg).

[b] After allowing for sales of the stillage coproduct.

upscaled to produce in principle 83 L of 95% ethanol per hour was predicted to have production costs of $1.80 per gallon; the single largest contributor (61%) to production costs was the sorghum feedstock.[15]

6.3.2 CORN ETHANOL IN THE 1980s: RISING INDUSTRIAL ETHANOL PRICES AND THE DEVELOPMENT OF THE INCENTIVE CULTURE

A key change in the pricing structure of industrial alcohol in the United States occurred in the decade after 1975: The price of petrochemical ethylene showed an increase of nearly 10-fold, and this steep rise in feedstock costs pushed the price of synthetic industrial alcohol from $0.15 per liter ($0.57 per gallon) to $0.53 per liter ($2.01 per gallon).[16] Corn prices fell significantly (from $129 per ton to $87 per ton) between 1984 and early 1988. Because the coproduct costs were increasing as a percentage of the corn feedstock cost at that time, the net corn cost for ethanol

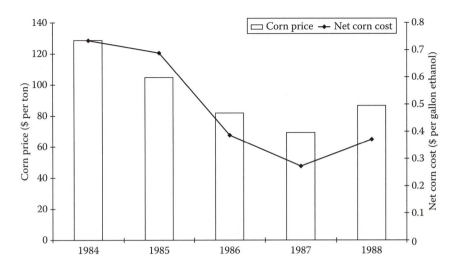

FIGURE 6.3   Corn price movements in the United States in the mid-1980s. (Data from Keim, C. R., and Venkatasubramanian, K. 1989. *Trends in Biotechnology* 7:22.)

production (the net cost as delivered to the ethanol production plant minus the revenue obtained by selling the coproducts) and the net corn cost per unit volume of ethanol were both halved (Figure 6.3).

By 1988, the costs involved in corn-derived ethanol production were entirely competitive with those of synthetic industrial alcohol (Table 6.5). The major

TABLE 6.5
Production Costs for Corn-Derived Ethanol in the United States in 1988

Manufacturing Input	Production Cost ($ per Liter)[a]	Production Cost ($ per Gallon)[a]	Production Cost (% of Total)
Direct:			
Grain	0.098	0.37	31.4
Steam and electric power	0.040	0.15	12.6
Enzymes	0.010	0.04	3.2
Yeast	0.010	0.04	3.2
Labor	0.010	0.04	3.2
Investment related	0.145	0.55	46.4
Total	**0.313**	**1.18**	

Source: Data from Keim, C. R., and Venkatasubramanian, K. 1989. *Trends in Biotechnology* 7:22.

[a] Average values calculated from the quoted range of values.

concern was that unexpectedly high investment costs could place a great strain on the economics of the process if the selling price for ethanol dipped. In general, such costs could be minimized by adding on anhydrous ethanol capacity to an existing beverage alcohol plant or adding an ethanol production process to a starch or corn syrup plant; however, expensive grassroots projects could face financial problems. Across the whole range of production facilities (small and large, new or with added capacity, and with varying investment burdens), a manufacturing price for ethanol could be as low as $0.18 per liter ($0.68 per gallon) or as high as $0.42 per liter ($1.59 per gallon).[16] By 1988, the average fuel ethanol selling price had fallen below $0.30 per liter ($1.14 per gallon, equivalent to $2.07 in 2009)—an economic movement that would have placed severe pressures on farm-scale production business plans.

As an incentive to fuel ethanol production, a federal excise tax concession of $0.16 per liter ($0.15 per gallon) and direct payments by states to producers amounting to as much as $0.11 per liter ($0.42 per gallon)—in conjunction with loan guarantees and urban development grants—encouraged the development of production by grassroots initiatives.[16] Industrial-size facilities, built without special incentives, were already reaching capacities in excess of 1 billion gallons per year as large corporations began to realize the earning potential of fuel ethanol.

STEM TOPIC 6.1: ETHANOL PRODUCTION WITHOUT COPRODUCT SALES OR SUBSIDIES

A study of candidate biofuels crops in Oregon determined that neither switchgrass nor soybean cultivation was practical on climatic grounds; corn is grown in the state, although in insufficient quantities to act as a feedstock, but could be imported from the Midwest.[17] Forestry is abundant locally, however, and could be a basis for a biomass fuel ethanol industry.

Included in the financial analyses made for the report were data devoid of local or national incentives; in addition, coproduct sales were also separately itemized so that a "clean" picture of the relative economics of different feedstocks could be obtained (see Table STEM 6.1.)

Even with the costs of shipping corn from the Midwest, the production costs of corn ethanol were only 2% greater than from local lignocellulosic sources: forest thinning, clearing invasive juniper, and waste from wood processing. The higher costs of trucking woody biomass and of wood-to-ethanol conversion narrowed the economic gap between imported corn and local-source timber.

Canola biodiesel was also considered in the report because of the potential for increasing canola planting in the state, but this would require an attractively high rate of return; this is the set of circumstances if blenders were induced to source canola biodiesel by high federal credits.

TABLE STEM 6.1
Unsubsidized Production Costs for Fuel Ethanol

	Cost (US$/Gallon, 2009 Basis)		
Cost	Corn (Ethanol)	Canola (Biodiesel)	Cellulosic (Ethanol)
Agriculture	1.00	2.53	0.77
Feedstock transport	0.30	0.10	0.37
Conversion to ethanol	1.35	0.71	1.47
Total	2.65	3.34	2.61

6.3.3 WESTERN EUROPE IN THE MID-1980S: ASSESSMENTS OF BIOFUELS PROGRAMS MADE AT A TIME OF FALLING REAL OIL PRICES

European commentators and analysts were far less sanguine on the desirability of fuel ethanol as a strategic industry for the future in the mid-1980s. This was a time of steeply falling oil prices, expressed in real terms and in the actual selling price (Figure 1.3, Chapter 1). In 1987, two independent assessments of ethanol production from agricultural feedstocks were published in the United Kingdom and Europe.[18–20] Across Europe, the introduction of lead-free fuels heralded an important new market for ethanol and other additives; the competition among these compounds (including MTBE and methanol) was likely to be intense for the estimated 2 million-tonne-per-year market by 1998.[18]

The UK survey included wheat grain and sugar beet as possible local sources of carbohydrates; in both cases, raw plant materials dominated the production cost analysis (Table 6.6). The monetary value of coproducts was important, although only

TABLE 6.6
Cost Estimates for Wheat- and Sugarbeet-Derived Ethanol in the United Kingdom in 1987

Manufacturing Input	Production Cost from Wheat ($ per Liter)[a]	Production Cost from Wheat ($ per Gallon)[a]	Production Cost from Sugar Beet ($ per Liter)[a]	Production Cost from Sugar Beet ($ per Gallon)[a]
Raw materials	0.44	1.68	0.44	1.68
Operating costs	0.09	0.34	0.13	0.50
Capital cost	0.04	0.17	0.10	0.39
Coproduct credit	0.16	0.62	0.10	0.39
Total	**0.41**	**1.57**	**0.58**	**2.18**

Source: Data from Marrow, J. E., Coombs, J., and Lees, E. W. 1987. *An assessment of bio-ethanol as a transport fuel in the UK.* London: Her Majesty's Stationery Office.

[a] Currency exchange values used from January 1987 to convert the original pound sterling data.

animal feeds were considered as viable sources of income to offset ethanol production costs.[18] The market prices for all major agricultural products were determined by the price support policies of the Common Agricultural Policy (CAP), an essential part of the Treaty of Rome (March 1957) under which the European Economic Community (EEC) was set up and regulated. Among its many provisions, the CAP was designed to ensure a fair standard of living for farmers and reasonable consumer prices. The CAP operated to guarantee a minimum price for basic agricultural products through intervention prices and protected the community's internal markets against fluctuations in world prices through the establishment of threshold prices.

However, technical progress was also a goal of the CAP to increase agricultural productivity. The CAP has been controversial inside the EEC and, subsequently, in the European Community (EC) and European Union (EU) as individual member states have received varying benefits from the CAP; however, it has the advantage of enabling commodity prices to be more predictable—a useful factor when calculating possible trends in feedstock prices for the production of biofuels. In 1987, the likely costs of ethanol from wheat and sugar beet were greatly in excess of the refinery price of petrol (gasoline), with a cost ratio of 3.2–4.4:1, allowing for the lower energy content of ethanol.[18] The continuing influence of the CAP was, moreover, highly unlikely to reduce feedstock costs for fuel ethanol to be price competitive with conventional fuels.

The second volume of the UK study gave outline production cost summaries for ethanol derived from wood (no species was specified) using acid and enzymatic hydrolysis for the liberation of glucose from cellulose (Table 6.7). Additionally, straw residues were considered from cereals (wheat, barley, and oats), field beans, and oil seed rape (canola) using acid and enzymic hydrolysis; electricity and lignin were modeled as saleable coproducts. No source for ethanol could yield a product with a production cost less than three times that of conventional fuels (Figure 6.4).

TABLE 6.7

Cost Estimates for Wood-Derived Ethanol in the United Kingdom in 1987

Manufacturing Input	Production Cost Acid Hydrolysis ($ per Liter)[a]	Production Cost Acid Hydrolysis ($ per Gallon)[a]	Production Cost Enzymic Hydrolysis ($ per Liter)[a]	Production Cost Enzymic Hydrolysis ($ per Gallon)[a]
Raw materials	0.30	1.13	0.27	1.04
Operating costs	0.18	0.67	0.27	1.04
Capital cost	0.14	0.53	0.29	1.08
Coproduct credit	0.14	0.52	0.03	0.11
Total	**0.48**	**1.82**	**0.81**	**3.05**

Source: Data from Marrow, J. E., Coombs, J., and Lees, E. W. 1987. *An assessment of bio-ethanol as a transport fuel in the UK.* London: Her Majesty's Stationery Office.

[a] Currency exchange values used from January 1987 to convert the original pound sterling data.

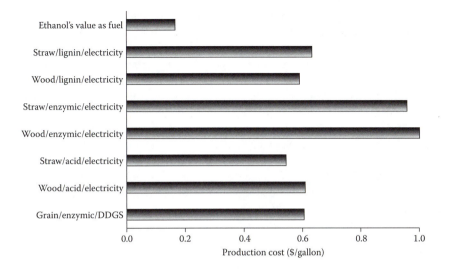

FIGURE 6.4 Production prices of ethanol from grain, wood, and straw feedstocks by acid or enzymic hydrolysis and with saleable coproducts in the UK. (Data recalculated from Marrow, J. E., and Coombs, J. 1987. *An Assessment of Bio-ethanol as a Transport Fuel in the UK—Volume 2.* London: Her Majesty's Stationery Office.)

These poor economics resulted in the authors being unable to recommend initiating a large program of work directed toward fuel ethanol production in the United Kingdom, although continued support of existing research groups was favored to enable the United-Kingdom to be able to take advantage of fundamental breakthroughs, especially in lignocellulose conversion. A return to the high oil prices experienced in 1973 and 1974 and 1978–1980 was considered unlikely until well into the twenty-first century, in any case, because transport costs were inevitably an important element of feedstock costs in the United Kingdom, and rising oil process would tend to increase the total costs for fuel ethanol production.[19]

The European study presented less detailed economic data but considered a wider range of feedstocks, not all of which were (or are) major agricultural products across the whole of Europe, but did represent potential sources for expanded agricultural production or as dedicated energy crops. For production facilities capable of manufacturing in excess of 150 million L (40 million gallons) of ethanol per year, wheat grain was the cheapest source, with a production cost of the feedstock equivalent to $0.36 per gallon (converting the now obsolete European currency unit values to US$ at the exchange rate prevailing in early January 1987), followed by corn ($0.62 per gallon), sugar beet ($0.83 per gallon), Jerusalem artichoke ($0.87 per gallon), potatoes ($1.89 per gallon), and wine ($2.73 per gallon).[20] The estimated cost of ethanol production from wheat was 0.49 ECU per liter (equivalent to $0.53 per liter or $2.01 per gallon); in comparison, the Rotterdam price refinery price for premium gasoline in late 1986 was approximately $0.10 per liter. The consultants who assembled the report concluded the following:

- Encouragement of a bioethanol program was not in the economic interests of the EC.
- A large reduction of the feedstock costs would be required to show a net economic benefit from such ethanol production.
- Alternatively, an oil price in the range of $30–40 per barrel would be required to achieve economic viability for fuel ethanol.

Although the conversion of ethanol to ethylene was technically feasible, a cost analysis of this route indicated that it would be even less viable economically than ethanol production as a fuel additive.

The European study included an analysis of the development of fuel ethanol industries in Brazil and the United States, noting that the bulk of the financial incentives in the U.S. corn ethanol sector benefited the large producers rather than the small operators or the corn farmers. Even more disturbing to European decision makers was the conclusion that blenders had benefited disproportionately, enjoying effectively cost-free ethanol as a gasoline additive during 1986 and using the available subsidies to start a price-cutting war between ethanol producers rather than promoting total sales. The prospect of subsidies being essential for establishing and maintaining a fuel ethanol program in Europe was nevertheless consistent with the EC and its long-established strategic approach to agricultural development.

6.3.4 BRAZILIAN SUGARCANE ETHANOL IN 1985: AFTER THE FIRST DECADE OF THE PROÁLCOOL PROGRAM TO SUBSTITUTE FOR IMPORTED OIL

The year 1985 marked the end of the first decade of the national program to use sugarcane-derived ethanol as an import substitute for gasoline. Rival estimates of the cost of Brazilian fuel ethanol varied widely, from $35 to $90 per barrel of gasoline replaced. Assessing the economic impact of the various subsidies available to alcohol producers was difficult, but indicated a minimum unsubsidized price of $45 per barrel on the same gasoline replacement basis.[21] Assuming that gasoline was mostly manufactured from imported petroleum, the overall cost comparison between gasoline and nationally produced ethanol was close to achieving a balance—clearly so if the import surcharge then levied on imported oil were taken into consideration (Table 6.8).

Anticipating future sentiments expressed for biofuels, justifying full gasoline replacement at then current world oil prices was held to be dependent on the additional benefits of ethanol production. These include employment creation, rural development, increased self-reliance, and reduced vulnerability to crises in the world oil market.[21]

6.3.5 ECONOMICS OF U.S. CORN AND BIOMASS ETHANOL ECONOMICS IN THE MID-1990s

In retrospect, it is perhaps surprising that the ground-breaking 1996 monograph on ethanol contained no estimate of production costs from lignocellulosic substrates,

TABLE 6.8

Production Costs for Sugar-Derived Ethanol in Brazil by 1985

Cost Component	Cost
Ethanol	
Sugarcane	$10–12/ton
Ethanol yield	65 L/ton
Distillation	$0.09–0.11/L
Production cost	$0.264–0.295/L
Replacement ratio 100%	1.2 L ethanol per liter of gasoline
Replacement ratio 20%	1.0 L ethanol per liter of gasoline
Ethanol cost 100%	$50–65 per barrel of gasoline replaced
Ethanol cost 20%	$42–47 per barrel of gasoline replaced
Gasoline	
Imported petroleum	$29/barrel
Shipping costs	$2/barrel
Import surcharge	$6/barrel
Refining cost	$10/barrel
Total	$47/barrel
Total – surcharge	$41/barrel

Source: Geller, H. S. 1985. Ethanol fuel from sugarcane in Brazil. *Annual Reviews in Energy* 10:135.

but rather advanced only a brief economic analysis of corn-derived ethanol.[22] U.S. production of corn ethanol had, after the relative doldrums of 1985–1990, begun to surge—a trend only briefly halted in 1995 (see Figure 1.16, Chapter 1). Wet and dry milling processes had also become competitive as ethanol added a major product to the long established mix of feed additives, corn oil, etc.

Table 6.9 summarizes *average* production costs from both wet and dry milling. As in the late 1980s, the mean values masked a wide range of variation: $1.06–1.40 per gallon (equivalent to $1.93–2.55 in 2009 prices) for dry milling and $0.232–0.338 per liter ($0.88–1.28 per gallon) for wet milling. The wet milling process was highly sensitive to the prices of corn and the process coproducts, even to the point of net corn prices becoming negative at times of lower feedstock but higher coproduct prices.[22] Further technical advances, including the introduction of corn hybrids with properties tailored for wet milling (e.g., accelerated steeping) and improved fermentor designs allowing cell entrapment were estimated to offer cost reductions of $0.4–0.7 per gallon.

The missing analysis of lignocellulosic ethanol in the mid-1990s was supplied by two reviews also published in 1996.[23,24] Two scenarios were considered in the cost modeling: a base case and an advanced technology option (although without a precise date for implementation); the data are summarized in Table 6.10. Using plausible

TABLE 6.9

Production Costs for Corn-Derived Ethanol in the United States by the Mid-1990s

Manufacturing Input	Production Cost ($ per Liter)[a]	Production Cost ($ per Liter)[a]	Production Cost (% of Total)
	Dry Corn Milling		
Net corn costs	0.120	0.45	51.1
Other operating costs	0.105	0.40	44.7
Annualized capital costs	0.010	0.04	4.3
Total	**0.235**	**0.89**	
	Wet Corn Milling		
Net corn costs	0.097	0.37	41.3
Other operating costs	0.103	0.39	43.8
Annualized capital costs	0.085	0.32	36.2
Total	**0.285**	**1.08**	

Source: Data from Elander, R. T., and Putsche, V. L. 1996. In *Handbook on Bioethanol: Production and Utilization,* ed. Wyman, C. E., Chap. 15. London: Taylor & Francis.

[a] Average values calculated from the quoted range of values (in 1993 dollars).

technology for the mid-1990s, a production cost of $1.18 per gallon was computed (equivalent to $2.15 in 2009 prices); a fourfold increase in the capacity of the facility, together with innovative bioprocess technologies, was predicted to reduce the production costs to approximately $0.50 per gallon.

6.3.6 CASE STUDY: THE VIEW FROM SWEDEN

Although a few pioneering studies attempted cost estimates of wood-derived ethanol from the 1980s onward, they focused on aspects of the technological processes required rather than making firm conclusions about market prices.[25,26] Swedish studies appear to have been the first to present detailed cost breakdowns for ethanol production from accessible large-scale woody biomass sources.[27,28]

The first of the two to be published was highly unusual in that it used recent advances in pentose utilization by recombinant bacteria to model a pentose stream process (i.e., using the solubilized sugars from the pretreatment of wood [willow] feedstock—the procedure involved the impregnation of the material with SO_2 and subsequent steaming) and included a detoxification procedure to reduce the levels of inhibitors from the hydrolysate.[27] The fermentation with *Escherichia coli* KO11 was assumed to consume 96% of the pentose sugars (and all of the much smaller amount of hexoses) to generate over 67 million L (1.8 million gallons) of ethanol per year. The final production cost of 95% aqueous alcohol was equivalent to $1.82 per gallon (equivalent to $2.62 in 2009) after allowance for financial costs and assuming a small net income from CO_2 as a coproduct (Table 6.11).

TABLE 6.10

Production Costs for Cellulosic Ethanol in the United States by the Mid-1990s

Manufacturing Input	Capital, Labor, and Related Items (Cents per Liter)	Energy (Cents per Liter)	Production Cost (Cents per Liter)	Production Cost (% of Total)
		Base Case[a]		
Feedstock			45.97	39.0
Other raw materials			9.78	8.3
Pretreatment	13.75	6.55	20.30	17.2
Cellulase preparation	1.55	1.67	3.22	2.7
SSF	13.83	3.34	17.17	14.6
Pentose conversion	3.22	0.99	4.21	3.6
Distillation	2.74	5.10	7.84	6.7
Power cycle	28.61	−26.96	1.65	1.4
Other	7.34	0.36	7.70	6.5
Total			**117.84**	
		Advanced Technology[b]		
Feedstock			35.84	71.3
Other raw materials			0.95	1.9
Pretreatment	3.22	5.63	8.85	17.6
Fermentation	1.95	1.00	2.95	5.9
Distillation	1.79	2.83	4.62	9.2
Power cycle	14.06	−22.03	−7.97	−15.8
Other	4.74	0.32	5.06	10.1
Total			**50.30**	

Source: Data from Lynd, L. R. 1996. *Annual Reviews of Energy and Environment* 21:403.

[a] 658,000 dry tons/year; 60.1 million gallons/year; installed capital $150.3 million.

[b] 2,738,000 dry tons/year; 249.9 million gallons/year; installed capital $268.4 million.

An essential parameter was that of to what extent the cells could be recycled: Single batch use of the cells increased the production cost to $0.64 per liter ($2.42 per gallon). The single largest contributor to the production cost, however, was the financial burden of repaying the investment in the plant (i.e., over 37% of the total annual production cost outlay) (Table 6.11).

A very similar analysis of three different approaches to utilizing the full carbohydrate potential of pine wood—digesting the cellulose component with concentrated acid, dilute acid, and enzymatic methods—calculated full manufacturing costs for ethanol of $1.89–2.01 per gallon (equivalent to $2.72–2.89 in 2009 prices).[28] The bulk of the production cost (up to 57.5%) was accounted for by the financial costs of installing the hardware for generating fermentable carbohydrates (hexoses as well as pentoses) in more complex total processes with longer cycle times.

TABLE 6.11

Cost Estimates for Ethanol Production from Pentose Stream from Willow

Manufacturing Input	Annual Capacity/Output (Tonne/Year)	Annual Cost ($ × 10³)	Cost (% of Total)
Wood hydrolysate	13,877	439	13.1
Ammonia (25%)	370	101	3.0
Phosphoric acid (80%)	179	90	2.7
Magnesium oxide	4.8	1	0.03
Sodium sulfite	252	142	4.2
Calcium oxide	1,750	236	7.1
Sulfuric acid (37%)	214	18	0.5
Electricity		80	2.4
Steam	12,885	226	6.8
Distillation		391	11.7
Maintenance		73	2.2
Labor		300	9.0
Annual capital costs		1,247	37.3
Working capital		8	0.2
Coproduct credit for CO_2	5,883	8	−0.2
Total		**3,344**	
Ethanol (m³/year)	6,906		
Ethanol production cost ($/liter)	0.48		

Source: Data from von Sivers, M. et al. 1994. *Biotechnology Progress* 10:555.

STEM TOPIC 6.2: EFFECTS OF SCALE OF PRODUCTION ON CELLULOSIC ETHANOL COSTS

A claim frequently made is that increasing the scale of production of cellulosic ethanol would bring cost benefits in the form of greatly reduced production costs. In 2007, Swedish researchers tested this assertion, compiling data published between 1995 and 2005 using all the various forms of production options devised or suggested for cellulosic ethanol biomanufacture[29] (see Figure STEM 6.2.1).

If all the variants can be considered as comparable, the trend to reduced production costs is marked, lowering costs by nearly 80% (the data have been converted to a 2009 dollar basis from those presented in the 2007 publication). As expected, capital costs rise in a positive correlation with the scale of production (see Figure STEM 6.2.2).

Therein resides the problem: Only by investing in larger production units can the costs be proved to decrease; however, finding the necessary investment requires proof that the larger units are economically viable and beneficial. Resolving this conundrum for unproved technologies is rendered even more complex by the competing biotechnologies (ethanologen, pretreatment, biomass substrate, etc.) implicit in the data.

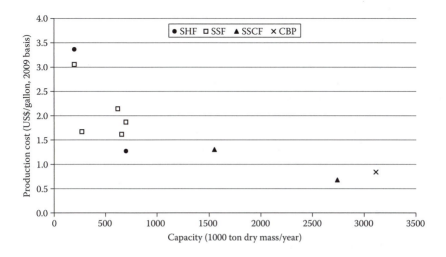

FIGURE STEM 6.2.1 Variation of production cost with cellulosic ethanol plant capacity with various fermentation options involving enzymic hydrolysis: SHF (separate hydrolysis and fermentation), SSF (simultaneous hydrolysis and fermentation), SSCF (simultaneous hydrolysis and cofermentation), and CBP (consolidated bioprocessing).

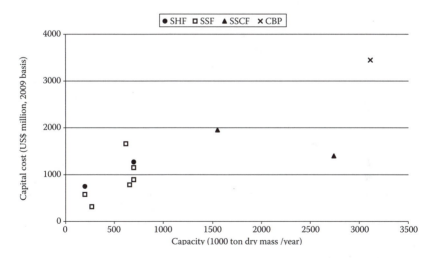

FIGURE STEM 6.2.2 Variation of capital cost with cellulosic ethanol plant capacity: SHF (separate hydrolysis and fermentation), SSF (simultaneous hydrolysis and fermentation), SSCF (simultaneous hydrolysis and cofermentation), and CBP (consolidated bioprocessing).

6.3.7 SUBSEQUENT ASSESSMENTS OF LIGNOCELLULOSIC ETHANOL IN EUROPE AND THE UNITED STATES

6.3.7.1 Complete Process Cost Models

Acknowledging the great uncertainties in establishing guideline costs for lignocellulosic feedstocks, a Swedish review of the cellulosic ethanol in 1999 noted that large-scale processes and improved overall ethanol yield would be highly desirable for future economic production of biofuels.[30] The first trend probably heralded the demise of the farm-scale ethanol plant (Section 6.3.1.4) because it is a production model probably only relevant to local and private consumption of transportation fuels as the market for fuel ethanol imposes competitive pricing. The second point is a natural conclusion from the vast efforts invested in developing recombinant producing organisms and bioprocesses.

The Swedish authors have continued to explore cost models for ethanol from lignocellulosic substrates:

- The SSF bioprocesses (Chapter 5, Section 5.6) offer improved economics over standard separate hydrolysis and fermentation because of higher ethanol yields and reduced capital costs. With softwood biomass sources, there are also significant advantages if either process can be operated with higher levels of insoluble material and if the stillage is recycled after distillation ("backsetting")—in principle reducing the production cost to $0.42 per liter or $1.59 per gallon (Table 6.12).[31]
- Operating steam pretreatment of softwoods in two steps (at lower and higher temperatures) to maximize the recovery of hemicellulose sugars and cellulosic glucose, respectively, has a higher overall ethanol yield and reduced requirement of enzymes but is more capital intensive and has a higher energy demand. The net result is no reduction in the production cost of ethanol (Table 6.13); further improvement to the process, including a higher insoluble solids content for the second step, might reduce the production cost by 5 or 6%.[32]

The National Renewable Energy Laboratory, in association with a consultant engineers, presented an outline cost model for the industrial-scale production of cellulosic ethanol (2000 tonnes per day consumption of feedstock, 52 million gallons of ethanol per year) from a hardwood yellow poplar biomass source.[33] Operating costs were calculated to be approximately $0.62 per gallon of ethanol (Table 6.14). Assuming a discount rate of 10%, discounted cash flow analysis indicated a minimum selling price of $1.44 per gallon (equivalent to $1.84 in 2009) for a capital investment of $234 million.

On the technical level, the key features of the envisaged process included:

- acid pretreatment of the biomass substrate (19% of the installed equipment cost)
- on-site generation of cellulase

TABLE 6.12

Cost Estimates for Ethanol Production from Softwood Using Different Bioprocess Technologies

Manufacturing Input	SSF[a] Base Case ($ per Liter)	SSF[a] Base Case ($ per Gallon)	SHF[b] Base Case ($ per Liter)	SHF[b] Base Case ($ per Gallon)	SSF[a] 8% Solids ($ per Liter)	SSF[a] 8% Solids ($ per Gallon)	SHF[b] 8% Solids ($ per Liter)	SHF[b] 8% Solids ($ per Gallon)
Wood[c]	0.16	0.61	0.19	0.73	0.16	0.61	0.19	0.70
Enzymes	0.08	0.31	0.06	0.21	0.08	0.31	0.06	0.21
Yeast	0.06	0.22	0.00	0.00	0.04	0.14	0.00	0.00
Other operating costs	0.04	0.16	0.05	0.19	0.03	0.13	0.04	0.16
Labor, maintenance, insurance	0.07	0.26	0.10	0.37	0.06	0.24	0.09	0.32
Capital costs	0.16	0.62	0.25	0.93	0.14	0.53	0.20	0.76
Coproduct credits[d]	0.01	0.03	0.01	0.03	0.06	0.22	0.07	0.27
Total	**0.57**	**2.14**	**0.64**	**2.41**	**0.46**	**1.74**	**0.50**	**1.87**

Source: Data from Wingren, A., Galbe, M., and Zacchi, G. 2003. *Biotechnology Progress* 19:1109.

[a] Simultaneous saccharification and fermentation, 63 billion L per year (base case).

[b] Separate hydrolysis and fermentation, 55 billon L per year (base case).

[c] 195,600 tonnes raw material per year, operated continuously (8,000 h per year), notionally located in northern Sweden.

[d] CO_2 and solid fuel.

TABLE 6.13

Cost Estimates for Ethanol Production from Softwood Using Different Pretreatment Options

Manufacturing Input	Steam Pretreatment One-Step[a] ($ per Liter)	Steam Pretreatment One-Step[a] ($ per Gallon)	Steam Pretreatment Two-Step[b] ($ per Liter)	Steam Pretreatment Two-Step[b] ($ per Gallon)
Wood[c]	0.19	0.71	0.18	0.69
Chemicals	0.11	0.43	0.11	0.42
Utilities	0.03	0.10	0.03	0.10
Other operating costs	0.09	0.33	0.09	0.33
Capital costs	0.21	0.78	0.21	0.79
Coproduct credits[d]	0.07	0.27	0.06	0.24
Total	**0.55**	**2.08**	**0.55**	**2.09**

Source: Data from Wingren, A. et al. 2004. *Biotechnology Progress* 20:1421.

[a] 215°C, residence time 5 min; SO_2 added to 2% of the water content of the wood; 47 billion L of ethanol per-year capacity.

[b] 190°C, residence time 2 min, then 210°C for 5 min; SO_2 added to 2% of the water content of the wood; 49 billion L of ethanol per-year capacity.

[c] 200,000 tonnes per year; plant operating time of 8,000 h per year.

[d] CO_2 and solid fuel.

TABLE 6.14

Operating Costs for Yellow Poplar Sawdust Ethanol

Input	Production Cost (Cents per Gallon)[a]	Production Cost (% of Total)
Feedstock[a]	37.0	60.0
Chemicals	8.0	13.0
Nutrients	6.2	10.0
Fossil fuels	0.9	1.5
Water	0.9	1.5
Utility chemicals	1.2	1.9
Solid waste disposal	1.2	1.9
Fixed costs	13.5	21.9
Electricity credit[b]	7.2	−11.7
Total	**61.7**	

Source: Data from Wooley, R. et al. 1999. *Biotechnology Progress* 15:794.

[a] Poplar sawdust at $25 per tonne.

[b] Excess electricity sold to grid at $0.04 per kilowatt-hour.

- SSF of the pretreated substrate with a *Zymomonas mobilis* capable of utilizing only glucose and xylose
- wastewater treatment via anaerobic digestion to methane
- utilization of three available waste fuel streams (methane, residual lignin solids, and a concentrated syrup from evaporation of the stillage) in a fluidized bed combustor, burner, and turbogenerator (33% of the installed equipment cost)

Although the complete array of technology in the model was unproven on a large scale, much of the process could be accurately described as "near term" or "based on the current status of research that is complete or nearly so."[33] The computed minimum selling price for ethanol was 20% higher than that of corn-derived ethanol, with a more than twofold greater investment cost ($4.50 per gallon as compared with approximately $2 per gallon).

6.3.7.2 Reviews of "Gray" Literature Estimates and Economic Analyses

Outside primary scientific journals, data from a range of sources (including reports prepared for governments and conference proceedings) were compiled on the basis of 2003 costs as a baseline for future modeling.[34] Ethanol produced from sugar, starch (grain), and lignocellulosic sources covered production cost estimates from less than $1 per gallon to over $4 per gallon (Table 6.15). Even with the lower production costs for lignocellulosic ethanol in the United States, taking into account financial outlays and risks ($260 million for a 50 million-gallon annual production plant), an ethanol price of $2.75 per gallon (equivalent to $2.94 in 2009) would be more realistic.[35]

The International Energy Agency's assessment of sugar- and starch-derived ethanol (2005 reference basis) is that Brazil enjoys the lowest unit costs ($0.20 per liter, or $0.76 per gallon); starch-based ethanol in the United States costs (after production subsidies) an average of around $0.30 per liter, or $1.14 per gallon, and a European

---

TABLE 6.15
Estimated Production Costs for Ethanol in 2003

Source of Ethanol	Production Cost (€/GJ)	Production Cost ($/L)[a]	Production Cost ($/Gallon)
Sugarcane (Brazil)	10–12	0.24–0.29	0.91–1.10
Starch and sugar (United States and Europe)	16.2–23	0.39–0.55	1.48–2.08
Lignocellulosic (United States)	15–19	0.36–0.46	1.36–1.74
Lignocellulosic (Europe)	34–45	0.82–1.08	3.10–4.09

Source: Hamelinck, C. N., van Hooijdonk, G., and Faaij, A. P. C. 2005. *Biomass & Bioenergy* 28:384.

[a] Higher heating value of ethanol = 83,961 Btu/gallon (24 GJ/L).

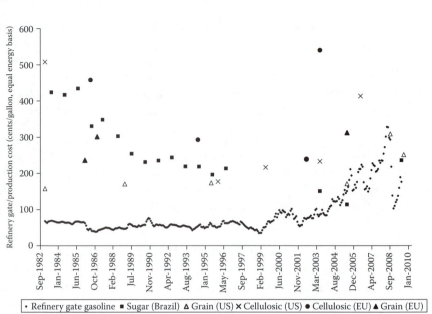

FIGURE 6.5 Refinery gate price of U.S. oil and estimates of fuel ethanol production costs between 1980 and 2006.

cost (including all subsidies) is $0.55 per liter, or $2.08 per gallon.[36] Brazilian production costs for fuel alcohol, close to $100 per barrel in 1980, decreased rapidly in the 1980s and then more slowly, but only a severe shortage of sugarcane or a marked rise in sugar prices would interrupt the downward trend in production costs.[37,38]

With due allowance of the lower fuel value of ethanol, therefore, the historical trend of fuel ethanol production costs versus refinery gate price of gasoline is showing some degree of convergence (Figure 6.5). In particular, the real production costs of both sugar- and corn-derived ethanol have fallen so that the production costs (with all tax incentives in place, where appropriate) are now probably competitive with the production cost of gasoline, as predicted for biomass ethanol in 1999.[39]

However, critics of the corn ethanol program have argued that the price of fuel ethanol is artificially low because total subsidies amount to $0.79 per gallon for production costs of $1.21 per gallon (i.e., some $3 billion are expended in subsidizing the substitution of only 1% of the total oil use in the United States).[40] Although incentives for domestic ethanol use in Brazil were discontinued by 1999 (except as part of development policies in the northeast region), a cross-subsidy was created during the 1990s to subsidize ethanol production through taxation on gasoline and diesel; this was operated via tight government control of the sale prices of gasoline, diesel, and ethanol, and the monopoly represented in the country by PETROBRÁS.[38]

Brazilian sugarcane ethanol has reached the stage of being an importable commodity to the United States, promoting the development of shipping, port handling, and distribution network infrastructures. In Europe, grain alcohol *might* be cost competitive; with the much higher tax rates prevailing in Europe, the scope for regulating the end user price is much higher. In contrast, the economics of cellulosic ethanol remain problematic, although it is possible that, at least in the United States, production costs may become competitive with gasoline within the next 5–10 years unless crude oil prices decrease significantly again.

6.4 PILOT PLANT AND INDUSTRIAL EXTRAPOLATIONS FOR CELLULOSIC ETHANOL

6.4.1 NEAR-FUTURE PROJECTIONS FOR CELLULOSIC ETHANOL PRODUCTION COSTS

The persistently high projected production costs of cellulosic ethanol (particularly in Europe) have catalyzed several attempts to predict trends in 5-year, 10-year, and longer scenarios. The implicit or explicit rationale is that only lignocellulosic biomass is sufficiently abundant to offer a means of substituting a sizeable proportion of the gasoline presently used for transportation.

In 1999, the National Renewable Energy Laboratory published a projection of the economic production costs for lignocellulosic ethanol that, starting from a baseline of $1.44 per gallon (in 1997; equivalent to $1.92 in 2009), computed a decrease to $1.16 per gallon with a 12% increase in yield (to 76 gallons per ton of feedstock), with a 12% increase in plant production capacity together with a 12% reduction in new capital costs.[33] A price trajectory envisaged this price deceasing to below $0.80 per gallon by 2015 based on developments in cellulase catalytic efficiency and production and in ethanologenic production organisms:

- improved cellulose-binding domain, active site, and reduced nonspecific binding
- improved cellulase producers genetically engineered for higher enzyme production
- genetically engineered crops as feedstocks with high levels of cellulases
- ethanologens capable of producing ethanol at temperatures in excess of 50°C
- ethanologens capable of consolidated bioprocessing of cellulose to ethanol

These topics were covered in Chapters 2–5; they remain research topics, although cellulase production and costs have certainly been improved greatly at the industrial scale of production.

A European study suggested that by 2010 cellulosic ethanol production costs could decrease to $2.00 per gallon, to $1.18 per gallon by 2020, and to $0.79 per gallon after 2025.[34] These improvements in process economics were considered to result from the combined effects of higher hydrolysis and fermentation efficiencies, lower specific capital investment, increases of scale, and cheaper biomass feedstock costs. However, the prospect of ethanol ever becoming cost competitive with gasoline was considered to be unlikely, although much of the analysis was undertaken during the early stages of the great surge in oil prices and gasoline production costs after 2002 (Figure 6.5).

A more recent publication by the same research group (Utrecht, the Netherlands) computed a 2006 production cost for cellulosic ethanol of €22/GJ ($2.00 per gallon, assuming exchange parity between the currencies) that was anticipated to fall

TABLE 6.16

Estimated Production Costs for Bioethanol from Biofuel Crops in 2005

Biofuel Crop	Production Cost (€/GJ)	Production Cost ($/L)[a]	Production Cost ($/Gallon)
Rape (canola)	20	0.48	0.91–1.10
Sugar beet	12	0.29	1.48–2.08
Willow	3–6	0.07–0.14	0.26–0.53
Poplar	3–4	0.07–0.10	0.26–0.38
Miscanthus	3–6	0.07–0.14	0.26–0.53

Source: Data from International Energy Agency. 2006. *World energy outlook,* Chap. 14. Paris.

[a] Higher heating value of ethanol = 83,961 Btu/gallon (24 GJ/L).

to €11/GJ ($1.00 per gallon) by 2030. The single largest contributor to the production cost in 2006 was capital related (46%), but this was expected to decrease in both absolute and relative terms so that biomass costs predominated by 2030.[41]

An analysis by the International Energy Agency for cellulosic ethanol (from willow, poplar, and *Miscanthus* biomass sources) predicted production costs below €0.05 per liter (€0.18 per gallon) by 2030 with achieved biomass yield increases per unit land area of between 44 and 100% (Table 6.16). On this basis, biomass-derived ethanol would enjoy lower production costs than other sources in Europe (cereal grain, sugar beet, etc.).[36] These three biomass options each target different areas:

- Willow already has considerable commercial experience in Sweden, the United Kingdom, etc. and has strong potential in Eastern Europe, where growing conditions and economics are favorable.
- Poplar is already grown for pulp production, with typical rotation cycles of 8–10 years.
- Elephant grass (*Miscanthus*) is a perennial crop suited to warmer climates (where high yields are possible), but its production potential in Europe in general is uncertain.

Transport costs are important for all biomass options; the energy used in transportation is equal to 1 or 2% of that contained in the biomass for short distances and 6–10% for long-distance shipping. Transport, handling, and storage costs were considered to add 10% to ethanol production costs when the feedstocks were shipped by road to a local plant.[36]

6.4.2 SHORT- TO MEDIUM-TERM TECHNICAL PROCESS IMPROVEMENTS AND THEIR ANTICIPATED ECONOMIC IMPACTS

Each of the exercises in predicting future cost and price trends discussed in Section 5.4.1 suffers from a high content of speculative thought because the costs of many key technologies are speculative at industrial scales of production. Indeed, many aspects have not progressed beyond the laboratory bench. However, some aspects of upstream (feedstock provision and preparation) and fermentation strategies have progressed to the stage that economic implications can be calculated across a range of cellulosic processes.

The ultimate starting point for cellulosic ethanol production is the collection and delivery of the biomass substrate and the associated costs. A central assumption for cellulosic ethanol is that a large supply of such material not only can be sourced but also can be provided cheaply. With the much more advanced starch ethanol industry, a much larger body of data is available for analyzing the effect of feedstock cost increase on ethanol production costs. For example, the Agricultural Research Service of USDA has developed a detailed process and cost model for the dry milling of corn showing that most (>85%) of any increase in the price of corn is transferred to the final product cost; a relatively small decrease in the starch content (from 59.5 to 55%, w/w) can similarly have a negative impact on the productivity of an ethanol production unit.[42]

Projecting forward from a 1995 baseline, feedstock costs for a mature biomass ethanol technology were anticipated to be within the price range of $34–38.60 per dry ton.[23] Switchgrass farming has been estimated to cost $30–36 per dry tonne. In comparison with straw or corn stover, the collection of switchgrass is probably less expensive because of the high yield of a denser biomass; nevertheless, delivered costs of the switchgrass to a production plant could be as low as $37 per dry tonne of compacted material to $47 per tonne of bales.[43] Similarly, corn stover is difficult to handle on account of its low bulk density, and chopped corn stover can be compacted into briquettes that can reach a density of 950 g/L; these more easily transported briquettes are more durable if produced from stover with low water content (5–10%).[44]

A more radical option is the pipeline transport of corn stover (e.g., at 20% wet solids concentration); this is cheaper than trucking at >1.4 million dry tonnes per year. It also allows the possibility of conducting a partial saccharification during transport if enzymes are added, thus reducing the need for investment in the fermentation plant and lowering production costs by $0.07 or 0.08 per gallon.[45]

In general, as cellulosic ethanol plant capacity is increased to cut unit production costs, the land area required for collection of sufficient biomass feedstock increases; as biomass supplies are sought from larger and larger distances, the costs of moving the raw material increases, introducing possibly diseconomies into production models. One solution is to introduce more flexibility into the feedstock "diet," taking advantage of whatever surpluses of other biomass material may seasonally occur.

For example, a California study investigated what biomass supplies could be considered for a 40 million-gallon facility in the San Joaquin Valley; locally grown corn was significantly more expensive than Midwestern corn ($1.21 per gallon of ethanol versus $0.92 per gallon), but surplus raisins and tree fruit and (although much more expensive) grapes and citrus fruit might all be included in a composite biomass harvest.[46] The proximate example of feedstock diversity, however, is that of cane sugar bagasse: the Dedini Hidrólise Rápida (rapid hydrolysis, www.dedini.com.br) process uses organic solvent extraction of sugarcane bagasse as a pretreatment method and aims to double the alcohol production per hectare of sugarcane harvested.

Using paper sludge as a feedstock for ethanol production has been claimed to be profitable because it provides a near-ideal substrate for cellulase digestion after a comparatively easy and low-cost pretreatment; even without xylose conversion to ethanol, such a technology may be financially viable at small scales, perhaps as low as 15 tonnes of feedstock processed per day.[47] This is a genuinely low-cost feedstock because, in the absence of any productive use, paper sludge goes into landfills—at a cost to its producer.

This has led to several attempts to find viable means of converting such waste into fuel. The pulp and paper industry in Canada is estimated to produce at least 1.3 million tonnes of sludge every year; at up to 70% cellulose, this raw material is economically competitive with cereal grain as a substrate for ethanol production.[48] Hungarian researchers have also identified paper sludge and other industrial cellulosic wastes as cost-effective routes to ethanol.[49]

This illustrates the proposition that biomass ethanol facilities might be designed and constructed to utilize a range of biomass substrates adventitiously when they become available. As in the California example noted previously, investigation of wastes from fresh and processed vegetables defined sizeable resources of plant material (450,000 tonnes per year) in Spain. Easily pretreated with dilute acid, such inputs could be merged with those for starch or lignocellulosic production lines at minimal (or even negative) cost.[50]

Substantial cost savings in cereal-based ethanol production can be achieved by a more integrated agronomic approach. Although fermentor stillage could be used as a substitute for mineral fertilizer, total ethanol production costs were 45% lower if cereals were grown after a previous nitrogen-fixing legume crop. Intensifying cereal yields certainly increased crops per unit of land area, but ethanol production costs per liter dropped because the ethanol yield per unit area outweighed the other costs. Field trials also suggest that, under German conditions, barley may be economically favorable in comparison with wheat and rye.[51]

All these conclusions may be directly applicable to cereal *straw* as well as cereal grain. Moreover, because one of the strongest candidates for cellulosic ethanol production (wheat straw) also has one of the lowest ethanol yields per unit of dry mass, the ability to mix substrates flexibly has capacity advantages if feedstock handling and processing regimes can be harmonized (Figure 6.6).[52]

Once in the ethanol facility, biomass pretreatment and hydrolysis costs are important contributors to the total production cost burden; for example, with hardwoods and softwoods, enzymes represent 18–23% of the total ethanol production costs. Combining enzyme recycling and doubling the enzyme treatment time might improve the economic cost by 11%.[53] Bench-scale experiments strongly indicated that production costs could be reduced if advanced engineering designs could be adopted, in particular improving the efficiency of biomass pretreatment with dilute acid in sequential co- and countercurrent stages.[54]

Different pretreatment techniques (dilute acid, hot water, ammonia fiber explosion, ammonia recycle percolation, and lime) are all capital intensive because low-cost reactors are counterbalanced by higher costs associated with catalyst recovery or ethanol recovery. As a result, the five rival pretreatment options exhibited very similar production cost factors.[55] Microbial pretreatments could

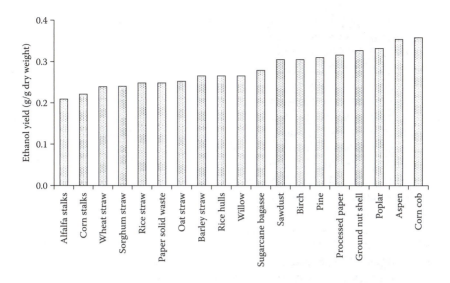

FIGURE 6.6 Maximum ethanol yield from lignocellulosic biomass sources. (Data from Chandel, A. K. et al. 2007. *Biotechnology and Molecular Biology Review* 2:14.)

greatly reduce the costs and energy inputs required by biomass hydrolysis techniques because enzymic digestibility is increased and hydrodynamic properties improved in stirred bioreactors. A full economic analysis remains to be undertaken in industrial ethanol facilities.[56] Continued optimization of chemical pretreatments has resulted in a combined phosphoric acid and organic solvent option that has the great advantage of requiring a relatively low temperature (50°C) and only atmospheric pressure.[57]

The ability to generate higher ethanol concentrations in the fermentation step would improve overall economic performance by reducing the costs of ethanol recovery; one solution is to run fermentations at higher biomass substrate loadings (i.e., accomplish biomass liquefaction and saccharification at high solids concentrations). Wheat straw could be processed to a paste/liquid in a reactor system designed for high solids content; the material would then be successfully fermented by *Saccharomyces cerevisiae* at up to 40% (w/v) dry matter in the biological step.[58]

What is the optimum processing of the various chemical streams (soluble sugars and oligosaccharides, pentoses and hexoses, cellulose, and lignin) resulting from the pretreatment and hydrolysis of lignocellulosic biomass? The utilization of pentose sugars for ethanol production is certainly beneficial for process economics—a conclusion reached as early as 1989 in a joint U.S.–New Zealand study of pine as a source of woody biomass. Ethanol production costs of $0.75 per liter ($2.83 per gallon) were calculated, decreasing by 5% if the pentose stream was used for fermentable sugars.[59] The stillage after distillation is also a source of carbohydrates as well as nutrients for yeast growth; replacement of up to 60% of the fresh water in the fermentation medium was found to be possible in a softwood process, with consequential reductions in production costs of as much as 17%.[60]

Because larger production fermentors are part of the drive toward economies-of-scale savings in production costs, reformulating media with cheaper ingredients becomes more important. In the fermentation industry at large, devising media to minimize this operating cost parameter has had a long history; as recombinant ethanologens are increasingly engineered (Chapters 3 and 4), suitable media for large scales of production are mandatory. With *E. coli* KO11, for example, laboratory studies showed that expensive media could be substituted by a soya hydrolysate-containing medium, although the fermentation would proceed at a slightly slower rate. Both this and a corn steep liquor-based medium could contribute as little as $0.06 per gallon of ethanol produced from biomass hydrolysates.[61,62]

Operating a membrane bioreactor (in a demonstration pilot plant of 7,000 L capacity) showed that the yearly capital costs could be reduced to $0.18–0.13 per gallon, with total operating costs for the unit of $0.017–0.034 per gallon.[63] Another advanced engineering design included continuous removal of the ethanol product in a gas stream; compared with a conventional batch process, ethanol stripping gave a cost savings of $0.03 per gallon with a more concentrated substrate being used, thus resulting in less water to remove downstream.[64] Simply concentrating the fermented broth, if the ethanol concentration is low, in a conventional process is feasible if reverse osmosis is employed, but not if the water is removed by evaporation with its high energy requirement.[65]

Incremental savings in cellulosic ethanol production costs are therefore entirely possible as processes are evolved. Many of the steps involved obviously require higher initial investment when compared with basic batch fermentation hardware, and it is unlikely that radical innovations will be introduced until a firm set of benchmark costing is achieved in semi-industrial- and full production-scale units.

On the other hand, efficient utilization and realization of the sales potential of coproducts remain an immediate possibility. Coproduct credits have long been an essential feature of estimates of ethanol production (Section 6.3); among these, electricity generation has been frequently regarded as readily engineered into both existing and new ethanol production facilities—especially with sugarcane as the feedstock for ethanol production. For example, in Brazil, steam turbines powered by combustion of sugarcane bagasse can generate 1 MWh per cubic meter (1,000 L) of ethanol, and economic analysis shows that this is viable if the selling price for electricity is >$30 per MWh (the sales price in Brazil in 2005).[66] Pyrolysis of sugarcane wastes to produce "bio-oil" could yield 1.5 ton of saleable product per ton of raw sugar used, but the selling cost of the product will be crucial for establishing a viable coproduction process.[67]

6.4.3 BIOPROCESS ECONOMICS: A CHINESE PERSPECTIVE

China's demand for oil as a transportation fuel is forecast to increase more than 10-fold between 1990 and 2030 (from 30 to 396 million tonnes), reaching 50% of that of the United States by that date.[68] Economic analysis has shown that sweet sorghum and its bagasse as well as rice hulls and corn stover have extensive availability in northern China and could represent attractive feedstocks for cellulosic ethanol production.[69]

Investigations into gasoline supplementation with endogenously produced ethanol began in 1999 and, by 2004, E90 grades were available in eight provinces; a renewable energy law and a national key R&D program for cellulosic ethanol were applied to the energy sector during 2005.[70] Other primary factors in China's stated interest in cellulosic ethanol include:

- Fuel ethanol production was initiated from cereal grain feedstocks, but a government-set selling price of $1.65 per gallon required large subsidies because the production costs were over $1.80 per gallon.
- Total biomass production could exceed the International Energy Agency's prediction for transportation fuel needs by 2030 at a low feedstock cost ($22 per dry ton).
- Assuming successful implementation of the types of cellulosic ethanol technologies on which recent U.S. and European cost models have been based, production cost estimates for Chinese production sites would be in the range $0.43–0.95 per gallon.

Building and operating commercial cellulosic ethanol plants in China thus appears very feasible and would generate exactly the kind of practical experience and knowledge that would induce other nations to invest. The Chinese government announced the allocation of $5 billion in capital investment in the coming decade for ethanol production capacity with a focus on noncereal feedstocks.[70]

STEM TOPIC 6.3: CAN EUROPEAN CELLULOSIC ETHANOL EVER COMPETE FOR COST?

European assessments have tended to be the highest for production costs with both cereal ethanol and cellulosic ethanol. Constructing a complete train of events for the biomanufacture of cellulosic ethanol considered all events from growing the biological feedstock to the retail price of ethanol on the forecourt[71] (see Figure STEM 6.3.1).

Using actual oil prices and realistic estimates for biomass sources, the results can be presented in the form of a trend line of the breakeven oil price versus the biomass feedstock price (see Figure STEM 6.3.2). With softwoods, the breakeven oil price was the range of $125–150 per barrel, a figure approximated by the peak oil price in 2008. With cereal straw, only a much higher oil price ($175 per barrel) would give economic competitiveness.

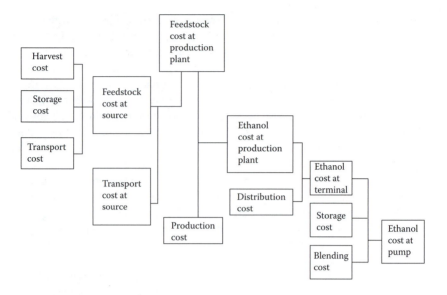

FIGURE STEM 6.3.1 Schematic cost structure evolution for cellulosic ethanol under European conditions.

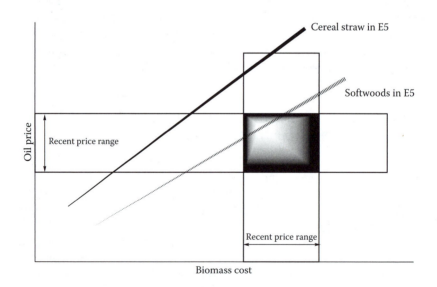

FIGURE STEM 6.3.2 Breakeven line for oil price and biomass feedstock price for cellulosic ethanol in Europe.

6.5 GOVERNMENTAL AND MACROECONOMIC FACTORS

6.5.1 MANDATORY BIOFUELS TARGETS

National targets for the use of biofuels became common before 2005 (Table 6.17). Subsequently, the dates for target achievement and even their desirability have been questioned more in Europe, where environmental issues have increasingly dominated the debate; cellulosic ethanol and other sustainable biofuels have been the beneficiaries—at least, in terms of funding programs—but this has not hastened the onset of industrial production. The most far reaching, in terms of impact on advanced biofuels programs, is the renewable fuel standard (RFS)—a provision of the U.S. Energy Policy Act of 2005 and the Energy Independence and Security Act of 2007.

The RFS mandated the increase of renewable fuels use to 36 billion gallons annually by 2022. The mechanism is the blending of biofuels (mostly corn ethanol) with conventional gasoline; this percentage-based standard is calculated by the Environmental Protection Agency: 7.76% in 2008, increased to 10.21% in 2009 (equivalent to 11 billion gallons). Funding programs were also instituted for cellulosic

TABLE 6.17
National Targets for Biofuels Use

Country	Target (% of Biofuels in Total Road Fuel Consumption)	Target Deadline	Production Incentives?	Consumption Incentives?
		Ethanol		
United States	2.78[a]	2006	√	√
Brazil	40% Rise in production	2010	√	√
Japan	500 million L	2010	×	×
Canada	3.5	2010	×	√
		Biofuels		
European Union	5.75	2010	√	√
Sweden	3	2005	√	√
France	10	2015	√	√
Germany	2	2005	√	√
United Kingdom	5	2020	√	√
India	5	(Unspecified)	√	√
China	15 (Total renewables)	2020	√	×
Thailand	2	2010	√	√

Source: Data from International Energy Agency. 2006. *World energy outlook,* chap. 14. Paris.

[a] Four billion gallons (2006) rising to 7.5 billion gallons by 2012.

ethanol and for sugar-to-ethanol production processes in Hawaii, Texas, Louisiana, and Florida.

The great advantage of the RFS and similar programs for actual and potential manufacturers of biofuels is the prospect of guaranteed markets; however, by themselves, they do not *guarantee* new production to meet the timelines. Anecdotal evidence and media reports in late 2009 indicated that the contribution of cellulosic ethanol to meeting RFS demands would experience serious shortfalls as early as 2010—reaching little more than 10% of the 100 million-gallon annual production target. Lack of financial credit and slow approval rates for federal funding programs have been blamed. More optimistically, corn ethanol production could increase to 25 billion gallons annually by 2030, doubling 2009's expected figure.

Corn has its own problems, however, in the form of massive price inflation, which peaked in mid-2008 (http://www.ams.usda.gov/mnreports/lswethanol.pdf); this caused a spike in corn ethanol prices (Figure 6.5). Corn prices in late 2009 were approximately 80% of those a year earlier, but ethanol prices remained considerably higher than gasoline prices (Figure 6.5). Corn ethanol is a mature industry that is now subject to market factors other than those of yearly crop yield; the 2009 crop was estimated as 7% greater than in 2008 and the second largest harvest in recorded history (on USDA figures). Nevertheless, attempts in 2008 to delay the implementation of the RFS because it could entail severe economic harm—high corn ethanol prices compared with rapidly declining gasoline prices—may presage future problems if oil markets remain depressed.

National targets imply incentives for consumption as well as for production, as discussed briefly in Section 6.3.2. Globally, the use of such incentives has proved to be highly variable; in the United States, with corn ethanol production centered in a few Midwestern states, production and consumption incentives are more predictable but still fail to present a totally coherent use of these policy tools. A summary prepared in late 2006 captured some of this complexity[72]:

- Production incentives were highest in Illinois (Illinois Renewable Fuels Development Program), Iowa (Iowa Renewable Fuel Funds Assistance program), and Minnesota, South Dakota, and Nebraska (each state had its ethanol production incentive); these could be expressed in terms of maximum amounts of funding available per volume capacity or additional capacity created, ranging from $0.026 to 0.052 per liter ($0.099–0.197 per gallon).
- Other states in the region had lower incentives (Wisconsin, Montana, Wyoming, Missouri, and Indiana).
- A number of states offered tax exemptions in addition to the federal exemption; these were highest in Idaho (then with no active fuel ethanol production). Only two of the main corn ethanol producers (South Dakota and Iowa) offered additional exemptions on state excise taxes over and above federal tax exemptions.

The use of these incentives to support biofuels production has attracted criticism and prompted calls for the removal of the federal domestic ethanol tax credit

($0.45/gallon in 2009), but it is often forgotten that the oil and gas industries have enjoyed subsidies for much longer than have biofuels. When the U.S General Accountability Office's assistant director for energy issues, natural resources, and the environment made a presentation to a conference in September 2006, he gave an oversight of the data.[73] Tax incentives to the U.S. ethanol industry between 1981 and 2005 amounted to only 12% of those to oil and gas between 1968 and 2005 (Figure 6.7). These figures underestimate the full sums expended in incentives for oil and gas provision because they date only from when full records of revenue losses were kept by the U.S. Treasury, not when an incentive was implemented (the Tariff Act of 1913, in the case of the oil and gas industry).

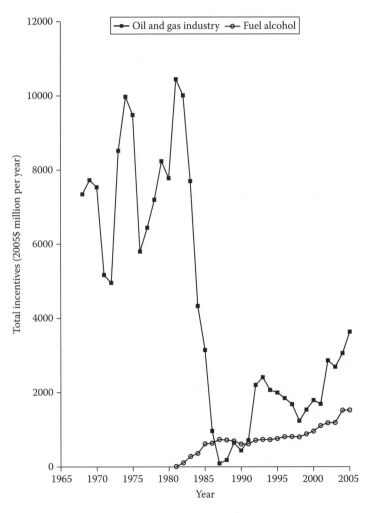

FIGURE 6.7   U.S. tax incentives and subsidies to the oil/gas and fuel alcohol industries. (Data from Agbara, G. M. 2006. *Federal Energy Tax Incentives and Subsidies and the Current State of Biomass Fuels: A View from Congressional Oversight.* Available from http://www.rice.edu/energy/events/past/biofuels2006.html.)

Although the magnitudes of some subsidies for conventional fuels are much reduced presently as compared to the 1970s and 1980s, they still outweigh the sums laid out to support biofuels. An incentives culture therefore has a long history in shaping and managing energy sectors. This supports continuing assistance for biofuels production R&D and its industrial implementation if biofuels can deliver on environmental and GHG emission targets.

STEM TOPIC 6.4: ETHANOL TAX INCENTIVES FOR BIOFUELS USE

Data presented in 2007 demonstrated the gross disparities in national tax policies for biofuels (see Table STEM 6.4).[74]

The case of Norway (with its strong disincentives for biodiesel use) is particularly instructive. Norway has abundant offshore oil and natural gas resources plus so many hydroelectricity schemes that its natural gas can be almost entirely exported. Its immediate neighbors, Sweden and Finland,

TABLE STEM 6.4
Excise Tax Rates for Conventional and Biofuels (2006)

Nation	Unleaded Gasoline	Excise Tax (US Cents/L)		
		E10	Diesel	Biodiesel
Incentives				
United States	4.9	3.5		
Canada	8.6	7.8		
Russia	35%[a]	25%[a]	35%[a]	25%[a]
France	73.1	65.8	48.5	48.5
Germany	73.9	72.9	51.1	51.1
The Netherlands	72.3	71.7	36.2	36.2
Spain	46.3	41.7	33.1	33.1
Austria	50.8	80.8	35.3	31.9
Switzerland	56.8	56.8	59.0	56.8
Neutral				
United Kingdom	94.0	94.0	94.0	94.0
Czech Republic	36.8	36.8	27.7	27.7
Disincentives				
Norway	95.6	95.6	64.1	95.6
Mexico	66.6%[a]	78.9%[a]		
Greece	37.1	37.1	31.4	37.1

[a] Variable dependent on world oil price.

lack these natural resources but have massive tracts of forests and have been among world leaders in cellulosic ethanol R&D. Interest in biofuels is widespread in Norway, but biofuels research there is minor in comparison to development in the rest of Scandinavia, according to SINTEF, the largest independent scientific research organization in the region (www.sintef.no).

Russia, in contrast, has no public involvement with biofuels but has a track record of promoting their use. Greece's prejudice against biodiesel could result from fears that olive oil markets will be distorted.

6.5.2 Impact of Fuel Economy on Ethanol Demand for Gasoline Blends

Improvements in automobile fuel economy would unambiguously improve the chances of an easier and better managed introduction of biomass-based fuel alcohols. For example, doubling the mileage achieved by gasoline-fueled vehicles in the United States would reduce the demand for ethanol by 45% or more at ethanol/gasoline blends of 10% or higher (Figure 6.8).[75] Mandatory fuel economy standards and voluntary agreements with automobile manufacturers in OECD and other countries aim at varying degrees of improved mileage in passenger cars and light commercial vehicles (Table 6.18).

The cumulative effects of these measures inevitably run counter to those of incentives that are intended to encourage biofuels consumption by reducing the size of novel fuel markets. This is unavoidable and no more a threat than that posed by any alternative energy source derived from physics and geology (solar, wind, wave, tidal,

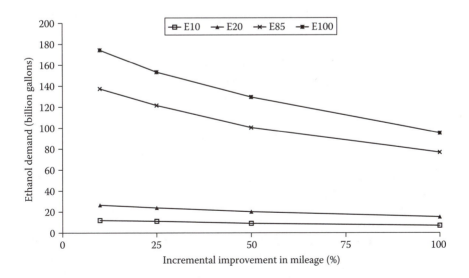

FIGURE 6.8 The impact of fuel economy on projected demand for ethanol in various blends with gasoline. (Data from Morrow, W. R., Griffin, W. M., and Matthews, H. S. 2006. *Environmental Science & Technology* 40:2877.)

TABLE 6.18

Selected Policies on Light-Duty Vehicle Fuel Economy

Country	Target	Target Deadline	Policy Basis
United States	20.7–22.2 mpg	2007	Mandatory
	24 mpg	2011	Mandatory
Japan	23% Reduction in fuel consumption (cars)	Progressive	Mandatory
	13% Reduction in fuel consumption (light trucks)	Progressive	Mandatory
China	10% Reduction in fuel consumption	2005	Mandatory
	20% Reduction in fuel consumption	2008	Mandatory
Australia	18% Reduction in fuel consumption (cars)	2010	Voluntary
Canada	Increase in corporate average fuel economy in line with U.S. standards	2007–2011	Voluntary

Source: Modified from International Energy Agency. 2006. *World energy outlook,* Chap. 14. Paris.

geothermal, etc.)—all of which will be essential if electric vehicles are to be powered sustainably (see Chapter 9).

6.5.3 Biofuels Pricing in the Era of Carbon Taxation

The Stern Report was commissioned by the UK government and collected data on the economic costs of global warming caused by the accumulation of atmospheric CO_2.[76] The headline figures were widely quoted:

- Extreme weather could reduce global gross domestic product (GDP) by up to 1% per year.
- A 2–3°C rise in temperatures could reduce global economic output by 3%.
- If temperatures rise by 5°C, up to 10% of global output could be lost, and the poorest countries would lose more than 10%.
- In the worst-case scenario, global consumption per head would fall 20%.
- To stabilize at manageable CO_2 levels, emissions would need to stabilize by 2025–2030 and fall between 1 and 3% in the succeeding years. This would cost 1% of GDP annually.

The methodology employed took average global temperature rises of between 1 and 6°C and defined physical changes consequent on these rises that could then be transformed into economic costs of the ensuing damage. At the extreme, entire global regions would experience major declines in crop yields (e.g., up to a third in Africa). The effects of intermediate temperature rises are more difficult to calculate accurately; they include falling crop yields in developed countries and more frequent floods, forest fires, droughts, and heat waves. Recent historical events can be used in the computations, adding in real sums from after-the-event assessments; for example, the unusual heat waves in 2003 in Europe resulted in 35,000 deaths and agricultural losses reaching $15 billion.

Chapter 22 of the report was entitled "Creating a Global Price for Carbon"—a prescient title for subsequent events. Recent years have introduced carbon footprints and carbon neutrality as everyday events. The most likely move for the next decade is to tax carbon outputs. This could move the debate about incentives for biofuels into its converse (i.e., taxing nonrenewables); a broadly similar price of carbon is necessary and will be a severe challenge for international collective action. As part of this developing carbon economy, corn ethanol, with its failure to convince scientists of its minor dependence on conventional fuels, might fare only a little less badly than gasoline and diesel. Conversely, sugarcane ethanol and cellulosic ethanol could be competitively priced no matter what their actual costs of production were.

Individual ethanol facilities (corn, sugarcane, or cellulosic) might be mandated to prepare eco audits on energy use on which to base their "factory gate" prices. Energy-inefficient processes or sites would simply disappear or—as happened during 2009—be retrofitted to biomanufacture other and more advanced biofuels.

6.6 SUMMARY

Detailed cost models for ethanol produced from corn, sugarcane, and cellulosic biomass have been prepared and discussed since the late 1970s. For most of the three succeeding decades, biofuels production cost estimates were 2–10 times higher than those for conventional gasoline, allowing for differences in energy content. Starting in 2004 or 2005, however, corn and sugarcane ethanol were competitive with gasoline, and even cellulosic ethanol began to overlap in this era of rapidly inflating global oil prices.

That change in the economic case for biofuels has unfortunately been forgotten, replaced by another effect of rapidly rising oil prices: large increases in corn prices that distorted corn ethanol production costs and—more importantly—generated the public perception that biofuels themselves caused food prices worldwide to increase and penalize the poor.

The final retail price of biofuels (corn and sugarcane ethanol in 2009) is affected by federal and state tax regimes and a confused global picture of consumption incentives. Only Brazil has developed a mature market place where ethanol blends and gasoline can be purchased selectively on the basis of daily price movements.

National targets for biofuels use have also created markets for ethanol and biodiesel. Production incentives have encouraged expansion of corn and sugarcane production, but cellulosic ethanol production remains at small or experimental scales.

Existing and future strategies to restrain fossil fuel use and slow global atmospheric CO_2 accumulation will focus on taxing carbon. Some—but not all—forms of ethanol will acquire price competitiveness with respect to conventional gasoline. International agreements on carbon taxation will be necessary for a coherent economic framework for the use of biofuels and other renewable forms of energy to emerge.

REFERENCES

1. Cox, S., ed. 2006. *Economics. Making sense of the modern economy.* London: *The Economist Newspaper* Ltd. in association with Profile Books Ltd.
2. Holden, C. 1985. Is bioenergy stalled? *Science* 227:1018.
3. Herrera, S. 2006. Bonkers about biofuels. *Nature Biotechnology* 24:755.
4. Anonymous. 2006. The dragon and the eagle. In *Economics. Making sense of the modern economy,* ed. Cox, S., 89. London: *The Economist Newspaper* Ltd. in association with Profile Books Ltd.
5. Wheals, A. E. et al. 1999. Fuel ethanol after 25 years. *Trends in Biotechnology* 17:482.
6. Pimentel, D. 1991. Ethanol fuels: Energy security, economics, and the environment. *Journal of Agricultural and Environmental Ethics* 4:1.
7. Patzek, T. W. et al. 2005. Ethanol from corn: Clean, renewable fuel for the future, or drain on our resources and pockets? *Environment, Development and Sustainability* 7:319.
8. Zechendorf, B. 1999. Sustainable development: How can biotechnology contribute? *Trends in Biotechnology* 17:219.
9. Martines-Filho, J., Burnquist, H. L., and Vian, C. E. F. 2006. Bioenergy and the rise of sugarcane-based ethanol in Brazil. *Choices* 21(2):91.
10. Moreira, J. R., Noguiera, L. A. H., and Parente, V. 2005. Biofuels for transport, development, and climate change: Lessons from Brazil. In *Growing in the greenhouse: Protecting the climate by putting development first,* ed. Bradley, R. and Baumert, K. A., chap. 3. Washington, D.C.: World Resources Institute.
11. Hahn-Hägerdal, B. et al. 2006. Bio-ethanol—The fuel of tomorrow from the residues of today. *Trends in Biotechnology* 24:549.
12. Paul, J. K., ed. 1980. *Large and small scale ethyl alcohol processes from agricultural raw materials.* Park Ridge, NJ: Noyes Data Corporation.
13. Wilke, C. R. et al. 1981. Raw materials evaluation and process development studies for conversion of biomass to sugars and ethanol. *Biotechnology and Bioengineering* 23:1681.
14. Westby, C. A., and Gibbons, W. R. 1982. Farm-scale production of fuel ethanol and wet grain from corn in a batch process. *Biotechnology and Bioengineering* 24:1681.
15. Gibbons, W. R., Westby, C. A., and Dobbs, T. L. 1984. A continuous, farm-scale, solid-phase fermentation process for fuel ethanol and protein feed production from fodder beets. *Biotechnology and Bioengineering* 26:1098.
16. Keim, C. R., and Venkatasubramanian, K. 1989. Economics of current biotechnological methods of producing ethanol. *Trends in Biotechnology* 7:22.
17. Jaeger, W. K., Cross, R., and Egelkraut, T. M. 2007. *Biofuel potential in Oregon: Background and evaluation of options.* Oregon State University, available at http://extension.oregonstate.edu/catalog/pdf/sr/sr1078.pdf.
18. Marrow, J. E., Coombs, J., and Lees, E. W. 1987. *An assessment of bio-ethanol as a transport fuel in the UK.* London: Her Majesty's Stationery Office.
19. Marrow, J. E., and Coombs, J. 1987. *An assessment of bio-ethanol as a transport fuel in the UK—Volume 2.* London: Her Majesty's Stationery Office.
20. Commission of the European Communities. 1987. *Cost/benefit analysis of production and use of bioethanol as a gasoline additive in the European Community.* Luxembourg: Office for Official Publications of the European Communities.
21. Geller, H. S. 1985. Ethanol fuel from sugar cane in Brazil. *Annual Review of Energy* 10:135.
22. Elander, R. T., and Putsche, V. L. 1996. Ethanol from corn: Technology and economics. In *Handbook on bioethanol: Production and utilization,* ed. Wyman, C. E., chap. 15. London: Taylor & Francis.

23. Lynd, L. R. 1996. Overview and evaluation of fuel ethanol from cellulosic biomass: technology, economics, the environment, and policy. *Annual Review of Energy and the Environment* 21:403.
24. Lynd, L. R., Elander, R. T., and Wyman, C. E. 1996. Likely features and costs of mature biomass ethanol technology. *Applied Biochemistry and Biotechnology* 57–58:741.
25. Wayman, M., and Dzenis, A. 1984. Ethanol from wood: Economic analysis of an acid hydrolysis process. *Canadian Journal of Chemical Engineering* 62:699.
26. Ladisch, M. R., and Svarckopf, J. A. 1991. Ethanol production and the cost of fermentable sugars from biomass. *Bioresource Technology* 36:83.
27. von Sivers, M. et al. 1994. Cost analysis of ethanol from willow using recombinant *Escherichia coli. Biotechnology Progress* 10:555.
28. von Sivers, M., and Zacchi, G. 1995. A techno-economical comparison of three processes for the production of ethanol from wood. *Bioresource Technology* 51:43.
29. Galbe, M. et al. 2007. Process engineering economics of bioethanol production. *Advances in Biochemical Engineering/Biotechnology* 108:303.
30. von Sivers, M., and Zacchi, G. 1999. Ethanol from lignocellulosics: A review of the economy. *Bioresource Technology* 56:131.
31. Wingren, A., Galbe, M., and Zacchi, G. 2003. Techno-economic evaluation of producing ethanol from softwood: Comparison of SSF and SHF and identification of bottlenecks. *Biotechnology Progress* 19:1109.
32. Wingren, A. et al. 2004. Process considerations and economic evaluation of two-step steam pretreatment for production of fuel ethanol from softwood. *Biotechnology Progress* 20:1421.
33. Wooley, R. et al. 1999. Process design and costing of bioethanol technology: A tool for determining the status and direction of research and development. *Biotechnology Progress* 15:794.
34. Hamelinck, C. N., van Hooijdonk, G., and Faaij, A. P. C. 2005. Ethanol from lignocellulosic biomass: Techno-economic performance in short, middle, and long term. *Biomass & Bioenergy* 28:384.
35. Bohlmann, G. M. 2006. Process economic considerations for production of ethanol from biomass feedstocks. *Industrial Biotechnology* 2:14.
36. International Energy Agency. 2006. *World energy outlook,* chap. 14. Paris.
37. Goldemberg, J. 1996. The evolution of ethanol costs in Brazil. *Energy Policy* 24:1127.
38. Moreira, J. R., and Goldemberg, J. 1999. The alcohol program. *Energy Policy* 27:229.
39. Wyman, C. E. 1999. Biomass ethanol: Technical progress, opportunities, and commercial challenges. *Annual Review of Energy and the Environment* 24:189.
40. Pimentel, D., Patzek, T., and Cecil, G. 2007. Ethanol production: Energy, economic, and environmental losses. *Reviews of Environmental Contamination & Toxicology* 189:25.
41. Hamelinck, C. N., and Faaij, A. P. C. 2006. Outlook for advanced biofuels. *Energy Policy* 34:3268.
42. Kwiatkowski, J. R. et al. 2006. Modeling the process and costs of fuel ethanol production by the corn dry-grind process. *Industrial Crops and Products* 23:288.
43. Kumar, A., and Sokhansanj, S. 2007. Switchgrass (*Panicum virgatum* L.) delivery to a biorefinery using integrated biomass supply analysis and logistics (BSAL) model. *Bioresource Technology* 98:1033.
44. Mani, S., Tabil, L. G., and Sokhansanj, S. 2006. Specific energy requirement for compacting corn stover. *Bioresource Technology* 97:1420.
45. Kumar, A., Cameron, J. B., and Flynn, P. C. 2005. Pipeline transport and saccharification of corn stover. *Bioresource Technology* 96:819.
46. Burnes, E. I. et al. 2004. Economic analysis of ethanol production in California using traditional and innovative feedstock supplies. *Applied Biochemistry and Biotechnology* 113–116:95.

47. Fan, Z., and Lynd, L. R. 2007. Conversion of paper sludge to ethanol. II. Process design and economic analysis. *Bioprocess Biosystems and Engineering* 30:35.
48. Ag-West biotech, Inc. http://www.agwest.sk.ca/publications/infosource/inf_may98.php.
49. Kádár, Z., Szengyel, Z., and Réczey, K. 2004. Simultaneous saccharification and fermentation (SSF) of industrial wastes for the production of ethanol. *Industrial Crops and Products* 20:103.
50. del Campo, I. et al. 2006. Diluted acid hydrolysis pretreatment of agri-food wastes for bioethanol production. *Industrial Crops and Products* 24:214.
51. Rosenberger, A. et al. 2002. Costs of bioethanol production from winter cereals: The effect of growing conditions and crop production intensity levels. *Industrial Crops and Products* 15:91.
52. Chandel, A. K. et al. 2007. Economics and environmental impact of bioethanol production technologies: An appraisal. *Biotechnology and Molecular Biology Review* 2:14.
53. Gregg, D. J., Boussaid, A., and Saddler, J. N. 1998. Techno-economic evaluations of a generic wood-to-ethanol process: effect of increased cellulose yields and enzyme recycle. *Bioresource Technology* 63:7.
54. Nagle, N., Ibsen, K., and Jennings, E. A. 1999. Process economic approach to develop a dilute-acid cellulose hydrolysis process to produce ethanol from biomass. *Applied Biochemistry and Biotechnology* 77–79:595.
55. Eggman, T., and Elander, R. T. 2005. Process and economic analysis of pretreatment technologies. *Bioresource Technology* 96:2019.
56. Keller, F. A., Hamilton, J. E., and Nguyen, Q. A. 2003. Microbial pretreatment of biomass: potential for reducing severity of thermochemical biomass pretreatment. *Applied Biochemistry and Biotechnology* 105–108:27.
57. Zhang, Y.-H. P. et al. 2007. Fractionating recalcitrant lignocellulose at modest reaction conditions. *Biotechnology and Bioengineering* 97:214.
58. Jorgensen, H. et al. 2007. Liquefaction of lignocellulose at high-solids concentrations. *Biotechnology and Bioengineering* 96:862.
59. Manderson, G. J. et al., Price sensitivity of bioethanol produced in New Zealand from *Pinus radiata* wood. *Energy Sources* 11:135.
60. Alkasrawi, M., Galbe, M., and Zacchi, G. 2002. Recirculation of process streams in fuel ethanol production from softwood based on simultaneous saccharification and fermentation. *Applied Biochemistry and Biotechnology* 98–100:849.
61. York, S. W., and Ingram, L. O. 1996. Soy-based medium for ethanol production by *Escherichia coli*. *Journal of Industrial Microbiology and Biotechnology* 16:374.
62. Ashgari, A. et al. 1996. Ethanol production from hemicellulose hydrolysates of agricultural residues using genetically engineered *Escherichia coli* strain KO11. *Journal of Industrial Microbiology and Biotechnology* 16:42.
63. Escobar, J. M., Rane, K. D., and Cheryan, M. 2001. Ethanol production in a membrane bioreactor: Pilot-scale trials in a wet corn mill. *Applied Biochemistry and Biotechnology* 91–93:283.
64. Taylor, F. et al. 2000. Dry-grind process for fuel ethanol by continuous fermentation and stripping, *Biotechnology Progress* 16:541.
65. Zacchi, G., and Axelsson, A. 1989. Economic evaluation of preconcentration in production of ethanol from dilute sugar solutions. *Biotechnology and Bioengineering* 34:223.
66. van Haandel, A. C. 2005. Integrated energy production and reduction of the environmental impact at alcohol distillery plants, *Water Science and Technology* 52:49.
67. Alonso Pippo, W., Garzone, P., and Cornacchia, G. 2007. Agro-industry sugarcane residues disposal: The trends of their conversion into energy carriers in Cuba. *Waste Management* 27:869.
68. International Energy Agency. 2006. *World energy outlook,* annex A. Paris.

69. Gnansounou, E., Dauriat, A., and Wyman, C. E. 2005. Refining sweet sorghum to ethanol and sugar: Economic trade-offs in the context of North China. *Bioresource Technology* 96:985.
70. Yang, B., and Lu, Y. 2007. The promise of cellulosic ethanol production in China. *Journal of Chemical Technology and Biotechnology* 82:6.
71. Slade, R., Bauen, A., and Shah, N. 2009. The commercial performance of cellulosic ethanol supply-chains in Europe. *Biotechnology for Biofuels* 2:3.
72. Mabee, W. E. 2007. Policy options to support biofuels production. *Advances in Biochemical Engineering/Biotechnology* 108:329.
73. Agbara, G. M. 2006. *Federal energy tax incentives and subsidies and the current state of biomass fuels: A view from Congressional oversight.* Available from http://www.rice.edu/energy/events/past/biofuels2006.html.
74. Mabee, W. E. 2007. Policy options to support biofuel production. *Advances in Biochemical Engineering/Biotechnology* 108:329.
75. Morrow, W. R., Griffin, W. M., and Matthews, H. S. 2006. Modeling switchgrass derived cellulosic ethanol distribution in the United States. *Environmental Science & Technology* 40:2877.
76. Stern, N. 2007. *The economics of climate change.* Downloadable from http://www.hm-treasury.gov.uk/stern_review_report.htm.

7 Advanced Biofuels
The Widening Portfolio of Alternatives to Ethanol

7.1 INTRODUCTION

Ethanol has a long history of use as a combustible fuel. What other low molecular weight alcohols are biosynthesized by microbial cells and have any advantages over ethanol? Glycerol is a well-known side product of ethanol fermentations with yeasts and strategies for its suppression have long been explored. Turning the equation on its head, can glycerol production be easily maximized as a route to biofuels?

Moving beyond alcohols, are other volatile organic compounds involved in the central metabolism of microorganisms and that could be produced efficiently from cellulosic feedstocks?

One simple gaseous alkane, methane, is the staple product of biogas, used throughout the world in low-technology applications. Less obvious is the capability of both microorganisms and plants to biomanufacture hydrogen gas as a product that is the one unambiguous carbonless fuel when it is combusted. Can biohydrogen be the long-term solution to greenhouse gas emissions and provide an alternative to gasoline in internal combustion engines or in more advanced engineering solutions?

At the extreme of the technological spectrum, do we need combustible fuels or are there already electrochemical engineering solutions that harness the energy metabolism of microbes to provide electrical power?

7.2 BIOBUTANOL AND ABE

The ABE (acetone, butanol, and ethanol in various proportions) fermentation-producing solvents have a long history of use for industrial production in Europe, where it was one of the largest industrial fermentation processes early in the twentieth century. It was later entirely superseded by petrochemical routes; with the oil crises of the 1970s, renewed interest in this bioprocess was aided by the accelerating advance of microbial physiology and genetics at that time.[1-3] The process therefore predates fuel ethanol biomanufacture; even more bizarrely, process development continued in the U.S.S.R. until at least the 1980s and in China until the 1990s.[4,5]

Rather than discussing a novel bioprocess, therefore, ABE production is an example of a fermentation route that can be rediscovered or resurrected. The microbial species capable of this multiproduct biosynthesis are clostridia, which also have remarkable appetites for cellulosic and hemicellulosic polymers, that are able to metabolize hexose sugars and pentoses (usually, xylose and arabinose).[6,7]

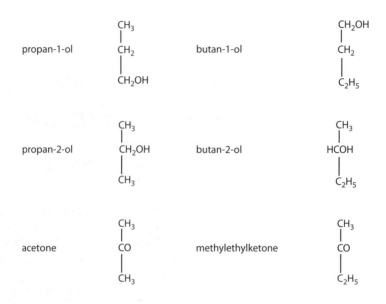

FIGURE 7.1   Candidate advanced C3 and C4 biofuels.

The neologism "biobutanol" (for *n*-butanol and its isomers with the structural formula C_4H_9OH; see Figure 7.1) appeared when DuPont (www2.dupont.com) and British Petroleum announced their intention to produce and market butanol as a biofuel in 2007, arguing the case for the superiority of butanol over ethanol:

- Butanol has a higher energy content than ethanol and can be blended with gasoline at higher concentrations for use in standard vehicle engines (11.5% in the United States, with the potential to increase to 16%).
- Suitable for transport in pipelines, butanol has the potential to be introduced into gasoline easily and without additional supply infrastructure.
- Butanol/gasoline mixtures are less susceptible to separation in the presence of water than ethanol/gasoline blends and demand no essential modifications to blending facilities, storage tanks, or retail station pumps.
- Butanol's low vapor pressure (lower than gasoline) means that vapor pressure specifications do not need to be compromised.
- Production routes from conventional agricultural feedstocks (corn, wheat, sugarcane, beet sugar, cassava, and sorghum) are all possible, supporting global implementation.
- Lignocellulosics from fast growing energy crops (e.g., grasses) or agricultural wastes (e.g., corn stover) are also feasible feedstocks.

The principal hurdles to process optimization are in manipulating cultures and strains to improve product specificity (Figure 7.2) and yield and in reducing the toxicity of butanol and O_2 (the fermentation must be strictly anaerobic) to producing cells.[7,8] This toxicity is not remarkable because simple alcohols (including ethanol) are widely used as surface disinfectants. In a solventogenic

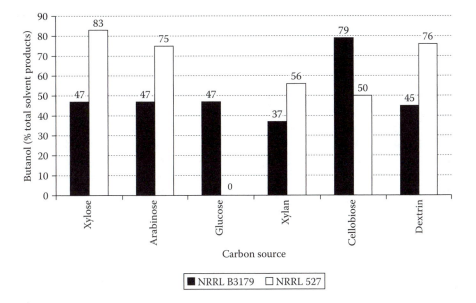

FIGURE 7.2 Variation in butanol production with two strains of *Clostridium acetobutylicum* grown on six different carbon sources. (Data from Singh, A., and Mishra, P. 1995. *Microbial Pentose Utilization. Current Applications in Biotechnology (Progress in Industrial Microbiology,* vol. 33), chap 7. New York: Elsevier Science & Technology.)

fermentation, however, product toxicity is a serious problem, causing sporulation as a drastic adaptation.[9]

Commencing in 2000, several notable advances were made in the biotechnology of butanol production:

- Hyperproducing strains can be isolated. *Clostridium beijerinckii* BA101 expresses high activities of amylase when grown in starch-containing media, accumulating solvents up to 29 g/L and as high as 165 g/L when adapted to a fed-batch fermentation with product recovery by pervaporation using a silicone membrane.[10–12]
- Gas stripping has also been developed as a cost-effective means to remove butanol and reduce any product inhibition.[13]
- At the molecular level, the high product yields with hyperproducing strains can be ascribed to a defective glucose transport system exhibiting poor regulation and a more efficient use of glucose during the solventogenic stage.[14]
- It has been demonstrated that the ABE fermentation can utilize corn fiber sugars (glucose, xylose, arabinose, and galactose) and is not inhibited by major sugar degradation products of pretreated lignocellulosic substrates.[15,16]
- A single clostridial gene can be overexpressed to increase both solvent production and producer cell tolerance of product accumulation.[17]
- Understanding of the molecular events causing loss of productivity in solventogenic strains spontaneously or during repeated subculturing or continuous fermentation has improved.[18,19]

A technoeconomic evaluation of a production facility with an annual capacity of 153,000 tonnes published in 2001 estimated production costs for butanol of $0.29 per kilogram ($0.89 per gallon, equivalent to $1.08 using 2009 prices). It assumed a conversion efficiency of 0.50 g of product per gram of glucose, with corn as the feedstock.[20]

The downstream processing operations for the ABE fermentation are necessarily more complex than for fermentations with a single product (e.g., ethanol). Not only can the insoluble materials from the harvested fermentation be used as a source for animal feed, but the fermentation broth must also be efficiently fractionated to maximize the economic returns possible from three saleable solvent products. Detailed analysis of a conventional downstream process modeled solvent extraction, solvent stripping, and two distillation steps to recover 96% of the butanol from a butanol-dominated mix of products.[21] An optimal arrangement of these downstream steps could reduce the operating costs by 22%.

Advanced bioprocess options have been devised to avoid product toxicity:

- a continuous two-stage fermentation design to maintain the producing cells in the solventogenic stage[22]
- packed bed biofilm reactors with *C. acetobutylicum* and *C. beijerinckii*[23]
- a continuous production system with a high cell density obtained by cell recycling and capable of operation for more than 200 h without strain degeneration or loss of productivity[24]
- investigation of simultaneous saccharification and fermentation processes by adding exogenous cellulase to poorly cellulolytic strains[25]
- investigation of a continuous flash fermentation process for the production of butanol consisting of three interconnected units: fermentor, cell-retention system using microfiltration, and vacuum flash vessel for the continuous recovery of butanol[26]

A novel feedstock for butanol production is sludge (i.e., the waste product in activated sludge processes for wastewater treatments); this material is generated at the rate of 4×10^7 m^3 per annum in Japan, and most of it is discharged by dumping.[27] Adding glucose to the sludge supported growth and butanol production and a marked reduction in the content of suspended solids within 24 h. Domestic organic waste (i.e., food residues) has been tested as a substrate for the clostridial ABE fermentation, using chemical and enzymic pretreatments. Growth and ABE formation were mainly supported by soluble sugars, and steam pretreatment produced inhibitors of growth and solvent formation.[28,29]

7.3 BACTERIAL PRODUCTION OF C3–C7 ALCOHOLS AND RELATED COMPOUNDS

Given the known limitations of clostridial strains (especially their slow growth rates), attention has begun to be paid to the metabolic engineering of butanol synthesis in other bacteria. The expression of a nonclostridial butanol-producing pathway in *Escherichia coli* is a promising strategy.[10,30] Low butanol titers in the

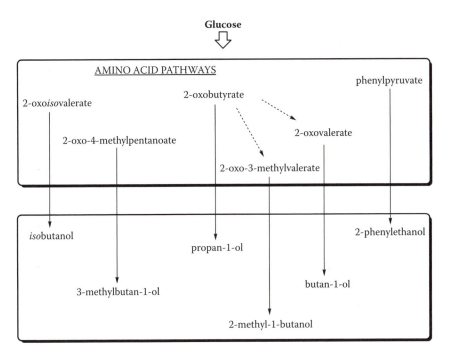

FIGURE 7.3 Engineered pathways for higher alcohol production.

fermentation broth are a problem, and both SSF and product removal by gas stripping have been investigated to reduce the cost of butanol recovery.[10] *Pseudomonas putida* strains have been adapted to grow in up to 6% (v/v) butanol, the highest reported butanol concentration tolerated by a microbe, and might be alternative hosts for butanol production.[31]

The genetic engineering of *E. coli* can be surprisingly versatile; alcohols containing between three and seven carbon atoms are biosynthesized with a 2-oxo acid decarboxylase and an alcohol dehydrogenase with relatively lax substrate selectivity (Figure 7.3).[32] Substrates for these novel pathways are intermediates of bacterial amino acid biosynthesis.

To produce propan-2-ol (isopropanol), the heterologous expression of a clostridial metabolic pathway was accomplished[33]:

- gene *thl* (acetyl-CoA acetyltransferase from *C. acetobutylicum*)
- gene *atoAD* (acetoacetyl-CoA transferase from *E. coli*)
- gene *adc* (acetoacetate decarboxylase from *C. acetobutylicum*)
- gene *adh* (secondary alcohol dehydrogenase from *C. beijerinckii*)

Such an engineered strain produced isopropanol with a yield of 43.5% (mole/mole basis); when the alcohol dehydrogenase was not used, the strain produced acetone at 73.5% (mole/mole basis).

Butan-2-one, or methyl ethyl ketone (MEK), is a close chemical relative of butan-2-ol (Figure 7.1). MEK could be biosynthesized by omitting the last step of the

2-butanol pathway, and a genetically engineered bacterium (*E. coli*) is the subject of a patent application.[34]

STEM TOPIC 7.1: CAN BACTERIA SYNTHESIZE GASOLINE HYDROCARBONS?

A study from Japan published in 2001 reported the isolation of a bacterium from the sludge of a sewage disposal plant that had the unusual ability to accumulate lipid material (including triglycerides) in the extracellular phase; even more remarkably, so much lipid material was produced that it formed a surface layer that contained not only the expected components but also easily detectable quantities of alkane hydrocarbons.[35] The bacterial species was identified as a strain of *Vibrio furnessi* on the basis of morphological, physiological, and DNA information.

A further curiosity was that the carbon chain lengths of the alkanes did not match those expected from simple decarboxylation reactions from the fatty acids that were also accumulated by the bacterium (see Figure STEM 7.1.1):

$$C_nH_{2n+1}.COOH \rightarrow C_nH_{2n+1} + CO_2$$

This prompted an evaluation of other biosynthetic pathways, and both radioactively labeled and unlabeled precursors were fed to membrane fractions from the bacterium and converted into alkanes without decarboxylation.[36] This implies a triple reduction pathway (see Figure STEM 7.1.2.)

Genetic engineering of this microbe could generate strains able to synthesize lower carbon alkanes (including octane) that would be directly analogous to gasoline.

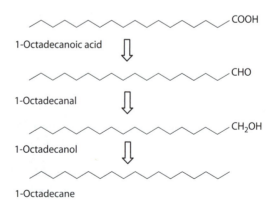

FIGURE STEM 7.1.1 Biosynthetic reduction of carboxylic acids to alkanes in *Vibrio furnessi* M1.

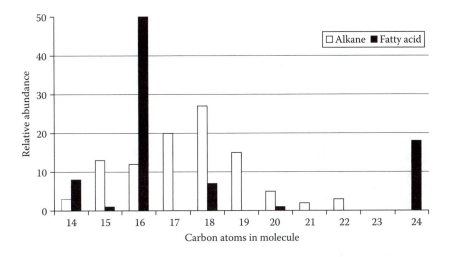

FIGURE STEM 7.1.2   Distributions of carboxylic acids and alkanes in *Vibrio furnessi* M1. (Data from Park, M.-O. 2005. *Journal of Bacteriology* 187:1426.)

7.4   GLYCEROL

Unlike ethanol, which is a major fermentation product of relatively few microbial species, glycerol is ubiquitous because of its incorporation into the triglycerides that are essential components of cellular membranes, as well as being accumulated in vegetable oils. Glycerol has a well-characterized biosynthetic route from glucose and other sugars (Figure 7.4). In fermentations for potable ethanol, the priority is to regulate glycerol accumulation because its formation is a waste of metabolic potential in

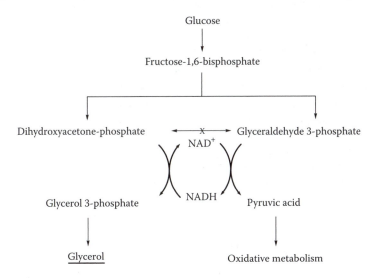

FIGURE 7.4   Glycerol biosynthesis in yeast genetically modified to eliminate interconversion of triose phosphates.

fuel alcohol production. Much research effort has therefore been devoted to *minimizing* glycerol formation by yeasts—for example, by regulating the glucose feeding rate to maintain an optimal balance of CO_2 production and O_2 consumption.[37]

Conversely, *maximizing* glycerol production by yeasts is also straightforward; successful strategies include[38]:

- adding bisulfite to trap acetaldehyde (an intermediate in the formation of ethanol), thus inhibiting ethanol production and forcing glycerol accumulation to restore the balance of intracellular redox cofactors
- growing yeast cultures at much higher pH values (7 or above) than traditionally used for ethanol fermentations
- using naturally osmotolerant yeasts—glycerol is often accumulated inside yeast cells to counteract the adverse effects of high osmotic pressures

Suitable osmotolerant strains can accumulate 13% (w/v) glycerol within 4 or 5 days. Even more productive is the osmophilic yeast *Pichia farinosa* that has been reported to produce glycerol at up to 30% (w/v) within 192 h in a fed-batch fermentation with glucose as the carbon source as a molar yield of glycerol from glucose of 0.90.[39] Fixing a maximum theoretical yield of glycerol from glucose is difficult because it is highly dependent on the totality of biochemical routes available to the producer cells. When *Saccharomyces cerevisiae* was genetically manipulated to overproduce glycerol, the glycerol yield was 0.50 g/g of glucose (i.e., a molar yield of 0.98).[40] The genetic background of this strain included deletion of the gene encoding the enzyme (triose phosphate isomerise) that normally would equilibrate the two possible metabolic routes of triose phosphates (Figure 7.4).

The yield of 1 mol of glycerol per mole of glucose consumed in the engineered strain may be the maximum obtainable because, in comparison with ethanol, production is too far up the glycolytic pathway for sufficient NADH to be available for further reduction. Although an osmotolerant *Saccharomyces* strain isolated from sugarcane molasses could accumulate high levels of glycerol (up to 260 g/L), the molar yield was still only 0.92.[41]

However, an osmotolerant *Candida glycerinogenes* was reported to produce glycerol with a molar yield of 1.20 ± 0.10.[42] This unusually high molar yield did not scale up from a 30–50,000 L fermentor (1.01 molar yield); a mutant strain with high glycerol 3-phosphate dehydrogenase activity in the early part of the fermentation also achieved a molar yield of only 1.03 (an unreplicated piece of data). It appears that 1 mol of glycerol per mole of glucose consumed remains the ceiling for productivity.[43]

Glycerol does not possess the physicochemical properties suitable for use as a fuel in internal combustion engines, but can be chemically converted to useful fuels for a range of applications (discussed in Section 8.5, Chapter 8). Glycerol is also a valuable carbon substrate for microbial fermentations (see Section 10.4.3, Chapter 10).

7.5 THE MIXALCO PROCESS

In this mixed fermentation and chemical process, microbial biochemistry is utilized to first generate a mixture of carboxylic acids, including propionic acid; the biocatalysts are anaerobic microbes.[44,45]

As with glycerol, carboxylic acids are not biofuels per se; after extraction and chemical reduction with hydrogen, a mixture of volatile alcohols can be generated, predominantly propan-2-ol (Figure 7.1). The MixAlco process is a patented technology able to convert *any* biodegradable material (including municipal solid waste, sewage sludge, manure, agricultural residues, and energy crops) into biofuels and is being commercialized to a pilot plant (www.terrabon.com). Furthermore, using zeolite catalysts, the alcohols can be converted into a hydrocarbon mix approximating gasoline with favorable economics[46]:

- A 40-tonne/hour plant processing municipal solid waste (incurring a $45/tonne tipping fee) and using hydrogen from a pipeline or refinery ($2.00/kg) could produce alcohols for $1.13 per gallon or gasoline for $1.75 per gallon.
- An 800-tonne/hour plant processing high-yield biomass (costing $60/tonne) and gasifying fermentation residues and waste biomass to hydrogen ($1.42/kg) could produce alcohols for $1.33 per gallon or gasoline for $2.04 per gallon.

7.6 BIOHYDROGEN

7.6.1 THE HYDROGEN ECONOMY AND FUEL CELL TECHNOLOGIES

The International Energy Agency in its 2006 review of world energy trends forecast that, by 2030, hydrogen-powered vehicles would have begun to "decarbonize" transportation—if, that is, production from low- and zero-carbon sources develops, if there are breakthroughs in hydrogen storage, and if the necessary infrastructure (requiring huge investments) develops.[47] The chemistry of hydrogen combustion entirely avoids greenhouse gas emissions:

$$2H_2 + O_2 \rightarrow 2H_2O$$

whether this occurs in thermal power generation or in any of the presently developed types of hydrogen fuel cell (Table 7.1).[48]

TABLE 7.1
Hydrogen Fuel Cells: Types, Fuels, and Power Ranges

Fuel Cell Type	Operating Temperature (°C)	Electric Efficiency (%)	Power Range (kW)
Alkaline	60–120	35–55	<5
Proton exchange membrane	50–100	35–45	5–120
Phosphoric acid	Approx. 220	40	200
Molten carbonate	Approx. 650	>50	200–MW
Solid oxide	Approx. 1,000	>50	2–MW

Source: Information from Hoogers, G. 2003. In *Fuel Cell Technology Handbook,* ed. Hoogers, G., chap.1-1. Boca Raton, FL: CRC Press.

STEM TOPIC 7.2: THERMODYNAMICS
OF WATER ELECTROLYSIS

In principle, generating H_2 from the most abundant potential source—water—is eminently straightforward (i.e., the electrolysis of water). The simplest demonstration of the electrolytic production of H_2 and O_2 (first accomplished in 1800) requires only an electrical power source connected to two electrodes placed in water. H_2 gas will appear at the cathode while O_2 gas appears at the anode, and both can be collected in inverted glass tubes placed over the electrodes.

A more sophisticated engineering utilizes a solid membrane electrolyzer with electrodes constructed from perfluorosulfonic acid polymer. Most importantly, such a construct is the inverse of a proton exchange membrane fuel cell in which H_2 and O_2 are recombined to *produce* electrical energy[49] (see Figure STEM 7.2.)

At the anode, water is split to O_2, electrons, and hydrated protons (hydroxonium ions):

$$3H_2O \rightarrow 2H_3O^+ + 2e^- + \tfrac{1}{2}O_2$$

The hydroxonium ions migrate through the membrane to the cathode, where they recombine with electrons to form H_2 and H_2O:

$$2H_3O^+ + 2e^- \rightarrow 2H_2O + H_2$$

Thermodynamics calculations show that the electrical energy required to electrolyze water decreases as the temperature is increased, but increases if the device is operated at elevated pressures.

How efficient is water electrolysis? A report published by the National Renewable Energy Laboratory in 2006 concluded that 9 kWh of electricity

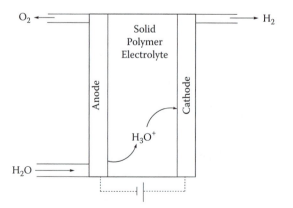

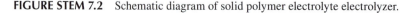

FIGURE STEM 7.2   Schematic diagram of solid polymer electrolyte electrolyzer.

and 8.9 L of water would be minimally required to produce 1 kg of H_2 under standard physical conditions (25°C and 1 atm pressure). However, commercial electrolyzer system efficiencies were only $64.5 \pm 8.5\%$, corresponding to 70.1–53.4 kWh/kg H_2.[50] Energy losses as heat generated in electrolyzers are the main culprits in limiting energy capture via electrolysis.

The Division of Technology, Industry and Economics of the United Nations Environment Program (UNEP) noted in its 2006 review that publicly funded research into hydrogen technologies was intensive in Organization for Economic Cooperation and Development (OECD) nations (Figure 7.5).[51] Both OECD countries and a growing number of developing economies have active hydrogen economy targets:

- Japan was the first country to undertake an ambitious fuel cell program: 10 years of R&D funded at $165 million and completed in 2002. Following that, the New Hydrogen Project focused on commercialization; funding reaching $320 million in 2005, with the aims of producing and supporting 50,000 fuel-cell-powered vehicles by 2010 and 5 million by 2020 (with 4,000 H_2 refueling stations by then); 2,200 MW of stationary fuel cell cogeneration systems by 2010; and 10,000 MW by 2020.
- The transition to the hydrogen economy envisaged by the U.S. government (the Hydrogen Fuel Initiative) is set to proceed via four phases: technology

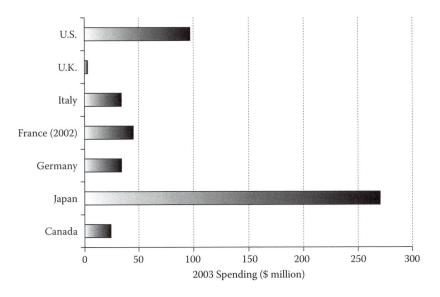

FIGURE 7.5 Publicly funded research into hydrogen technologies in the largest OECD economies, 2003. (Data from United Nations Environment Program. 2006. *The hydrogen economy. A nontechnical review*. Paris.)

development, initial market penetration, infrastructural investment, and full realization to begin by 2025.

- Funding for the hydrogen economy in the EU was provided by the Renewable Energy Sixth Framework Program from 2003 to 2006, and subsequent plans are expected to generate combined public and private funding of approximately $2.8 billion by 2011.
- Canada's H_2 R&D focused on the Ballard PEM fuel cell and the Hydrogenics alkaline water electrolyzer, with public funding of over $25 million per year.
- Korea budgeted $586 million for hydrogen-related projects through to 2011, aiming at the introduction of 10,000 fuel cell vehicles, the development of H_2 production from renewable resources, and the development of a stationary fuel cell with a capacity of 370 MW.
- India allocated $58 million from 2004 to 2007 for projects in universities and governmental research laboratories, with car manufacturers expected to contribute $116 million by 2010.
- Russia began to fund a joint project on fuel cell development between the Russian Academy of Science and the Norlisk Nikel Company at $30 million in 2005.
- Brazil's Hydrogen Roadmap focused on production from water electrolysis, reforming of natural gas, reforming or gasification of ethanol and other biofuels, storage technologies (including metal hydrides), and fuel cells.
- In 2007, the first liquid H_2-dispensing fuel pump was installed in Norway as the first step in providing the H_2 highway, a 360-mile route from Stavanger to Oslo expected to be complete by 2009. California launched a H_2 highway network to include up to 200 fueling stations by 2010.

7.6.2 BIOPRODUCTION OF GASES: METHANE AND H_2 AS PRODUCTS OF ANAEROBIC DIGESTION

"Biogas" is a mixture of methane (CH_4) and CO_2 prepared usually from the anaerobic digestion of waste materials by methanogenic bacterial species (*Methanosarcina, Methanosaeta, Methanobrevibacter,* etc.); this is a technology applied globally because it is ideally suited for local use in rural communities in developing economies as a cheap source of nonbiomass direct fuel.[52-54] As a low-technology but established approach to wastewater treatment, it is applicable on an industrial scale; its only disadvantages is the need to remove malodorous volatile sulfur compounds. Anaerobic digestion is an effective means of capturing the energy present in biological materials (Figure 7.6). Biogas production is also ideal for purifying wastewater from ethanol facilities for detoxification and recirculation, thus reducing production costs by generating locally an input for combined heat and power or steam generation.[55]

Much less widely known are the bacteria that can form H_2 as an end product of carbohydrate metabolism. Included in the vast number of species capable of some kind of biological fermentation are a wide array of microbes from anaerobic environments (including *E. coli*) that were known as active research topics as far back as the

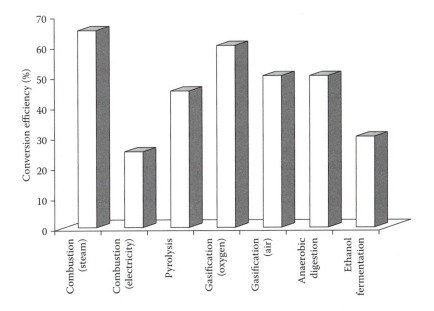

FIGURE 7.6 Bioconversion efficiencies of thermochemical and microbiological processes for biomass. (Data from Lewis, C. 1976. *Process Biochemistry* 11 (part 9, November):29.)

1920s; some of these were discovered by Louis Pasteur in the nineteenth century.[56,57] The ability of microorganisms isolated from the digestive tract to produce H_2 from *cellulosic* substrates is another scientific research subject with a surprisingly long history.[58]

7.6.2.1 Heterotrophic Microbes Producing H_2 by Hydrogenase Activity

The best taxonomically and physiologically characterized examples of H_2 producers are clostridia. However, other genera (including bacilli) as well as a microbial flora from anoxic marsh sediments and other environments are known to be capable of the H_2 production and either the ABE fermentation or the accumulation of one or more of the ABE trio, carboxylic acids (acetic, butyric, etc.), and/or other products (acetoin, 2,3-butanediol, etc.).[59]

A wide spectrum of carbon sources support H_2 production at rates up to 1,000 mL per hour per gram of cells at a maximum yield of 4 mol of H_2 per mole of glucose with the stoichiometry given by the following reaction:

$$C_6H_{12}O_6 + 2H_2O \rightarrow 2CH_3COOH + 2CO_2 + 4H_2$$

This reaction is sufficiently exothermic to support microbial growth. The yield of H_2, however, is subject to feedback inhibition by H_2, requiring that the partial pressure of the gas be kept low to avoid problems with growth rate or a shift to acid production.

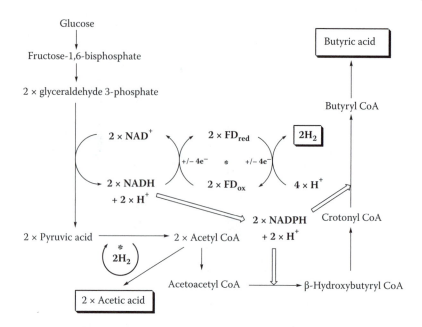

FIGURE 7.7 Hydrogenase and the reoxidation of redox cofactors in acid- and H_2-producing clostridial species. (Modified from Moat, A. G., and Foster, J. W. 1995. *Microbial Physiology*, 3rd ed., chap 7. New York: Wiley-Liss.)

Another H_2-forming fermentation has butyric acid as its major acidic product, although the molar production of H_2 is only half that of acetate-accumulating strains:

$$C_6H_{12}O_6 \rightarrow C_3H_7COOH + 2CO_2 + 2H_2$$

The key enzyme in heterotrophic H_2 producers is hydrogenase, an enzyme that catalyzes the reoxidation of reduced ferredoxin (Fd), an iron-containing protein reduced by ferredoxin-NAD and pyruvate-ferredoxin oxidoreductases, with the liberation of molecular hydrogen (Figure 7.7)[60]:

$$2Fd^{2+} + 2H^+ \rightarrow 2Fd^{3+} + H_2$$

A summary of hydrogenase-containing bacteria is given in Table 7.2.

Hydrogenases are a diverse group of enzymes and are often cataloged on the basis of the metal ion they contain as an essential component of the active site.[61] The fastest H_2-evolving species under laboratory conditions, *Clostridium acetobutylicum*, produces two different hydrogenases[62–64]:

1. an iron-containing enzyme whose gene is located on the chromosome
2. a dual-metal (nickel, iron) enzyme whose gene is located on a large plasmid

The iron-dependent hydrogenase from *C. acetobutylicum* has a specific activity eightfold higher than similar enzymes from green algae, even when all three

TABLE 7.2

Biochemistry of Hydrogenase-Containing Bacteria

Microbe	Examplary Species	Substrate	H_2 Productivity (mole/mole)
	Anaerobes		
Clostridia	*Clostridium butyricum, C. beijerincki*	Glucose	Up to 4
Methylotrophs	*Methylomonas albus, M. trichosporium*	Formic acid	Up to 2.45
Methanogens	*Methanobacterium soehgenii*	Formic acid	1
Rumen bacteria	*Ruminococcus albus*	Glucose	Up to 0.59
Archaea	*Pyrococcus furiosus*	Glucose etc.	?
	Facultative Anaerobes		
Enterobacteria	*Escherichia coli, Enterobacter aerogenes*	Glucose	Up to 0.9
	Aerobes		
Alcaligens	*Alcaligenes eutrophus*	Gluconic acid, formic acid, etc.	?
Bacilli	*Bacillus licheniformis*	Glucose	Up to 1.5

Source: Information from Nandi, R., and Sengupta, S. 1998. *Critical Reviews in Microbiology* 24:61.

enzymes are expressed in and purified from the clostridial host.[65] The active sites of iron-dependent hydrogenases may be the simplest such structures yet studied at the molecular level and of enormous potential importance for the industrial development of hydrogenases is the finding that even simple complexes of iron sulfide and CO mimic hydrogenase action.[66] The crucial structure involves two iron atoms with different valency states at different stages of the reaction mechanism (Figure 7.8). These findings raise the possibility of rational design of improved hydrogenases by the binding of novel metal complexes with existing protein scaffolds from known enzymes.

In contrast, nickel–iron bimetallic hydrogenases possess complex organometallic structures with CO and cyanide (CN^-) as additional components, the metal ions bound to the protein via multiple thiol groups of cysteine resides, and an important coupling between the active site and iron–sulfur clusters.[67,68] Multigene arrays are required for the biosynthesis of mature enzyme.[69] Nevertheless, progress has been impressive in synthesizing chemical mimics of the organometallic centers that contain elements of the stereochemistry and atomic properties of the active site.[70] The enzyme kinetics of nickel–iron hydrogenases remain challenging, and it is possible that more than one type of catalytic activation step is necessary for efficient functioning in vivo.[71,72]

Such advances in basic understanding will open the door to replacing expensive metal catalysts (e.g., platinum) in hydrogen fuel cells by iron- or iron/nickel-based

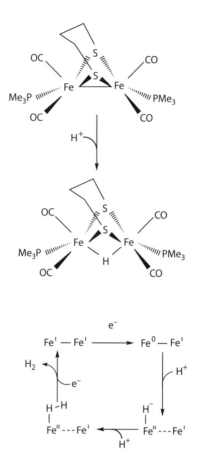

FIGURE 7.8 Simple organometallic complexes as biochemical mimics of hydrogenase enzymes. (Modified from Darensbourg, M. Y. et al. 2003. *Proceedings of the National Academy of Sciences USA* 100:3683.)

biocatalysts. The sensitivity of many hydrogenases to inhibition by O_2 is a serious drawback but, of the many organisms known to produce hydrogenases, some contain forms with no apparent sensitivity to O_2 and can function under ambient levels of the gas.[73]

7.6.2.2 Nitrogen-Fixing Microorganisms

The fixation of gaseous molecular nitrogen, N_2, to biologically utilizable nitrogen is performed as an essential part of the global nitrogen cycle by bacteria that may be free living or exist in symbiotic associations with plants, bivalves, or marine diatoms.[74] The central enzymic reduction of nitrogen to ammonia is that catalyzed by nitrogenase:

$$N_2 + 8H^+ + 8e^- \rightarrow 2NH_3 + H_2$$

Molecular hydrogen is an obligatory product of the overall reaction. However, it is so energy expensive that most nitrogen fixers rarely evolve H_2—instead employing another hydrogenase (uptake hydrogenase) to recycle the H_2 via oxidation by O_2 and conservation of part of the potentially released energy to support nitrogenase action. For this reason, nitrogenase-containing microbes are not viewed as likely sources of H_2 for biofuels.

7.6.2.3 Development of "Dark" H_2 Production Systems

The major challenge in developing commercial H_2 generation by "dark" biotechnology (i.e., by fermentation processes) has been the low productivity of natural H_2-evolving microbes. This can be resolved into two distinct limitations: the molar yield from a fermentable substrate and the expected low growth rates and cell densities of microbial producers under energy-poor environmental conditions.

On the basis of data presented in Table 7.2, a maximum of 4 mol of H_2 per mol of fermentable glucose substrate equates (on a mass balance) to only 8 g per 180 g of sugar consumed. In cell-free biotransformations, using mixtures of enzymes, a nearly threefold higher productivity has been demonstrated.[75] If the pentose phosphate pathway (Figure 3.2, Chapter 3) can be run in a cyclic manner to oxidize glucose 6-phosphate completely to CO_2 and H_2O, each mole of glucose consumed (in six turns of the cycle) can generate 12 mol of reduced cofactor:

$$6C_6H_{12}O_6 + 12\ NADP^+ \rightarrow 5C_6H_{12}O_6 + 6CO_2 + 12NADPH + 12H^+$$

Coupling the reoxidation of NADPH to the hydrogenase from *Pyrococcus furiosus* (Table 7.2)—one of only a few hydrogenases known to accept NADPH as a reducing agent—generated 11.6 mol of H_2 per mole of phosphorylated glucose oxidized.

A genetically modified strain of *E. coli* overexpressed the gene for formate hydrogen lyase, an enzyme catalyzing the following reaction:

$$HCOOH \rightarrow CO_2 + H_2$$

By growing the cells to high cell densities under glucose-supported aerobic conditions before transfer to an anaerobic fermentor, a high catalytic potential for formate transformation to H_2 was established, reaching 300 L of H_2 per hour per liter of culture.[76] This rate of H_2 production could support a 1-kW fuel cell operating at 50% efficiency using only 2 L of culture medium maintained under continuous conditions by a feed of formic acid. Further strain construction (deleting lactate dehydrogenase and fumarate reductase genes) has improved the induction of the formate hydrogen lyase activity.[77] The same genetic manipulations have eliminated side reactions of (phosphoenol)pyruvic acid in glucose-grown *E. coli*, maximizing the transformation of pyruvate to formate via the pyruvate formate lyase-catalyzed step that involves coenzyme A (CoASH):

$$CH_3COCOOH + CoASH \rightarrow CH_3CO\text{-}SCoA + HCOOH$$

Such a genetic background (with formate hydrogen lyase as the next step) increases the production of H_2 from glucose as the fermentation substrate, although only rates of approximately 20 L per hour per liter of culture have been achieved.[78]

Thermophilic and hyperthermophilic microbes are obvious choices for production strains to accelerate H_2 formation. A strain of the bacterium *Klebsiella oxytoca* isolated from a hot spring in China could produce H_2, even in the presence of 10% O_2

in the gas phase, but it had a low molar yield (1 mol/mol of glucose consumed).[79] The extreme thermophile *Caldicellulosiruptor saccharolyticus* shows up to 92% of the theoretical H_2 yield from glucose (4 mol/mol) at low growth rates at 72 or 73°C, indicating possible applications in long-term free or immobilized cultures.[80] This organism also can produce H_2 from hydrolyzed paper sludge industrial waste as the sole carbon source, and is unusual in that it can utilize xylose faster than glucose.[81,82]

An advanced bioprocess option for H_2 production utilizes a membrane bioreactor to maintain the bacteria inside the reactor while allowing fluids to exit.[83] This design could be the optimal methodology to restrict the growth of methanogenic bacteria that consume H_2 and generate CH_4, a gas with only 42% of the energy content of H_2.[84] Restricting the residence time of materials in a continuous flow reactor system allows H_2 producers to outcompete the slower growing methanogens. With a 12-h residence time, glucose could be utilized as a substrate for H_2 production with an overall consumption of 98% and a 25% efficiency (assuming 4 mol of H_2 per mole of glucose), accumulating H_2 at a concentration of 57–60% (volume basis) in the headspace.[83]

The production of H_2 need not be based on pure bacterial cultures; mixed cultures, even ones with only indirect evidence of the microbial flora present, are suitable for wastewater treatment or for local production sites in isolated rural localities. A clostridial population (on the basis of the spectrum of metabolites produced in parallel to H_2) provided a system capable of stable and prolonged production, with H_2 reaching 51% in the gas phase and with no methanogenesis observed.[85] Some process control, however, is unavoidable to maintain H_2-evolving capacity, particularly pH; maintaining a pH of 6.0 may inhibit the growth of lactobacilli in a mixed culture of *Clostridium* and *Coprothermobacter* species that could utilize untreated sludge and lake sediment material as substrates.[86]

The choice of pH regulant may be crucial, and the accumulation of sodium to toxic levels has been noted to limit continuous biohydrogen systems from sucrose-supplemented anaerobic sewage sludge.[86] At a constant pH, the combination of substrate-material retention time and temperature (at 30 or 37°C) can have a marked effect on the balance between different clostridial species, the appearance of non-clostridial bacterial species, and the overall molar yield of H_2 from carbohydrates.[87] In a molasses wastewater treatment plant, the H_2 production rate was highest in an ethanol-forming stage of the process, and at least six types of H_2-producing microbes were found to be present (predominantly, a novel species, *Ethanoligenes harbinese*).[88] Such a complex microbial ecology may be highly adaptable to differing types and compositions of carbon sources during production cycles or when seasonally available.

The use of advanced reactor types has been explored; for example, the fluidized bed reactor design has been utilized with a mixed community that rapidly established H_2 production from *C. butyricum*. Instability developed over the course of time as propionate producers gradually took over, and biofilm types of reactors may not be the optimal design because of the efficient adhesion of H_2-consuming microbial species to the carrier.[89] A trickle-bed reactor packed with glass beads inoculated with a pure culture of *C. acetobutylicum* certainly gave high H_2 gas concentrations, but soon (60–72 h) clogged because of bacterial growth.[90]

Irrespective of the long-term prospects for the industrial production of H_2, dark fermentations are very likely to be permanent features of wastewater treatment technologies and as an alternative to methane for local "biogas" production. Indeed, an obvious application of the trickle-bed reactor may be for the treatment of high-carbohydrate wastewaters requiring no energy input for stirring a conventional mixed tank and producing H_2 as a recoverable fuel gas as well as a more dilute stream for conventional biogas production.[90] The use of water streams with lower organic loading, however, may be advantageous for H_2 production because supersaturation of the gas space inside bioreactors may feed back to inhibit H_2 synthesis.[91]

Removing CO_2 (e.g., by the use of alkali to absorb the gas) is also beneficial to H_2 production, probably by minimizing the flow of utilizable substrates to acetogenic bacteria capable of synthesizing acetic acid from H_2 and CO_2.[92] Animal-waste-contaminated water can also be made acceptable to biological H_2 producers if the ammonia concentration can be reduced and maintained below toxic concentrations in continuous flow systems, especially if the microbial community can be gradually adapted to increased ammonia levels.[93] Food-processing aqueous streams with high chemical oxygen demands can support biohydrogen production at 100 times the rates possible with domestic wastewaters—often reaching commercially viable amounts of H_2 if used on site as a heating fuel.[94]

STEM TOPIC 7.3: CHEMICAL AND BIOLOGICAL OXYGEN DEMAND

Chemical oxygen demand (COD) and biological oxygen demand (BOD) are essential in environmental sciences. They appear in this text because biofuels production facilities generate waste effluents, but those aqueous wastes can be cleaned up using microbes known to biotechnology.

COD is measured as an indicator of organic pollutants in water. In a typical industrial wastewater treatment plant, COD measured on both influent and effluent water indicates the efficiency of the treatment process. Due to its unique chemical properties, the dichromate ion ($Cr_2O_7^{2-}$) is a commonly specified oxidant in the majority of applications and is reduced to the chromic ion (Cr^{3+}). COD can be measured by titrimetric and colorimetric methods.

The BOD is the amount of oxygen required by aerobic microorganisms to decompose the organic matter in a sample of polluted water. BOD is also termed biochemical oxygen demand.

The ratio of COD and BOD is indicative of the source of the pollution. An example is a study of the lagoon of Abidjan in Nigeria, West Africa.[95] This large lagoon receives five major effluents (I, II, III, IV, and V) from Abidjan City, an urban area that expanded greatly in the later twentieth century (see Table STEM 7.3).

Effluent II derived from a big health center and, with its efficient wastewater treatment, its COD and BOD were both low. Because COD/BOD ratios for domestic waste water approximated 2.0, domestic effluents dominated other sources of pollution and could be amenable to removal of biodegradable material by biotechnological means.

TABLE STEM 7.3
COD and BOD Values and Ratios

Effluent	COD (mg O_2/L)	BOD (mg O_2/L)	COD/BOD
I	239	102	2.3
II	102	28	3.6
III	202	108	1.9
IV	1632	740	2.2
V	215	153	1.4

As an excellent example of the opportunistic use of H_2-producing microbes in biofuels production, a strain of *Enterobacter aerogenes* was shown to be highly adept at producing both H_2 and ethanol from glycerol-containing wastewaters from biodiesel production. Continuous production in a packed bed reactor using porous ceramic support material maximized the H_2 production rate.[96] But, as a final twist, H_2 producers may have an unexpected role in assisting a microbial community of methanogens to achieve full productivity (i.e., a syntropic relationship may be established to provide the methanogens with a readily utilizable substrate). Adding mesophilic or thermophilic H_2-producing cultures increases biogas production from animal manure slurry, and the added species persist for several months of semicontinuous operation.[97]

7.6.3 PRODUCTION OF H_2 BY PHOTOSYNTHETIC ORGANISMS

In comparison with fermentor-based H_2 production, the use of photosynthetic organisms has received wider publicity because envisaged bioprocesses have convincing environmental credentials—that is, the ability to produce a carbonless fuel using only water, light, and air (CO_2) as inputs. However, hydrogen photobiology is highly problematical because of the incompatibility of the two essential steps:

1. In the first stage, water is split to produce O_2.
2. In the second stage, the photoproduced electrons are combined with protons to form H_2 by either a hydrogenase or a nitrogenase—and O_2 is a potent inhibitor of such an anaerobic system.

Nature provides two related solutions to this dilemma.[98] First, filamentous cyanobacteria (e.g., *Anabaena cylindrica*) compartmentalize the two reactions into different types of cells: vegetative cells for generating O_2 from water and using the reducing power to fix CO_2 into organic carbon compounds, which then pass to specialized nitrogenase-containing heterocyst cells that evolve H_2 when N_2 reduction is blocked by low ambient concentrations of N_2. The second scenario is that of nonheterocystous cyanobacteria that separate O_2 and H_2 evolution temporally (in day and night cycles), although the same overall effect could be achieved using separate light and dark reactors. With either type of nitrogen-fixing organism, however, the high energy requirement of nitrogenase would lower solar energy conversion efficiencies to unacceptably low levels.

Hydrogenase is the logical choice of biocatalyst for H_2 production. More than 30 years have now passed since the remarkable experimental demonstration that simply mixing chloroplasts isolated from spinach leaves with hydrogenase and ferredoxin isolated from cells of *Clostridium kluyveri* generated a laboratory system capable of direct photolysis of water and H_2 production.[99] The overall reaction sequence in that hybrid biochemical arrangement was the following:

$$H_2O + light \rightarrow \frac{1}{2}O_2 + 2e^- + 2H^+ \rightarrow ferredoxin \rightarrow hydrogenase \rightarrow H_2$$

Light-induced photolysis of water produced electrons that traveled via the photosystems of the chloroplast preparations to reduce ferredoxin prior to hydrogenase catalyzing the reunion of electrons and protons to form molecular hydrogen. For decades, therefore, the nagging knowledge that direct photolytic H_2 production is technically feasible has both tantalized and spurred on research into solar energy conversion.

Thirty years previously, the abilities of some unicellular green (chlorophyll-containing) algae (i.e., microalgae) to generate H_2 under unusual (O_2-free) conditions where hydrogenase was synthesized had already been defined.[100] When illuminated at low light intensities in thin films (5–20 cellular monolayers), such microalgae can show conversion efficiencies of up to 24% of the photosynthetically active radiation.[101]

What is the biological function of hydrogenase in such highly aerobic organisms? An induction period with darkness and anaerobiosis appears to be essential.[102] Photosynthetic H_2 production is also enhanced if the concentration of CO_2 is low, suggesting that the hydrogenase pathway is competitive with the normal CO_2-fixing activity of chloroplasts.[103] Because the electron transport via the hydrogenase pathway is still coupled to bioenergy conservation (photosynthetic phosphorylation), hydrogenase may represent an emergency strategy in response to adverse environmental conditions—for example, in normally aboveground plant parts subject to water logging and anaerobiosis where essential maintenance and cellular repair reactions can still operate with a continuing source of energy.

It naturally follows that reintroduction of O_2 and CO_2 would render such a function of hydrogenase superfluous; this explains the inhibitory effect of O_2 on hydrogenase and the ability of even background levels of O_2 to act as an electron acceptor in direct competition with hydrogenase-mediated H_2 production.[104] Genetic manipulation and directed evolution of algal hydrogenases with reduced or (in the extreme case) no sensitivity to O_2 is therefore unavoidable if maximal and sustained rates of photohydrogen production can be achieved in microalgal systems.

Progress has begun to be made on the molecular biology of microalgal hydrogenases, including the isolation and cloning of the two genes for the homologous iron hydrogenases in the green alga *Chlamydomonas reinhardtii*.[105] Random mutagenesis of hydrogenase genes could rapidly isolate novel forms retaining activity in the presence of O_2 and/or improved hydrogenase kinetics. Screening mutants of *Chlamydomonas reinhardtii*, however, has revealed unexpected biochemical complexities—in particular, the requirement for functional starch metabolism in

H_2 photoproduction.[106] Several changes were indeed identified during the successful improvement of H_2 photoproduction by this alga[107]:

- There was rational selection of mutants with altered electron transport activities with maximized electron flow to hydrogenase.
- Isolates were then screened for increased H_2 production rates, leading to a mutant with reduced cellular O_2 concentrations and thus having less inhibition of hydrogenase activity.
- The most productive mutant also had large starch reserves.

Using the conventional representation of electron transport inside chloroplast membrane systems, the possible interactions of photohydrogen production and other photosynthetic activities can be visualized (Figure 7.9).[108] Active endogenous metabolism could remove photoproduced O_2 by using O_2 as the terminal electron acceptor in mitochondria. However, the problem of different spatial sites for O_2 production and O_2 utilization still requires a reduced sensitivity of hydrogenase to O_2 because the gas cannot be removed instantaneously. Only in "test tube" systems can O_2-removing chemicals be supplied, for example, as glucose plus glucose oxidase to form gluconic acid by reaction between glucose and O_2.[99]

The obvious implication of the redox chemistry of Figure 7.9 is that the normal processes of photosynthesis, involving reduction of NADP for the subsequent

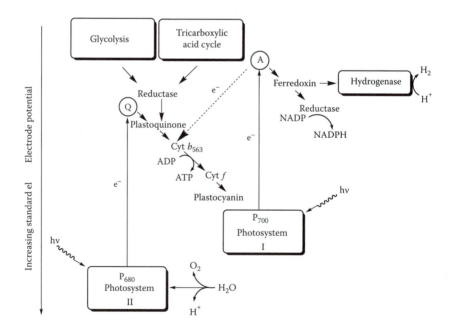

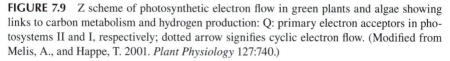

FIGURE 7.9 Z scheme of photosynthetic electron flow in green plants and algae showing links to carbon metabolism and hydrogen production: Q: primary electron acceptors in photosystems II and I, respectively; dotted arrow signifies cyclic electron flow. (Modified from Melis, A., and Happe, T. 2001. *Plant Physiology* 127:740.)

reduction of CO_2 to sugars, can be separated in time, with light-dependent O_2 evolution and dark H_2 production, or if H_2 production can proceed with inhibited O_2 evolution (i.e., indirect biophotolysis).[98] The particular advantage of this arrangement is that the light-dependent stage can be operated in open pools to maximize productivity at minimal cost.

Sustained H_2 production could be achieved over approximately 100 h after transfer of light-grown *C. reinhardtii* cells to a medium deficient in sulfur; these conditions reversibly inactivated photosystem II and O_2 evolution, while oxidative respiration *in the continued light* depleted O_2, thus inducing hydrogenase.[109] The subsequent H_2 production *only* occurred in the light and was probably a means of generating energy by photosystem I activity (Figure 7.9). Starch and protein were consumed while a small amount of acetic acid was accumulated.

This was the first reported account of a single-organism, two-stage photobiological production process for H_2, although a prototype light/dark device using three stages (one light, two dark) with a marine microalga and a marine photosynthetic bacterium was tested in Japan in the 1990s.[110] How much H_2 could a microalgae-based approach produce? With *C. reinhardtii* cells given an average irradiance of 50 mol of photons per square meter per day (a possible value in temperate latitudes, although highly variable on a day-to-day and seasonal basis), the maximum H_2 production would be 20 g per square meter per day—equivalent to 80 kg per acre per day (or 200 kg per hectare per day). However, the likely value, allowing for low yields of H_2 production measured under laboratory conditions, the far from complete absorption of incident light, and other factors, is only 10% of this.[108]

In a further refinement of this approach, the sulfate-limited microalgae were shown to form a stable process for 4,000 h; two automated photobioreactors were coupled to grow the cells aerobically before being continuously delivered to the second, anaerobic stage.[111] Until all the biological and physical limitations can be overcome, large infrastructural investments in high and predictable sunlight regions would be required, and the capital costs for such solar power stations would be high. However, the technical complexity may only approximate that of installing extensive photovoltaic cell banks for the direct production of electricity—an option vigorously advocated by critics of biofuels programs.[112,113]

Cyanobacteria (blue-green algae) are prokaryotes but share with higher photosynthetic organisms the basic electron transport chains of photosystems I and II (Figure 7.9). The molecular biology and biochemistry of hydrogenases in cyanobacteria are well understood, the complete genomes of several such organisms have been sequenced, and interspecies gene transfer is established.[114] Much of the research has unfortunately concentrated on nitrogenase as a source of H_2, but many cyanobacteria contain hydrogenases catalyzing the reversible formation of H_2—a route with far more biotechnological potential for commercial H_2 generation. Protein engineering has begun to reduce the O_2 sensitivities of cyanobacterial hydrogenases.[115]

The physiological role of hydrogenase in cyanobacteria has been debated for decades; hydrogenase may have a safety valve function under low O_2 conditions when a light to dark transition occurs. Inactivating quinol oxidase (an enzyme with a similar hypothetical biochemical function) and nitrate reductase (a third electron sink) increase photohydrogen evolution rates.[116]

Thermophilic cyanobacteria are known to be capable of H_2 photoproduction at up to 50°C in open-air cultures maintained for over 3 weeks.[117,118] If a fermentable carbon source is supplied, a sustained photoevolution of H_2 can be achieved, with photolysis of water (a photosystem II activity; see Figure 7.9), while carbohydrate-mediated reduction of the plastoquinone pool continues independently.[119] This H_2 production system has been termed "photofermentation"; in principle, relatively little light energy is required to drive the reaction because of the energy input from the fermentable substrate.[98] The green alga *C. reinhardtii* shares this pattern of metabolism with cyanobacteria, behaving under photofermentative conditions much like an enteric bacterium such as *E. coli,* exhibiting pyruvate formate lyase activity, and accumulating formate, ethanol, acetate, CO_2, and H_2 as well as glycerol and lactate.[120]

The overlapping molecular structures of cyanobacteria and nonphotosynthetic bacteria were exemplified by the coupling (both in vivo and in vitro) between cyanobacterial photosynthetic electron transport components with clostridial hydrogenase; even more remarkable was the expression in a *Synechococcus* strain of the hydrogenase gene from *C. pasteurianum,* the enzyme being active in the cyanobacterial host.[121] As a possible pointer to the future of designing an improved photosynthetic organism for H_2 production, the "hard wiring" of a bacterial hydrogenase with a peripheral subunit of a photosystem I subunit of the cyanobacterium *Thermosynechococcus elongatus* resulted in a fusion protein that could associate functionally with the rest of the photosystem I complex in the cyanobacterium and display light-driven H_2 evolution.[122]

Photosynthetic bacteria differ from other photosynthetic organisms in using bacteriochlorophyll rather than chlorophyll as the central pigment for light-induced electron transport. They also lack photosystem II (Figure 7.9) and perform anoxygenic photosynthesis, requiring electron donors more reduced than water, including reduced sulfur and organic compounds.[123] Being able to fix gaseous nitrogen, the photosynthetic bacteria contain nitrogenase in addition to hydrogenase, and they occur globally in widely different habitats, including fresh, brackish, and sea waters; hot sulfur springs; paddy fields; wastewaters; and even in Antarctica.

Hydrogen can be photoproduced in the presence of an organic substrate, sometimes with high efficiencies deduced from the maximum theoretical H_2 production on a molar basis (Table 7.3). Both free and immobilized cells have been used to produce H_2 over extended periods of time (Table 7.4). All photosynthetic bacteria can use H_2 as a reductant for the fixation of CO_2 into organic carbon, and considerable reengineering of the molecular biochemistry is unavoidable if the cells are to be evolved into biological H_2 producers.[60] Photofermentations are also known, and *Rhodobacter capsulatus* has been used as a test organism to evaluate photobioreactor designs potentially reaching 3.7% conversion efficiency of absorbed light energy into H_2 fuel energy.[124]

Photosynthetic bacteria may have the additional capability of catalyzing the water shift reaction that is usually considered only in thermochemistry:

$$CO + H_2O \rightarrow CO_2 + H_2$$

However, unlike the thermochemical process, the biologically catalyzed reaction can proceed at moderate temperatures.[99] A continuous process was devised

TABLE 7.3
Photofermentative Hydrogen Production by Photosynthetic Bacteria

Species	Substrate	Stoichiometry	Conversion Efficiency (%)
Rhodobacter capsulatus	Acetate	$C_2H_4O_2 + 2H_2O = 2CO_2 + 4H_2$	57–100
Rhodobacter capsulatus	Butyrate	$C_4H_8O_2 + 6H_2O = 4CO_2 + 10H_2$	23–80
Rhodopseudomonas sp.7	Ethanol	$C_2H_5O + 3H_2O = 2CO_2 + 6H_2$	45
Rhodobacter sphaeroides	Glucose	$C_6H_{12}O_6 + 6H_2O = 6CO_2 + 12H_2$	99
Rhodomicrobium vannielii	Lactate	$C_3H_6O_3 + 3H_2O = 3CO_2 + 6H_2$	78–100
Rhodobacter sphaeroides	Malate	$C_4H_6O_5 + 3H_2O = 4CO_2 + 6H_2$	57–100
Rhodopseudomonas sp.7	Propanol	$C_3H_8O + 5H_2O = 3CO_2 + 9H_2$	36
Rhodopseudomonas palustris	Pyruvate	$C_3H_4O_3 + 3H_2O = 3CO_2 + 5H_2$	52
Rhodobacter capsulatus	Succinate	$C_4H_6O_4 + 4H_2O = 4CO_2 + 7H_2$	72

Source: Data from Sasikala, K. et al. 1993. *Advances in Applied Microbiology* 38:211.

for *Rhodospirillum rubrum* with illumination supplied by a tungsten light.[125] With biomass as the substrate for gasification, a substantially (if not entirely) biological process for H_2 production can be envisaged. A biological reactor might be larger and slower but could achieve comparable efficiencies of heat recovery in integrated systems. The most likely niche market use would occur in facilities where the water–gas shift was an option occasionally (but gainfully) employed, but where the start-up time for a thermal catalytic step would be undesirable.[126]

TABLE 7.4
Photofermentative Hydrogen Production by Immobilized Cells

Species	Electron Donor	Immobilization Method	H_2 Evolution Rate (mL/h/g dry weight)
Rhodobacter sphaeroides	Malate	Alginate	16.2
Rhodospirillum rubrum	Acetate	Alginate	9.0
Rhodospirillum rubrum	Lactate	Alginate	30.7
Rhodopseudomonas sp.7	Starch	Alginate	80.0
Rhodospirillum rubrum	Lactate	Agar beads	57.3
Rhodospirillum molischianum	Wastewater	Agar blocks	139.0
Rhodopseudomonas palustris	Malate	Agar blocks	41.0
Rhodobacter capsulatus	Lactate	Carrageenan	111.0
Rhodospirillum rubrum	Acetate	Agar cellulose fiber	15.6
Rhodospirillum rubrum	Lactate	Agarose	22.9
Rhodospirillum rubrum	Lactate	Pectin	21.0

Source: Data from Sasikala, K. et al. 1993. *Advances in Applied Microbiology* 38:211.

TABLE 7.5
US Patents Covering Photosynthetic and Fementative H₂ Production

Date	Title	Assignee/Inventor(s)	Patent
Photoproduction			
10/4/1984	Method for producing hydrogen and oxygen by use of algae	U.S. Department of Energy	US 4,442.211
1/30/1985	Process for producing hydrogen by alga in alternating light/dark cycle…	Miura et al.	US 4,532,210
3/31/1992	Measurement of gas production of algal clones	Gas Research Institute (Chicago, IL)	US 5,100,781
2/16/1999	Process for selection of oxygen-tolerant algal mutants that produce H₂ under aerobic conditions	Midwest Research Institute (Kansas City, MO)	US 5,871,952
6/25/2002	Molecular hydrogen production by direct electron transfer	McTavish, H.	US 6,410,258
1/24/2006	Hydrogen production using hydrogenase-containing oxygenic photosynthetic organisms	Midwest Research Institute (Kansas City, MO)	US 6,989,252
2/13/2007	Modulation of sulfate permease for photosynthetic hydrogen production	University of California (Oakland, CA)	US 7,176,005
6/12/2007	Fluorescence techniques for online monitoring of state of hydrogen-producing microorganisms	Midwest Research Institute (Kansas City, MO)	US 7,229,785
Fermentation			
9/10/2002	System for rapid biohydrogen phentotypic screening of microorganisms using chemochromic sensor	Midwest Research Institute (Kansas City, MO)	US 6,448,068
3/1/2005	Method of producing hydrogen gas by using hydrogen bacteria	Japan Science and Technology Corporation	US 6,860,996
5/3/2005	Method and apparatus for hydrogen production from organic wastes and manure	Gas Technology Institute (Des Plaines, IL)	US 6,887,692
8/1/2006	Method for hydrogen production from organic wastes using a two-phase bioreactor system	Gas Technology Institute (Des Plaines, IL)	US 7,083,956
6/19/2007	Process for enhancing anaerobic biohydrogen production	Feng Chia University (Taiwan)	US 7,232,669

Patents describing processes for H₂ production using photosynthetic microbes cover two decades, with topics as diverse as their basic biology, molecular, and enzymic components, and analytical methodologies. After 2000, patents also appeared that focus on biohydrogen production by fermentative organisms (Table 7.5).

7.7 MICROBIAL FUEL CELLS: ELIMINATING THE MIDDLEMEN OF ENERGY CARRIERS

Hydrogen ions (protons, H^+) can accept reducing equivalents (conventionally represented as electrons, e^-) generated either photosynthetically or by the oxidation of organic and inorganic substrates inside microbial cells:

$$2e^- + 2H^+ \rightarrow H_2$$

The terminal electron donor (e.g., reduced ferredoxin) could donate electrons to the anode of a battery. Protons could then, in the presence of O_2, complete the electric circuit at the cathode by the reaction:

$$O_2 + 4e^- + 4H^+ \rightarrow 2H_2O$$

This forms a highly environmentally friendly source of electric power (a battery), fueled by microbial metabolic activity. That, in essence, is the definition of a microbial fuel cell (MFC).[127-129]

At its simplest, an MFC is a dual-chamber device with an electrolyte, a cation exchange membrane to separate anodic and cathodic compartments, a supply of O_2 for the cathode, and an optional sparge of inert gas for the anode (Figure 7.10). The transfer of electrons to the anode may be directly (via unknown terminal electron donors on the cell surface) or by employing redox-active "mediators" that can be reduced by the cells and reoxidized at the anode (e.g., Neutral Red reduced by hydrogenase).[130,131] A wide spectrum of microbial species has been tested in MFC environments; the species are usually anaerobes or facultative anaerobes chosen to function in the O_2-deficient anode compartment; examples include:

- immobilized cells of the yeast *Hansenula anomala*[132]
- a mixed microbial community of *Proteobacterium*, *Azoarcus*, and *Desulfuromonas* species with ethanol as the fuel source[133]
- *Desulfitobacterium hafniense* with humic acids or the humate analog anthraquinone-2,6-disulfonate added as an electron-carrying mediator with formic acid, H_2, lactate, pyruvate, or ethanol as the fuel[134]
- *E. coli* in MFCs as power sources for implantable electronic devices[135]

The first use of the term "microbial fuel cell" appears to date from the early 1960s in studies by research scientists with the Mobil Oil Company with hydrocarbon-metabolizing *Nocardia* bacteria, but the basic concepts may date back 30 or even 50 years earlier.[136] Development of MFCs as commercial and industrial functionalities are methods of water treatment and as power sources for environmental sensors. The power produced by these systems is currently limited, primarily by high internal (ohmic) resistance; however, improvements in system architecture might result in power generation that is more dependent on the bioenergetic capabilities of the microorganisms.[137]

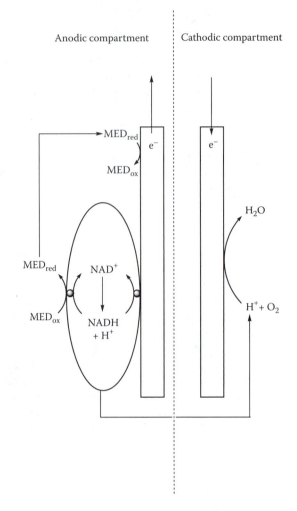

FIGURE 7.10   Redox reactions occurring in a microbial fuel cell: MED is the soluble mediator reduced by the microbial terminal electron donor at the microbial cell surface.

Creating a scalable architecture for MFCs is essential to provide large surface areas for oxygen reduction at the cathode and bacteria growth on the anode. A tubular ultrafiltration membrane with a conductive graphite coating and a nonprecious metal catalyst can be used to produce power in an MFC; this is a promising technology that is intrinsically scalable for creating larger systems.[138] For the anodes, highly conductive, noncorrosive materials are needed that have a high specific surface area (i.e., surface area per volume) and an open structure to avoid biofouling. Graphite fiber brush anodes have high surface areas and a porous structure can produce high power densities—qualities that make them ideal for scaling up MFC systems.[139]

Because rumen bacteria have been shown to generate electricity from cellulosic materials, potentially immense substrate supplies could be available for MFC arrays.[140] An even greater flexibility can be designed—for example, coproducing H_2

and ethanol production from glycerol-containing wastes discharged from a biodiesel fuel production plant with *Enterobacter aerogenes* in bioelectrochemical cells with thionine as the exogenous electron transfer mediator.[141]

The first generation of MFCs have proved highly interesting research tools; the immediate target remains that of increasing MFC power output to support a wide spectrum of potential activities in a cost-effective manner.[142] Treatment of domestic and industrial wastewaters is only one of an emerging group of topics discussed on the MFC community research Web site (www.microbialfuelcell.org).[143]

7.8 SUMMARY

Any alcohol of low molecular mass can function as a combustible fuel. Microbial production of butan-1-ol was an industrial process in the early twentieth century and has been reborn as biobutanol a century later. Such alcohols as butan-1-ol have superior physical properties to ethanol but their biosynthesis requires anaerobic bacteria or extensive genetic engineering of bacteria to generate suitable biocatalysts.

Glycerol, in contrast, is a known coproduct of yeast ethanol processes. By manipulation of strains and fermentation conditions, high-level production of glycerol is possible as a prelude to the chemical transformation of glycerol to advanced biofuels.

A parallel process converts sugars to a mixture of carboxylic acids by fermentation. These acids (propanoic acid, etc.) can be converted chemically to fuel alcohols.

A more radical option is to biosynthesize H_2 gas as the ideal alternative to carbon-based fuels. Both plants and microbes can bioproduce H_2, but the biochemistry is complicated by an inhibition by O_2. Selected photosynthetic and nonphotosynthetic producer cells are being developed for large-scale H_2 generation, but the required engineering is often sophisticated and yields remain low. However, the long-term aim of bioproduced H_2 utilized by onboard fuel cells to power automobiles remains highly attractive for sustainable transportation in the future.

At the farthest extreme from liquid biofuels, microbial fuel cells are a means by which to generate electricity directly from organic compounds. Higher power outputs are being extensively sought, although even prototype devices may be readily adapted for purifying domestic wastewaters, simultaneously reducing COD (STEM Topic 7.3) while generating electricity.

REFERENCES

1. Awang, G. M., Jones, G. A., and Ingledew, W. M. 1988. The acetone-butanol-ethanol fermentation. *Critical Reviews in Microbiology* 15(suppl. 1):S33.
2. Woods, D. R. 1995. The genetic engineering of microbial solvent production. *Trends in Biotechnology* 13:259.
3. Lee, S. Y. et al. 2008. Fermentative butanol production by *Clostridia. Biotechnology and Bioengineering* 101:209.
4. Zverlov, V. V. et al. 2006. Bacterial acetone and butanol production by industrial fermentation in the Soviet Union: Use of hydrolyzed agricultural waste for biorefinery. *Applied Microbiology and Biotechnology* 71:587.

5. Ni, Y., and Sun, Z. 2009. Recent progress on industrial fermentative production of acetone–butanol–ethanol by *Clostridium acetobutylicum* in China. *Applied Microbiology and Biotechnology* 83:415.

6. Mitchell, W. J. 1998. Physiology of carbohydrate to solvent conversion by clostridia. *Advances in Microbial Physiology* 39:31.

7. Singh, A., and Mishra, P. 1995. *Microbial pentose utilization. Current applications in biotechnology (Progress in industrial microbiology,* vol. 33), chap 7. New York: Elsevier Science & Technology.

8. Ezeji, T. C., Qureshi, N., and Blaschek, H. P. 2007. Bioproduction of butanol from biomass: From genes to reactors. *Current Opinion in Biotechnology* 18:220.

9. Qureshi, N., and Blaschek, H. P. 1999. Production of acetone–butanol–ethanol (ABE) by a hyper-producing mutant strain of *Clostridium beijerinckii* BA101 and recovery by pervaporation. *Biotechnology Progress* 15:594.

10. Zheng, Y. N. et al. 2009. Problems with the microbial production of butanol. *Journal of Industrial Microbiology and Biotechnology* 36:1127.

11. Qureshi, N., and Blaschek, H. P. 2000. Butanol production using *Clostridium beijerinckii* BA101 hyper-producing mutant strain and recovery by pervaporation. *Applied Biochemistry and Biotechnology* 84–86:225.

12. Qureshi, N., and Blaschek, H. P. 2001. Recent advances in ABE fermentation: Hyper-butanol producing *Clostridium beijerinckii* BA101. *Journal of Industrial Microbiology and Biotechnology* 27:287.

13. Ezeji, T. C., Qureshi, N., and Blaschek, H. P. 2004. Butanol fermentation research: Upstream and downstream manipulations. *The Chemical Record* 4:305.

14. Lee, J. et al. 2005. Evidence for the presence of an alternative glucose transport system in *Clostridium beijerinckii* NCIMP 8052 and the solvent-hyperproducing mutant BA101. *Applied and Environmental Microbiology* 71:3384.

15. Qureshi, N. et al. 2006. Butanol production from corn fiber xylan using *Clostridium acetobutylicum*. *Biotechnology Progress* 22:673.

16. Ezeji, T. C., Qureshi, N., and Blaschek, H. P. 2007. Butanol production from agricultural residues: Impact of degradation products on *Clostridium beijerinckii* growth and butanol fermentation. *Biotechnology and Bioengineering* 97:1460.

17. Tomas, C. A., Welker, N. E., and Papoutsakis, E. P. 2003. Overexpression of *groESL* in *Clostridium acetobutylicum* results in increased solvent production and tolerance, prolonged metabolism, and changes in the cell's transcriptional program. *Applied and Environmental Microbiology* 69:4951.

18. Scotcher, M. C., and Bennett, G. N. 2005. *SpoIIE* regulates sporulation but does not directly affect solventogenesis in *Clostridium acetobutylicum* ATCC 824. *Journal of Bacteriology* 187:1930.

19. Stim-Herndon, K. P. et al. 1996. Analysis of degenerate variants of *Clostridium acetobutylicum* ATCC 824. *Anaerobe* 2:11.

20. Qureshi, N., and Blaschek, H. P. 2001. ABE production from corn: A recent economic evaluation. *Journal of Industrial Microbiology and Biotechnology* 27:292.

21. Liu, J. et al. 2004. Downstream process synthesis for biochemical production of butanol, ethanol, and acetone from grains: Generation of optimal and near-optimal flowsheets with conventional operating units. *Biotechnology Progress* 20:1518.

22. Mutschlechner, O., Swoboda, H., and Gapes, J. R. 2000. Continuous two-stage ABE-fermentation using *Clostridium beijerinckii* NRRL B592 operating with a growth rate in the first stage vessel close to its maximal value. *Journal of Molecular Microbiology and Biotechnology* 2:101.

23. Qureshi, N. et al. 2005. Biofilm reactors for industrial bioconversion processes: Employing potential of enhanced reaction rates. *Microbial Cell Factories* 4:24.

24. Tashiro, Y. et al. 2005. High production of acetone–butanol–ethanol with high cell density culture by cell-recycling and bleeding. *Journal of Biotechnology* 120:197.
25. Claassen, P. A. M. et al. 1999. Utilization of biomass for the supply of energy carriers. *Applied Microbiology and Biotechnology* 52:741.
26. Pinto Mariano, A. et al. 2009. Optimization strategies based on sequential quadratic programming applied for a fermentation process for butanol production. *Applied Biochemistry and Biotechnology* 159:366.
27. Kobayashi, G. et al. 2005. Utilization of excess sludge by acetone–butanol–ethanol fermentation employing *Clostridium saccharoperbutylacetonicum* NRR N1-4 (ATTC 13564). *Journal of Bioscience and Bioengineering* 99:517.
28. Claassen, P. A. M. et al. 2000. Acetone, butanol and ethanol production from domestic organic waste by solventogenic clostridia. *Journal of Molecular Microbiology and Biotechnology* 2:39.
29. Lopez-Contreras, A. M. et al. 2000. Utilization of saccharides in extruded domestic organic waste by *Clostridium acetobutylicum* ATCC 824 for production of acetone, butanol and ethanol. *Applied Microbiology and Biotechnology* 54:162.
30. Atsumi, S. et al. 2007. Metabolic engineering of *Escherichia coli* for 1-butanol production. *Metabolic Engineering* 10:1016.
31. Rühl, J., Schmid, A., and Blank, L. M. 2009. Selected *Pseudomonas putida* strains able to grow in the presence of high butanol concentrations. *Applied and Environmental Microbiology* 75:4653.
32. Atsumi, S., Hana, T., and Liao, J. C. 2008. Non-fermentative pathways for synthesis of branched-chain higher alcohols as biofuels. *Nature* 451:86.
33. Hanai, T., Atsumi, S., and Liao, J. C. 2007. Engineered synthetic pathway for isopropanol production in *Escherichia coli*. *Applied and Environmental Microbiology* 73:7814.
34. Larossa, R. A. et al. Strain for butanol production. USPTO application 2009016291.
35. Park, M.-O. et al. 2001. Isolation and characterization of a bacterium that produces hydrocarbons extracellularly which are equivalent to light oil. *Applied Microbiology and Biotechnology* 56:448.
36. Park, M.-O. 2005. New pathway for long-chain *n*-alkane synthesis via 1-alcohol in *Vibrio furnessi* M1. *Journal of Bacteriology* 187:1426.
37. Bideaux, C. et al. 2006. Minimization of glycerol production during high-performance fed-batch fermentation process in *Saccharomyces cerevisiae,* using a metabolic model as a prediction tool. *Applied and Environmental Microbiology* 72:2134.
38. Wang, Z. et al. 2001. Glycerol production by microbial fermentation. A review. *Biotechnology Advances* 19:201.
39. Vijaikishore, P., and Karanth, N. G. 1987. Glycerol production by fermentation: A fedbatch approach. *Biotechnology and Bioengineering* 30:325.
40. Overkamp, K. M. et al. 2002. Metabolic engineering of glycerol production in *Saccharomyces cerevisiae*. *Applied and Environmental Microbiology* 68:2814.
41. Gong, C. S. et al. 2000. Coproduction of ethanol and glycerol. *Applied Biochemistry and Biotechnology* 84–86:543.
42. Zhuge, J. et al. 2001. Glycerol production by a novel osmotolerant yeast *Candida glycerinogenes*. *Applied Microbiology and Biotechnology* 55:686.
43. Zhuge, B. et al. 2005. A novel *Candida glycerinogenes* mutant with high glycerol productivity in high phosphate concentration medium. *World Journal of Microbiology and Biotechnology* 21:453.
44. Holtzapple, T. M. et al. 1999. Biomass conversion to mixed alcohol fuels using the MixAlco process. *Applied Biochemistry and Biotechnology* 79:609.

45. Holtzapple, T. M., and Granda, C. B. 2009. Carboxylate platform: The MixAlco process. 1. Comparison of three biomass conversion platforms. *Applied Biochemistry and Biotechnology* 156:95.

46. Granda, C. B. et al. 2009. Carboxylate platform: the MixAlco process. 2. Process economics. *Applied Biochemistry and Biotechnology* 156:107.

47. International Energy Agency. 2006. *World energy outlook,* chap. 10. Paris.

48. Hoogers, G. 2003. Introduction. In *Fuel cell technology handbook,* ed. Hoogers, G., chap.1-1. Boca Raton, FL: CRC Press.

49. Laoun, B. 2007. Thermodynamics aspect of high-pressure hydrogen production by water electrolysis. *Revue des Energies Renouvelables* 10:435.

50. Kroposki, B. et al. 2006. Electrolysis: Information and opportunities for electrical power utilities. Technical report NREL/TP-581-40605, National Renewable Energy Laboratory, Golden, Colorado.

51. United Nations Environment Program. 2006. *The hydrogen economy. A non-technical review*. Paris.

52. Lettinga, G. 1996. Sustainable integrated biological wastewater treatment. *Water Science and Technology* 33:85.

53. Claassen, P. A. M. et al. 1999. Utilization of biomass for the supply of energy carriers. *Applied Microbiology and Biotechnology* 52:741.

54. Lewis, C. 1976. Energy relationships of fuel from biomass. *Process Biochemistry* 11(part 9, November):29.

55. Torry-Smith, M., Sommer, P., and Ahring, B. K. 2003. Purification of bioethanol effluent in a USAB reactor system with simultaneous biogas formation. *Biotechnology and Bioengineering* 84:7.

56. Thimann, K. V. 1963. *The life of bacteria,* 2nd ed., chap. 14. New York: Macmillan.

57. Thimann, K. V. 1963. *The life of bacteria,* 2nd ed., chap. 15. New York: Macmillan.

58. Thimann, K. V. 1963. *The life of bacteria,* 2nd ed., chap. 17. New York: Macmillan.

59. Moat, A. G., and Foster, J. W. 1995. *Microbial physiology,* 3rd ed., chap 7. New York: Wiley-Liss.

60. Nandi, R., and Sengupta, S. 1998. Microbial production of hydrogen: An overview. *Critical Reviews in Microbiology* 24:61.

61. Vignais, P. M., Billoud, B., and Meyer, J. 2001. Classification and phylogeny of hydrogenases. *FEMS Microbiology Reviews* 25:455.

62. Soni, B. K., Soucaille, P., and Goma, G. 1987. Continuous acetone butanol fermentation: Influence of vitamins on the metabolic activity of *Clostridium acetobutylicum. Applied Microbiology and Biotechnology* 27:1.

63. Gorwa, M. F., Croux, C., and Soucaille, P. 1996. Molecular characterization and transcriptional analysis of the putative hydrogenase gene of *Clostridium acetobutylicum* ATCC 824. *Journal of Bacteriology* 178:2668.

64. Nölling, J. et al. 2001. Genome sequence and comparative analysis of the solvent-producing bacterium *Clostridium acetobutylicum. Journal of Bacteriology* 183:4823.

65. Girbal, L. et al. 2005. Homologous and heterologous overexpression in *Clostridium acetobutylicum* and characterization of purified clostridial and algal Fe-only hydrogenases with high specific activities. *Applied and Environmental Microbiology* 71:2777.

66. Darensbourg, M. Y. et al. 2003. The organometallic active site of [Fe]hydrogenase: Models and entatic states. *Proceedings of the National Academy of Sciences USA* 100:3683.

67. Bagley, K. A. et al. 1995. Infrared-detectable groups sense changes in charge density on the nickel center in hydrogenase from *Chromatium vinosum. Biochemistry* 34:5527.

68. Volbeda, A. et al. 1996. Structure of the [NiFe] hydrogenase active site: Evidence for biologically uncommon Fe ligands. *Journal of the American Chemical Society* 118:12989.

69. Lenz, O. et al. 2005. Requirements for heterologous production of a complex metalloenzyme: The membrane-bound [NiFe] hydrogenase. *Journal of Bacteriology* 187:6590.
70. Zhu, W. et al. 2005. Modulation of the electronic structure and the Ni–Fe distance in heterobimetallic models for the active site in [NiFe]hydrogenase. *Proceedings of the National Academy of Sciences USA* 102:18280.
71. Armstrong, F. A. 2004. Hydrogenases: Active site puzzles and progress. *Current Opinion in Chemical Biology* 8:133.
72. Ösz, J et al. 2005. Theoretical calculations on hydrogenase kinetics: Explanation of the lag phase and the enzyme concentration dependence of the activity of hydrogenase uptake. *Biophysical Journal* 89:1957.
73. Tye, J. W. et al. 2005. Better than platinum? Fuel cells energized by enzymes. *Proceedings of the National Academy of Sciences USA* 102:16911.
74. Woodward, J. et al. 2000. Enzymatic production of biohydrogen. *Nature* 405:1014.
75. Yoshida, A. et al. 2005. Enhanced hydrogen production from formic acid by formate hydrogen lyase-overexpressing *Escherichia coli* strains. *Applied and Environmental Microbiology* 71:6762.
76. Yoshida, A. et al. 2007. Efficient induction of formate hydrogen lyase of aerobically grown *Escherichia coli* in a three-step biohydrogen production process. *Applied Microbiology and Biotechnology* 74:754.
77. Yoshida, A. et al. 2006. Enhanced hydrogen production from glucose using *ldh*- and *frd*-inactivated *Escherichia coli* strains. *Applied Microbiology and Biotechnology* 73:67.
78. Minnan, L. et al. 2005. Isolation and characterization of a high H_2-producing strain *Klebsiella oxytoca* HP1 from a hot spring. *Research in Microbiology* 156:76.
79. de Vrije, T. et al. 2007. Glycolytic pathway and hydrogen yield studies of the extreme thermophile *Caldicellulosiruptor saccharolyticus*. *Applied Microbiology and Biotechnology* 74:1358.
80. Kadar, Z. et al. 2003. Hydrogen production from paper sludge hydrolysate. *Applied Biochemistry and Biotechnology* 105–108:557.
81. Kadar, Z. et al. 2004. Yields from glucose, xylose, and paper sludge hydrolysate during hydrogen production by the extreme thermophile *Caldicellulosiruptor saccharolyticus*. *Applied Biochemistry and Biotechnology* 113–116:497.
82. Oh, S.-E. et al. 2004. Biological hydrogen production using a membrane bioreactor. *Biotechnology and Bioengineering* 87:119.
83. Liu, H., and Fang, H. P. 2002. Hydrogen production from wastewater by acidogenic granular sludge. *Water Science and Technology* 47:153.
84. Cheong, D. Y., Hansen, C. L., and Stevens, D. K. 2007. Production of bio-hydrogen by mesophilic anaerobic fermentation in an acid-phase sequencing batch bioreactor. *Biotechnology and Bioengineering* 96:421.
85. Kawagoshi, Y. et al. 2005. Effect of inoculum conditioning on hydrogen fermentation and pH effect on bacterial community relevant to hydrogen production. *Journal of Bioscience and Bioengineering* 100:524.
86. Kyazze, G. et al. 2007. Performance characteristics of a two-stage dark fermentative system producing hydrogen and methane continuously. *Biotechnology and Bioengineering* 97:759.
87. Iyer, P. et al. 2004. H_2-producing bacterial communities from a heat-treated soil inoculum. *Applied Microbiology and Biotechnology* 66:166.
88. Ren, N. et al. 2007. Microbial community structure of ethanol type fermentation in bio-hydrogen production. *Environmental Microbiology* 9:1112.
89. Koskinen, P. E. P., Kaksonen, A. H., and Puhakka, J. A. 2007. The relationship between instability of H_2 production and compositions of bacterial communities within a dark fermentation fluidized-bed reactor. *Biotechnology and Bioengineering* 97:742.

90. Zhang, H., Bruns, M. A., and Logan, B. E. 2006. Biological hydrogen production by *Clostridium acetobutylicum* in an unsaturated flow reactor. *Water Research* 40:728.

91. Van Ginkel, S. W., and Logan, B. E. 2005. Increased biological hydrogen production with reduced organic loading. *Water Research* 39:3819.

92. Park, W. et al. 2005. Removal of headspace CO_2 increases biological hydrogen production. *Environmental Science & Technology* 39:4416.

93. Salerno, M. B. et al. 2006. Inhibition of biohydrogen production by ammonia. *Water Research* 40:1167.

94. Van Ginkel, S. W., Oh, S.-E., and Logan, B. E. 2005. Biohydrogen gas production from food processing and domestic wastewaters. *International Journal of Hydrogen Energy* 30:1535.

95. Briton Bi, G. H., Yao, B., and Ado, G. 2006. Evaluation of the Abidjan Lagoon pollution. *Journal of Applied Sciences & Environmental Management* 10:175.

96. Ito, T. et al. 2005. Hydrogen and ethanol production from glycerol-containing wastes discharged after biodiesel manufacturing process. *Journal of Bioscience and Bioengineering* 100:260.

97. Bagi, Z. et al. 2007. Biotechnological intensification of biogas production. *Applied Microbiology and Biotechnology* 76:473.

98. Benemann, J. 1996. Hydrogen biotechnology: Progress and prospects. *Nature Biotechnology* 14:1101.

99. Benemann, J. et al. 1973. Hydrogen evolution by a chloroplast-ferredoxin-hydrogenase system. *Proceedings of the National Academy of Sciences USA* 70:2317.

100. Gaffron, H., and Rubin, J. 1942. Fermentative and photochemical production of hydrogen in algae. *Journal of General Physiology* 26:219.

101. Greenbaum, E. 1988. Energetic efficiency of hydrogen photoevolution by algal water splitting. *Biophysical Journal* 54:365.

102. Roessler, P. G., and Lien, S. 1984. Activation and de novo synthesis of hydrogenase in *Chlamydomonas*. *Plant Physiology* 76:1086.

103. Kessler, E. 1973. Effect of anaerobiosis on photosynthetic reactions and nitrogen metabolism in the green alga *Chlamydomonas reinhardtii*. *Archives of Microbiology* 93:91.

104. Lee, J. W., and Greenbaum, E. 2003. A new oxygen sensitivity and its potential application in photosynthetic H_2 production. *Applied Biochemistry and Biotechnology* 105–108:303.

105. Melis, A., Seibert, M., and Happe, T. 2004. Genomics of green algal hydrogen research. *Photosynthesis Research* 82:277.

106. Posewitz, M. C. et al. 2004. Hydrogen photoproduction is attenuated by disruption of an isoamylase gene in *Chlamydomonas reinhardtii*. *Plant Cell* 16:2151.

107. Kruse, O. et al. 2005. Improved photobiological H_2 production in engineered green algal cells. *Journal of Biological Chemistry* 280:34170.

108. Melis, A., and Happe, T. 2001. Hydrogen production. Green algae as a source of energy. *Plant Physiology* 127:740.

109. Melis, A. et al. 2000. Sustained photobiological hydrogen gas production upon reversible inactivation of oxygen evolution in the green alga *Chlamydomonas reinhardtii*. *Plant Physiology* 122:127.

110. Akano, T. et al. 1996. Hydrogen production by photosynthetic microorganisms. *Applied Biochemistry and Biotechnology* 57–58:677.

111. Fedorov, A. S. et al. 2005. Continuous hydrogen production by *Chlamydomonas reinhardtii* using a novel two-stage, sulfate-limited chemostat system. *Applied Biochemistry and Biotechnology* 121–124:403.

112. Patzek, T. W. 2004. Thermodynamics of the corn-ethanol biofuel cycle. *Critical Reviews in Plant Sciences* 23:519.

113. Patzek, T. W., and Pimentel, D. 2006. Thermodynamics of energy production from biomass. *Critical Reviews in Plant Sciences* 24:329.
114. Tamagnini, P. et al. 2002. Hydrogenases and hydrogen metabolism of cyanobacteria. *Microbiology and Molecular Biology Reviews* 66:1.
115. McTavish, H., Sayavedra-Soto, L. A., and Arp, D. J. 1995. Substitution of *Azotobacter vinelandii* hydrogenase small-subunit cysteines by serines can create insensitivity to inhibition by O_2 and preferentially damages H_2 oxidation over H_2 evolution. *Journal of Bacteriology* 177:3960.
116. Gutthann, F. et al. 2007. Inhibition of respiration and nitrate assimilation enhances photohydrogen evolution under low oxygen concentrations in *Synechocystis* sp. PCC 6803. *Biochimica et Biophysica Acta* 1767:161.
117. Miyamoto, K., Hallenbeck, P. C., and Benemann, J. R. 1979. Nitrogen fixation by thermophilic blue-green algae (cyanobacteria): temperature characteristics and potential use in biophotolysis. *Applied and Environmental Microbiology* 37:454.
118. Miyamoto, K., Hallenbeck, P. C., and Benemann, J. R. 1979. Hydrogen production by the thermophilic alga *Mastigocladus laminosus:* Effects of nitrogen, temperature, and inhibition of photosynthesis. *Applied and Environmental Microbiology* 38:440.
119. Cournac, L. et al. 2004. Sustained photoevolution of molecular hydrogen in a mutant of *Synechocystis* sp. strain PCC 6803 deficient in the type I NADPH-dehydrogenase complex. *Journal of Bacteriology* 186:1737.
120. Hemschemeier, A., and Happe, T. 2005. The exceptional photofermentative hydrogen metabolism of the green alga *Chlamydomonas reinhardtii. Biochemical Society Transactions* 33:39.
121. Asada, Y., and Miyake, J. 1999. Photobiological hydrogen production. *Journal of Bioscience and Bioengineering* 88:1.
122. Ihara, M. et al. 2006. Light-driven hydrogen production by a hybrid complex of a [NiFe]-hydrogenase and the cyanobacterial photosystem I. *Photochemistry and Photobiology* 82:676.
123. Sasikala, K. et al. 1993. Anoxygenic phototrophic bacteria: Physiology and advances in hydrogen production technology. *Advances in Applied Microbiology* 38:211.
124. Hoekema, S. et al. 2006. Controlling light-use by *Rhodobacter capsulatus* continuous cultures in a flat-panel photobioreactor. *Biotechnology and Bioengineering* 95:613.
125. Najafpour, G. et al. 2004. Hydrogen as a clean fuel via continuous fermentation by anaerobic photosynthetic bacteria. *Rhodospirillum rubrum* Afr. *Journal of Biotechnology* 3:503.
126. Amos, W. A. 2004. Biological water–gas shift conversion of carbon monoxide to hydrogen. Milestone completion report, NREL-MP-560-35592, National Renewable Energy Laboratory, Golden, Colorado.
127. Tayhas, G., and Palmore, R. 2004. Bioelectric power generation. *Trends in Biotechnology* 22:99.
128. Trabaey, K., and Verstraete, W. 2005. Microbial fuel cells: Novel biotechnology for energy generation. *Trends in Biotechnology* 23:291.
129. Logan, B. E., and Regan, J. M. 2006. Microbial fuel cells: Challenges and applications. *Environmental Science & Technology* 40:5172.
130. Chaudhuri, S. K., and Lovley, D. R. 2003. Electricity generation by direct oxidation of glucose in mediatorless microbial fuel cells. *Nature Biotechnology* 21:1229.
131. McKinlay, J. B., and Zeikus, J. G. 2004. Extracellular iron reduction is mediated in part by neutral red and hydrogenase in *Escherichia coli. Applied and Environmental Microbiology* 70:3467.
132. Prasad, D. et al. 2007. Direct electron transfer with yeast cells and construction of a mediatorless microbial fuel cell. *Biosensors and Bioelectronics* 22:2604.
133. Kim, J. R. et al. 2007. Electricity generation and microbial community analysis of alcohol powered microbial fuel cells. *Bioresource Technology* 98:2568.

134. Milliken, C. E., and May, H. D. 2007. Sustained generation of electricity by the spore-forming, Gram-positive, *Desulfitobacterium hafniense* strain DCB2. *Applied Microbiology and Biotechnology* 73:1180.

135. Justin, G. et al. 2004. Biofuel cells: A possible power source for implantable electronic devices. *Conference Proceedings of IEEE Engineering in Medicine and Biology Society* 6:4096.

136. Davis, J. B., and Yarborough, H. F. 1962. Preliminary experiments on a microbial fuel cell. *Science* 137:615.

137. Logan, B. E., and Regan, J. M. 2006. Electricity-producing bacterial communities in microbial fuel cells. *Trends in Microbiology* 14:512.

138. Zuo, Y. et al. 2007. Tubular membrane cathodes for scalable power generation in microbial fuel cells. *Environmental Science & Technology* 41:3347.

139. Logan, B. et al. 2007. Graphite fiber brush anodes for increased power production in air-cathode microbial fuel cells. *Environmental Science & Technology* 41:3341.

140. Rismani-Yazdi, H. et al. 2007. Electricity generation from cellulose by rumen microorganisms in microbial fuel cells. *Biotechnology and Bioengineering* 97:1398.

141. Sakai, S., and Yagishita, T. 2007. Microbial production of hydrogen and ethanol from glycerol-containing wastes discharged from a biodiesel fuel production plant in a bio-electrochemical reactor with thionine. *Biotechnology and Bioengineering* 98:340.

142. Hou, H. et al. 2009. Microfabricated microbial fuel cell arrays reveal electrochemically active microbes. *PLoS ONE* 4(8):e6570.

143. Ahn, Y., and Logan, B. E. 2010. Effectiveness of domestic wastewater treatment using microbial fuel cells at ambient and mesophilic temperatures. *Bioresource Technology* 101:469.

8 Chemically Produced Biofuels

8.1 INTRODUCTION

To most chemists, fermentation-based manufacture is a slow means to generate dilute aqueous solutions—a strategy that works best for high-value products (monoclonal antibodies, enzymes, and multiply chiral molecules). Chemical processes are fast and robust and result in highly concentrated products. Extracting useful molecules from plants is an ancient science, but can chemistry and chemical engineering transform plant biomass to useful biofuels?

Plant seed oils are major global sources for foods and represent global resources of relatively pure materials. Are microbes superior sources of such feedstocks for industrial chemistry?

Most importantly, can chemical reactions generate hydrogen in sufficient quantities to transform energy supply radically to support hydrogen fuel cells in the coming decades?

8.2 BIODIESEL: CHEMISTRY AND PRODUCTION PROCESSES

Practical interest in the oils extracted from plant seeds as sources of usable transportation fuels has a historical lineage back to Rudolf Diesel and Henry Ford. Minimally refined vegetable oils can be blended with conventional diesel fuels. A 10% lower energy content of widely available oils (on a volume basis) is incurred, with the consequent reduction of maximum fuel energy but without any modification of the injection system being required; such diesel fuel extenders remain cheap and plentiful.[1]

Biodiesel is not a biotechnological product; it is manufactured with any suitable vegetable oil from crops with no history of plant biotechnology (and even from animal fats) by an entirely chemical procedure. Commentators include biodiesel in the portfolio of emerging biofuels because of its principal biological origin as a plant seed oil.

8.2.1 VEGETABLE OILS AND CHEMICALLY PROCESSED BIOFUELS

Industrial production of biodiesel initially focused on transforming vegetable oils into a mixture of fatty acid esters by a process of transesterification of triglycerides with low molecular weight alcohols (almost always methanol because methanol provides the most volatile fatty acid esters; see Figure 8.1). Transesterification generates a product with physicochemical properties similar to those of conventional diesel fluids (Table 8.1).

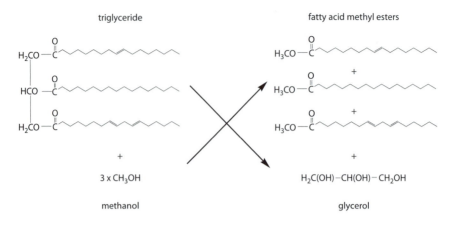

triglyceride

fatty acid methyl esters

3 x CH₃OH

methanol

H₂C(OH)−CH(OH)−CH₂OH

glycerol

FIGURE 8.1 Transesterification of triglycerides to fatty acid methyl esters (biodiesel).

By 2005, world production of biodiesel had reached 2.91 million tons of oil equivalent, of which 87% was manufactured in the EU (62% in Germany); only the United States (7.5%) and Brazil (1.7%) were other major producers. At that time, this total supply amounted to less than 20% of that of global fuel ethanol production.[2] World biodiesel supply increased by threefold between 2000 and 2005, and a marked acceleration in the United States as well as in Europe up to 2030 was predicted by the International Energy Agency.

Soybean oil dominated U.S. biodiesel production initially but production sites designed for multiple feedstocks now make up over 70% of total capacity; a much smaller group of production sites uses other plant oils, animal fats, and recycled cooking oils as feedstocks (Figure 8.2). Data supplied by the National Diesel Board (www.nbb.org) for October 2009 indicate 193 sites in the United States with a total

TABLE 8.1
Canola Seed Oil, Biodiesel, and Diesel

Physical Parameter	Diesel	Semirefined Oil	Methyl Esters (Biodiesel)
Onset of volatilization (°C)	70	280	70
End of volatilization (°C)	260	520	250
90% Distillation temperature (°C)	220	500	242
Density (g/mL)	0.83–0.85	0.92	0.88
Cetane number	45–54	32–40	48–58
Viscosity (mm²/sec at 20°C)	5.5	73	7
Cloud point (°C)	1 (seasonal)	−11	−1

Source: Data from McDonnell, K. et al. 1999. *Journal of the American Oil Chemists Society* 76:539; and Culshaw, F., and Butler, C. 1993. *A review of the potential of biodiesel as a transport fuel.* Department of Trade and Industry Energy Technology Support Unit Report ETSU-R-71. London: Her Majesty's Stationery Office.

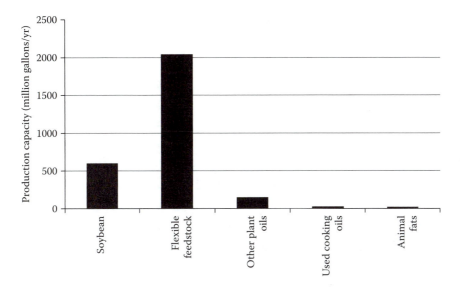

FIGURE 8.2 Breakdown of feedstock utilization by U.S. biodiesel producers. (Data from the National Biodiesel Board, October 2009.)

production capacity of 2.8 billion gallons per year. The range of sizes of the sites is enormous; at one extreme, a site capable of producing 180 million gallons per annum dwarfs the smallest site (annual capacity of 37,000 gallons). Almost as striking is the geographical spread of U.S. biodiesel producers, with 44 states represented in late 2009. In addition to commercial sites, unknown numbers of private producers are using home/garage kits to produce biodiesel whose chemical properties and suitability for motor vehicle use can easily be assessed.[3]

In the EU, rape (canola) is the most abundant suitable monoculture crop; it has the particular advantage of being readily cultivated in the relatively cold climates of northern Europe.[4] However, it is the sheer variety of single or mixed sources of oil and fat that could be transformed into biodiesel that has attracted both large-scale and niche market industrial interest; at one extreme, even used cooking oil (manufactured initially from corn, sunflower, etc.) can serve as the biological input—a widely publicized example of exemplary social recycling.[5]

8.2.2 Biodiesel Composition and Production Processes

Biodiesel is unique among biofuels in that it is not a single, defined chemical compound, but rather a variable mixture, even from a monoculture crop source. The triglycerides in any plant oil are a mixture of unsaturated and saturated fatty acids esterified to glycerol; fatty materials from land animals have much higher contents of saturated fatty acids (Table 8.2).[6] This variability has one far reaching implication: Reducing the content of saturated fatty acid methyl esters in biodiesel reduces the cloud point (the temperature below which crystallization becomes sufficiently advanced to plug fuel lines). A diesel suitable for winter use may have a cloud point

TABLE 8.2

Fatty Acid Composition of Plant Oils and Beef Tallow

Material	Lauric	Myristic	Saturated Palmitic (% Total Fatty Acids)	Stearic	Arachidic	Palmitoleic	Oleic	Unsaturated Linooleic (% Total Fatty Acids)	Liolenic	Other[a]
Corn oil		1.4	10	3.2	5.8	1.5	49.6	34.3		
Canola oil			1				32	15	1	50
Soybean oil	0.2	0.1	9.8	2.4	0.9	0.4	28.9	50.7	6.5	0.1
Sunflower oil			5.6	2.2	0.9		25.1	66.2		
Beef tallow		6.3	27.4	14.1			49.6	2.5		

Source: Data from Lide, D. R., ed. 1992. *CRC Handbook of Chemistry and Physics*, 73rd ed., section 7-29. Boca Raton, FL: CRC Press.

[a] Erucic (canola); C14 monoethenoic (soybean).

below −11°C, and winterization (treatment at low temperature and removal of solidi-
fied material) of biodiesel generates a product with similarly improved operability
and start-up characteristics.[7,8]

The idiosyncratic fatty acid content of canola seed oil, with its preponderance of
the very long chain erucic acid, has a quite different significance. Erucic acid has been
known since the 1950s to stimulate cholesterol synthesis by animals.[9] The potential
adverse health effects (increased risk of circulatory disease) led to legislation on the
erucic acid content of edible oils and the development of low erucic acid cultivars.

By contrast, high erucic acid oils have a market because erucic acid and its derivatives are feedstocks for the manufacture of specialized chemicals.[10] High-erucic acid
oils would be either desirable or neutral for biodiesel production, but low erucic
cultivars are higher yielding. In any case, legal requirements were in place in the EU
as early as 1992 to separate the two types of oilseed rape cultivation geographically
to minimize cross-pollination and contamination of agricultural products intended
for human consumption.[4]

The majority of the biodiesel producers continue to employ a base-catalyzed
reaction with sodium or potassium hydroxide (Figure 8.3).[11] This straight forward
technology has economic attractions: use of low temperatures and pressures in the
reaction, high conversion efficiencies in a single step, and no requirement for exotic
materials in the construction of the chemical reactor. The liberation of glycerol
(sometimes referred to as "glycerine" or "glycerin") in the transesterification reaction generates a potentially saleable coproduct.

The generation of fatty acid methyl esters (FAMEs) is the same reaction as that
needed to form volatile derivatives of fatty acids prior to their analysis by gas liquid chromatographic methods. The key parameters for optimization are the reaction
time, temperature, and the molar ratio of oil to alcohol, although choices of the type
of catalyst used and the short-chain alcohol coreactant can also be made.[12] Different
oil types of plant origin have been the subject of intense process optimization studies; examples from 2005 to 2009, exemplifying the global nature of R&D activities
with biodiesels, are summarized in Table 8.3.[13–23] Although elevated temperatures
are common, lower temperatures (<40°C) are possible.

More subtle factors include differential effects on product yield and purity; for
example, temperature has a significant positive effect on biodiesel purity but a negative influence on biodiesel yield, and the alcohol-to-oil molar ratio is significant only
for biodiesel purity (with a positive influence).[24] Whereas the biodiesel yield may
increase at decreasing catalyst concentration and temperature, the methanol-to-oil
ratio may not affect the material balance.[25]

A variety of novel catalysts have been explored, partly to avoid the use of caustic
materials but also to facilitate catalyst recovery and reuse—for example:

- sulfonated amorphous carbon[26]
- ion-exchange resins[27]
- amorphous SiO_2-ZrO_2[28]
- K_2CO_3 on an alumina/silica support[29]
- zinc dodecatungstophosphate ($Zn_{1.2}H_{0.6}PW_{12}O_{40}$) nanotubes[30]

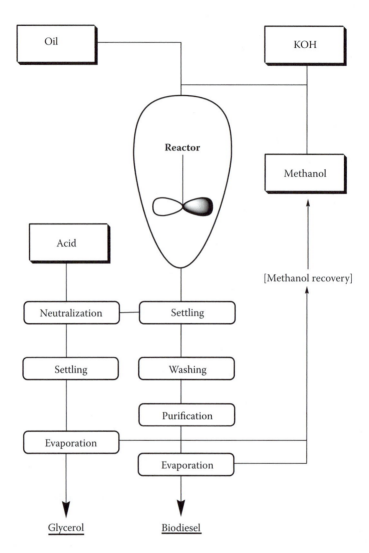

FIGURE 8.3 Schematic of typical biodiesel production process with alkaline catalysis.

Indeed, the requirement for a catalyst can be eliminated if high temperatures and pressures are used to generate supercritical fluid conditions. Under such conditions, alcohols can react directly with triglycerides or (in two-stage procedures) with fatty acids liberated from triglycerides.[31]

Crude oils can give poor transesterification rates because of their contents of free fatty acids and other components lost during refining; the free fatty acids (up to 3% of the oils) react with the alkaline catalysts and form saponified products during the transesterification. Extra steps must be devised to overcome the problems posed by some feedstocks. For example, a two-step pretreatment process was devised for an Indonesian seed oil (*Jatropha curcas*) exhibiting very high free fatty acids (15%). The first step was carried out with sulfuric acid as catalyst in a reaction at 50°C and

TABLE 8.3

Transesterification Optimization for Production of Biodiesel from Different Plant Oils

Plant Seed Oil (Country)	Alcohol	Temperature (°C)	Reaction Time (h)	Molar Ratio (alcohol/oil)	Catalyst	Ref.
Soybean and castor (Brazil)	Ethyl	70	3	9	NaOH	13
Pongamia pinnata (India)	Methyl	60		10	KOH	14
Waste frying oils (Portugal)	Methyl		1	4.8	NaOH	15
Rapeseed (Korea)	Methyl	60	0.33	10	KOH	16
Sunflower (Spain)	Methyl	25		6	KOH	17
Moringa oleifera (Pakistan)	Methyl	60		6	KOH	18
Rapeseed + hexane (China)	Methyl	55–60	2	9	NaOH	19
Zanthoxylum bungeanum (China)	Methyl	60	1.33	24	H_2SO_4	20
Cottonseed (China)	Methyl	40	3	135	NaOH	21
Guizotia abyssinica (India)	Methyl	65			NaOH	22
Palm (Malaysia)	Methyl	65		15	CaO	23

removal of the methanol–water layer; the second step was a conventional transesterification using an alkaline catalyst to produce biodiesel at 65°C.[32]

Rice bran stored at room temperature can show extensive (>75%) hydrolysis of triglycerides to free fatty acids. The successful processing of the oil fraction also required a two-step methanolysis process (with both steps catalyzed by acid), resulting in a 98% methyl ester formation in less than 8 h and the coproduction of residue with high contents of such nutraceuticals such as γ-oryzanol and phytosterols.[33] Supercritical methanol treatment (without any catalyst) at 350°C can generate esters from both triglycerides and free fatty acids, thus giving a simpler process with a higher total yield of biodeisel.[34]

Experimental innovations in biodiesel production include:

- A six-stage continuous reactor for transesterification of palm oil is claimed to produce saleable biodiesel within residence time of 6 min in a laboratory prototype with a production capacity of 17.3 L per hour.[35]
- A continuous process takes place for the manufacture of biodiesel from crude vegetable oils under high-power, low-frequency ultrasonic irradiation.[36]
- A two-phase membrane reactor developed to produce biodiesel from canola oil and methanol offers an immiscible combination, providing a mass-transfer challenge in the early stages of the transesterification. This design of reactor is particularly useful in removing unreacted oil from the product, yielding high-purity biodiesel and shifting the reaction equilibrium to the product side.[37]
- A continuous reactor process with a high reaction rate is achieved by atomizing the heated oil before spraying it into a reaction chamber filled with methanol vapor in a countercurrent flow arrangement. The separation of product and the excess methanol stream in the reactor are continuous.[38]
- A conventional stirred tank reactor replaced with a microchannel reactor can arrive at a compact and mini-fuel-processing plant design with extraordinarily short residence times (<0.5 min).[39]

Following biodiesel production, the fuel's thermal properties have been improved—in this case to reduce the onset of volatilization (Table 8.1) of soybean-derived biodiesel to below that of conventional diesel. Ozonolysis is one known method in which the onset freezing temperature of ozonated methyl soyate is reduced from –63 to –86°C.[40]

8.2.3 BIODIESEL ECONOMICS

The International Energy Agency's 2006 analysis concluded that biodiesels would not be price competitive with conventional diesels if all subsidies to crops and production were excluded; if the biodiesel source were animal fat, however, the derived biodiesel would be competitive even at crude oil prices below $50–55 per barrel.[2] By 2030, assuming various process improvements and economies of scale, biodiesel from vegetable oils was also predicted to be competitive at crude oil prices at $50–55 per barrel; European biodiesel would continue to be more expensive than U.S. biodiesel, with feedstock costs the largest contributor (Figure 8.4). This continues a tradition of cost assessments that commenced in the 1990s.

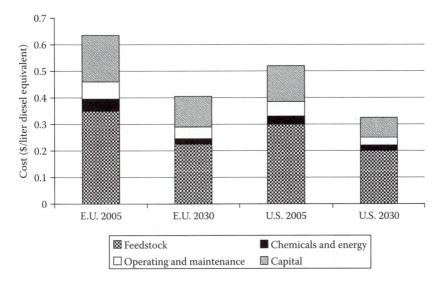

FIGURE 8.4 Production costs of biodiesel in the EU and United States (including subsidies to crop production) for 2005 and 2030 reference scenario. (Data from International Energy Agency. 2006. *World Energy Outlook,* chap. 14. Paris.)

Rapeseed oil-derived biodiesel was estimated to require a total subsidy of between 10 and 186% of the price of conventional diesel in 1992; the variation reflected the price of the seeds sown to grow the crop. This was equivalent to the cost of biodiesel being between 11 and 286% of the refinery gate cost of conventional diesel; the contemporary price for seeds would have resulted in biodiesel being 243% of the conventional diesel cost.[4] These calculations required both coproducts (glycerol and rapeseed meal) to be income generators; in addition, the spring-sown crop (although with lower yields per unit of land) had lower production costs.

An evaluation of costs from U.S. soybean and sunflower in 2005 concluded that soybean biodiesel would have production costs 2.8-fold those of conventional diesel, whereas sunflower-derived biodiesel was over fivefold more expensive to produce than petroleum diesel fuel.[41] Sunflower seeds had a higher oil content (25.5%) than soybeans (18%), but were estimated to have a lower crop productivity (1,500 kg/ha versus 2,700 kg/ha); in both cases, oil extraction was calculated to be highly energy intensive.

The production costs for biodiesels also depend on the production route. With waste cooking oil as the feedstock, although an alkali-catalyzed process using virgin vegetable oil had the lowest fixed capital cost, an acid-catalyzed process using the waste oil was more economically feasible overall. It provided a lower total manufacturing cost, a more attractive after-tax rate of return, and a lower biodiesel breakeven price; in addition, plant capacity was found to be a significant factor affecting the economic viability of biodiesel manufacture.[42,43]

The U.S. Department of Agriculture's Agricultural Research Service has developed a computer model to estimate the capital and operating costs of a moderately sized industrial biodiesel production facility (annual production capacity of 10 million gallons)[44]:

- Facility construction costs were calculated to be $11.3 million.
- The largest contributors to the equipment cost (accounting for nearly one third of expenditures) were storage tanks to contain a 25-day capacity of feedstock and product.
- At a value of $0.52 per kilogram for feedstock soybean oil, a biodiesel production cost of $0.53 per liter ($2.00 per gallon) was predicted.
- The single greatest contributor to this value was the cost of the oil feedstock, which accounted for 88% of total estimated production costs. An analysis of the dependence of production costs on the cost of the feedstock indicated a direct linear relationship between the two.
- Process economics included the recovery of coproduct glycerol generated during biodiesel production and its sale into the commercial glycerol market, which reduced production costs by approximately 6%.

Waste cooking oils, restaurant grease, and animal fats are inexpensive feedstocks; they represent 30% of total U.S. fats and oil production, but are currently devoted mostly to industrial uses and animal feed; because the free fatty acids may represent >40% of the material, the production process may be complex.[45] Nevertheless, such unconventional feedstocks may become increasingly important because soybean oil prices surged during 2007 and reached an unsustainably high peak in mid-2008; this caused a near collapse in soybean oil use for biodiesel production, although there was some recovery during 2009 (Figure 8.5). With its main feedstock unpredictably expensive, U.S. biodiesel has struggled to be competitive on price with diesel fuels

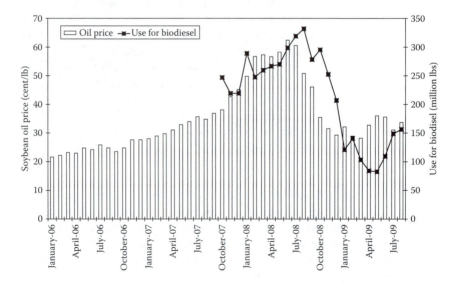

FIGURE 8.5   Soybean oil prices and monthly use for biodiesel, 2006–2009. (Data from U.S. Department of Agriculture.)

or heating oil; tax incentives may be necessary to overcome these production price issues.

8.2.4 Energetics of Biodiesel Production and Effects on Greenhouse Gas Emissions

Inevitably, an essential facet of the public discussions on costs and subsidies of biodiesel production has been that of its potential amelioration of greenhouse gas emissions. If significant, this would augment the case for production and consumption incentives to offset production costs higher than those for conventional diesel.

In the early 1990s, net energy balance values of up to 3.8:1 were calculated for rapeseed-derived biodiesel, depending on how the coproducts and crop straw were assessed in the calculations (Figure 8.6).[4] Unpublished reports and communications quoted in that report were from 1.3 to 2.1 without coproduct credits and from 2 to 3 if thermal credits for the meal and glycerol coproducts were included. Radically different conclusions were reached in a 2005 publication: Biodiesel production from soybean oil required 27% more *fossil* energy than the biodiesel energy content, while sunflower oil was even less viable (requiring 118% more *fossil* energy than in the product).[41]

Midway (in time) between these conflicting estimates was a report from the National Renewable Energy Laboratory (NREL), whose main conclusion was that, whereas conventional diesel yielded only 0.83 of a unit per unit of fossil fuel consumed, biodiesel (from soybean oil) yielded 3.2 units of fuel product energy for every unit of fossil energy consumed in its life (i.e., biodiesel was eminently renewable).[46] Direct comparison of these conflicting results shows that the

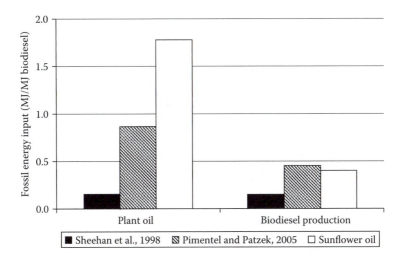

FIGURE 8.6 Estimated fossil energy inputs to biodiesels manufactured from soybean or sunflower oil. (Data from Pimentel, D., and Patzek, T. W. 2005. *Natural Resources Research* 14:65; and Sheehan, J. et al. 1998. Final report, NREL/SR-580-24089 UC Category 1503, National Renewable Energy Laboratory, Golden, Colorado.)

disagreements are major for the stages of soybean cultivation and for biodiesel production (Figure 8.6).

As occurs so often in biofuel energy calculations, part of the discrepancy resides in how the energy content of coproducts is allocated and handled in the computations. As the authors of the 2005 study pointed out, if the energy credit of soybean meal is subtracted, then the excess energy required for biodiesel production falls to 2% of the biodiesel energy content. Other factors have also been implicated[47]:

- Lime application was a major agricultural input in the 2005 study but may have been overestimated by a factor of five; that is, the application rate should have been spread over 5 years rather than every season. Making this change reduces the energy required to only 77% of the biodiesel energy content.
- In the NREL study, the energy input to produce soybeans and then extract the oil was divided 18:82 in favor of soybean meal (i.e., following the weight split of oil and residual material). Allocating only 18% of the input energy to soybean oil changes the energy balance dramatically to 5.3 times *more* biodiesel energy than fossil energy input.
- Even adding in energy requirements for oil transport and transesterification as well as biodiesel transport produces a favorable energy balance of 2.9:1.

In any of these scenarios, the energy balance is highly dependent on viewing the process as a biorefinery producing coproducts as well as biodiesel. If the energetic (and economic) value of the soybean meal cannot be realized, then the balance will be negative; even using the soybean meal as a "green manure" spread on the soybean fields would only partially offset the major loss of replaced fossil energy in the total process. A focus of future attention may be that of realizing an economic return on the greatly increased amounts of seed meal and of finding a viable route for glycerol. Refining the glycerol coproduct to a chemically pure form is expensive and alternative uses of glycerol for small- and medium-scale biodiesel facilities are being explored (e.g., its use as an animal feed supplement).[48]

As the number of industrial units producing biodiesel increases, assessments of energy balances should be possible from collected data rather than from calculations and computer simulations. A report on activities in six Brazilian and Colombian biodiesel facilities using palm oil as the agricultural input attempted to do precisely this.[49] Net energy balances were in the range of 6.7–10.3; the differences arose because of important variables:

- Rates of fertilizer application were different.
- There were different uses of plant residues as fertilizers or as boiler fuel for electricity production.
- Some sites had on-site electricity generation, whereas others were entirely dependent on purchased electricity.
- Efficiencies in the generation of coproducts and the recovery of unused palm oil differed.

Taken together as a group, these palm oil biodiesel producers were assessed as more energy efficient than reference manufacturers in Europe or the United States. A detailed estimate of biodiesel from soybean oil in the United States, however, arrived at a net energy balance of 1.93:1, although this conclusion was critically dependent on full credits taken for soybean meal and glycerol coproducts (without them, the balance decreases to only 1.14:1).[50]

The energy balance is an important parameter in defining to what extent biodiesel can reduce greenhouse gas emissions; in the extreme case, if biodiesel requires more fossil energy in its production than can be usefully recovered in the product, no savings could possibly accrue.[41] With a favorable energy balance for soybean biodiesel, its use could displace 41% of the greenhouse gas emissions relative to conventional diesel.[50] As headline statements, the NREL study on biodiesel use for public transport concluded[46]:

- Substituting 100% biodiesel (B100) for petroleum diesel reduced the life cycle consumption of petroleum by 95% while a 20% blend (B20) reduced consumption by 19%.
- B100 reduced CO_2 emissions by 74.5%, B20 by 15.7%.
- B100 completely eliminated tailpipe emissions of sulfur oxides and reduced life cycle emissions of CO, sulfur oxides, and total particulate matter by 32, 35, and 8%, respectively.
- Life cycle emissions of NO_x and hydrocarbons were higher (13.4 and 35%, respectively) with B100, but there were small reductions in methane emissions.

Earlier assessments indicated that, if biodiesel were to be used, only 55% of the CO_2 emitted from fossil diesel could be saved because of the CO_2 emissions inherent in the production of biodiesel and that, other than a marked reduction in sulfur oxides, effects on CO, hydrocarbons, NO_x, and polyaromatic hydrocarbons were inconsistent.[4] As the use of biodiesel has widened globally, the number of publications exploring individual pollutants or groups of greenhouse gas emissions has expanded (Table 8.4).[51–59] The most recent of these reports, however, found that

TABLE 8.4
Studies on Biodiesels and Their Impact on Aerial Pollutants

Biodiesel Source	Pollutants Investigated	Ref.
Neem oil (Bangladesh)	$CO\downarrow$, $No_x\uparrow$, smoke$\downarrow$	51
Soybean oil (Turkey)	$CO\downarrow$, $No_x\uparrow$, particulates$\downarrow$, hydrocarbons$\downarrow$	52
Rapeseed oil (Korea)	$CO\uparrow$, $No_x\uparrow$, smoke$\downarrow$, $CO_2\uparrow$	53
Soybean oil (U.S.)	Particulates$\downarrow$	54
Waste cooking oil (Spain)	Particulates$\downarrow$, smoke$\downarrow$	55
Soybean, rapeseed oil (Germany)	Mutagenicity of particulates$\uparrow$	56
Palm oil (China)	$CO\downarrow$, polyaromatics$\downarrow$, particulates$\downarrow$, hydrocarbons$\downarrow$	57
Brassica carinata (Italy)	$No_x\uparrow$, particulates$\downarrow$	58
Soybean oil (U.S.)	$CO\downarrow$, $No_x\downarrow$, particulates$\downarrow$, hydrocarbons$\downarrow$	59

although local tailpipe emissions were reduced, a complete life cycle analysis indicated higher hydrocarbons and NO_x; assessments of impacts of biodiesel blends may therefore arrive at different conclusions depending on whether rural or urban trials are compared.[59]

Increased mutagenicity in particulate emissions with biodiesel has been reported.[60] This was unusual because two earlier reports from the same research group in Germany found *reduced* mutagenicity with rapeseed oil- and soybean-derived biodiesels.[56,61] A high sulfur content of the fuel and high engine speeds (rated power) and loads were associated with an increase in mutagenicity of diesel exhaust particles. This is in accord with the desirability of biodiesels because of their very low sulfur contents: zero or barely detectable as compared with up to 0.6% (by weight) in conventional diesels.[4]

There are suggestions that exhaust emissions from biodiesels are less likely to present any risk to human health relative to petroleum diesel emissions. However, the recommendation has been made that the speculative nature of a reduction in health effects based on chemical composition of biodiesel exhaust needs to be followed up with thorough investigations in biological test systems.[62]

STEM TOPIC 8.1: HOW COST EFFECTIVE ARE FIRST-GENERATION BIOFUELS IN REDUCING ATMOSPHERIC CO_2 ACCUMULATION?

European options for biofuels are quite restricted: rapeseed (canola) oil for biodiesel and sugar beet and cereal grains for ethanol. European Union targets for replacing fossil fuels with renewable energy sources (including biofuels) have cost implications. A German study[63] focused on assessing the costs of abating greenhouse emissions by various strategies (see Figure STEM 8.1).

Both European biodiesel and ethanol (from sugar beet) emerged as expensive means of reducing greenhouse gas emissions. Much cheaper alternatives were to improve energy efficiency in gas and coal (lignite) power stations and to use woody biomass for power generation. Brazilian ethanol, on the other hand, was by far the best option because its production costs were then so low.

This argument was not, as the authors admitted, an exhaustive cost-benefit analysis. The primary reason for using biodiesel would be to power diesel engines; electricity-generating options such as wood-burning power stations would be highly relevant, given sufficient electric or gasoline/electric hybrids on European roads, but the costs of achieving that status would be enormous. Similarly, the study estimated that total EU tax revenue losses by incentive measures for biofuels would reach €1 billion ($1.47 billion) annually by 2010. Taxes are, of course, questionable; however, given the highly socialized nature of the major European societies, it is unlikely that tax-raising powers could ever not be counted into the abatement equation.

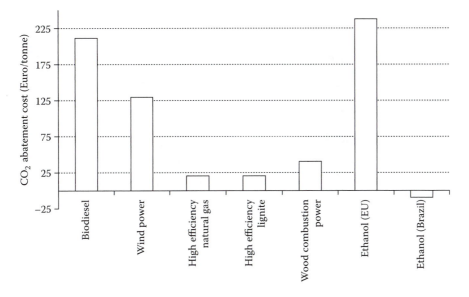

FIGURE STEM 8.1 Comparative CO_2 abatement costs of biodiesel, ethanol, and means of improving power generation technologies in Europe. (Data from Frondel, M., and Peters, J. 2005. Discussion paper 36, Rheinisch-Westfälisches Institut für Wirtschaftforschung, Essen, Germany. Available from http://repec.rwi-essen.de/files/DP_05_036.pdf.)

8.2.5 CASE STUDY 1: HYDROGENATED PLANT OILS AND "GREEN DIESEL"

The petrochemical industry has evidenced intense interest in biofuels for a number of years. In trade journals, the key word phrase is often "clean fuels," and there is no doubt that environmental issues have been dominant in the attitude of the conventional oil industry in its broadest sense (from oil extraction to refinery operations) toward developments in the international production and trade in biofuels. The two major achievements in the conventional clean fuels agenda have been the elimination of lead forms of gasoline and the production of ultralow-sulfur diesel.[64] Since 2005, an innovation in oil refineries has been to introduce hydrogenated vegetable oils that can be produced on site by combining technologies:

- "Cracking" of oil alkanes (saturated hydrocarbons such as octane) to unsaturated hydrocarbons (alkenes) generates H_2 as a coproduct.
- This H_2 can be used to hydrotreat vegetable oils to produce the second-generation renewable diesel fuel or green diesel or hydrotreated vegetable oils (HVOs).

The biggest challenge faced by the manufacturers of the HVOs is that of differentiating the product from first-generation biodiesel because the new fuels have important advantages over earlier biodiesels.[65] The key feature of HVOs is an energy density higher than that of biodiesel, although not quite equaling that of conventional

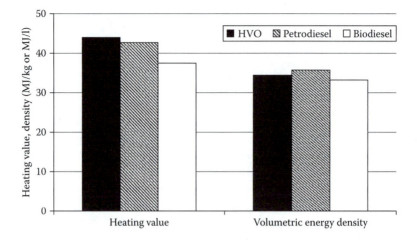

FIGURE 8.7 Energy values of hydrogenated vegetable oil, conventional diesel, and first-generation biodiesel. (Data from Mikkonen, S. 2008. *Hydrocarbon Processing* February: 63–66.)

low-sulfur diesel (Figure 8.7). With a diesel engine, a gallon of HVO could power a vehicle for approximately twice the distance achieved with an internal combustion engine fueled with E85.

The high-energy density of HVOs means that they could be introduced into the transportation fuel chain by three distinct routes:

1. by blending small amounts into petrodiesel (B2–B5)
2. by blending in larger amounts, perhaps up to B30, without compromising fuel quality and engine operation
3. pushing the blend to its limit with city bus fleets when fuel injection systems will need recalibrating because of the lower density of HVOs and their higher cetane numbers compared to petrodiesel

HVOs have no lower storage stabilities than conventional diesels, thus encouraging a wide rage of applications with minimal financial implications.

Finally, HVOs emit fewer pollutants. In comparison with low-sulfur diesel, HVOs have reduced tailpipe emissions when tested with bus and truck engines: up to 40–50% less for particulates, hydrocarbons, and CO and approximately 10% less for NO_x.[65]

8.2.6 CASE STUDY 2: ENZYMES FOR BIODIESEL PRODUCTION

A biotechnological approach to biodiesel production has slowly emerged that employs enzyme catalysts, lipases, to carry out transesterification (sometimes described as alcohololysis) rather than straightforward hydrolysis of triglycerides. Therefore, rather than liberating free fatty acids and glycerol, transesterification generates biodiesel methyl fatty acid esters in one step (with glycerol as the coproduct).[66] The principal process advantage of the enzyme-based approach is the ability to use

TABLE 8.5
Enzyme Catalysts for the Transesterification of Oils to Biodiesel

Plant Oil	Alcohol	Conditions	Lipase	Enzyme Activity	Ref.
Soybean	Several	Continuous batch operation at 30°C	Lipozyme TL IM	95% after 10 batches	66
Soybean, sunflower, etc.	Ethyl	Quantitative conversion within 7 h at 25°C	Novozyme 435	85% after 9 batches	67
Olive	Methyl	Stepwise addition of methanol at 60°C	Novozyme 435	70% after 8 batches	68
Sunflower, etc.	Ethyl acetate	Immobilized enzyme at 25°C	Novozyme 435	85% after 12 batches	69

low to moderate temperatures and atmospheric pressure in the reaction vessel while ensuring little or no chemical decomposition (i.e., a high product purity). The main drawbacks are the high costs of enzymes and the much longer incubation times to achieve >90% conversion of the triglycerides—up to 120 h.[67]

The barrier to full commercialization is that of maintaining the (relatively expensive) enzyme active during repeated batch use. Rival enzyme products show differing stability, and methanol appears to induce a faster loss of activity than does ethanol.[68] Examples of enzyme-catalyzed processes using oils of plant origin and with prolonged survival of the lipases are summarized in Table 8.5.[67–70] Even immobilized lipases are inhibited by methanol and glycerol; the use of *tert*-butanol as solvent, continuous removal of glycerol, and the stepwise addition of methanol reduce the inhibitory effects and increase the cost effectiveness of lipase-catalyzed processes.[71]

Exploration of lipases from a wider range of microbial sources than have been commercialized has begun to accelerate. An interesting example is a lipase from *Pseudomonas cepacia* used in an immobilized form within a chemically inert, hydrophobic sol-gel support. Under optimal conditions with soybean oil, high methyl and ethyl ester formations were achieved within a 1-h reaction, and the immobilized lipase was consistently *more* active than the free enzyme, losing little activity when subjected to repeated use.[72]

Immobilized lipases from *Enterobacter aerogenes, Rhizopus oryzae,* and *Candida rugosa* have each shown high survivability during repeated rounds of biodiesel production.[73,74] A lipase from *Photobacterium lipolyticum* is highly tolerant to methanol.[75] A lipase-producing bacterium strain screened from soil samples in China, identified as *Pseudomonas fluorescens,* exhibited a novel psychrophilic lipase (with a temperature optimum of only 20°C); this enzyme may represent a highly competitive energy-saving biocatalyst because lipase-mediated biodiesel production is normally carried out at 35–50°C.[76]

The most recent trend in lipase technologies has been to immobilize whole cells rather than isolated lipases; by eliminating the requirement to extract active enzyme before immobilization, significant cost benefits are expected. Fungal cells often inhabit al environments akin to polymeric support surfaces. Immobilization of fungal mycelia within biomass support particles, and the expression of lipase activity on the surfaces of the cells, could generate highly efficient whole-cell biocatalysts for industrial applications.[77] Bacteria are also amenable to this approach; a recombinant *Escherichia coli* expressing a *Proteus* lipase gave a fast biodiesel process (a yield of nearly 100% after 12 h) and could function well at a low temperature (15°C) that was thought to be the lowest temperature among all known catalysts for biodiesel production.[78]

8.3 FISCHER–TROPSCH DIESEL: CHEMICAL BIOMASS–LIQUID FUEL TRANSFORMATIONS

8.3.1 THE RENASCENCE OF AN OLD CHEMISTRY FOR BIOMASS-BASED FUELS?

Direct combustion of plant biomass is only one of three thermochemical routes for biomass utilization. Gasification (incomplete combustion) yields different mixes of products depending on the conditions used[79,80]:

- With pure oxygen as the combustant, a synthesis (or producer) gas with a high CO content is generated.
- The use of air rather than oxygen reduces the heating value because nitrogen dilutes the mixture of gases.
- If water is present and high temperatures are reached, hydrogen may also be formed, but excess water tends to result in high CO_2 concentrations and greatly reduces the heating value of the gaseous product.

Synthesis/producer gases resulting from gasification technologies generally have low heating values (4–10 MJ per liter) and are best suited to in situ power and heat generation. The third thermochemical method, pyrolysis (i.e., heating in the absence of air or oxygen), can be an efficient means of generating a gas high in hydrogen and CO.

A mixture of CO and H_2 gases can be converted chemically into a mixture of hydrocarbons. The chemistry of gas-to-liquid fuel transformations was developed in the first quarter of the twentieth century and utilized extensively in Germany during World War II. Further evolution led to commercial production processes being initiated for peacetime purposes in the 1990s.[81,82] The essential step, known as the Fischer–Tropsch reaction, can be written as

$$nCO + (2n + 1)H_2 \rightarrow C_n H_{2n + 2} + nH_2O$$

where $[CH_2]_n$ represents a range of hydrocarbons from low molecular weight gases ($n = 1$, methane) by way of gasoline ($n = 5$–12), diesel fuel ($n = 13$–17), and as far as solid waxes ($n > 17$). The reaction requires catalysts for realistic rates to be achieved, usually iron or cobalt (although transition metals will function effectively) at high temperatures (180–350°C) and high pressures. The higher the temperature is, the higher the proportion of gas and liquid hydrocarbon products will be.

To date, no process has been commercialized from plant biomass feedstocks, and the FT technology could be described as "radical" or "*n*th" generation for biofuels were it not that the key elements of the chemistry and production options are well established

TABLE 8.6

Operating and Planned FT Plants Based on Methane (Natural Gas) Feedstock

Country	Company or Companies	Production Level (Barrels Per Day)	Start-Up Year
South Africa	PetroSA	20,000	1992
Malaysia	Shell	15,000	1993
Qatar	Sasol, Qatar Petroleum, Chevron	34,000	2005
Nigeria	Chevron Nigeria, Nigeria National Petroleum	34,000	2007
Qatar	Shell, Qatar Petroleum	140,000	2009
Qatar	ExxonMobil, Qatar Petroleum	154,000	2011

in industrial processes with fossil inputs. In a climate of high crude oil prices, the environmental desirability of low-sulfur diesel and the drive to commercialize otherwise unmarketable natural gas in remote locations are important synergies (Table 8.6).[83]

FT biomass–liquid fuel (FT–BtL) from lignocellulosic sources is particularly attractive because of the high CO_2 emission reduction potential (up to 90% when substituting conventional gasoline and diesel) and the ability to use woody materials from low-grade land, thus avoiding the pressures on land use in OECD countries contemplating agriculture-based fuel ethanol or biodiesel production on a large scale. The principal barrier to large-scale biomass FT–BtL is the suboptimal mixture of gases in syngas as prepared from plant materials: The lower the molar ratio of H_2/CO is, the more the proportion of high molecular weight products formed in the FT reaction will be. However, biomass gasification results in a wide range of H_2/CO ratios—often with an excess of CO—together with appreciable amounts of CO_2, methane, and higher hydrocarbons as well as smaller amounts of condensable tars and ammonia.[84]

STEM TOPIC 8.2: COMBUSTION, GASIFICATION, AND PYROLYSIS OF BIOMASS

The complete combustion of the carbon in biomass requires sufficient O_2:

$$C + O_2 \rightarrow CO_2$$

A molar ratio of O_2/C of less than 1 results in incomplete combustion or gasification, with varying amounts of CO, CO_2, and C as products—for example:

$$C + nO_2 \rightarrow (2n-1)CO_2 + 2(1-n)CO$$

$$C + nO_2 \rightarrow 2nCO + (1-2n)C$$

Pyrolysis is an anaerobic process with the complete exclusion of air:

$$\text{Biomass carbon} \rightarrow \text{coke} + \text{hydrocarbon volatiles}$$

Methane can be transformed to CO and H_2 by a number of different reactions, including the uncatalyzed (but, again, high-temperature and high-pressure) processes[85]:

$$CH_4 + O_2 \rightarrow CO_2 + 2H_2O$$

$$CH_4 + H_2O \rightarrow CO + 3H_2$$

Partial removal of CO (and formation of additional H_2) is possible by the water-shift reaction:

$$CO + H_2O \rightarrow CO_2 + H_2$$

Finally, the physical removal (adsorption) of CO_2 (an inert gas for FT reactions) is relatively straightforward, but a higher yielding process can be devised (at least, in principle) by including a catalytic reduction of the CO_2 to using multiple FT reactors in series with an intermediate water removal step[86]:

$$CO_2 + 3H_2 \rightarrow [-CH_2-] + 2H_2O$$

Complete wood-based FT–BtL production therefore involves a multistage process incorporating biomass pretreatment, syngas purification, and optional syngas recycling, plus gas turbine power generation for unused syngas and, for FT diesel, a hydrocracking step to generate a mixture of diesel, naphtha, and kerosene (Figure 8.8).[84,85]

8.3.2 ECONOMICS AND ENVIRONMENTAL IMPACTS OF FT DIESEL

In comparison with natural-gas-based FT syntheses, biomass requires more intensive engineering, and gas-cleaning technology has been slow to evolve for industrial purposes, even though it is essential for the successful use of biomass because of the sensitivity of FT catalysts to contaminants. In terms of U.S. dollars in 2000, investment costs of $200–340 million would be required for an industrial facility offering conversion efficiencies of 33–40% for atmospheric gasification systems and 42–50% for pressured systems.

However, the estimated production costs for FT diesel were high—in excess of 10 times those of conventional diesel.[86] This conclusion has been worded differently: Unless the environmental benefits of FT diesel are valued in economic terms, the technology can become viable only if crude oil prices rise substantially.[87] This did occur (Figure 6.1, Chapter 6) and the cost differential has undoubtedly narrowed, although with no signs of a surge in investor confidence.

If it can be produced economically using an energy crop such as switchgrass as the substrate, FT diesel rates better than E85 (from corn-derived ethanol) as a biofuel to replace fossil fuel use in assessments performed by the Argonne National Laboratory (Figure 8.9).[88,89] FT diesel greatly outperformed E85 for total fossil fuel savings and also exhibited much reduced emissions of total particulates, sulfur oxides, and nitric oxides, although it fared worse than E85 using the criterion of total CO. Compared with conventional diesel fuel, FT diesel had higher total missions of volatile organic carbon, CO, and nitric oxides (Figure 8.9).

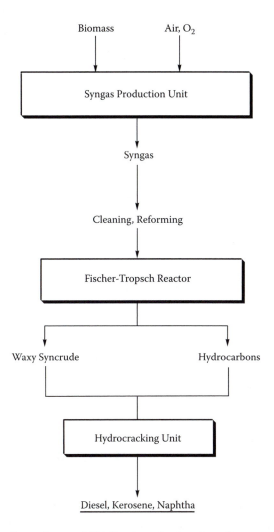

FIGURE 8.8 Outline scheme of FT diesel production from biomass. (Combined from Morales, F., and Weckhuysen, B. M. 2006. *Catalysis* 19:1; and Tijmensen, M. J. A. et al. 2002. *Biomass & Bioenergy* 23:129.)

An experimental biofuels conversion technology is for the production of HydroThermalUpgrading® (HTU) diesel, originally devised by Royal Dutch Shell.[90] At high temperature (300–350°C) and pressure, wet biomass feedstocks such as beet pulp, sludge, and bagasse can be converted to a hydrocarbon-containing liquid (biocrude). Biocrude is a heavy organic liquid, immiscible with water, that contains 10–18% (w/w) oxygen. The biocrude can then be upgraded by a catalyzed reaction with H_2 to remove the oxygen and the HTU diesel can be blended with conventional diesel in any proportion without engine adjustments. A pilot plant for HTU diesel is presently operated in the Netherlands.

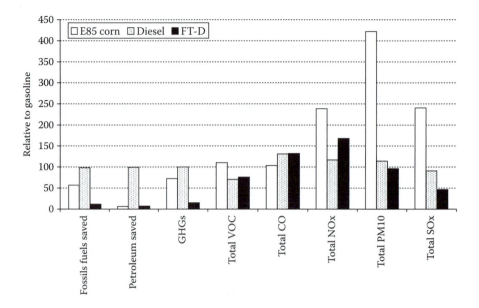

FIGURE 8.9 Well-to-wheel energy use and emissions for E85 from corn, diesel, and FT diesel produced from switchgrass, relative to conventional gasoline (= 100). (Data from Wu, M., Wu, Y., and Wang, M. 2005. Mobility chains analysis of technologies for passenger cars and light-duty vehicles fueled with biofuels: Application of the GREET model to the Role of Biomass in America's Future Energy (RBAFE) Project. Argonne National Laboratory, Argonne, IL, May 2005.)

8.4 BIODIESEL FROM MICROALGAE AND MICROBES

8.4.1 MARINE AND AQUATIC BIOTECHNOLOGY

Any source of triglycerides could act as a feedstock for the production of biodiesel or green diesel (Section 8.2); however, in 2007, a review concluded that microalgae appeared to be the *only* source of renewable biodiesel capable of meeting the global demand for petrodiesel transportation fuels.[91] The main argument advanced was that the oil productivity of microalgae can greatly exceed that of the best seed oil terrestrial plants; although both life forms utilize sunlight as their ultimate energy source, microalgae do so far more efficiently than do crop plants.

The same area of land that produces 1 tonne of biodiesel from a conventional crop plant (canola, soybean, etc.) might yield 100 tonnes of biodiesel grown from algae. The *entire world* petroleum demand (31 billion barrels of crude oil per annum) could therefore be met from algae grown on an area less than 5% the size of North America.[92] It is evident why so many start-up countries have been interested in "oilgae" and why ExxonMobil decided in 2009 to invest strategically in this biofuels sector.

This was a scenario very different from that detailed in the 1998 closeout report on two decades of research funded by the U.S. Department of Energy.[93] This program had set out to investigate the production of biodiesel from high-lipid algae grown in ponds and utilizing waste CO_2 from coal-fired power plants. The main achievements of the research were

- the establishment of a collection of 300 species (mostly green algae and diatoms), housed in Hawaii, that accumulated high levels of oils; some species were capable of growth under extreme conditions of temperature, pH, and salinity
- a much greater understanding of the physiology and biochemistry of intracellular oil accumulation—in particular, the complex relationships among nutrient starvation, cell growth rate, oil content, and overall oil productivity
- significant advances in the molecular biology and genetics of algae, including the first isolation from a photosynthetic organism of the gene encoding acetyl-CoA carboxylase, the first committed step in fatty acid biosynthesis[94]
- the development of large surface area (1,000 m²) pond systems capable of utilization of 90% of the injected CO_2

Although algal production routes had the enormous advantage of not encroaching on arable land or other agricultural resources for food crops, the perceived problem in the 1998 report was the high cost of algal biodiesel relative to conventional automotive fuels—up to $69 per barrel in 1996 prices. The higher the biological productivity was, the lower the production costs were, while using flue gas was more economical than buying CO_2 supplies (Figure 8.10). With crude oil prices then at $20 per barrel or less, such production costs were disappointingly high.

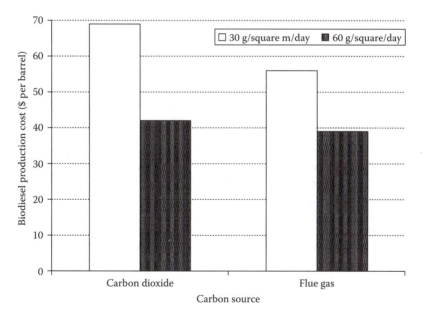

FIGURE 8.10 Estimated production costs of biodiesel from microalgae with two different carbon sources and at differing productivities. (Data from Sheehan, J. et al. 1998. NREL/TP-580-24190, National Renewable Energy Laboratory, Golden, Colorado.)

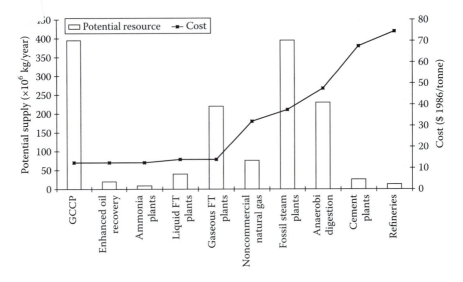

FIGURE 8.11 Potential supplies and costs of CO_2 for microalgal biodiesel production in the United States by 2010. (Data from Sheehan, J. et al. 1998. NREL/TP-580-24190, National Renewable Energy Laboratory, Golden, Colorado.)

The open pond technology was not only the simplest but also the cheapest production choice. Closed system production offered far more controllable growth environments for the algae. However, the cost of even the simplest tubular photobioreactors was projected to have capital costs 10 times higher than those of open pond designs.

In addition, open pond cultures had been commercialized for high-value algal chemical products and any attempt at large-scale (>1 ton per year) closed production systems had failed.[93] Choices of location and species had dramatically increased productivity during the lifetime of the program, from 50 to 300 tonnes per hectare per year—close to the calculated theoretical maximum for solar energy conversion (10%).

The report concluded, therefore, that microalgal fuel production was not limited by engineering issues, but rather by cultivation factors, including species control in large outdoor environments, harvesting methods, and overall lipid productivity. Encouragingly, the potential supply of industrial waste sources of CO_2 in the United States by 2010 was estimated to be as high as 2.25×10^6 tonnes per year. Fischer–Tropsch conversion plants from fossil fuels (Section 8.3) and gasification/combined cycle power (GCCP) facilities offered the largest amounts of CO_2 at low-cost prices (Figure 8.11).

In the subsequent era of high oil prices, interest and research efforts into algal sources of oils for biodiesel production have become more globally distributed. Typical of this recent change in scientific and technical priorities have been studies of *Chlorella protothecoides*—but with the cells grown heterotrophically (using chemical nutrients) rather than photosynthetically with a source of CO_2:

- Heterotrophically grown cells contained 57.9% oil—over three times higher than in autotrophic (photosynthetically CO_2-fixing) cells; chemical pyrolysis yielded an oil with a lower oxygen content, a higher heating value, a lower density, and a lower viscosity than autotrophic cell bio-oil.[95]

- With a corn powder hydrolysate as carbon source (rather than glucose), a high cell concentration could be achieved; the oil (55.2%) could be efficiently extracted with hexane as a solvent and converted to biodiesel by transesterification with an acid catalyst.[96]
- Optimization of the transesterification defined a temperature of 30°C and a methanol/oil molar ratio of 56:1, resulting in a process time of 4 h.[97]
- The process could be upscaled from 5 to 11,000 L, maintaining the lipid content; hexane-extracted oil could be transformed to methyl esters using an immobilized lipase and with an transesterification efficiency of over 98% within 12 h.[98]

The choice of a microalgal species for biodiesel production is potentially very wide, and it is likely that many optimal strains will be isolated. In the present discovery phase of algal biodiesel R&D, typical candidate species include:

- Lipids of the yellow-green microalgae *Nannochloropsis oculata* have been converted to biodiesel using conventional methanol transesterification chemistry and alumina-supported CaO and MgO catalysts.[99]
- Cultures of *Scenedesmus obliquus* reached 61.3% of the dry cell weight within 8 days in an optimized nutrient medium.[100]
- *Neochloris oleabundans* could reach a maximum lipid content of 56% after 6 days of nitrogen starvation.[101]

Other types of microalgae—dinoflagellates—have been investigated as lipid sources, but their relatively slow growth rates (with growth periods of over 30 days) might limit their economic usefulness without unusually high productivity parameters.[102]

Are there alternatives to the open pond system that might increase microalgal productivity and accelerate commercialization? A species of the alga *Chlorella* has been grown attached to polystyrene foam.[103] The attached cells were harvested by simply scraping them off, leaving residual colonies that served as inocula for regrowth; this mode of growth resulted in a higher biomass yield than that from surface-only growth. The system could utilize dairy manure wastewater as a highly heterotrophic growth medium. Over 10 days, the attached algal culture removed 61–79% total nitrogen and 62–93% total phosphorus from the wastewater and demonstrated good performance for biomass yield, biodiesel production potential, ease of harvesting biomass, and physical robustness for reuse.

8.4.2 MICRODIESEL

The cultivation of photosynthetic microalgae under dark conditions supplied with organic carbon closely resembles typical microbial fermentations. Because several bacterial species are well known as accumulators of triglycerides (oils) and esters of fatty acids with long chain alcohols (waxes), the logical conclusion was to combine these biosynthetic abilities with that of ethanol formation to generate the precursors of triglycerides in microbial production systems (i.e., "microdiesel" produced without any need for a chemically or enzyme-catalyzed transesterification).[104]

The simple bacterium *E. coli* was used as host for the *Zymomonas mobilis* pyruvate decarboxylase and alcohol dehydrogenase genes for ethanol production (Chapter 4, Section 4.2.2.1), together with the gene encoding an unspecific wax ester synthase/acyl-CoA:diacylglycerol transferase from a bacterial strain (*Acinetobacter baylyi*) known to accumulate lipid as an internal cell storage reserve. The resulting recombinant could accumulate ethyl esters of fatty acids at up to 26% of the cellular dry mass in fermentations fed with glucose. Insomuch as glucose is a fully renewable carbohydrate supply (via, for example, cellulose or starch), microdiesel is a genuinely sustainable source of preformed transportation fuel, although the chemical engineering aspects of its extraction from bacterial cells and the economics of its production systems require further definition.

A refinement preliminary to industrial feasibility studies would be to transfer to a host capable of higher endogenous accumulation of lipids; many of these hosts are Gram-negative species (like *E. coli*) with well-understood primary metabolism and molecular biology.[105] One obvious metabolic problem is that bacteria produce fatty acids as cell envelope precursors; the biosynthesis of free fatty acids (often regarded by microbial physiologists as toxic in high concentrations) is tightly regulated at multiple levels. When four genetic modifications to the *E. coli* genome were made, an efficient producer of fatty acids was engineered that could exhibit a maximal conversion efficiency of 5% of the carbon source into fatty acid products—at least 50% of which were present in the free acid form amenable to direct esterification.[106]

8.5 CHEMICAL CONVERSIONS OF GLYCEROL PRODUCED BY FERMENTATION

Glycerol represents 10% by weight of typical triglycerides, and biodiesel production generates large amounts of this coproduct (Figure 8.1). This has increasingly marginalized chemical routes to glycerol synthesis that are cost inefficient or otherwise suboptimal in a competitive manufacturing environment.[107] Produced by any manufacturing route (chemical or biological), glycerol provides a good input to Fischer–Tropsch conversion to liquid alkanes via syngas generated over platinum-based catalysts at relatively low temperatures (225–350°C).[108] The gas mixture from glycerol conversion at 300–450°C has a molar excess of H_2 over CO up to 1.83:1, a high ratio (up to 90:1) between CO and CO_2, and only traces of methane. With a subsequent Fischer–Tropsch step, the overall conversion of glycerol to hydrocarbons can be written as

$$25C_3O_3H_8 \rightarrow 7C_8H_8 + 19CO_2 + 37H_2O$$

This is a mildly exothermic process (enthalpy change of −63 kJ per mole of glycerol). High rates of conversion of glycerol into syngas were observed using aqueous glycerol concentrations of 20–30% (w/w).

8.6 CHEMICAL ROUTES FOR THE PRODUCTION OF MONOOXYGENATED C6 LIQUID FUELS FROM BIOMASS CARBOHYDRATES

5-Hydroxymethylfurfural (HMF) was discussed in Section 2.3.3.2 in Chapter 2 as a toxic product of acidic pretreatment techniques for biomass. The boiling point of HMF is too high (291°C) to be considered as a liquid fuel, but when HMF is subject to chemical hydrogenolysis, a more volatile product, 2,5-dimethylfuran (DMF) is formed (Figure 8.12).[109] DMF has a boiling point of 93°C—20°C higher than ethanol—and has a research octane number (RON) of 119. The by-product 2-methylfuran has an even higher RON (131) but is more water soluble than DMF.

HMF is most readily formed by the dehydration of fructose, a naturally occurring sugar and a straightforward isomerization product of glucose. Mineral acids such as hydrochloric acid can be used to catalyze the reaction; 88% conversion can be achieved at 180°C.[109] Solvent extraction with butan-1-ol can be employed to HMF prior to a hydrogenolysis over a mixed Cu–Ru catalyst at 220°C.

Such production routes, beginning with enzymic conversion of glucose to fructose and proceeding via entirely thermochemical processes, have been described as hybrid.[110] They have the advantage of avoiding reliance on large fermentation

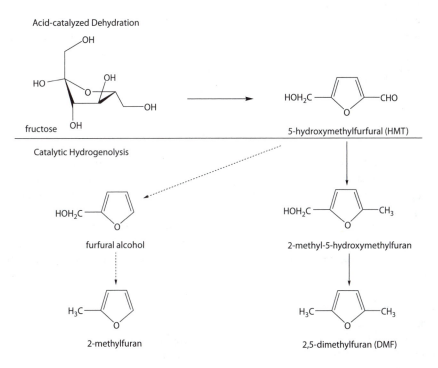

FIGURE 8.12 Thermochemical production of C6 furans and C5 by-products from fructose. (Modified from Román-Leshkov, Y. et al. 2007. *Nature* 447:982.)

vessels for production steps and therefore being potentially much more rapid. Their economics could be similar to, or an improvement on, those for Fischer–Tropsch liquid fuels.

The conversion of glucose to fructose, catalyzed by the enzyme glucose isomerase, has been a major industrial application of enzymology since the 1960s. The product (high-fructose corn syrup) was introduced as a substitute for Cuban sugar in the reduced-calorie sweetener market in the United States.[111] The enzyme technology has been continuously improved, evolving to immobilized forms of the enzyme. The potential of enzymes from hyperthermophilic microbes has now been explored, and their stability at 80°C rivals that of conventional enzyme processes operating at 55–65°C.[112] Rapid and efficient processing of glucose solutions to high concentrations of fructose is feasible if desirable biocatalytic and thermostability properties of suitable enzymes can be realized.

8.7   BIOMETHANOL AND BIODIMETHYLETHER

The versatility of the Fischer–Tropsch process is that almost any hydrocarbon produced that can be derived from petroleum can be made from syngas—not only alkanes and alkenes but also oxygenated compounds. The exact mixture of products obtained can be varied by choices of catalyst, pressure, and temperature, and straight-chain alcohols are produced in the synthol reaction at 400–450°C and 14 MPa pressure in the presence of an iron catalyst.[113] Industrial production from natural gas, however, has been dominated since the 1960s by a lower temperature and pressure process invented by Imperial Chemical Industries in which CO, CO_2, and H_2, derived by steam reforming, are reacted over a mixed $Cu/ZnO/Al_2O_3$ catalyst at 250°C and 50–10 MPa when two reactions occur[114]:

$$CO + 2H_2 \rightarrow CH_3OH$$

$$CO_2 + 3H_2 \rightarrow CH_3OH + H_2O$$

A recent development has been to combine syngas production from methane with the reduction of ZnO to metallic zinc in a metallurgical plant; the syngas has a H_2/CO ratio of approximately 2:1, which is highly suitable for methanol production.[115] A renewable-resource route for methanol (one of the largest bulk chemicals in the contemporary world) via biomass gasification is entirely feasible as an intermediate step, however. This would be entirely appropriate given methanol's older name of "wood alcohol," which is indicative of its historical provenance by incomplete combustion.

As an energy carrier, methanol is inferior to ethanol, with an energy content only 75% (on either a weight or a volume basis) that of ethanol and approximately 50% that of conventional gasoline.[116] Blends of methanol with conventional gasoline up to 20% can be tolerated without the need for engine modifications (i.e., as a fuel extender); the corrosive effect of methanol on some engine materials limits the extent of this substitution.[117]

Methanol would have been an excellent replacement for methyl tertiary butyl ether as a gasoline oxygenate additive (Chapter 1, Section 1.5), but its acute neurotoxicity is well known and a barrier to several potential uses. One notable exception is as an at-site (or on-board) source of hydrogen for fuel cells; between 1983 and 2000, nearly 50 patents were granted to automobile producers such as General Motors Corporation and Honda, chemical multinationals (DuPont, Inc., BASF AG, etc.), and major oil companies (CONOCO, Inc., etc.) for catalytic methanol "reforming" systems.[114]

Combined reforming with liquid water and gaseous oxygen has been intensively investigated for use in mobile applications for transportation:

$$(s+p)CH_3OH \text{ (l)} + sH_2O \text{ (l)} + 0.5pO_2 \rightarrow (s+p)CO_2 + (3s+2p)H_2$$

because the composition of the reactant feed can be varied and the process carried out under a wide range of operating conditions.

The first pilot plant for testing and evaluating the production process for biomethanol was established in a program that commenced in 2000 between the Ministry of Agriculture, Forestry and Fisheries of Japan and Mitsubishi Heavy Industries at Nagasaki (Japan); various feedstocks were investigated, including wood, rice husks, rice bran, and rice straw.[118] The test plant consisted of

- a drier and grinder for the biomass input (crushed waste wood)
- a syngas generator
- a gas purifier
- a methanol synthesis vessel (with an unspecified catalyst)

The pilot plant was designed for a capacity of 240 kg per day, with a methanol yield (weight of methanol produced per unit of dry weight of material) of 9–13%.

No economic analysis of the Japanese pilot facility has been published, but a theoretical study of methanol production via the syngas route suggested that methanol from biomass (by 2002) had production costs approximately twice those of conventional gasoline on an equal energy basis.[119] Large-scale methanol production via plant biomass gasification in South Africa has been estimated to yield purified methanol with production costs down to $0.38 per liter ($1.66 per gallon).[120] This encouraging result is timely because direct methanol fuel cell (DMFC) technology has also been developed: The Toshiba Corporation in Japan announced in 2004 the development of a micro-DMFC suitable for powering MP3 players, and a U.S. patent covering aspects of DMFC construction was issued in March 2007.[121]

The Japanese invention utilizes a polymer electrolyte membrane device with the following electrochemical reactions:

$$CH_3OH + H_2O \rightarrow CO_2 + 6H^+ + 6e^- \text{ (anode)}$$

$$1.5O_2 + 6H^+ + 6e^- \rightarrow 3H_2O \text{ (cathode)}$$

The inputs are concentrated methanol and air (O_2); the only outputs are water, CO_2, and electricity (100 mW) sufficient to power a portable device for 20 h on a 2-cm^3 charge of solvent. The prospects for large DMFCs for heavier duty use are unclear.

)ther thermochemical route has been explored to convert biomass-derived to dimethylether (DME), $(CH_3)_2O$—a highly volatile liquid that is a suitable fuel for diesel engines because of its low self-ignition temperature and high cetane number.[122] Although bioDME has only half the energy content of conventional diesel, diesel engines can easily be retrofitted for bio-DME use. Well-to-wheel analyses showed that bioDME was a little inferior to FT diesel for total fossil fuel substitution and pollutant emissions.[88,89] The bioDME project, funded by the EU from September 1, 2008, and with a total budget of €28.4 million ($41.7 million), aims to take bioDME to the pilot plant stage and develop automobile engineering for its use in collaboration with Volvo in Sweden; the feedstock will be black liquor effluents from pulping mills.

8.8 CHEMISTRY AND THE EMERGENCE OF THE HYDROGEN ECONOMY

It is highly doubtful that industrial biohydrogen processes will be the entry points for the widespread use of H_2 as a fuel. Despite a number of major national and international initiatives and research programs, fossil-fuel-based and alternative energy processes are widely considered to be essential before 2030, or even as late as 2050. Of these technologies, H_2 production by coal gasification is clearly the worst alternative in terms of fossil energy use and greenhouse gas emissions (Figure 8.13).[123] Nevertheless, gasification and electricity-powered electrolytic routes to H_2 offer the promise of production costs rivaling or even less than those of conventional gasoline

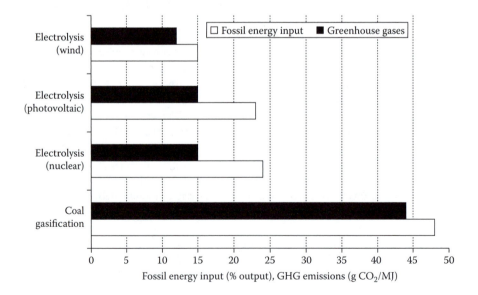

FIGURE 8.13 Alternative nonbiological production routes for H_2: primary (fossil) energy inputs and associated greenhouse gas emissions. (Data from Mason, J. E. 2007. *Energy Policy* 35:1315.)

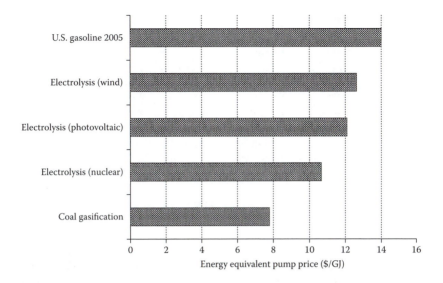

FIGURE 8.14   Predicted retail (pump) prices for H_2 and gasoline on an equal energy basis and assuming that fuel cell vehicles are more efficient than conventional internal combustion engine vehicles. (Data from Mason, J. E. 2007. *Energy Policy* 35:1315.)

for use in fuel-cell-powered vehicles; an anticipated fuel economy is approximately twice that of conventional internal combustion engines (Figure 8.14).

As a carbonless production route, the internationally accepted route map is the sulfur–iodine cycle based on the following three chemical reactions:

$$H_2SO_4 \rightarrow SO_2 + H_2O + \frac{1}{2}O_2 \ [850°C]$$

$$I_2 + SO_2 + 2H_2O \rightarrow 2HI + H_2SO_4 \ [120°C]$$

$$2HI \rightarrow H_2 + I_2 \ [220–330°C]$$

The high temperatures required for the first reaction have prompted research programs investigating solar-furnace splitting of sulfuric acid—for example, in the five-nation project HYTHEC (hydrogen thermochemical cycles), which involves research teams from France, Germany, Spain, Italy, and the UK in the search for a long-term massive hydrogen production route that would be sustainable and independent of fossil fuel reserves (www.hythec.org).

In Japan, all the major automobile manufacturers—Toyota, Honda, Nissan, Mazda, Daihatsu, Mitsubishi, and Suzuki—are active in the development of fuel-cell-powered vehicles.[124] In Europe, HYVOLUTION is a program with partners from 11 EU countries, Russia, and Turkey that is funded by approximately $9.5 million. It aims to establish decentralized H_2 production from biomass, to maximize the number and diversity of H_2 production routes, and to increase energy security of supply at both local and regional levels (www. biohydrogen.nl/hyvolution). The approach is based on combined bioprocesses with thermophilic and phototrophic bacteria to provide H_2 production with high efficiencies in small-scale, cost-effective industries to reduce H_2 production costs to \$10 per gigajoule by 2020. With production costs in the \$5–7 per gigajoule range, biomass-derived H_2 would be highly competitive with conventional fuels or biofuels.[119] Principal objectives for HYVOLUTION include:

- pretreatment technologies to optimize biodegradation of energy crops
- maximized conversion of biomass to H_2
- assessment of installations for optimal gas cleaning
- minimum energy demand and maximal product output
- identification of market opportunities for a broad feedstock range

---

STEM TOPIC 8.3: COST/POWER PARAMETERS OF HYDROGEN FUEL CELLS

The International Energy Agency's 2005 study on the hydrogen economy[125] assembled technical and economic data from a global range of sources and strategies, including a cost analysis for the manufacture of protein exchange membrane (PEM) fuel cells (see Figure STEM 8.3).

The platinum catalyst was estimated to contribute only 1% to the total cost; the largest single cost input was the elaboration of bipolar plates from milled graphite or gold-coated stainless steel.

Plausible cost movements by 2030 were estimated to reduce the cost from \$2,000 per kilowatt to \$100 per kilowatt, although further reductions to \$50 per kilowatt were assessed as essential for fuel cell vehicles to be cost competitive with conventional vehicles.

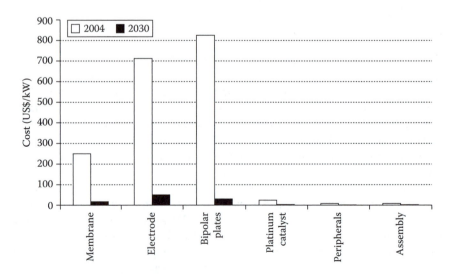

FIGURE STEM 8.3 Production costs of PEM hydrogen fuel cells. (Data from International Energy Agency. 2005. *Prospects for Hydrogen and Fuel Cells*. Paris.)

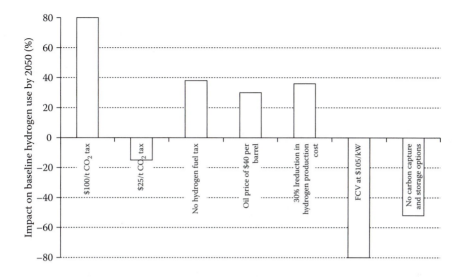

FIGURE 8.15 Sensitivity analysis of predicted H_2 use by 2050. (Data from International Energy Agency. 2005. *Prospects for Hydrogen and Fuel Cells*. Paris.)

Based in Sweden, the SOLAR H program links molecular genetics and biomimetic chemistry to explore radically innovative approaches to renewable H_2 production, including artificial photosynthesis in manmade systems (www.fotmol.uu.se). Japanese research has already explored aspects of this interface between industrial chemistry and photobiology—for example, incorporating an artificial chlorophyll (with a zinc ion replacing the green plant choice of magnesium) in a laboratory system with sucrose and the enzymes invertase and glucose oxidase, together with a platinum colloid to photoevolve H_2.[126]

The size of the investment required to bring the hydrogen economy to fruition remains daunting: from several billion to a few trillion dollars over several decades.[125] The International Energy Agency also estimates that H_2 production costs must be reduced 3- to 10-fold and fuel cell costs 10- to 50-fold. Stationary fuel cells could represent 2–3% of global generating capacity by 2050, and total H_2 use could reach 15.7 EJ by then. There are appreciated risks in these prognostications; governments are holding back from imposing fuel taxes on H_2 but imposing high CO_2 penalties (strongly positive for increasing the possible use of H_2) while high fuel cell prices for automobiles will be equally negative (Figure 8.15).

8.9 SUMMARY

Biodiesel is a biofuel produced chemically from plant seed oils and methanol. First-generation biodiesel has been manufactured from major food seed crops (e.g., soybean), but novel nonfood sources are being intensively investigated. Biodiesel is much less geographically restricted than is ethanol from corn grain or cane sugar, and biodiesel has a strong claim to be truly global in its scope of production on an industrial scale.

_...e economics and fossil fuel requirements of biodiesel production have received detailed and often critical scrutiny; possible benefits of biodiesel use in ameliorating greenhouse gas and other atmospheric pollutants have also been challenged. A novel route to convert plant seed oil to biofuels is by catalytic hydrogenation; the resulting hydrotreated vegetable oils (green diesel) may have superior properties and environmental impacts while being closer structurally to conventional diesels.

Plant biomass can be converted thermochemically via incomplete combustion to a mixture of CO, H_2, and other gases known as syngas. Syngas is a gaseous biofuel in its own right and can be combusted for on-site steam and/or electricity generation or be incorporated in local or district heating schemes. The Fischer–Tropsch (FT) reaction converts syngas into a hydrocarbon mix approximating diesel. FT diesel is the leading contender of biomass-to-liquid fuel technologies that utilize chemical rather than biological production platforms. FT reactions can convert syngas from glycerol (produced by fermentation or as a coproduct of biodiesel manufacture) to renewable diesel. Other thermochemical reactions convert syngas to methanol or dimethylether as candidate biofuels (biomethanol and bioDME).

Returning to biodiesel, massive interest and some widely publicized financial investment in sources other than plant seed oils—in particular, microalgae and bacteria—have been evident. Microbial lipids can be generated in large amounts but require much reduced land areas in comparison with terrestrial plants.

Modern developments in catalytic chemistry point to routes from carbohydrates to liquid biofuels such as dimethylfuran as the end products of combined (hybrid) enzymic and chemical processes.

Although syngas-based and biologically produced H_2 are long-term supports for the hydrogen economy, chemistry offers more immediate possibilities—in particular, the sulfur–iodine cycle. Advanced bioorganic chemistry has explored H_2 generation by mimicking elements of photosynthetic biochemistry.

Other than biodiesel from plant seed oils, no chemically produced biofuel has reached full commercial production.

REFERENCES

1. McDonnell, K. et al. 1999. Properties of rapeseed oil for use as a diesel fuel extender. *Journal of the American Oil Chemists Society* 76:539.
2. International Energy Agency. 2006. *World energy outlook,* chap. 14. Paris.
3. Gerpen, J. V. 2009. Biodiesel: Small scale production and quality requirements. *Methods in Molecular Biology* 581:281.
4. Culshaw, F., and Butler, C. 1993. A review of the potential of biodiesel as a transport fuel. Department of Trade and Industry Energy Technology Support Unit report ETSU-R-71. London: Her Majesty's Stationery Office.
5. Mittelbach, M., and Enzelsberger, H. 1999. Transesterification of heated rapeseed oil for extending diesel fuel. *Journal of the American Oil Chemists Society* 76:545.
6. Lide, D. R., ed. 1992. *CRC handbook of chemistry and physics,* 73rd ed., section 7-29. Boca Raton, FL: CRC Press.
7. Lee, I., Johnson, A., and Hammond, E. G. 1996. Reducing the crystallization temperature of biodiesel by winterizing methyl soyate. *Journal of the American Oil Chemists Society* 73:631.

8. Dunn, R. O. 1999. Thermal analysis of alternative diesel fuels from vegetable oils. *Journal of the American Oil Chemists Society* 76:109.

9. Carroll, K. K. 1953. Erucic acid as the factor in rape oil affecting adrenal cholesterol in the rat. *Journal of Biological Chemistry* 200:287.

10. Mietkiewska, E. et al. 2004. Seed-specific heterologous expression of a *Nasturtium FAE* gene in *Arabidopsis* results in a dramatic increase in the proportion of erucic acid. *Plant Physiology* 136:2665.

11. Mittelbach, M. et al. 1983. Diesel fuel derived from vegetable oils: Preparation and use of rape oil methyl ester. *Energy and Agriculture* 2:369.

12. Jeong, G. T. et al. 2004. Production of biodiesel fuel by transesterification of rapeseed oil. *Applied Biochemistry and Biotechnology* 113–116:747.

13. de Oliveira, D. et al. 2005. Optimization of alkaline transesterification of soybean oil and castor oil for biodiesel production. *Applied Biochemistry and Biotechnology* 121–124:553.

14. Karmee, S. K., and Chadha, A. 2005. Preparation of biodiesel from crude oil of *Pongamia pinnata*. *Bioresource Technology* 96:1425.

15. Felizardo, P. et al. 2006. Production of biodiesel from waste frying oils. *Waste Management* 26:487.

16. Jeong, G. T., and Park, D. H. 2006. Batch (one- and two-stage) production of biodiesel fuel from rapeseed oil. *Applied Biochemistry and Biotechnology* 129–132:668.

17. Vicente, G., Martínez, M., and Aracil, J. 2007. Optimization of integrated biodiesel production. I. A study of the biodiesel purity and yield. *Bioresource Technology* 98:1724.

18. Rashid, U. et al. 2008. *Moringa oleifera* oil: A possible source of biodiesel. *Bioresource Technology* 99:8175.

19. Shi, H., and Bao, Z. 2008. Direct preparation of biodiesel from rapeseed oil leached by two-phase solvent extraction. *Bioresource Technology* 99:9025.

20. Zhang, J., and Jiang, L. 2008. Acid-catalyzed esterification of *Zanthoxylum bungeanum* seed oil with high free fatty acids for biodiesel production. *Bioresource Technology* 99:8995.

21. Qian, J. et al. 2008. In situ alkaline transesterification of cottonseed oil for production of biodiesel and nontoxic cottonseed meal. *Bioresource Technology* 99:9009.

22. Sarin, R., Sharma, M., and Khan, A. A. 2009. Studies on *Guizotia abyssinica* L. oil: Biodiesel synthesis and process optimization. *Bioresource Technology* 100:4187.

23. Boey, P. L., Maniam, G. P., and Hamid, S. A. 2009. Utilization of waste crab shell (*Scylla serrata*) as a catalyst in palm olein transesterification. *Journal of Oleo Science* 58:499.

24. Vicente, G., Martínez, M., and Aracil, J. 2007. Optimization of integrated biodiesel production. II. A study of the material balance. *Bioresource Technology* 98:1754.

25. Toda, M. et al. 2005. Green chemistry: Biodiesel made with sugar catalyst. *Nature* 438:178.

26. Shibasaki-Kitakawa, N. et al. 2007. Biodiesel production using anionic ion-exchange resin as heterogeneous catalyst. *Bioresource Technology* 98:416.

27. da Silva, N. L. et al. 2006. Optimization of biodiesel production from castor oil. *Applied Biochemistry and Biotechnology* 129–132:405.

28. Zaccheria, F. et al. 2009. Esterification of acidic oils over a versatile amorphous solid catalyst. *ChemSusChem* 2:535.

29. Lukić, I. et al. 2009. Alumina/silica supported K_2CO_3 as a catalyst for biodiesel synthesis from sunflower oil. *Bioresource Technology* 100:4690.

30. Li, J. et al. 2009. $Zn_{1.2}H_{0.6}PW_{12}O_{40}$ nanotubes with double acid sites as heterogeneous catalysts for the production of biodiesel from waste cooking oil. *ChemSusChem* 2:177.

31. Demirbas, A. 2008. Studies on cottonseed oil biodiesel prepared in noncatalytic SCF conditions. *Bioresource Technology* 99:1125.
32. Berchmans, H. J., and Hirata, S. 2008. Biodiesel production from crude *Jatropha curcas* L. seed oil with a high content of free fatty acids. *Bioresource Technology* 99:1716.
33. Zullaikah, S. et al. 2005. A two-step acid-catalyzed process for the production of biodiesel from rice bran oil. *Bioresource Technology* 96:1889.
34. Kusidiana, D., and Saka, S. 2001. Methyl esterification of free fatty acids of rapeseed oil as treated in supercritical methanol. *Journal of Chemical Engineering Japan* 34:383.
35. Leevijit, T. et al. 2008. Performance test of a 6-stage continuous reactor for palm methyl ester production. *Bioresource Technology* 99:214.
36. Stavarache, C. et al. 2007. Ultrasonically driven continuous process for vegetable oil transesterification. *Ultrasonics Sonochemistry* 14:413.
37. Dubé, M. A., Tremblay, A. Y., and Liu, J. 2007. Biodiesel production using a membrane reactor. *Bioresource Technology* 98:639.
38. Behzadi, S., and Farid, M. M. 2009. Production of biodiesel using a continuous gas-liquid reactor. *Bioresource Technology* 100:683.
39. Wen, Z. et al. 2009. Intensification of biodiesel synthesis using zigzag micro-channel reactors. *Bioresource Technology* 100:3054.
40. Baber, T. M. et al. 2005. Application of catalytic ozone chemistry for improving biodiesel product performance. *Biomacromolecules* 6:1334.
41. Pimentel, D., and Patzek, T. W. 2005. Ethanol production using corn, switchgrass, and wood; biodiesel production using soybean and sunflower. *Natural Resources Research* 14:65.
42. Zhang, Y. et al. 2003. Biodiesel production from waste cooking oil. 1. Process design and technological assessment. *Bioresource Technology* 89:1.
43. Zhang, Y. et al. 2003. Biodiesel production from waste cooking oil. 2. Economic assessment and sensitivity analysis. *Bioresource Technology* 90:229.
44. Haas, M. J. et al. 2006. A process model to estimate biodiesel production costs. *Bioresource Technology* 97:671.
45. Canakci, M. 2007. The potential of restaurant waste lipids as biodiesel feedstocks. *Bioresource Technology* 98:183.
46. Sheehan, J. et al. 1998. Life cycle inventory of biodiesel and petroleum diesel for use in an urban bus. Final report, NREL/SR-580-24089 UC Category 1503, National Renewable Energy Laboratory, Golden, Colorado.
47. Van Gerpen, J., and Shrestha, D. 2006. Biodiesel energy balance. Available from http://uidaho.edu/bioenergy.
48. Thompson, J. C., and He, B. 2006. Characterization of crude glycerol from biodiesel production from multiple feedstocks. *Applied Engineering in Agriculture* 22:261.
49. da Costa, R. E. et al. 2007. The energy balance in the production of palm oil biodiesel—Two case studies: Brazil and Colombia. Swedish Bioenergy Association (available from http://svebio.se/attachments/33/295.pdf).
50. Hill, J. et al. 2006. Environmental, economic, and energetic costs and benefits of biodiesel and ethanol biofuels. *Proceedings of the National Academy of Sciences USA* 103:11206.
51. Nabi, M. N., Akhter, M. S., and Zaglul Shahadat, M. M. 2006. Improvement of engine emissions with conventional diesel fuel and diesel-biodiesel blends. *Bioresource Technology* 97:372.
52. Canakci, M. 2007. Combustion characteristics of a turbocharged DI compression ignition engine fueled with petroleum diesel fuels and biodiesel. *Bioresource Technology* 98:1167.
53. Jeong, G.-T., Oh, Y.-T., and Park, D.-H. 2006. Emission profile of rapeseed methyl ester and its blend in a diesel engine. *Applied Biochemistry and Biotechnology* 129–132:165.

54. Jung, H., Kittelson, D. B., and Zachariah, M. R. 2006. Characteristics of SME biodiesel-fueled diesel particle emissions and the kinetics of oxidation. *Environmental Science & Technology* 40:4949.

55. Lapuerta, M., Rodríguez-Fernández, J., and Agudelo, J. R. 2008. Diesel particulate emissions from used cooking oil biodiesel. *Bioresource Technology* 99:731.

56. Bünger, J. et al. 2007. Strong mutagenic effects of diesel engine emissions using vegetable oil as fuel. *Archives of Toxicology* 81:599.

57. Yuan, C. S. et al. 2007. A new alternative fuel for reduction of polycyclic aromatic hydrocarbon and particulate matter emissions from diesel engines. *Journal of the Air and Waste Management Association* 57:465.

58. Cardone, M. et al. 2002. *Brassica carinata* as alternative oil crop for the production of biodiesel in Italy: Engine performance and regulated and unregulated exhaust emissions. *Environmental Science & Technology* 36:4656.

59. Pang, S. H., Frey, H. C., and Rasdorf, W. J. 2009. Life cycle inventory energy consumption and emissions for biodiesel versus petroleum diesel fueled construction vehicles. *Environmental Science & Technology* 43:6398.

60. Bünger, J. et al. 2000. Mutagenicity of diesel exhaust particles from two fossil and two plant oil fuels. *Mutagenesis* 15:391.

61. Bünger, J. et al. 1998. Mutagenic and cytotoxic effects of exhaust particulate matter of biodiesel compared to fossil diesel fuel. *Mutation Research* 415:13.

62. Swanson, K. J., Madden, M. C., and Ghio, A. J. 2007. Biodiesel exhaust: The need for health effects research. *Environmental Health Perspectives* 115:496.

63. Frondel, M., and Peters, J. 2005. Biodiesel: A new Oildorado? Discussion paper 36, Rheinisch-Westfälisches Institut für Wirtschaftforschung, Essen, Germany. Available from http://repec.rwi-essen.de/files/DP_05_036.pdf.

64. Simmons, M. R. 2008. Can motors fuels be redesigned? *Hydrocarbon Processing* February:42–44.

65. Mikkonen, S. 2008. Second-generation renewable diesel offers advantage. *Hydrocarbon Processing* February:63–66.

66. Linko, Y. Y. et al. 1998. Biodegradable products by lipase biocatalysis. *Journal of Biotechnology* 66:41.

67. Du, W., Xu, Y., and Liu, D. 2003. Lipase-catalysed transesterification of soya bean oil for biodiesel production during continuous batch operation. *Biotechnology and Applied Biochemistry* 38:103.

68. Hernández-Martín, E., and Otero, C. 2008. Different enzyme requirements for the synthesis of biodiesel: Novozym® 435 and Lipozyme® TL IM. *Bioresource Technology* 99:277.

69. Sanchez, F., and Vasudevan, P. T. 2006. Enzyme catalyzed production of biodiesel from olive oil. *Applied Biochemistry and Biotechnology* 135:1.

70. Modi, M. K. et al. 2007. Lipase-mediated conversion of vegetable oils into biodiesel using ethyl acetate as acyl acceptor. *Bioresource Technology* 98:1260.

71. Ranganathan, S. V., Narasimhan, S. L., and Muthukumar, K. 2008. An overview of enzymatic production of biodiesel. *Bioresource Technology* 99:3975.

72. Noureddini, H., Gao, X., and Philkana, R. S. 2005. Immobilized *Pseudomonas cepacia* lipase for biodiesel fuel production from soybean oil. *Bioresource Technology* 96:769.

73. Kumari, A. et al. 2009. Enzymatic transesterification of *Jatropha* oil. *Biotechnology for Biofuels* 2:1.

74. Lee, J. H. et al. 2008. Optimization of the process for biodiesel production using a mixture of immobilized *Rhizopus oryzae* and *Candida rugosa* lipases. *Journal of Microbiology and Biotechnology* 18:1927.

75. Yang, K. S., Sohn, J. H., and Kim, H. K. 2009. Catalytic properties of a lipase from *Photobacterium lipolyticum* for biodiesel production containing a high methanol concentration. *Journal of Bioscience and Bioengineering* 107:599.

76. Fukuda, H., Kondo, A., and Noda, H. 2001. Biodiesel fuel production by transesterification of oils. *Journal of Bioscience and Bioengineering* 92:405.
77. Fukuda, H. et al. 2008. Whole-cell biocatalysts for biodiesel fuel production. *Trends in Biotechnology* 26:668.
78. Gao, B. et al. 2009. Development of recombinant *Escherichia coli* whole-cell biocatalyst expressing a novel alkaline lipase-coding gene from *Proteus* sp. for biodiesel production. *Journal of Biotechnology* 139:169.
79. Malte, P. C. 1989. Combustion. In *Biomass handbook,* ed. Kitani, O. and Hall, C. W., chap. 2.3.1. New York: Gordon and Breach Science Publishers.
80. Li, C.-Z. 2006. Gasification: A route to clean energy. *Process Safety and Environmental Protection (Transactions of IChemE.)* 84:407.
81. Müller, S. 2007. Sunshine, sand and soybean king. Brazil's rise to agricultural superpower. *BioForum Europe* 5(11):17.
82. Schulz, H. 1999. Short history and present trends of FT synthesis. *Applied Catalysis A* 186:3.
83. Morales, F., and Weckhuysen, B. M. 2006. Promotion effects in Co-based Fischer–Tropsch catalysis. *Catalysis* 19:1.
84. Campbell, I. 1983. *Biomass, catalysts and liquid fuels,* chap. 5. London: Holt, Rinehart and Winston.
85. Tijmensen, M. J. A. et al. 2002. Exploration of the possibilities for production of Fischer–Tropsch liquids and power via biomass gasification. *Biomass & Bioenergy* 23:129.
86. Srinivas, S., Malik, R. K., and Mahajani, S. M. 2006. Fischer-Tropsch synthesis using bio-syngas and CO_2. *Advances in Energy Research (AER—2006): Proceedings of the 1st National Conference on Advances in Energy Research,* Department of Energy Systems Engineering, Indian Institute of Technology, Bombay, p. 317.
87. Hamelinck, C. N. et al. 2004. Production of FT transportation fuels from biomass; technical options, process analysis and optimization, and development potential. *Energy* 29:1743.
88. Wu, M., Wu, Y., and Wang, M. 2005. Mobility chains analysis of technologies for passenger cars and light-duty vehicles fueled with biofuels: Application of the GREET model to the Role of Biomass in America's Future Energy (RBAFE) Project. Argonne National Laboratory, Argonne, IL, May 2005.
89. Wu, M., Wu, Y., and Wang, M. 2006. Energy and emission benefits of alternative transportation liquid fuels derived from switchgrass: A fuel life cycle assessment. *Biotechnology Progress* 22:1012.
90. HTU diesel. http://www.refuel.eu/biofuels/htu-diesel/.
91. Chisti, Y. 2007. Biodiesel from microalgae. *Biotechnology Advances* 25:294.
92. Rhodes, C. J. 2009. Oil from algae; salvation from peak oil? *Science Progress* 92:39.
93. Sheehan, J. et al. 1998. A look back at the U.S. Department of Energy's Aquatic Species Program, biodiesel from algae. NREL/TP-580-24190, National Renewable Energy Laboratory, Golden, Colorado.
94. Roessler, P. G. et al. 1994. Characteristics of the gene that encodes acetyl-CoA carboxylase in the diatom *Cyclotella cryptica. Annals of the New York Academy of Sciences* 721:250.
95. Miao, X., and Wu, Q. 2004. High yield bio-oil production from fast pyrolysis by metabolic controlling of *Chlorella protothecoides. Journal of Biotechnology* 110:85.
96. Xu, H., Miao, X., and Wu, Q. 2006. High quality biodiesel production from a microalga *Chlorella protothecoides* by heterotrophic growth in fermenters. *Journal of Biotechnology* 126:499.
97. Miao, X., and Wu, Q. 2006. Biodiesel production from heterotrophic microalgal oil. *Bioresource Technology* 97:841.
98. Li, X., Xu, H., and Wu, Q. 2007. Large-scale biodiesel production from microalgal *Chlorella protothecoides* through heterotrophic cultivation in fermenters. *Biotechnology and Bioengineering* 98:764.

99. Umdu, E. S., Tuncer, M., and Seker, E. 2009. Transesterification of *Nannochloropsis oculata* microalga's lipid to biodiesel on Al_2O_3 supported CaO and MgO catalysts. *Bioresource Technology* 100:2828.

100. Mandal, S., and Mallick, N. 2009. Microalga *Scenedesmus obliquus* as a potential source for biodiesel production. *Applied Microbiology and Biotechnology* 84:281.

101. Gouveia, L. et al. 2009. *Neochloris oleabundans* UTEX #1185: A suitable renewable lipid source for biofuel production. *Journal of Industrial Microbiology and Biotechnology* 36:821.

102. Fuentes-Grünewald, C. et al. 2009. Use of the dinoflagellate *Karlodinium veneficum* as a sustainable source of biodiesel production. *Journal of Industrial Microbiology and Biotechnology* 36:1215.

103. Johnson, M. B., and Wen, Z. 2009. Development of an attached microalgal growth system for biofuel production. *Applied Microbiology and Biotechnology,* published online July 28.

104. Kalscheuer, R., Stölting, T., and Steinbüchel, A. 2006. Microdiesel: *Escherichia coli* engineered for fuel production. *Microbiology* 152:2529.

105. Alvarez, H. M., and Steinbüchel, A. 2002. Triacylglycerols in prokaryotic microorganisms. *Applied Microbiology and Biotechnology* 60:367.

106. Lu, X., Vora, H., and Khosla, C. 2008. Overproduction of free fatty acids in *E. coli*: Implications for biodiesel production. *Metabolic Engineering* 10:333.

107. Pagliaro, M. et al. 2007. From glycerol to value-added products. *Angewandte Chemie International Edition* 46:4434.

108. Soares, R. R., Simonetti, D. A., and Dumesic, J. A. 2006. Glycerol as a source for fuels and chemicals by low-temperature catalytic processing. *Angewandte Chemie International Edition* 45:3982.

109. Román-Leshkov, Y. et al. 2007. Production of dimethylfuran for liquid fuels from biomass-derived carbohydrates. *Nature* 447:982.

110. Schmidt, L. D., and Dauenhauer, P. J. 2007. Hybrid routes to biofuels. *Nature* 447:914.

111. Bhosale, S. H., Rao, M. B., and Deshpande, V. V. 1996. Molecular and industrial aspects of glucose isomerase. *Microbiological Reviews* 60:280.

112. Bandish, R. K. et al. 2002. Glucose-to-fructose conversion at high temperature with xylose (glucose) isomerases from *Streptomyces murinus* and two hyperthermophilic *Thermotoga* species. *Biotechnology and Bioengineering* 80:185.

113. Schobert, H. H. 1990. *The chemistry of hydrocarbon fuels,* chap. 11. London: Butterworth.

114. Agrell, J. et al. 2002. Catalytic hydrogen generation from methanol. *Catalysis* 16:67.

115. Ale Ebrahim, A., and Jamshidi, E. 2004. Synthesis gas production by zinc oxide reaction with methane: Elimination of greenhouse gas emission from a metallurgical plant. *Energy Conversion Management* 45:345.

116. Campbell, I. 1983. *Biomass, catalysts and liquid fuels,* chap 1. London: Holt, Rinehart and Winston.

117. Biomethanol. http://www.refuel.eu/biofuels/biomethanol/.

118. Nakagawa, H. et al. 2007. Biomethanol production and CO_2 emission reduction from forage grasses, trees, and crop residues. *Japan Agricultural Research Quarterly* 41:173.

119. Hamelinck, C. N., and Faaij, A. P. C. 2002. Future prospects for production of methanol and hydrogen from biomass. *Journal of Power Sources* 111:1.

120. Amiguna, B., Gorgens, J., and Knoetze, H. 2010. Biomethanol production from gasification of non-woody plant in South Africa: Optimum scale and economic performance. *Energy Policy* 38:312.

121. Izenson, M. G., Crowley, C. J., and Affleck, W. H. 2007. Lightweight direct methanol fuel cell. US Patent 7,189,468.

122. Bio-DME. http://www.biodme.eu.
123. Mason, J. E. 2007. World energy analysis: H_2 now or later? *Energy Policy* 35:1315.
124. International Energy Agency. 2004. *Hydrogen and fuel cells: Review of national R&D programs.* Paris.
125. International Energy Agency. 2005. *Prospects for hydrogen and fuel cells.* Paris.
126. Takeuchi, Y., and Amao, Y. 2005. Biohydrogen production from sucrose using the light-harvesting function of zinc chlorophyll-*a. Bulletin of the Chemical Society of Japan* 78:622.

9 Sustainability of Biofuels Production

9.1 INTRODUCTION

How much plant biomass can be generated annually in nations or in the global community? If insufficient biomass can be sourced, no large-scale cellulosic ethanol industry can be supported to replace 20% or more of conventional gasoline consumption—and any advanced terrestrial biofuel would suffer from the same supply-side problem.

Can plant biomass be grown sustainably year after year without depleting the soil and causing yields to decline disastrously?

In tropical and semitropical regions, do crops intended for biofuels or bioenergy programs require irrigation? If so, can this supply be met?

Are major increases in plant yield mandatory for any global biofuels program and would this require genetic manipulation of crop and other plant species?

9.2 DELIVERING BIOMASS FEEDSTOCKS FOR CELLULOSIC ETHANOL PRODUCTION: THE LOGISTICS OF A NEW INDUSTRY

A study published in 2006 estimated that even if all U.S. corn production were to be dedicated to ethanol, only 12% of U.S. gasoline demand would be met.[1] Moving biofuels beyond "niche" markets therefore requires biomass-based ethanol. The feasibility of this had already been indicated by Canadian data in 2004. Although the total 2004 demand for fuel ethanol of 2,025 million L was met by ethanol produced from wheat, barley, corn, and potatoes, available nonfood crop supplies could have amounted to nearly 11,500 million L as corn stover, straw, wood residues, and forest residues.[2]

Although the tacit assumption in much biofuels literature is that abundant biomass supplies are at hand, little detailed information is available with which to test this hypothesis. The enormous scale of the potential supply of lignocellulose is frequently asserted; for example:

> Lignocellulose is the most abundant renewable natural resource and substrate available for conversion to fuels. On a worldwide basis, terrestrial plants produce 1.3×10^{10} metric tons (dry weight basis) of wood per year, which is equivalent to 7×10^9 metric tons of coal or about two thirds of the world's energy requirement.[3]

Some of this wood, however, represents trees grown or harvested as food crops and/or used directly as domestic fuel (e.g., in sub-Saharan Africa) or as industrial energy resources (in Scandinavia), while most—if not all—is involved in maintaining global ecological cycles and balances.

A useful data set was provided in 2000 in a review focused on sugarcane ethanol and especially as a means of estimating how biofuels could ameliorate global CO_2 emissions. Detailed computations were provided on how sunlight energy is converted to chemical energy inside plants on a limited amount of the Earth's surface.[4] For land use, the key numerical inputs (1996 data) were the following:

- Total land surface area is 14.9×10^9 ha.
- Only 4.9×10^9 ha was put to "productive" use as pastures and rangeland (67%), crop planting (29%), and settled land (4%).
- A further 4.4×10^9 ha was assessed as unfit for plant biomass production (rock, ice, tundra, and desert).
- Land unfit for highly productive plant biomass production (dry woods, mosaics, and taiga) amounted to 2.4×10^9 ha.
- The remaining 3.5×10^9 ha was potentially available for plant biomass production or for extending land used for grazing or crop planting (the types of land and landscape represented by this area are discussed in Section 9.2.2).

Only 9% of the total land area was used for crop planting in 1996 (Figure 9.1). With these estimates for actual and potential arable land, Brazilian sugarcane ethanol in that year was the sole industrial-scale biofuel on which to base modeling for energy use. The yearly production of ethanol amounted to 114 GJ per hectare; the estimated yearly fossil energy requirement of 3.2×10^{11} GJ could in principle be met with a land area devoted to sugarcane ethanol production of 2.8×10^9 ha—twice the global crop planting area (1.4×10^9 ha).

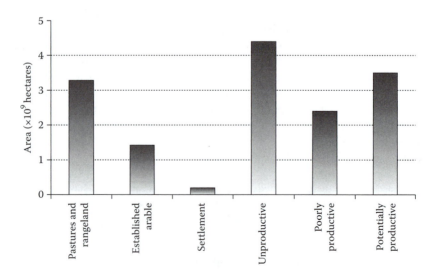

FIGURE 9.1 Global land use partition. (Data from Kheshgi, H. S., Prince, R. C., and Marland, G. 2000. *Annual Review of Energy and the Environment* 25:199.)

This would have been a daunting prospect but, given a further 3.5×10^9 ha potentially available for plant biomass, not totally inconceivable. If ethanol could be produced from the entire aboveground sugar plant with cellulosic ethanol technology, then ethanol energy of 309.8 GJ per hectare would equate to a land use of 1.03×10^9 ha—a high but more attainable dedication of land to biomass-derived biofuels. This not only provided a yardstick for biofuels use but also strongly reinforced the potential for cellulosic ethanol in the era when its use was still essentially a research topic.[5]

If only transportation fuels were to be considered, the land use requirements become more amenable—a consequence of the authors of the 2000 publication focusing on global CO_2 targets rather than the practical introduction of ethanol as a gasoline extender. Using data from the International Energy Agency, oil used for transport is expected to continue being approximately 20% of the yearly fossil fuel demand or to increase slowly to over 50% of the crude oil extraction rate (Figure 9.2).[6] To substitute the global demand for oil as a transport fuel, therefore, sugarcane grown for ethanol would need to occupy 40% of the world's arable land (stems only as a feedstock) or 15% if the entire harvestable biomass were to be used for ethanol production (using the 1996 Brazilian ethanol data for calculation).

These theoretical calculations demonstrated the potential conflict between land use for food crop production and for bioenergy—what has subsequently become widely known as the "food versus fuel" debate. Given the conclusion that biofuels could not be derived from biomass grown on a small percentage (<5%) of the world's available agricultural land, land use conflict was (and still is) highly likely to be a reality, and genetic engineering or other innovative technologies cannot avoid this without massive increases over current bioprocess-limited abilities to transform biomass into ethanol (or any other liquid biofuel).[7]

Given an ever growing human population and—more acutely—the rapidly industrializing "mega" nations of India and China, the likely dramatic increases in world

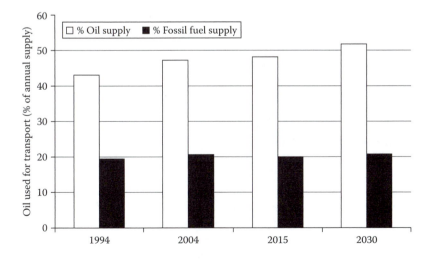

FIGURE 9.2 Oil used for transport as percentage of annual supply of oil and fossil fuels. (Data from International Energy Agency. 2006. *World Energy Outlook*, annex A. Paris.)

energy and transportation demand in the coming decades have given the land use issue for bioenergy consistently negative media coverage since 2006 and 2007. More positively, the perceived need to avoid land use conflicts has driven the search for radical alternatives to traditional plantation-based biofuels—for example, "oilgae," "microdiesel" (Chapter 8, Section 8.4), and recycling of food and other wastes for biofuels production.

A more balanced approach is to target only part of the 20% of fossil fuels (oil and liquefied natural gas) used for transportation, bearing in mind the plethora of alternative energy routes for electrical power generation: wind, wave and tidal, geothermal, solar (photovoltaic), hydroelectric, and microbial (Chapter 7, Section 7.7). Massive generation of renewable electricity obviously would cut fossil fuel use and power hybrid gasoline/electric vehicles; similarly, H_2 production (however costly in monetary terms or environmentally if dependent on fossil-fuel-based routes) will contribute significantly to the global energy landscape during the twenty-first century in a complex road map of evolving industrial-scale options. Biofuels will therefore only be called upon as part of a portfolio of energy technologies, and this redefines the argument to what provision of plant biomass for ethanol and other biofuels is *possible*.

9.2.1 Upstream Factors: Biomass Collection and Delivery

A study provided estimates of land area available for biomass energy crops and of the utilization of wood industry, agricultural, and municipal solid wastes for the United States in 2005.[8] A total of 1.3–2.3 billion tons of cellulosic biomass was estimated to be the annual input to ethanol production—potentially equivalent to a biofuel supply matching 30% of U.S. gasoline consumption. Ignoring questions of cost and efficiency of bioconversion, this study identified forestry and agricultural resources for use in biofuels production. Three types of forest resources were quantified:

1. primary (logging residues, removal of excess biomass in timberland fuel treatments, and fuel wood extracted from forestlands)
2. secondary (mill residues)
3. tertiary (urban wood residues from construction/demolition and recycling)

Similarly, lignocellulosic agricultural resources were divided among

1. primary (crop residues, perennial grasses, and perennial woody crops)
2. secondary (food/feed processing residues)
3. tertiary (municipal solid waste recycling)

Approximately 280 million tons (dry weight) annually of such resources was estimated to be available by the time of the report (Figure 9.3). This figure is somewhat shy of the headline billion tons, and careful scrutiny of the data is important to understand the dynamics of possible change in the biomass supply.

Two initial scenarios were considered for agricultural sources: moderate and high crop yield increases (Figure 9.4). By the middle of the twenty-first century, these were expected to increase from a baseline of 143 million tons annually (Figure 9.3) to

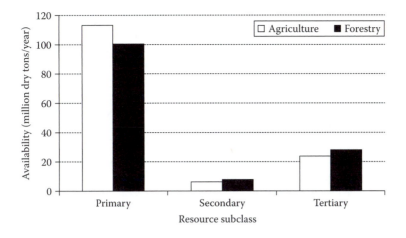

FIGURE 9.3 Projected mid twenty-first century U.S. plant biomass resources. (Data from Perlack, R. D. et al. 2005. Biomass as a feedstock for a bioenergy and bioproducts industry: The technical feasibility of a billion-ton annual supply. U.S. Department of Energy, April 2005. Available at http.//www.osti.gov/bridge.)

322 million tons (moderate yield increase) or 455 million tons (high-yield increase). The forestry resources were also expected to increase to 171 million tons from a baseline of 136 million tons. Moving above a maximum of 620–630 million tons annually would require additional resources—most importantly, changes in land use (discussed in the next section).

Augmentation of this restricted supply can arise from programs to thin native forests strategically so as to reduce fire hazards—in California, for example, in excess of 0.75 million tons per year could be generated by such activities.[9] With an excess

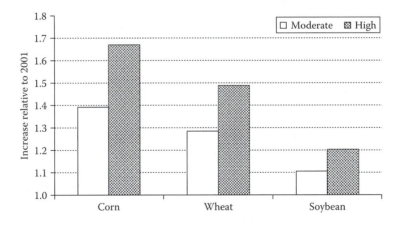

FIGURE 9.4 Projected increases in yield for major U.S. crops by middle of the twenty-first century. (Data from Perlack, R. D. et al. 2005. Biomass as a feedstock for a bioenergy and bioproducts industry: The technical feasibility of a billion-ton annual supply. U.S. Department of Energy, April 2005. Available at http.//www.osti.gov/bridge.)

of 600 million tons annually, gasoline substitution rates could reach at least 15%; in addition to this, the contribution from corn and other grain ethanol sources adds 5% of national gasoline usage. Corroborating data from Sweden (with its relatively low population density but high degree of forestation) suggest that 25% of that country's gasoline requirements could be substituted by lignocellulose-derived ethanol from existing biomass resources.[10]

The forestry component in the "billion ton vision" has been subjected to a critical analysis, with the following results[11]:

- Of the estimated 159 million dry tons of industrial wood by-products, 95% is already used for wood or paper products (38%) or for energy generation (only the residual unused fraction is included in Figure 9.3) and would have a high bark content and a corresponding low potential conversion to ethanol.
- Any fuel wood counted into the total biomass supply represents existing uses (domestic and/or industrial).
- Fuel trimmings could contribute 60 million tons annually, but a viable use for this resource would need to have minimal transportation costs (i.e., a nearby ethanol conversion plant); high bark contents would also reduce potential conversion to ethanol.
- Logging residues could be collected efficiently because they are derived from commercially viable operations, although, again, they have high bark contents.

A conservative estimate of total woody biomass availability could be as low as 64 million tons per year—equating to 2.56 billion gallons of ethanol, or 1.8% of 2006 gasoline usage. Even the secure and reliable provision of this quantity of forestry-related biomass was described as "challenging."

9.2.2 LIMITATIONS IMPOSED BY LAND AVAILABILITY AND LAND USE

Even in the United States, achieving a billion tons of biomass annually would require some changes in land use (Figure 9.5).[8] Compared to the baseline, the major changes were the following:

- reduction in food crop planting (by up to 7%)
- reduction in cropland pasture (by 37%)
- reduction in idle land (by 27%)
- new plantings of perennial bioenergy crops such as switchgrass

As an energy crop feedstock in North America, a grass such as switchgrass (*Panicum virgatum*) has persuasive economic, social, and agricultural advantages.[12] With such land use changes, the calculated biomass supplies from crop planting increased to 480 million annually (moderate yield increase) or 865 million annually (high yield increase). Adding in forestry and some timber processing resources gives a biomass supply of 651–1,036 million tons per year. Furthermore, counting in manure and other products, grains used for bioenergy, and all timber processing residual materials and fuel wood gives a total of over 1,300 million tons per year.

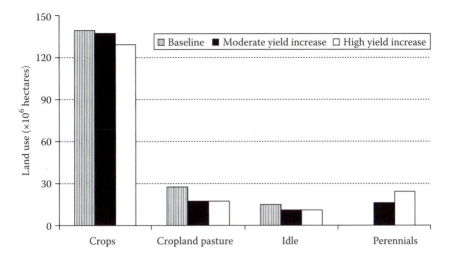

FIGURE 9.5 Projected changes in U.S. land use by middle of the twenty-first century with differing crop yield increases. (Data from Perlack, R. D. et al. 2005. Biomass as a feedstock for a bioenergy and bioproducts industry: The technical feasibility of a billion-ton annual supply. U.S. Department of Energy, April 2005. Available at http.//www.osti.gov/bridge.)

To achieve a 30% substitution of fossil fuels by biofuels requires new crops (perennial bioenergy crops) and more land use for crops and plant biomass. No encroachment of bioenergy crops on arable land was envisaged because technological increases in major crop yields (corn, wheat, and soybean) were assumed to be achievable; therefore, food versus fuel conflicts could be avoided. Because North America has an advanced and scientific agriculture, several drivers can synergize to ensure sufficiently high future crop yields, in particular:

- crop yield rates of increase for corn that were 1.2% per year in 2005 and were expected to decline only slowly (to 0.9% per year) by 2030
- wheat yield increases that were in the range 1.3–1.7% per year between 1997 and 2002
- decreasing fertilizer usage
- new soybean varieties that offer high biomass-to-seed ratios while maintaining seed yield
- (more controversially) genetically modified (GM) plants that have been developed for herbicide and insect resistance
- R&D programs for maximizing biomass removal and recovery from fields while managing soil conservation and no-till planting systems that are used on more than 24 million ha as the favored form of conservation tillage

Outside North America, agriculture is often less scientifically and technically developed and supported; where population densities are highest, farms may be unviable, small units operated on a subsistence basis. In the broader world, therefore, close attention must be paid to available land that could be harnessed for biofuels

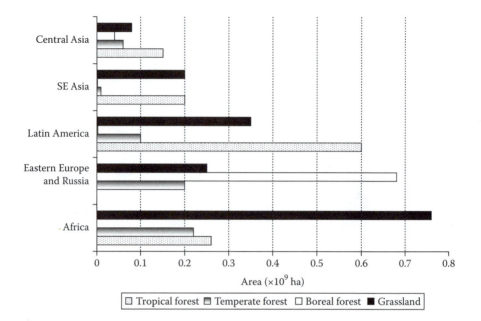

FIGURE 9.6 Potentially arable land by global region. (Data from Kheshgi, H. S., Prince, R. C., and Marland, G. 2000. *Annual Review of Energy and the Environment* 25:199.)

programs. The major land masses for analysis are (in decreasing order) Africa, Eastern Europe and Russia, Latin America, Southeast Asia (excluding China), and the former Central Asian Republics of the Soviet Union. Potential land for biomass and energy crop cultivation could be found in tropical, temperate, and boreal forests and in grasslands (Figure 9.6).[4]

These are mostly regions where environmental issues in land use are at their most extreme. The public face of soybean-oil-derived biodiesel in Brazil, for example, is that of sustainable symbiotic nitrogen fixation with the bacterium *Bradyrhizobium japonicum* and soybean cultivars selected to grow in the arid savannah in the hinterland state of Mato Grosso. However, environmentalist concerns about "deforestation diesel" remain acute: Brazil already exports 20 million tonnes of soybean annually and plans to increase this to 32 million tonnes by 2015.[13] Brazil launched a national program in 2002; targets under "probiodiesel" include 2% of all transport diesel to be biodiesel by 2008 (when all fuel distributors will be required to market biodiesel) and 5% by 2013.[14]

The debate about biodiesel in particular has become highly charged, but the issues affect all biofuels that use plant biomass resources. When, in early 2008, a new challenge to the environmental benefits of biofuels was raised on the basis of land use change involving extensive deforestation, this debate also rapidly became acerbic. A new analysis of biofuels production and greenhouse gas emissions included carbon emissions that would occur if forests and grasslands worldwide were to be converted to support a vastly increased biofuels production.[15]

Although a mathematical modeling approach was used, the essential argument was that when forested land or grassland was converted to biofuels production, a spike of CO_2 was generated (e.g., the burning of felled trees). This exceeded the CO_2 abatement resulting from biofuels use so much that, in effect, a carbon debt was incurred that would take decades or longer to repay. This is intuitively logical; the furor resulted from the *size* of the computed figures:

- Corn-based ethanol, instead of producing a 20% savings, would double greenhouse emissions over 30 years and increase greenhouse gas levels in the atmosphere for 167 years.
- Biofuels from switchgrass grown on U.S. corn lands would increase emissions by 50%.

The previous discussion (Sections 9.2 and 9.2.1) emphasized the enormous scale of biomass and bioenergy crop planting to substitute for significant fractions of global fossil fuel use in transportation. Extensive losses of primary rain forest are known to cause severe pollution events, including spectacular smogs covering hundreds or even thousands of square kilometers in Indonesia in 2000–2005. This is the climax of the process that did not commence with local deforestation for biofuels cropping, but rather can be traced back to U.S. corn ethanol[15]:

- U.S. corn is diverted from food uses to ethanol and this use will increase over the coming decade.
- Corn and other crop prices increase (corn by 40%, wheat by 17%, and soya by 20%).
- Worldwide, farmers react by changing land use to compensate for the diversion of crops to biofuels, taking advantage of changing economic conditions.
- Grassland, savannah, and forest areas are converted (with forests, this most obviously involves "slash and burn"), including 2.2×10^6 ha in the United States, 2.8×10^6 ha in Brazil, 1.1×10^6 ha in China, and 0.7×10^6 ha in Africa.

All these events *singularly* are far from new. They were discussed at length at the UN Climate Change Conference in Bali in 2007. From this emerged the Bali road map, which covered policy approaches and positive incentives required to reduce CO_2 emissions from deforestation and forest degradation in developing countries and the role of conservation, sustainable management of forests, and enhancement of forest carbon stock. Bali was an appropriate choice of venue because deforestation in Indonesia had reached alarming levels due to illegal logging, fires, and land conversion; the damaged forests are a threat to the biodiversity of Indonesian and global ecosystems.

The island of Bali is a paradigm for the contrary view. Before colonization by mankind, the island was mostly covered by rain forest, with some dry tropical forest; now, very little of the indigenous forest remains and the remnants are protected in a national park. Land clearance was caused by settlement and the introduction of rice farming after 300 BC (3 million people inhabit an island of 5,600 km², 1.5 times the area of Long Island).[16] Biodiversity has been severely compromised: The elephant is already extinct and the last tiger was shot in 1937; many native species are endangered by or vulnerable to extinction. Yet, in time, the settlers introduced the world-renowned Balinese culture so highly prized by tourists that 30-day visas have been enforced. Without land clearance for irrigated paddy fields that supported the vibrant community, its temples, dance, and music would not have evolved on the now densely populated island. Is there, consequently, good and bad deforestation?

The arguments raised in *Science* by the 2008 paper continue. The situation in late 2009 could be summarized thusly[17]:

- Methodologies and numerical data inputs have been questioned.
- The detailed workings of the complete mathematical model cannot be explored and replicated by others.
- Parameters of indirect land use changes are uncertain but must be taken into account, and more detailed work should be dedicated to this topic.

Another conclusion is that corn-derived ethanol was an easy target because even supporters of biofuels agree that corn is one of the worst feedstock choices for ethanol in terms of greenhouse gas emissions due to land change (Figure 9.7).

Some reduction in land use change impacts ("carbon payback time") would be gained by crop yield improvements and technology advances; this conclusion was reached from using a new and geographically detailed database of crop locations and yields and updated vegetation and soil biomass estimates in a more regionally defined model.[18] Nevertheless, no foreseeable changes in agricultural or energy technology may be able to deliver carbon benefits if crops grown for biofuels displace tropical forests; however, biofuels made from waste biomass or from perennial plant biomass grown on degraded or abandoned agricultural lands incur no significant carbon debt (Table 9.1).[19]

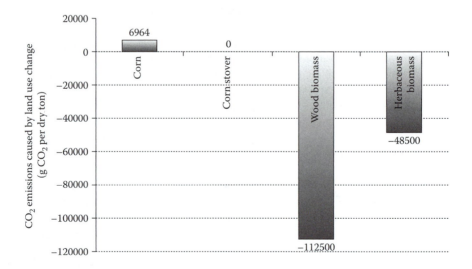

FIGURE 9.7 Estimated CO_2 emissions caused by land use change to ethanol production. (Data from Searchinger, T. et al. 2008. *Science* 319:1238.)

TABLE 9.1
Biofuel Carbon Debt on Land Use Change

Biofuel	Feedstock	Location	Previous Land Use	Time to Repay C Debt (Years)
Ethanol	Perennial biomass	United States	Marginal cropland	0
Ethanol	Perennial biomass	United States	Abandoned cropland	1
Ethanol	Sugarcane	Brazil	Cerrado woodland	17
Biodiesel	Soybean	Brazil	Cerrado grassland	37
Ethanol	Corn	United States	Abandoned cropland	48
Biodiesel	Palm	Southeast Asia	Tropical rainforest	86
Ethanol	Corn	United States	Central grassland	93
Biodiesel	Soybean	Brazil	Tropical rainforest	319
Biodiesel	Palm	Southeast Asia	Peatland rainforest	423

Source: Data from Fargione, J. et al. 2008. *Science* 319:1235.

STEM TOPIC 9.1: LAND AREA REQUIREMENTS FOR FIRST-GENERATION BIOFUELS

Ethanol and biodiesel derived from crop plants (mostly food crops) vary greatly in biofuel yields per unit area of land used for crop cultivation. This is one of the major determinants in economic cost as well as in strategic choices for biofuels production in regions and among nations. This variability is threefold:

1. nature of the biofuel
2. biological feedstock
3. location of production

Even with the same biofuel produced from the same feedstock, the variation in yield according to geographical region can be substantial (see Figure STEM 9.1).

Brazilian sugarcane ethanol retains its preeminence and is over 16 times more productive (on a land area basis) than soybean biodiesel in Brazil.[20]

Other marked differences include EU cereal feedstocks and palm oil biodiesel in comparison with any other biodiesel. Brazil intends to compete with Malaysia in palm oil production and will allow the establishment of oil palm plantations on degraded lands in the Amazon rainforest without contributing further to deforestation. Soybean biodiesel is a poor choice for Brazil, which grows soybean for export as a food source.

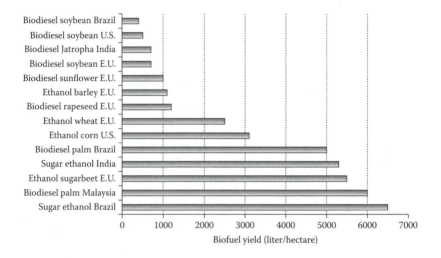

FIGURE STEM 9.1 Yield of ethanol and biodiesel from different crop species in different regions. (Data from Fulton, L., Howes, T., and Hardy, J. 2004. International Energy Agency, Paris. http://www.iea.org/textbase/nppdf/free/2004/biofuels2004.pdf.)

Land use issues in the European Union (EU) are less acute than in developing countries because of the set-aside provisions of the Common Agricultural Policy, which aim to provide financial encouragement to farmers *not* to overproduce agricultural surpluses, which otherwise would require a large financial outlay for their storage and disposal. Projections for 2020 include a baseline of 11% of arable land being set aside with no crops grown for biofuels production. This would drop to 3.3% with rapeseed for biodiesel and sugar beet and cereals used for ethanol if biofuels replaced 14% of transportation fossil fuels with a reliance on European production rather than imports of biofuels (Figure 9.8).[21] Total arable land would increase from 98.4 to 102.5 million ha.

For each additional million hectares used within the EU to produce biofuels feedstocks, land use change could be

- 370,000 ha of arable land diverted from exports to domestic production
- 400,000 ha taken out of set-aside
- 220,000 ha of land that would have found other uses remaining in arable use

Reduced arable land used for conventional food crops implies some improvement in crop yield but EU population growth is expected to be modest—1.2% between 2005 and 2020—unlike most of the world's nations and regions (Figure 9.9).[22] Parts of the EU will be subject to population loss due to migration; some nations with a high agricultural working population may even experience labor shortages unless net immigration is permitted from adjacent European nations or from developing countries. With or without such measures, however, crop yield improvements of 10% would be sufficient to allow anticipated land use changes.

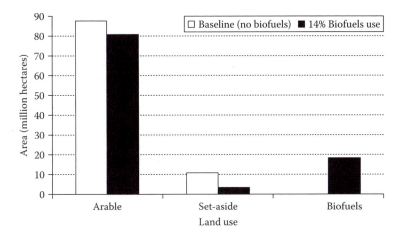

FIGURE 9.8 Projected land use changes in the European Union consequent on biofuels production. (Data from Department for Environment, Food and Rural Affairs. 2008. Report by AEA Technology plc, London.)

Beyond this relatively modest ambition (14%) for conventional fuel replacement, the estimates for land use become both uncertain and potentially very high. At 25% displacement of fossil transportation fuels, the land required using first-generation biofuels is in the wide range of 17–117% of total arable land.[21]

If increases in crop productivity do not keep pace with the world's increasing total population, will there be an inevitable conflict of food versus fuel? This appears to be most likely in the world's poorest regions, including Africa (Figure 9.9.). Can ethical

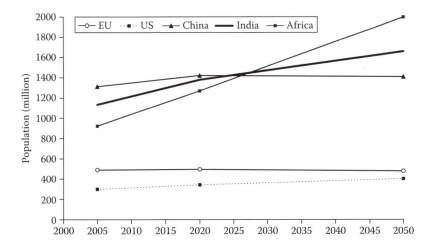

FIGURE 9.9 Projected world population change by region. (Data from Fésüs, G. et al. 2008. Regions 2020. Demographic challenges for European Regions. Commission of the European Communities. Brussels.)

questions be factored into policy making? Some ethical trade-offs related to land use in moving to a global bioenergy system include[23]:

- Emissions may be reduced, but added crop production may decrease the ability of the world's poor to feed themselves through increased demand.
- Environmentalists often value low-intensity crop production because it causes less environmental degradation at low usage of fertilizers and fossil fuels. However, more intensive agriculture can achieve higher food crop production and fewer land use changes.
- Although climate change affects biodiversity, the extensive use of biofuels could devastate local ecosystems, especially in less developed countries and regions.

Benefits of increasing rural employment opportunities must be balanced against these problems. How governmental and international bodies can coherently and convincingly explain such trade-offs remains speculative.

9.3   SUSTAINABLE DEVELOPMENT AND BIOMASS PRODUCTION

Massive demands for plant biomass have major implications for regional and global agriculture and greenhouse gas reduction strategies. After exploring "how much" questions, intensive plantation systems for bioenergy crops and biofuel feedstocks face questions posed by the plant physiologist and the soil scientist—broadly, "can it be done?" issues of ecological sustainability.

9.3.1   DEFINITIONS, SEMANTICS, AND ANALYSIS

Much of the public debate about biofuels has assumed that the production of ethanol and biodiesel as first-generation biofuels is inherently sustainable. This is mostly on the grounds that any agricultural activity is renewable, whereas the extraction of crude oil is necessarily a once-only activity, given the extremely long geological time scale of oil's generation. As a term, "sustainability" suffers from inexactness.[24,25] An exact definition can be deduced from thermodynamics and consequently defined in terms of mathematical and physical properties; that is, a cyclic process is sustainable *if and only if:*

- It is capable of being maintained indefinitely without interruption, weakening, or loss of quality.
- The environment on which this process depends and into which the process expels any waste material is itself equally renewable and maintainable.

These are very strict criteria and are not exemplified by, for example, an annual replanting of a crop plant such as maize, which depends on outlays of fossil fuel energy (for fertilizers, etc.) and may seriously deplete the soil or minerals and contribute to soil erosion. Even though such a system appears to be *renewed* every year, that is only within living memory or the history of agricultural production on Earth; on a geological timescale, this is a minuscule length of time.

It is a sad fact of human agricultural activity that periodic crises have accompanied large-scale, organized farming for millennia in the erosion and salt accumulation that caused the downfall of the Mesopotamian civilization, the overgrazing and poor cultivation practices that have caused the expansion of the Sahara, and the overintensive cultivation of fragile tropical rain forest soils that contributed to the collapse of the Mayan economy and society (among many other examples).[26]

Industrial agriculture and intensive farming are relatively recent arrivals (within the last century), and they have brought accelerated land degradation and soil erosion; agrochemical residues in water courses are a global problem, leading to eutrophication of freshwater fishing grounds and threatening the collapse of fragile ecosystems. To environmental lobbyists, therefore, the prospect of energy crop plantations is highly unwelcome if such agronomy requires and is dependent on the heavy use of fertilizers, insecticides, and pesticides, and the depletion of soil organic carbon.

A more analytically useful definition of the renewability of renewable energy could focus on a more analyzable timescale, perhaps 160 years (approximately the life span of the industrial activity that underpins modern society).[25] Perhaps even more appropriate would be a century (i.e., most of the time in which automobiles have been fueled by gasoline and diesel). For either option, generating a quantitative framework is crucial to any assessment of the practicalities of a biofuels program.

At the heart of this thermodynamic and mathematical framework—although using an intimidating formal language—is the principle that an industrial biomass facility can only be useful if the energy derived from its functioning exceeds that required to maintain operation (Figure 9.10). The maintenance requirement is equivalent to *restoration*. For a biomass plantation, the restoration energy is that necessary in fertilizers, herbicides, pesticides, electrical power, and fuel to process the biomass and convert the biomass to a useful output; the most energy-efficient output is electricity generated

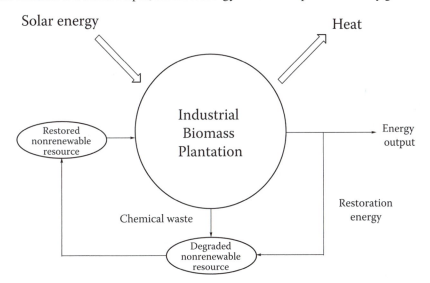

FIGURE 9.10   Energy flows into and from an industrial biomass plantation. (Modified from Patzek, T. W. 2004. *Critical Reviews in Plant Sciences* 23:519.)

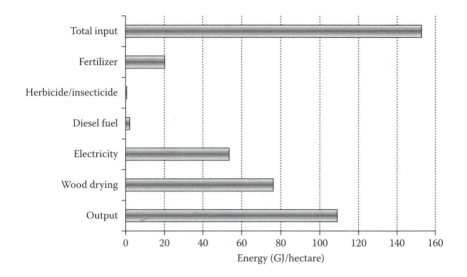

FIGURE 9.11   Energy uses and output energy from acacia biomass plantation. (Data from Patzek, T. W. 2004. *Critical Reviews in Plant Sciences* 23:519.)

on site by the combustion of wood pellets. For both acacia and eucalyptus species (fast growing plantation trees in Indonesia that have been studied in detail), the restoration energy has exceeded the maximum output energy (Figure 9.11).

On this evidence, these industrial biomass plantations were entirely unsustainable in thermodynamic terms—although they may be profitable (in monetary terms) and well managed—because the long-term requirements to maintain their operation exceed their energy yields. Such plantations could be sustainable if input energy requirements are minimized by resorting to sun drying the wood before processing to pellets; this appears to be incompatible with a high rate of biomass production.[25]

Similar arguments have long been used against any form of intensive agriculture, usually focused on soil degradation. Soil is a source of organic carbon as well as nitrogen, phosphate, and inorganic cations (potassium, calcium, and magnesium). Soil erosion by wind and rain reduces soil productivity and also affects runoff into aquatic ecosystems; soil compaction by heavy agricultural vehicles reduces productivity and water absorption and increases runoff. Organic matter decrease occurs if excessive amounts of biomass are removed.

Because of these effects, using agricultural residues for biofuels production has agronomic limitations: Although a worldwide potential residue harvest of 3.8×10^9 tonnes per year has been estimated, removing more than 30% of this could greatly increase soil erosion as well as deplete the soil organic carbon content.[27] Restoring soil carbon by adding "biochar" (prepared by the slow pyrolysis of biomass) has been suggested as a means of biosequestering carbon.[28]

The exploitation of adventitious resources—for example, weed infestations (water hyacinth is a major problem in lakes and waterways) and naturally occurring sugar-rich forest flowers—is a possible means of extending biomass supplies.[29,30] Of major candidate energy crops, switchgrass and other prairie grasses offer good balances between high net energy yields, low nutrient demands, and high soil and water conservation; with its high below-root mass, switchgrass can replace soil carbon lost during decades of prior tilling within (perhaps) 20 years.[31] Under U.S. farm conditions, evidence is mounting that dedicated energy crops would significantly reduce erosion and chemical runoff in comparison with conventional monoculture crops.[32–34] Replacing continuous cropping with crop rotation and no-till field operations might maximize agricultural residue availability.[35]

STEM TOPIC 9.2: SOIL CARBON AND THE GLOBAL CARBON CYCLE

Soil organic carbon is a significant pool of carbon, along with CO_2 in the atmosphere, CO_2 and dissolved carbonate in the oceans, and more organic carbon in the biosphere (living organisms and dead organisms not yet converted into soil organic carbon). The relative amounts are

biosphere (1) < air (1.3) < soil (2.7) < oceans (67.8)

The global carbon cycle consists of movements of carbon between these dynamic pools and one inelastic, depleting source (i.e., fossil fuels; see Figure STEM 9.2).

At present, the cycle is not a closed circle: 112.5×10^9 tonnes of carbon are removed from the atmosphere annually, but 118×10^9 tonnes are added (data from Soil Carbon Center, Kansas State University, www.soilcarboncenter.k-state.edu). The imbalance represents global CO_2 buildup and is accounted for by fossil fuel and deforestation emissions.

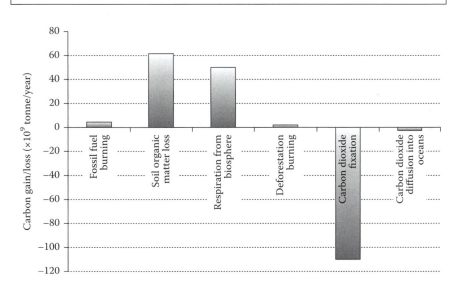

FIGURE STEM 9.2 Carbon gains and losses from terrestrial, oceanic, and atmospheric sources. (Data from www.soilcarboncenter.k-state.edu.)

An ironic twist is the impact of global climate change; global warming (at least for a time, perhaps 50–100 years) might *increase* plant growth to such an extent that biomass extraction targets would be more closely met. Even accurately predicting biomass yields from monocultures of fast growing tree species such as willow in the northeastern United States suffers from methodological weaknesses.[36] The effects of changes in daily minimum and maximum temperatures are complex because they differentially influence crop yield parameters; this is a major area of uncertainty for projecting yield responses to climate change.[37]

For three major cereal species (wheat, corn, and barley) and an increase in annual global temperatures since 1980 of approximately 0.4°C, there is evidence for *decrease* in yield. The magnitude of the effect is small in comparison with the technological yield gains over the same time period; however, it suggests that rising temperature might cancel out the expected increase in yield because of increased CO_2 concentrations.[38,39] Using the global environment as an uncontrolled experimental system has its methodological drawbacks but the results of the one long-term field trial to attempt to isolate the effects of temperature on rice growth indicate that a 15% reduction in yield could result from each 1°C rise in temperature, a much grater effect than predicted by simulation models.[40]

In contrast, the International Panel on Climate Change (IPCC) has predicted that CO_2 benefits will exceed temperature-induced yield reductions with a modest rise in temperature.[41] The detailed conclusions for food, fiber, and forest products are different:

- Crop productivity is projected to increase slightly at mid to high latitudes for local mean temperature increases of up to 1–3°C, depending on the crop, and then to decrease beyond that in some regions.
- At lower latitudes, especially seasonally dry and tropical regions, crop productivity is projected to decrease for even small local temperature increases (1 or 2°C), which would increase risk of hunger.
- Globally, the potential for food production is projected to increase with increases in local average temperature over a range of 1–3°C, but above this it is projected to decrease.
- Adaptations such as altered cultivars and planting times allow low and mid to high cereal yields to be maintained at or above baseline yields for modest warming.
- Increases in the frequency of droughts and floods are projected to affect local production negatively, especially in subsistence sectors at low latitudes.
- Globally, commercial timber productivity rises modestly with climate change in the short to medium term, with large regional variability around the global trend.

Taken together, these IPCC prognoses suggest some short-term improvement in the productivity of a range of plant species, including timber grown as feedstocks for cellulosic ethanol, but that the predictions become more unreliable as the geographical area narrows and related effects of climate are considered. This highlights the obvious conclusion that geostatistical surveys and controlled experiments should

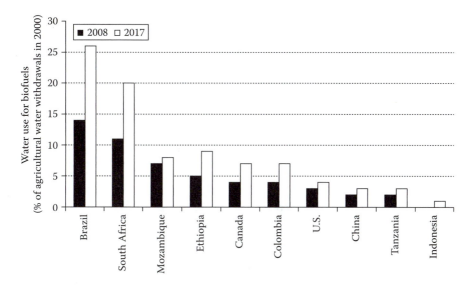

FIGURE 9.12   Global water use for irrigating biofuels crops. (Data from Hoogeveen, J., Faures, J.-M., and van de Giessen, N. 2009. *Irrigation and Drainage* 54:148.)

both be pursued vigorously as priority issues for agronomy and plant physiology in the coming decades.

Water resources are the final element in the biology of sustainability for producing biofuels. Up to 2008, however, surprisingly little (1%) of all the water withdrawn from water courses for irrigation was dedicated to crops grown for biofuels. As biofuels increase in scale, this percentage will increase, and concern has been expressed as to how this increased demand will place burdens on irrigation in countries outside North America and Europe with rapidly developing economies—for example, South Africa, Ethiopia, Mozambique, China, and Tanzania (Figure 9.12).[42] Augmented use of rain-fed plant species may counter this trend, but localized effects and competition will inevitably occur; this is most likely if irrigated sugarcane is expanded as the global primary source for fuel ethanol manufacture.

The UN's Department of Economic and Social Affairs included irrigation water use in its study of biofuels production in sub-Saharan Africa.[43] The use of crops requiring minimum irrigation was advocated. The use of *Jatropha curcas* and *J. moringa* as a source of seed oils for biodiesel is particularly attractive because of the low requirements for water, placing little increased pressure on frequently scarce and diminishing water resources.

Jatropha can be grown on unused and abandoned land, does not compete with food supply, and provides an income alternative to cotton (which has poor returns due to heavily subsidized global markets and high pesticide needs). Based on experience of *Jatropha* in India in rain-fed areas plantations, with poor soil can yield 2 kg of seed oil per tree; even in relatively poor desert soils yields can reach 1 kg per tree. Nevertheless, the quoted establishment costs for a *Jatropha* plantation included

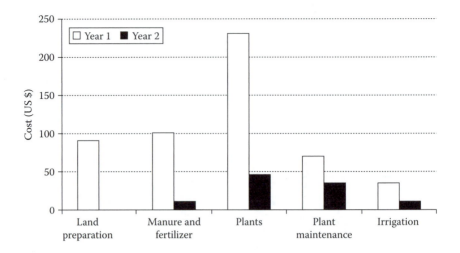

FIGURE 9.13 Indicative costs of establishing a small-scale *Jatropha* plantation for biodiesel in India. (Data from Energy and Transport Branch, Division for Sustainable Development, United Nations Department of Economic and Social Affairs. 2007. Background paper no. 2 DESA/DSD/2007/2, New York.)

irrigation as an item in both years 1 and 2 to supply water for seedling growth and vigor (Figure 9.13).

9.3.2 CASE STUDY: SUSTAINABILITY OF BRAZILIAN SUGARCANE ETHANOL

Much of the discussion on sustainability of cellulosic ethanol is speculative. Far more information is known about sugarcane ethanol—in particular, the large and expanding industry for its production in Brazil. This is a paradigm for the scenario discussed in the immediately preceding section—that is, a rapidly developing economy outside North America and Europe that uses irrigated plantations.

The productivity of the Brazilian sugarcane ethanol industry is highly impressive: The equivalent of 3% of global gasoline use is generated by the crop grown on only 2% of the total agricultural and pasture land area of Brazil.[44] There are nevertheless doubts from the environmental perspective: The first decade (1976–1985) of the national program probably achieved a reasonable soil balance by recycling fermentor stillage as fertilizer—a valuable source of minerals, particularly potassium.[45,46]

With the great expansion of the industry subsequently, significant pollution problems emerged. The volume of stillage that can be applied to agricultural areas varies from location to location, and in regions with near-surface groundwater much less stillage can be applied without contaminating the water supply.[46] In the case of the Ipojuca River in northeast Brazil, a joint German–Brazilian study found that sugar cultivation and adjacent ethanol production plants extensively use stillage for both fertilization and irrigation, and this has led to water heating, acidification, increased turbidity, O_2 imbalance, and increased coliform bacteria levels.[47] The study urged that a critical evaluation be made of the environmental status of the sugar alcohol industry, focusing on developing more environmentally friendly cultivation

TABLE 9.2

Selected Sustainability Criteria for Sugar Ethanol Production in Brazil

Criterion	Measurable Parameter	Expected Compliance
Greenhouse gas emissions	Net reduction 30% by 2007	Probable
Greenhouse gas emissions	Net reduction 50% by 2011	Probable
Competition with food supply	?	Uncertain
Biodiversity	No decline of protected areas in 2007	Very uncertain
	Active protection of local ecosystems by 2011	Very uncertain
Welfare	Compliance with treaties, declarations, etc.	Partial or unknown
Environment		
Waste management	Compliance with existing laws	Uncertain
Use of agrochemicals	Compliance with existing laws	Partial
Use of agrochemicals	Compliance with EU legislation by 2011	Uncertain
Prevention of soil erosion and nutrient depletion	Management plans	Unclear
Preservation of surface and ground water	Water use and treatment	Probably possible
Airborne emissions	Compliance with EU laws by 2011	Uncertain
Use of GMOs	Compliance with EU laws by 2011	Possible

Source: Modified from Smeets, E. et al. 2006. Sustainability of Brazilian bio-ethanol. Report NWS-E-2006-110 (ISBN 90-8672-012-9), Senternovem (The Netherlands Agency for Sustainable Development and Innovation), Utrecht, the Netherlands.

methods, waste-reducing technologies, and water recycling to protect the region's water resources.

The preservation of surface and ground water in Brazil in general as a consequence of the sugar alcohol industry's activities and development was ranked "uncertain, but probably possible" in a detailed study from the Netherlands (Table 9.2).[48] Sugarcane plantations have been found to rank well for soil erosion and runoff criteria in some locations in São Paolo state, although the experimental results date from the 1950s (Table 9.3). A subsequent study showed much poorer results for sugarcane in comparison with other monoculture crops (Figure 9.14).

Nevertheless, although Brazilian sugarcane alcohol (viewed as an industrial process) makes massive demands on the water supply (21 m³ per tonne of cane input), in principle, much of this water can be recycled. In addition, Brazil enjoys such a large natural supply of fresh water from its eight major water basins (covering an area of 8.5 million km²) that the ratio of water extracted to supply on a global basis is exceedingly small: approximately 1% per annum—equivalent to 30-fold less than comparable data for Europe. Local seasonal shortages may occur, however, and two of the four main sugar production regions have relatively low rainfall (Figure 9.15). Although sugar cultivation has mainly been rain fed, irrigation is becoming more common.

TABLE 9.3
Annual Soil Losses by Erosion and Runoff in Experimental Stations in Brazil

Crop	Soil Loss (tonne/ hectare)	Runoff (mm)	Crop	Soil Loss (tonne/ hectare)	Runoff (mm)
Fertile Soil, 9.4% Slope[a]			**Red Soil, 8.5% Slope[b]**		
Cassava	53	254	Castor beans	56.1	199
Cotton (in rotation)	38	250	Common beans	54.3	180
Soybean (continuous)	35	208	Cotton	51.4	183
Cotton (continuous)	33	228	cassava	42.6	170
Soybean (in rotation)	26	146	Upland rice	36.6	143
Sugarcane	23	108	Maize (residues incorporated)	30.9	144
Maize (in rotation)	19	151	Peanut	30.6	134
Maize + common beans	14	128	Maize (residues burned)	29.0	131
Maize (continuous)	12	67	Maize + macuna bean (incorporated)	28.2	133
Maize + macuna bean (incorporated)	10	100	Sugarcane	21.0	88
Maize + manure	6.6	97	Maize + lime	19.1	96
Maize + macuna bean (mulched)	3.0	42	Maize + manure	8.9	62
Gordura grass	2.6	46	Jaragua grass	5.5	45

Source: Data from Smeets, E. et al. 2006. Sustainability of Brazilian bio-ethanol. Report NWS-E-2006-110 (ISBN 90-8672-012-9), Senternovem (The Netherlands Agency for Sustainable Development and Innovation), Utrecht, the Netherlands.

[a] Average rainfall = 1,347 mm per year.
[b] Average rainfall = 1,286 mm per year.

Even before 1985, expanding sugar cultivation was putting pressure on the area of cultivatable land in Brazil used for food crops—although crops grown for export (including soybeans and coffee, as well as sugar) were more likely than sugarcane to occupy land traditionally used for food crops. Thus, from 1976 to 1982[45]:

- Land used for basic food crops in Brazil increased from 27.5 to 29.4 million ha.
- Land for export crops increased from 13.4 to 15.6 million ha.
- Land on which sugarcane was grown increased from 2.6 to 3.9 million ha.

The real competition is therefore a triple one among food, fuel, and export crops or, more accurately, between land used for food, fuel, and export crops and land cleared by deforestation. This is particularly pertinent in Brazil, where newly cultivatable land may well have previously been virgin rain forest.

Available data on the social impacts of land clearance are old and official figures are elusive; any overall effects of ethanol production on land use are intertwined with

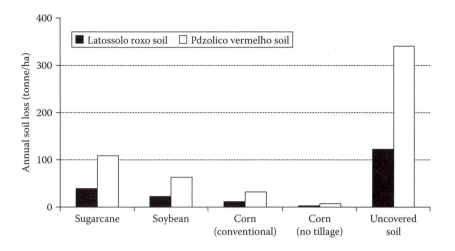

FIGURE 9.14 Soil erosion for two types of Brazilian soils. (1990 data from Smeets, E. et al. 2006. Report NWS-E-2006-110 [ISBN 90-8672-012-9], Senternovem [The Netherlands Agency for Sustainable Development and Innovation], Utrecht, the Netherlands.)

those of cash crops and social factors such as income distribution. Given the emotive nature of the subject matter, the debate tends to be acerbic:

> As it is currently developing, the Brazilian ethanol industry represents a direct challenge to food sovereignty and agrarian reform. Ethanol production to sustain the enormous consumption levels of the Global North will not lead the Brazilian countryside out of poverty or help attain food sovereignty for its citizens.[49]

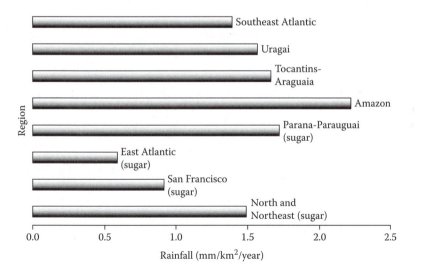

FIGURE 9.15 Annual rainfall in main sugar-producing and other regions of Brazil. (Data from Smeets, E. et al. 2006. Report NWS-E-2006-110 [ISBN 90-8672-012-9], Senternovem [The Netherlands Agency for Sustainable Development and Innovation], Utrecht, the Netherlands.)

On the other hand, to achieve poverty alleviation and the eradication of social exclusion and with support from environmentalists, Brazil proposed the Brazilian Energy Initiative at the 2002 World Summit on Sustainable Development (Johannesburg, South Africa) aiming at the establishment of global targets and time frames of minimum shares of energy from renewable sources.[50] Headline figures for the global numbers of malnourished people known to international agencies are another datum point with a large uncertainty: from below 1 billion to 3.7 billion.[39,51] As an economist from the Earth Policy Institute was quoted as saying: "The competition for grain between the world's 800 million motorists who want to maintain their mobility and its 2 billion poorest people who are simply trying to stay alive is emerging as an epic issue."[52]

Brazil became the global leader in ethanol exports in 2006, exporting 19% (3 billion L) of its production and plans to export 200 billion L annually by 2025, increasing sugarcane planting to cover 30 million ha.[49] Sugar for ethanol will increasingly be viewed by nations without a strong industrial base, but with suitable climatic conditions for sugarcane growth as a cash crop, in exactly the manner that Brazil regards coffee or soybeans. The example provided by Brazil in creating rural employment at low cost, reducing the economic burden of oil imports, and developing national industrial infrastructure will be one difficult to resist—especially if major sugar producers, including Brazil, India, Cuba, Thailand, South Africa, and Australia, unite to create an expanding alternative fuel market with sugar-derived ethanol.[46]

Economists and agronomists continue to call for an informed debate about land use in the context of increasingly large areas of highly fertile or marginal land being reallocated for energy crops.[53] Sugarcane-derived ethanol in Brazil shows the highest agricultural land efficiency in both replacing fossil energy for transportation and avoiding greenhouse gas emissions (Figure 9.16).[54] Impacts on acidification and

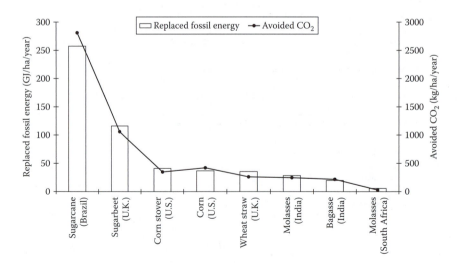

FIGURE 9.16 Potential fossil fuel replacements and CO_2 emission abatements with different fuel ethanol production technologies. (Data from von Blottnitz, H., and Curran, M. A. 2007. *Journal of Cleaner Production* 15:607.)

human and ecological toxicity, as well as deleterious environmental effects occurring mostly during the growing and processing of biomass, are more often ranked as unfavorable than favorable. Paradoxically, the principal economic drivers toward greater biofuel production in developing economies are precisely those widely accepted programs to reduce greenhouse gas emissions, increase energy security, and move to a scientifically biobased economy by promoting the use of biofuels.

9.3.3 Future Horizons for Cane Sugar Ethanol

Sugarcane ethanol is the leading first-generation biofuel and could herald the globalization of biofuels with an important split between the north (the United States, Canada, Europe, and Japan) and the south (Latin America, Africa, and South and Southeast Asia). The former increasingly has focused on the production of advanced biofuels (including hydrogen-based systems) while the south has been developing a long-term economic strategy based on sugar ethanol.[55] The premise for this radical split is that tropical countries "do it better" (i.e., can produce ethanol with positive energy gains) because they achieve at least twice the yield per hectare than can be demonstrated in temperate countries producing corn- or grain-derived ethanol. As locations for biodiesel crops, locales in the Southern Hemisphere simply cannot be bettered for land availability, climate, the types of plant species grown, or the yields extracted from them.

In this analysis, the status of India and China is ambiguous, straddling the boundary between agricultural and industrial nations: Both have ranked in the top 10 of nations with the least car ownership but now are in the top 10 of the world's largest economies. Like Brazil, India and China have the climate and rainfall (in some regions, at least, and seasonally) to establish large-scale ethanol and biodiesel programs, and both could include lignocellulosic ethanol in the range of feedstocks.

Based on the calculation that the creation of another 18 Brazils over the course of the next decade could produce sufficient biofuels to substitute for 20% of Organization for Economic Cooperation and Development (OECD) gasoline needs by 2020, the countries could include, in a swathe across the tropics: Brazil (doubling present national output), Argentina, Colombia, Ecuador, and Chile; Central American and Caribbean countries; Nigeria, Togo, Democratic Republic of Congo, Angola, Namibia, Ghana, Senegal, Gambia, Guinea, Togo, Mali, Chad, Zimbabwe, Zambia, Kenya, Swaziland, Rwanda, Uganda, Tanzania, and South Africa; Malaysia, Thailand, the Philippines, Indonesia, and Vietnam; Indian Ocean islands (Mauritius, etc.); China; and India.

Interrogating the UN land survey database reveals that large tracts of land have been classified as "potential arable land"—most obviously in sub-Saharan Africa and South and Central America (Table 9.4). In the heavily populated Asian Pacific region, considerable potential for finding unused land apparently exists. Skepticism can be expressed about the validity of this data use because the satellite imaging does not give full information about terrain: steepness of slope, land ownership and disputes, habitats for endangered species, etc.

In sub-Saharan Africa, as already discussed (Section 9.3.1), concerns are acute as to water sources and irrigation. The establishment of sugarcane ethanol production

TABLE 9.4
Global Land Use

Region	Actual Arable Land in Use (million hectares)	Potential Arable Land (million hectares)	Potential: Actually Used
Sub-Saharan Africa	157,608	1,109,851	7.0
South and Central America	143,352	1,028,473	7.2
Asia and the Pacific	477,706	777,935	1.6
Canada and the United States	233,276	479,632	2.1
Europe	213,791	384,220	1.8
North Asia	175,540	297,746	1.7

Source: Data from the U.N. Food and Agriculture Organization's Terrastat database.

on the scale of 20 additional Brazils would require only 40 million ha of land in suitable regions in tropical and semitropical zones in Africa, the Americas, and South and Southeast Asia—less than 0.002% of the potential arable land. A priority should therefore be to define precisely what area of land capable of rivaling the great fecundity of existing Brazilian sugarcane plantations could be accessed for ethanol production, commencing from the known major sugarcane producers and adjacent geographies (Figure 9.17).[56]

An extra 2,000 ethanol factories over the course of the next decade are easily assessed in technical and economic quantities: an output of 400 billion L of

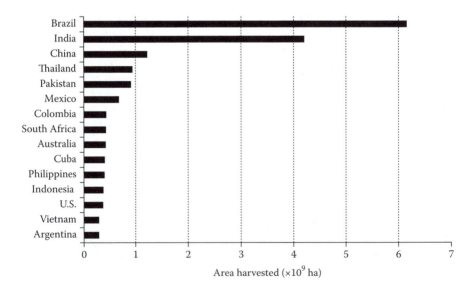

FIGURE 9.17 Major sugarcane producing nations. (Data from Goldemberg, J., and Guardabassi, P. 2009. *Energy Policy* 37:10.)

ethanol (and/or biodiesel), requiring investment of US$240 billion. Because the International Energy Agency expects $470 billion to be invested in the oil and gas industry in a single year by 2010, this sum is modest, but could only be undertaken by global investment funding agencies with OECD markets guaranteed by firm agreements.[57]

Climate change may significantly alter any predictions for Brazil as for elsewhere in the world because any alteration in rainfall patterns will have a great impact on areas presently or in future used for the cultivation of sugar, sweet sorghum, or other ethanol feedstock crops, as would hydroelectricity schemes and wind power turbine power generation.[58] If managed with due regard to ecological and environmental impacts, however, Southern Hemisphere nations have the potential to develop biofuels production processes; although their long-term sustainability can be questioned, inflammatory rhetoric might be avoided in the future:

> There has been much hyperbole voiced of late against biofuels. At the end of October [2007] the United Nations special rapporteur on the right to food, Jean Ziegler, stated at a press conference in New York that it was a crime against humanity to divert arable land to the production of crops which are then burned for fuel. These sentiments were then echoed by George Monbiot, in *The Guardian,* when he claimed, amongst other things, that biofuels could kill more people than the Iraq war.
>
> The oil lobby must be rubbing its hands with glee. Never in over a century of destructive use of fossil fuels have such charges been leveled against the internal combustion engine and the fossil fuels burnt that are actually creating the problem of global warming. Instead, it is the potentially clean substitutes that are attracting all the opprobrium.[59]

9.4   BIOENERGY CROPS AND GENETICALLY MANIPULATED PLANTS

Technical advances have been cited as essential to improve food and energy crop yields (Section 9.2.1). Wheat—the first biomass plant for cellulosic ethanol production—is grown as a monoculture and is inevitably subject to crop losses arising from pathogen infestation. Modern biotechnology and genetic manipulation offer novel solutions to the development of resistance mechanisms as well as yield improvements through increased efficiency of nutrient usage and tolerance to drought and other seasonal and unpredictable stresses.

But the deliberate release of any such genetic-manipulated (GM) species is contentious—highly so in Europe and India, where environmental campaigners are still skeptical that GM technologies offer any advantage over traditional plant breeding without the associated risks of monopoly positions adopted by international seed companies, the acquisition of desirable traits by weed species, and the horizontal transfer of antibiotic resistance genes to microbes.[60]

In contrast, the positive aspects of plant biotechnology have been succinctly expressed:

> Genetic transformation has offered new opportunities compared with traditional breeding practices since it allows the integration into a host genome of specific sequences leading to a strong reduction of the casualness of gene transfer.[61]

Because large numbers of insertional mutants have been collected in a highly manipulable model plant species (*Arabidopsis thaliana*), it has been possible for some years to inactivate any plant gene with a high degree of accuracy and certainty.

9.4.1 ENGINEERING RESISTANCE TRAITS FOR BIOTIC AND ABIOTIC STRESSES

After herbicide resistance in major crop species, the first target area for GM crop development was that of protection against plant pests.[61] Whereas tolerance of modern herbicides is usually located in the amino acid sequences of a handful of target genes in biosynthetic pathways, plants have multiple inducible mechanisms to fight back against microbial pathogens[62]:

- Permanent ("constitutive") expression of genes encoding chitinases or a ribosomal-inactivating protein confers partial protection against fungal attack.
- Enhancing lignin deposition in response to fungal or bacterial invasion is a possible multigene defense.
- Overexpressing genes encoding biosynthesis of phytoalexin antibiotics has been explored together with the introduction of novel phytoalexin pathways by interspecies gene transfer.
- Specific natural plant disease resistance genes are beginning to be identified and cloned for transfer into susceptible plants.

Wheat is prone to attack by the rust pathogens *Puccinia graminis* and *P. tritici;* stem and leaf rusts are considered to be major constraints to wheat production worldwide.[63] More immediately alarming from the perspective of any dependence on wheat straw biomass is that new and highly infectious variants of the pathogen have been noted in Africa for some years, and media reports in 2007 described its spread into Asia. A single gene (*Sr2*) has been identified as a broad-spectrum resistance locus for over 80 years; recently, this gene (or two tightly linked genes) conferred resistance and the associated dark pigmentation traits, pseudoblack chaff.[64] Stem rust-susceptible barley has been transformed into a resistant form by an *Agrobacterium* plasmid containing the barley resistance gene *Rpg1;* a single copy of the gene is sufficient to confer resistance against stem rust.[65]

Of the abiotic stresses that plants experience, drought is a serious limiting factor on growth and productivity even in the Northern Hemisphere, and cereal crops are highly prone to fluctuating yield depending on seasonal rainfall and average temperatures in the growing season. To be dependent on a monoculture crop such as wheat (for starch or straw) runs the risk of uncertain prices as well as variable feedstock availability.

With the evolution of domesticated cereal species over millennia, genetic diversity has been lost. Using the natural genetic diversity of wild species is an invaluable resource because wild types harbor very broad ranges of tolerance characteristics; other exploitable traits include those for salt tolerance (allowing saline water to be the source of irrigation) and—especially if global temperatures increase due to global warming—flowering times and other growth parameters more typical of Mediterranean regions for transfer to cultivars grown farther north.[66] Drought appears to exert physiological effects via oxidative stress signaling pathways, a property shared with freezing, heat, and salinity stresses. Protein kinases are often associated with signaling pathways, and expression of a protein kinase gene (*NPK1*) from tobacco in maize protects kernel weights when the water supply is reduced.[67]

9.4.2 BIOENGINEERING INCREASED CROP YIELD

How can *yield* be precisely defined with energy crops? This is both more flexible and less precise than, for example, crop yield as measured by grain size or weight per plant or per unit area. For a feedstock such as wheat straw, increasing plant biomass accumulation per plant may suffice. This implies a change in nutrient utilization or absorption from the soil, but may paradoxically reverse the trend toward "dwarf" crops (i.e., more grain, less stem/stalk) that has been emphasized in the green revolution type of agronomy.[68]

In principle, simply achieving higher leaf, stalk, and stem mass per plant is a straightforward target that is not limited by considerations of morphology for crops dedicated to energy supply and/or biofuels. Mass clonal propagation of commercial trees is now well advanced and gene transfer technologies have been devised for conifers and hardwoods; that is, forest biotechnology has emerged (probably irreversibly) out of the laboratory and into the global ecosystem.[69]

Traditional breeding and marker-assisted selection can identify genes involved in nutrient use efficiency that can then be used in gene transfer programs to improve features of plant nutrition. For crop plants in intensive agriculture, nitrogen assimilation and recycling within the plant over the stages of plant development are crucial.[70] Although plant biochemistry is increasingly well understood at the molecular level, what is much less clear is how to modulate gene expression (single genes or whole pathways) accurately to achieve harvestable yield increases.[71] As with other "higher" organisms, a greater understanding of how regulatory circuits and networks control metabolism at organ and whole-plant levels as an exercise in systems biology will be necessary before metabolic engineering for yield in crop plants becomes routine.[72]

Nevertheless, successes are now being reported for gene therapy with the goal of improving the assimilation of CO_2 into biomass—however this is defined:

- Transgenic rice plants with genes for phospho*enol*pyruvate carboxylase and pyruvate, *ortho*phosphate dikinase from maize, where the two enzymes are key to high photosynthetic carbon fixation under tropical conditions (C4 metabolism), increase photosynthetic efficiency and grain yield by up to 35% and have the potential to enhance stress tolerance.[73]
- Once CO_2 is fixed by green plants, some of the organic carbon is lost by respiratory pathways shared with microorganisms. Several studies report that partially disabling the oxidative pathways of glucose metabolism enhances photosynthetic performance and overall growth. For example, in transgenic tomato plants with targeted decreases in the activity of mitochondrial malate dehydrogenase, CO_2 assimilation rates increased by up to 11% and total plant dry matter by 19%.[74]

- Starch synthesis in developing seeds requires ADP-glucose phosphory-lase; expressing a mutant maize gene for this enzyme in wheat increases both seed number and total plant biomass. These effects are dependent on increased photosynthetic rates early in seed development.[75]

Much attention has been given to improving the catalytic properties of the primary enzyme of CO_2 fixation, ribulose 1,5-*bis*phosphate carboxylase (RuBisCO), the most abundant single protein on Earth and one with a chronically poor kinetic efficiency for catalysis. Although much knowledge has been garnered pertaining to the natural variation in RuBisCO's catalytic properties from different plant species and in developing the molecular genetics for gene transfer among plants, positive effects on carbon metabolism as a direct result of varying the amounts of the enzyme in leaves have proved very slow to materialize.[76]

A more radical approach offers far greater benefits. Accepting the inevitable side reaction catalyzed by RuBisCO (i.e., the formation of phosphoglyceric acid; Figure 9.18), transgenic plants were constructed to contain a bacterial pathway to recycle the lost carbon entirely inside the chloroplast rather than the route present in plant biochemistry that involved the concerted actions of enzyme in three plant cell organelles (the chloroplast, the mitochondrion, and peroxisome). Transgenic plants grew faster, produced more biomass (in shoot and roots), and had elevated sugar contents.[77]

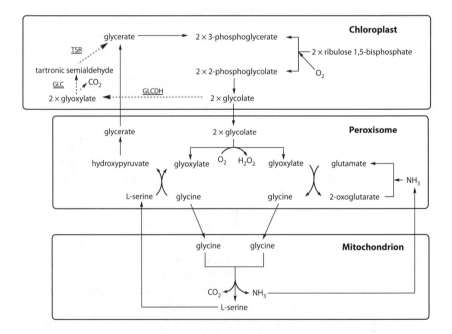

FIGURE 9.18 Intracellular C and N traffic for photorespiration transgenic bacterial glycolate catabolic pathways; the three bacterial gene-encoded enzymes (underlined) are GLCDH, glycolate dehydrogenase; GCL, glyoxylate carboxyligase; and TSR, tartronic semialdehyde reductase. (Modified from Kebeish, R. et al. 2007. *Nature Biotechnology* 25:593.)

Is there an upper limit to plant productivity? A temperate zone crop such as wheat is physiologically and genetically capable of much higher productivity and efficiency of converting light and CO_2 into biomass than can be achieved in a real environment (i.e., in hydroponics and with optimal mineral nutrition). Again, most studies focus on yield parameters such as grain yield (in mass per unit area); however, total leaf mass (a component of straw or stover) will also increase under such ideal conditions.[78]

The higher the light intensity is, the greater the plant response will be. However, natural environments have measurable total hours of sunshine per year, and the climate imposes average, minimum, and maximum rainfall and temperatures. Although supplementary lighting is expensive, with unlimited renewable energy resources substantial increases in plant productivity are theoretically possible. From a shorter-term perspective, however, the choice of biomass refuses to be erased from the agenda, and agricultural wastes simply cannot compete with the best energy crops. For example, using biofuel yield as the metric, the following ranking can be computed from relative annual yields (liters per hectare)[79]:

Corn stover (1.0) < poplar (2.9) < switchgrass (3.2) < elephant grass (4.4)

Rather than mobilizing molecular resources against the vagaries of climate, concentrating effort on maximizing biomass supply from a portfolio of crops other than those most presently abundant would pay dividends. Over-reliance of a few species (vulnerable to pests and climatic variability) would be minimized by expanding the number of such crops.

9.4.3 OPTIMIZING TRAITS FOR ENERGY CROPS INTENDED FOR BIOFUEL PRODUCTION

With a lignocellulosic platform for ethanol production, one obvious target is (as just discussed) the management of energy crop productivity to maximize the capture of solar energy and atmospheric CO_2. The chemical composition of the biomass is of great practical significance for the industrial bioprocessing of feedstocks[80,81]:

- developing crop varieties with reduced lignin contents (especially with softwoods)
- crops with increased cellulose (and, arguably) hemicellulose contents
- plants with the increased capability to degrade cellulose, hemicellulose, and lignin—*after harvest* (i.e., in a controlled manner capable of minimizing biomass pretreatment)

Of these, modifying lignin content has been the most successful. Classical genetics suggests that defining quantitative traits and their genetic loci is relatively easy and that (even better) some of these loci are those for *increased* cellulose biosynthesis.[82] As collateral, there is the confidence-building conclusion that lignin contents of commercial forest trees have been reduced to improve pulping for the paper industry; the genetic fine-tuning of lignin content, composition, or both is now technically feasible.[83]

Reductions in plant lignin content have been claimed using both single- and multiple-gene modifications (Figure 9.19):

- Downregulating either of the initial two enzymes of lignin biosynthesis—phenylalanine ammonia lyase and cinnamate 4-hydroxylase (C4H)—reduces lignin content and impairs vascular integrity in the structural tissues of plants.[84]
- Deletion of the second activity of the bifunctional C4H enzyme, coumarate 3-hydroxylase, results in reduced lignin deposition.[85]
- Later enzymes in the lignin pathway were considered to be less amenable for inhibiting lignification, but multiple gene downregulation could be effective.[85,86]
- Inactivating O-methyltransferase activity with an aspen gene incorporated into a transmissible plasmid in the antisense orientation reduced lignin formation by 28% in *Leucaena leucocephala* (a thornless tree included in the list of the 100 worst invaders that forms dense thickets and is difficult to eradicate once established). This genetic change also increased monomeric phenolic levels and increased the cellulose content by 9%, but did not visibly affect the plant phenotype.[87]

Are "lignin light" plants biologically viable for commercial cultivation? Altered stem lignin biosynthesis in aspen has a large effect on plant growth, reducing total leaf area and resulting in 30% less total carbon per plant; root growth was also compromised.[88] Vascular impairment can lead to stunted growth.[85] On the other hand, aspen wood in reduced-lignin transgenics was mechanically strong because less lignin was compensated for by increased xylem vessel cellulose.[86]

Smaller plants may be grown, as energy crops, in denser plantations; alternatively, plants with reduced stature may be easier to harvest, and various practical compromises between morphology and use can be imagined. This can be seen as analogous to the introduction of dwarfing rootstocks for fruit trees that greatly reduced plant height and canopy spread and facilitated manual and mechanical harvesting.

Also without obvious effects on plant growth and development was the introduction and heterologous expression in rice of the gene from *Acidothermus cellulolyticus* encoding a thermostable endo-1,4-β-glucanase. This protein constituted approximately 5% of the total soluble protein in the plant and was used to hydrolyze cellulose in ammonia fiber explosion-pretreated rice and maize.[89]

More ambitiously, enzymes of polysaccharide depolymerization are being actively targeted by plant biotech companies for new generations of crops intended for biofuels. A large number of genes are under strict developmental stage-specific transcriptional regulation for wood formation in species such as hybrid aspen; at least 200 genes are of unknown function, possibly undefined enzymes and transcription factors. This implies that heterologous glucanases and other enzymes could be produced during plant senescence to provide lignocellulose processing *inside the plant* before the preparation of substrates for ethanol fermentation.[90]

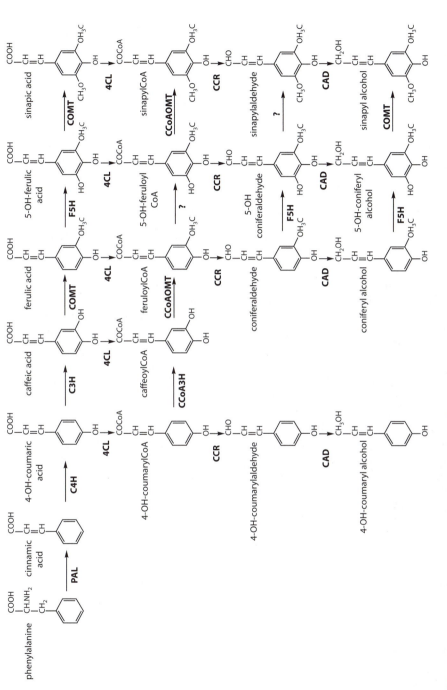

FIGURE 9.19 Outline of biosynthesis of lignin precursors: PAL, phenylalanine ammonia lyase; C4H, cinnamate 4-hydroxylase; C3H, 4-coumarate 3-hydroxylase; COMT, caffeate O-methyltransferase; CCoAOMT, caffeoyl-CoA O-methyltransferase; CCR, cinnamoyl-CoA reductase; CAD, cinnamyl alcohol dehydrogenase; F5H, ferulate 5-hydroxylase. (After Hertzberg, M. et al. 2001. *Proceedings of the National Academy of Sciences USA* 98:14732.)

9.4.4 GENETIC ENGINEERING OF DUAL-USE FOOD PLANTS AND DEDICATED ENERGY CROPS

The commercialization of GM technologies for staple food crops such as wheat has faced the obstacles of social and market opposition and resistance.[91] Fortunately (or fortuitously), gene transformation techniques for improving wheat yield have been challenging for plant biotechnology. Bread wheat (*Triticum aestivum* L.) has one of the largest and most complex plant polyploid genomes, 80% of which are noncoding sequences deriving from three ancestral genomes. New initiatives to analyze the minority expressed portion of the wheat genome are ongoing, but the structural complexity of the genome has been of enormous value to agronomy over millennia because major chromosomal rearrangements and deletions are well tolerated.[92]

STEM TOPIC 9.3: THE BIOTECHNOLOGY OF GM PLANTS

Plant biotechnology was born suddenly and dramatically in the early 1980s. Many commentators have dated the key event as the issuing of US Patent 4,940,835 (filed on July 7, 1986) to the Monsanto Company, Saint Louis (Missouri). The patent described the means of genetically reprogramming plants to express a gene encoding an herbicide-resistant enzyme that could be directed to its correct subcellular site in the cytoplasm. This technical and scientific achievement was the culmination of decades of work on plant biochemistry and understanding the molecular details of a plant disease.

The soil bacterium *Agrobacterium tumefaciens* is the causative agent of the crown gall disease of dicotyledonous plants, a condition that resembles neoplasm in animals but is linked to disturbances in plant hormone concentrations. The bacterium harbors a large plasmid (Ti) that encodes the crown gall tumorigenic genes and is stably incorporated in the plant chromosome; both desirable genes and genetic markers for antibiotic resistance can be readily inserted into disabled Ti plasmids (incapable of inducing grown gall symptoms). When incorporated in the host plant, foreign genes are stably expressed—hence the "Frankenstein food" epithet if the host plant species is, like soybean, destined for the human food chain.

The essential steps of US Patent 4,940,835[93] are shown in Figure STEM 9.3.

Public perceptions may be more favorable to GM technologies applied to dedicated energy crops, although even here the fear is that of the introduction and spread of unwanted genes into natural populations with consequences that are difficult to predict accurately. Geographical isolation of energy crops (as of GM plants designed to synthesize high-value biopharmaceuticals) is one extreme solution, but flies in the face of the limited land availability for nonfood crops. Even if the entire U.S. corn and soybean crop were to be devoted to ethanol and biodiesel production (with complete elimination of all direct and indirect food uses), only 5.3% of the 2005 gasoline and diesel requirement could be met.[94]

Step 1: Generate A Glyphosate-Resistant Enzyme

1. Plant cells growing in suspension culture
2. Gradual adaptation of cells to increasing glyphosate concentration
3. Isolation of highly resistant cell line
4. Purify glyphosate-resistant enzyme

⇩

Step 2: Construct A Chimeric Gene

1. Isolate gene encoding glyphosate-resistant enzyme
2. Remove nature gene promotor
3. Add strong promotor from Cauliflower Mosaic Virus 35S

⇩

Step 3: Construct A Ti Plasmid

1. Insert chimeric gene into disarmed Ti plasmid
2. Insert plasmid into *Agrobacterium tumefaciens*
3. Use bacterial cells to transform plants

⇩

Step 4: Isolate Glyphosate-Resistant Cells

1. Grow transformed cells in the presence of glyphosate
2. Select herbicide-resistant cells
3. Regenerate whole plants
4. Test plants for herbicide resistance

FIGURE STEM 9.3 Outline of Monsanto's procedure for generating glyphosate-resistant plants. (Modified from US Patent 4,940,835.)

Expressed another way, an area of land nearly 20-fold larger than that presently used for corn and soybean cultivation would be required for ethanol and biodiesel energy crops. Concerns relating to land delineation are highly probable even if a wheat straw/corn stover bioeconomy is used as the main supply of feedstocks for ethanol production.[95]

Equally inevitable, however, is that GM approaches will be applied to dual food–energy crops or to dedicated energy crops such as fast-growing willow and switchgrass. The USDA-ARS Western Regional Research Center (Albany, California) has created a gene inventory of nearly 8,000 gene clusters in switchgrass, 79% of which are similar to known protein or nucleotide sequences.[96] A plasmid system has also been developed for the transformation of switchgrass containing an herbicide resistance gene.[97]

9.5 SUMMARY

Estimates of land used for growing the most productive ethanol feedstock (sugarcane) led to the conclusion that substantial areas of the world's cultivable land would inevitably be dedicated to bioenergy. Fortunately, many other forms of alternative energy have been or are being actively developed; correspondingly, the global land area required for bioenergy to replace oil as a transportation fuel is much lower—as low as 15% with a highly efficient cellulosic feedstock industry for biofuels.

For the United States, a national estimate of plant biomass predicts a total availability in excess of 1 billion tons annually by the middle of the twenty-first century, with achievable increases in crop yield and some changes in land use that would minimize or eliminate "food versus fuel" issues. Extrapolating this conclusion to the rest of the world's agriculture and forestry resources is more difficult because land uses and suitability are more speculative; however, large amounts of potentially arable land have been claimed to be available.

The real conflicts in land use in developing economies are not solely between biofuels and food crops but also between cash crops grown for export, food crops (including those grown on a subsistence basis), and bioenergy crops. This triangle of interests will require careful management socially and politically and has major implications for ecology and the environment.

The sustainability of such massive bioenergy plantations has been questioned: The energy inputs necessary to grow and process the plant biomass with adequate replenishment of soil nutrients are high, and bioenergy based on plants suffers the same problems as does any intensive, energy-dependent agriculture. This topic requires much careful and detailed analysis in a variety of climates, soil types, and bioenergy crop scenarios.

The growing consensus that cellulosic biofuels offer the best means of ameliorating global CO_2 emissions stalled in 2008 when computations of CO_2 emissions caused by land clearance for feedstock crop planting were presented. Based on this rationale, the resulting surge in atmospheric CO_2 would result in a carbon debt that would require over a century to pay back by the successful implementation of cellulosic ethanol production. Although these results have been challenged, they remain an urgent focus for further analysis and modeling studies.

Less contentious is the conclusion that, although water use by irrigation has been minor for biofuels production, it will inevitably increase with augmented biofuels production. This is particularly important with sugarcane ethanol production, which is expanding from Brazil to become a major industry around the globe in tropical and semitropical regions. Localized water shortages and conflicts are likely to be exacerbated by predicted changes in rainfall patterns and as other climatic factors, such as global climate changes, unfold in the remainder of the twenty-first century.

Climate change and increased atmospheric CO_2 levels may actually boost plant productivity for some decades, but precise estimates remain elusive and are complicated by effects of region, altitude, and crop type. Inevitably, with a growing world population and challenging changes to the global ecosystem, technical advances in developing beneficial traits in major crop and other plant species to cope with biotic (plant pathogens) and abiotic (in particular, drought) stress will require greatly accelerated classical and/or modern plant biotechnology R&D. For nonfood crops, genetically modified plants may find vastly expanded uses and acceptability as aids to maximize the efficiency of photosynthetic carbon sequestration.

REFERENCES

1. Hill, J. et al. 2006. Environmental, economic, and energetic costs and benefits of biodiesel and ethanol biofuels, *Proceedings of the National Academy of Sciences USA* 103:11206.
2. Natural Resources Canada. 2004. *Economic, financial, social analysis and public policies for fuel ethanol phase 1.* Ottawa.
3. Demain, A. L., Newcomb, M., and Wu, J. H. D. 2005. Cellulase, clostridia, and ethanol. *Microbiology and Molecular Biology Reviews* 69:124.
4. Kheshgi, H. S., Prince, R. C., and Marland, G. 2000. The potential of biomass fuels in the context of global climate change: focus on transportation fuels. *Annual Review of Energy and the Environment* 25:199.
5. Putsche, V., and Sandor, D. 1996. Strategic, economic, and environmental issues for transportation fuels. In *Handbook on bioethanol: Production and utilization,* ed. Wyman, C. E., chap. 2. London: Taylor & Francis.
6. International Energy Agency. 2006. *World energy outlook,* annex A. Paris.
7. Torney, F. et al. 2007. Genetic engineering approaches to improve bioethanol production from maize. *Current Opinion in Biotechnology* 18:193.
8. Perlack, R. D. et al. 2005. Biomass as a feedstock for a bioenergy and bioproducts industry: The technical feasibility of a billion-ton annual supply. U.S. Department of Energy, April 2005. Available at http.//www.osti.gov/bridge.
9. Kadam, K. L. et al. 2000. Softwood forest thinnings as a biomass source for ethanol production: A feasibility study for California. *Biotechnology Progress* 16:947.
10. Wingren, A., Galbe, M., and Zacchi, G. 2003. Techno-economic evaluation of producing ethanol from softwood: Comparison of SSF and SHF and identification of bottlenecks. *Biotechnology Progress* 19:1109.
11. Maness, T. 2008. Forests as a potential feedstock for cellulosic ethanol. U.S. Forest Service, available from http://www.safnet.org/fp/documents/feedstock_for_cellulosic_ethanol_08.pdf.
12. McLaughlin, S. B. et al. 2002. High-value renewable energy from prairie grasses. *Environmental Science & Technology* 36:2122.
13. Müller, S. 2007. Sunshine, sand and soybean king. Brazil's rise to agricultural superpower. *BioForum Europe* 5(11):17.
14. Ramos, L. P., and Wilhelm, H. M. 2005. Current status of biodiesel development in Brazil. *Applied Biochemistry and Biotechnology* 121–124:807.
15. Searchinger, T. et al. 2008. Use of U.S. croplands for biofuels increases greenhouse gases through emissions from land-use change. *Science* 319:1238.
16. Hughes, J. D. 2001. *An environmental history of the world,* chap. 8. New York: Routledge.
17. Mathews, J. A., and Tan, H. 2009. Biofuels and indirect land use change effects: The debate continues. *Biofuels, Bioproducts and Biorefining* 3:305.

18. Gibbs, H. K. et al. 2008. Carbon payback times for crop-based biofuel expansion in the tropics: The effects of changing yield and technology. *Environmental Research Letters* 3:034001.

19. Fargione, J. et al. 2008. Land clearing and the biofuel carbon debt. *Science* 319:1235.

20. Fulton, L., Howes, T., and Hardy, J. 2004. Biofuels for transport. An international perspective. International Energy Agency, Paris. http://www.iea.org/textbase/nppdf/free/2004/biofuels2004.pdf.

21. Department for Environment, Food and Rural Affairs. 2008. Review of work on the environmental sustainability of international biofuels production and use. Report by AEA Technology plc, London.

22. Fésüs, G. et al. 2008. Regions 2020. Demographic challenges for European regions. Commission of the European Communities. Brussels.

23. Jordaan, S. M. 2007. Ethical risks of attenuating climate change through new energy systems: The case of a biofuel system. *Ethics in Science and Environmental Politics* 7:23.

24. Patzek, T. W. 2004. Thermodynamics of the corn-ethanol biofuel cycle. *Critical Reviews in Plant Sciences* 23:519.

25. Patzek, T. W., and Pimentel, D. 2005. Thermodynamics of energy production from biomass. *Critical Reviews in Plant Sciences* 24:327.

26. Holdren, J. P., and Ehrlich, P. R. 1974. Human population and the global environment. *American Scientist* 62:282.

27. Lal, R. 2005. World crop residues production and implications of its use as a biofuel. *Environment International* 31:575.

28. Mathews, J. A. 2008. Carbon-negative biofuels. *Energy Policy* 36:940.

29. Gunnarsson, C. C., and Petersen, C. M. 2007. Water hyacinths as a resource in agriculture and energy production: A literature review. *Waste Management* 27:117.

30. Swain, M. R. et al. 2007. Ethanol fermentation of mahula (*Madhuca latifolia* L.) flowers using free and immobilized yeast (*Saccharomyces cerevisiae*). *Microbiological Research* 162:93.

31. Downing, M., McLaughlin, S., and Walsh, M. 1995. Energy, economic, and environmental implications of production of grasses as biomass feedstocks. In *Proceedings of the Second Biomass Conference of the Americas: Energy, Environment, Agriculture, and Industry*, National Renewable Energy Laboratory, Golden, Colorado, p. 288.

32. Graham, R. L., Downing, M., and Walsh, M. E. 1996. A framework to assess regional environmental impacts of dedicated energy crop production. *Environmental Management* 20:475.

33. Sanderson, M. A. et al. 2001. Nutrient movement and removal in a switchgrass biomass-filter strip system treated with dairy manure. *Journal of Environmental Quality* 30:210.

34. Nelson, R. G., Ascough, J. C., and Langemeier, M. R. 2006. Environmental and economic analysis of switchgrass production for water quality improvement in northeast Kansas. *Journal of Environmental Management* 79:336.

35. Nelson, R. G. et al. 2004. Methodology for estimating removable quantities of agricultural residues for bioenergy and bioproduct use. *Applied Biochemistry and Biotechnology* 113–116:13.

36. Arevalo, C. B. M. et al. 2007. Development and validation of aboveground biomass estimations for four *Salix* clones in central New York. *Biomass Bioenergy* 31:1.

37. Lobell, D. B., and Ortiz-Monasterio, J. I. 2007. Impacts of day versus night temperatures on spring wheat yields. A comparison of empirical and CERES model predictions in three locations. *Agronomy Journal* 99:469.

38. Lobell, D. B., and Field, C. B. 2007. Global scale climate-crop yield relationships and the impacts of recent warming. *Environmental Research Letters* 2:014002.

39. Cassman, K. G. 2007. Climate change, biofuels, and global food security. *Environmental Research Letters* 2:011002.

40. Peng, S. et al. 2004. Rice yields decline with higher night temperature from global warming. *Proceedings of the National Academy of Sciences USA* 101:9971.

41. Intergovernmental Panel on Climate Change. 2007. Climate change 2007: Climate change impacts, adaptation and vulnerability. Summary for policymakers (available at: http://ipcc.ch/SPM6avr07.pdf).

42. Hoogeveen, J., Faures, J.-M., and van de Giessen, N. 2009. Increased biofuel production in the next decade: To what extent will it affect global freshwater resources. *Irrigation and Drainage* 54:148.

43. Energy and Transport Branch, Division for Sustainable Development, United Nations Department of Economic and Social Affairs. 2007. Small-scale production and use of liquid biofuels in sub-Saharan Africa: Perspectives for sustainable development. Background paper no. 2 DESA/DSD/2007/2, New York.

44. Goldemberg, J. 2008. The Brazilian biofuels industry. *Biotechnology for Biofuels* 1:6.

45. Geller, H. S. 1985. Ethanol fuel from sugar cane in Brazil. *Annual Review of Energy* 10:135.

46. Moreira, J. R., and Goldemberg, J. 1999. The alcohol program. *Energy Policy* 27:229.

47. Gunkel, G. et al. 2007. Sugar cane industry as a source of water pollution—Case study on the situation in Ipojuca River, Pernambuco, Brazil. *Water Air & Soil Pollution* 180:261.

48. Smeets, E. et al. 2006. Sustainability of Brazilian bio-ethanol. Report NWS-E-2006-110 (ISBN 90-8672-012-9), Senternovem (The Netherlands Agency for Sustainable Development and Innovation), Utrecht, the Netherlands.

49. Kenfield, I. 2007. Brazil's ethanol plan breeds rural poverty, environmental degradation. International Relations Center, Silver City (New Mexico), March 6, 2007(available from: http://americas.irc-online.org/am/4049).

50. Goldemberg, J., Coelho, S. T., and Lucon, O. 2004. How adequate polices can push renewables. *Energy Policy* 32:1141.

51. Pimentel, D., Patzek, T., and Cecil, G. 2007. Ethanol production: energy, economic, and environmental losses. *Reviews of Environmental Contamination & Toxicology* 189:25.

52. Brown, L. R., quoted in Daniel Howden. 2007. The big green fuel lie. *The Independent* (London) March 5, 2007: http://.independent.co.uk/environment/climate-change/the-big-green-fuel-lie-438937.html.

53. Hill, J. 2007. Environmental costs and benefits of transportation biofuel production from food- and lignocellulose-based energy crops. A review. *Agronomy for Sustainable Development* 27:1.

54. von Blottnitz, H., and Curran, M. A. 2007. A review of assessments conducted on bio-ethanol as a transportation fuel from a net energy, greenhouse gas, and environmental life-cycle perspective. *Journal of Cleaner Production* 15:607.

55. Mathews, J. A. 2007. Biofuels: What a biopact between North and South could achieve. *Energy Policy* 35:3550.

56. Goldemberg, J., and Guardabassi, P. 2009. Are biofuels a feasible option? *Energy Policy* 37:10.

57. Mathews, J. A. 2008. Biofuels, climate change and industrial development: can the tropical South build 2000 biorefineries in the next decade? *Biofuels, Bioproducts and Biorefining* 2:203.

58. Frossard Pereira de Lucenaa, A. et al. 2009. The vulnerability of renewable energy to climate change in Brazil. *Energy Policy* 37:879.

59. Mathews, J. A. 2008. Opinion: Is growing biofuel crops a crime against humanity? *Biofuels, Bioproducts and Biorefining* 2:97.

60. Paula, L., and Birrer, F. 2006. Including public perspectives in industrial biotechnology and the biobased economy. *Journal of Agricultural and Environmental Ethics* 19:253.
61. Lanfranco, L. 2003. Engineering crops, a deserving venture. *Rivista di Biologia* 96:31.
62. Lamb, C. J. et al. 1992. Emerging strategies for enhancing crop resistance to microbial pathogens. *Biotechnology* 10:1436.
63. Khan, R. R. et al. 2005. Molecular mapping of stem and leaf rust resistance in wheat. *Theories in Applied Genetics* 111:846.
64. Kota, R. et al. 2006. Fine genetic mapping fails to dissociate durable stem rust resistance gene *Sr2* from pseudo-black chaff in common wheat (*Triticum aestivum* L.). *Theoretical and Applied Genetics* 112:492.
65. Horvath, H. et al. 2003. Genetically engineered stem rust resistance in barley using the *Rpg1* gene. *Proceedings of the National Academy of Sciences USA* 100:364.
66. Ellis, R. P. et al. 2000. Wild barley: A source of genes for crop improvement in the 21st century? *Journal of Experimental Botany* 51:9.
67. Shou, H., Bordallo, P., and Wang, K. 2004. Expression of the *Nicotiana* protein kinase (NPK1) enhanced drought tolerance in transgenic maize. *Journal of Experimental Botany* 55:1013.
68. Nagano, H. et al. 2005. Genealogy of the "green revolution" gene in rice. *Genes & Genetic Systems* 80:351.
69. Merkle, S. A., and Dean, J. F. 2000. Forest tree biotechnology. *Current Opinion in Biotechnology* 11:298.
70. Good, A. G., Shrawat, A. K., and Muench, D. G. 2004. Can less yield more? Is reducing nutrient input into the environment compatible with maintaining crop production? *Trends in Plant Science* 9:597.
71. Sinclair, T. R., Purcell, L. C., and Sneller, C. H. 2004. Crop transformation and the challenge to increase yield. *Trends in Plant Science* 9:70.
72. Carrari, F. et al., Engineering central metabolism in crop species: Learning the system. *Metabolic Engineering* 5:191.
73. Ku, M. S. et al. 2001. Introduction of genes encoding C4 photosynthesis enzymes into rice plants: Physiological consequences. *Novartis Foundation Symposium* 236:100.
74. Nunes-Nesi, A. et al. 2005. Enhanced photosynthetic performance and growth as a consequence of decreasing mitochondrial malate dehydrogenase activity in transgenic tomato plants. *Plant Physiology* 137:611.
75. Smidansky, E. D. et al. 2007. Expression of a modified ADP-glucose pyrophosphorylase large subunit in wheat seeds stimulates photosynthesis and carbon metabolism. *Planta* 225:965.
76. Parry, M. A. et al. 2003. Manipulation of Rubisco: The amount, activity, function and regulation. *Journal of Experimental Botany* 54:1321.
77. Kebeish, R. et al. 2007. Chloroplastic photorespiratory bypass increases photosynthesis and biomass production in *Arabidopsis thaliana*. *Nature Biotechnology* 25:593.
78. Bugbee, B. G., and Salisbury, F. B. 1988. Exploring the limits of crop productivity. I. Photosynthetic efficiency of wheat in high irradiance environments. *Plant Physiology* 88:869.
79. Sanderson, K. 2006. A field in ferment. *Nature* 444:673.
80. McLaren, J. S. 2005. Crop biotechnology provides an opportunity to develop a sustainable future. *Trends in Biotechnology* 23:339.
81. Sticklen, M. 2006. Plant genetic engineering to improve biomass characteristics for biofuels. *Current Opinion in Biotechnology* 17:315.
82. Cardinal, A. J., Lee, M., and Moore, K. J. 2003. Genetic mapping and analysis of quantitative loci affecting fiber and lignin content in maize. *Theories in Applied Genetics* 106:866.
83. Baucher, M. et al. 2003. Lignin: Genetic engineering and impact on pulping. *Critical Reviews in Biochemistry and Molecular Biology* 38:305.

84. Anterola, A. M., and Lewis, N. G. 2002. Trends in lignin modification: A comparative analysis of the effects of genetic manipulations/mutations on lignification and vascular integrity. *Phytochemistry* 61:221.

85. Abbott, J. C. et al. 2002. Simultaneous suppression of multiple genes by single transgenes. Down-regulation of three unrelated lignin biosynthetic genes in tobacco. *Plant Physiology* 128:844.

86. Laigeng, L. et al. 2003. Combinatorial modification of multiple lignin traits in trees through multigene cotransformation. *Proceedings of the National Academy of Sciences USA* 100:4939.

87. Rastogi, S., and Dwivedi, U. N. 2006. Down-regulation of lignin biosynthesis in transgenic *Leucaena leucocephala* harboring *O*-methyltransferase gene. *Biotechnology Progress* 22:609.

88. Hancock, J. E. et al. 2007. Plant growth, biomass partitioning and soil carbon formation in response to altered lignin biosynthesis in *Populus tremuloides. New Phytologist* 173:732.

89. Oraby, H. et al. 2007. Enhanced conversion of plant biomass into glucose using transgenic rice-produced endoglucanase for cellulosic ethanol. *Transgenic Research* 16:739.

90. Hertzberg, M. et al. 2001. A transcriptional roadmap to wood formation. *Proceedings of the National Academy of Sciences USA* 98:14732.

91. Bhalla, P. L. 2006. Genetic engineering of wheat—Current challenges and opportunities. *Trends in Biotechnology* 24:305.

92. Francki, M., and Appels, R. 2002. Wheat functional genomics and engineering crop improvement. *Genome Biology* 3:1013.1.

93. Shah, D. M. et al. 1990. Glyphosate-resistant plants. U.S. Patent 4, 940, 835 awarded July, 10, 1990.

94. Hill, J. et al. 2006. Environmental, economic, and energetic costs and benefits of biodiesel and ethanol biofuels. *Proceedings of the National Academy of Sciences USA* 103:11206.

95. Torney, F. et al. 2007. Genetic engineering approaches to improve bioethanol production from maize. *Current Opinion in Biotechnology* 18:193.

96. Tobias, C. M. et al. 2005. Analysis of expressed sequence tags and the identification of associated short tandem repeats in switchgrass. *Theoretical and Applied Genetics* 111:956.

97. Somleva, M. N. 2006. Switchgrass (*Panicum virgatum* L.). *Methods in Molecular Biology* 344:65.

10 Biofuels as Products of Integrated Bioprocesses (Biorefineries)

10.1 INTRODUCTION

Corn ethanol factories are beginning to be renamed as "biorefineries." What is a biorefinery and how is it related conceptually to an oil refinery? Does it relate to biotech industry or industrial chemistry?

Are biorefineries a stage in the development of the biocommodity economy to replace oil refineries or will both types of refinery find a synergism?

In an era of fluctuating oil prices and questions about oil reserves and their ownership, is the timeline of oil depletion the best argument—sidelining arguments on climate change agendas—for massive investment in coproduction of biofuels and biochemicals in multiple sites using biomass feedstocks from surrounding catchment regions?

10.2 THE BIOREFINERY CONCEPT

As a neologism, "biorefinery" was coined in the early 1990s by Charles A. Abbas of the Archers Daniel Midland Company (Decatur, Illinois) from the practices implicit in the fractionation of corn and soybean; the wet milling process of corn was an excellent example of an ancestral biorefinery (Figure 1.20, Chapter 1). Certainly by the late 1990s, the word (or, in an occasional variant usage, "biomass refinery") was becoming increasingly popular.[1]

The concept has carried different meanings according to the user, but the central proposition has been that of a comparison with the petrochemical refinery that produces not only gasoline and other conventional fuels but also petrochemical feedstock compounds for the chemical industry. On this formal analogy, the fuels from a biorefinery would include ethanol, biodiesel, biohydrogen, and/or syngas products, and the range of fine chemicals is potentially enormous, reflecting the spectrum of materials that bacterial metabolism can fashion from carbohydrates and other monomers present in plant polysaccharides, proteins, and other macromolecules (Figure 10.1).

The capacity to process biomass material through to a mixture of products (including biofuels) for resale distinguishes a biorefinery from, for example, a traditional fermentation facility manufacturing acids, amino acids, enzymes, or antibiotics; from industrial sites that may use plant-derived inputs (corn steep liquor, soybean oil, soy protein, etc.); or from either of two modern polymer processes

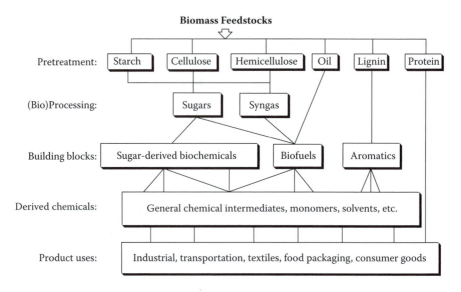

FIGURE 10.1   Material flow in an idealized biorefinery.

producing outputs from biomass resources that are often discussed in the context of biorefineries:

- Cargill Dow LLC's patented process for polylactic acid (PLA, "Natureworks PLA"), pioneered at a site in Blair, Nebraska, was the first commodity plastic to incorporate the principles of reduced energy consumption, waste generation, and emission of greenhouse gases and was awarded the 2002 presidential green chemistry award.[2]
- 1,3-Propanediol (1,3-PD) produced from glucose by highly genetically engineered *Escherichia coli* carrying genes from baker's yeast and *Klebsiella pneumoniae* in a process developed by a DuPont/Tate & Lyle joint venture is a building block for the polymethylene terphthalate polymers used in textile manufacture.[3]

However good these examples of the use of modern biotechnology to support generate or bulk chemistry industries are, they center on single-product fermentations (for lactic acid and 1,3-PD, respectively) that are not significantly different from many earlier bacterial bioprocesses. In particular, lactic acid has a very long history as a microbial ingredient of yogurts. It is used in the food industry to control pH, add flavor, and control microbial growth in products as diverse as alcoholic beverages, frozen desserts, and processed meat; the lactic acid production sector has major manufacturers in China, the United States, and Europe that utilize bacteria or *Rhizopus* molds in large-scale fermentations.

The following three definitions for biorefineries focus on the multiproduct, (usually) biofuel-associated nature of the envisaged successors to fossil-based units:

1. The U.S. Department of Energy: "A biorefinery is an overall concept of a processing plant where biomass feedstocks are converted and extracted into a spectrum of valuable products."[4]
2. The National Renewable Energy Laboratory: "A biorefinery is a facility that integrates biomass conversion processes and equipment to produce fuels, power, and chemicals from biomass. The biomass concept is analogous to today's petroleum refineries, which produce multiple fuels and products from petroleum."[5]
3. "Third generation (generation-III) and more advanced biorefineries...will use agricultural or forest biomass to produce multiple product streams, for example, ethanol for fuels, chemicals, and plastics."[1]

The concept provides a fascinating insight into how biofuels production facilities could develop in the coming decades as stepping stones toward the global production of chemical intermediates from biomass resources if lignocellulosic ethanol fails to meet commercial targets or if other developments (e.g., the successful emergence of a global hydrogen economy) render liquid biofuels such as ethanol and biodiesel short-lived experiments in bioindustrial innovation. The scale of chemical endeavor possible from biomass resources, as promulgated by enthusiasts for biorefineries, is arithmetically impressive[6]:

- By 2040, a world population of 10 billion could be supported by 2 billion ha of land for food production, leaving 800 million ha for nonfood crops.
- With a modest increase in agricultural productivity to 40 tonnes per hectare per year, land surplus to food production could yield 32 billion tonnes per year.
- Adding in 12 billion tonnes annually from forests and other agricultural waste streams yields a theoretical 50 billion tonnes.

Of this total, only 1 billion tonnes would be required to generate all the organics required as chemical feedstocks—leaving the rest for biofuels, including traditional biomass as a direct source of power and heat. These optimistic predictions, however, must be tempered with realism as to precisely how much land can be devoted to biorefinery feedstocks with conflicting requirements for food and with uncertainties as to land availability and suitability for intensive agriculture (Chapter 9, Section 9.2.2).

Calculations prepared from and for German industry show that only agricultural waste (i.e., cereal straw) could match the total demand for E10:gasoline blends as well as all the ethylene manufactured for the national chemical and plastics industries plus surplus ethanol for use in E85 blends and other chemical uses (Figure 10.2).[7]

The following are included in the range of roles proposed for biorefineries, as codified by biorefinery.nl, which is the umbrella organization in the Netherlands tasked with developing strategic aspects of biorefineries in Europe:

- primary processing units for waste streams from existing agricultural endeavors
- essential technologies for ensuring that biomass-derived ethanol and other biofuels can be produced at costs competitive with conventional fuels

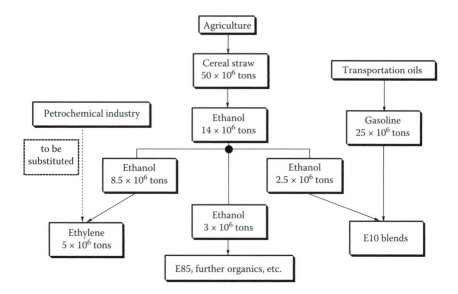

FIGURE 10.2 A possible biobased organic chemical industry in Germany: annual material flows. (Data from Kamm, B. et al. 2006. In *Biorefineries—Industrial Processes and Products*, vol. 1, ed. Kamm, B., Gruber, P. R., and Kamm, M., chap. 1. Weinheim, Germany: Wiley-VCH Verlag GmbH & Co. KGaA.)

- new additions to be integrated with the infrastructure of agricultural processing, which might include (in Europe) beet sugar refineries

As corollaries and (probably) axiomatic truths, biorefineries will only become interesting (as players in the industrial economy) when they reach large scales of operation and contribute significant amounts of materials to widely used and/or specialist chemistry platforms. They must be driven not essentially or solely as means to reduce greenhouse gas emissions, but also by considerations of the future depletion of fossil fuel reserves and the desire to broaden the substrate base. Governments would be instrumental in catalyzing these developments by favorable taxation regimes and economic subsidies.

10.3 BIOREFINERY ENTRY ROUTES

Producing low-volume, high-cost—or, at least, middle-volume and middle-price—products and coproducts from a biorefinery will be essential in establishing crop- and biomass-dependent biorefineries as components of the industrial landscape. The demise of the solvent (ABE) fermentation industry in the twentieth century (Chapter 7, Section 7.2) and the later failure of attempts to biomanufacture single-cell proteins in Europe are salutary reminders that biotechnological processes can fail due to economic and other factors unrelated to pure science.[8] At its heart, the biorefinery has the idea that plant biomass can be utilized in different ways and broaden the scope of biochemical engineering and of the spectrum of outputs rather than relying on single products whose markets may disastrously shift.

10.3.1 Fermentation of Biomass-Derived Substrates

The presence of up to 500 biorefineries in North America and 500–1,000 across Europe brings the option of large numbers of small- to medium-capacity sugar streams in widely scattered locations, often in agricultural or afforested regions. Each site could house a fermentation facility for fine chemicals as well as serve as a site for biofuels. The costs of transporting biomass and high-volume substrate solutions long distances by road or rail are unlikely to appeal to accountants or environmentalists. Ethanol alone not only has become the single largest biotechnological commodity as measured in terms of product volume or weight but also has demonstrated how metabolic engineering has meshed with traditional fermentation biology.[9]

The implication is that large numbers of production sites will arise at which many commercial products for the chemical, food, and other industries will be synthesized; in turn, this suggests a reversal of the process by which large-scale fermentation production for antibiotics, enzymes, vitamins, food flavors, and acids has been exported from the United States and Europe to areas with lower labor and construction costs (India, China, etc.). This has been described as "restructuring the traditional fermentation industry into viable biorefineries."[10] Such a vision will challenge the imaginations of industrial fermentation companies and demand that keener attention be paid to the practical economics of the biorefinery concepts; many synergies are to be expected.

Strategic research around the world has begun to explore what might be achieved with such an abundant supply of pentose sugars from lignocellulosic biomass and of the metabolic and biosynthetic uses that might result. *Corynebacterium glutamicum* has been much used—especially in Japan, Korea, and China—for the industrial production of amino acids such as glutamic acid and lysine for the food and agricultural feed sectors. To broaden its substrate utilization range to include xylose, a two-gene xylose catabolic pathway was constructed using the *E. coli xylA* gene (encoding xylose isomerase) with either the *E. coli xylB* gene (for xylulokinase) or a corynebacterial gene for this enzyme. Recombinants could grow in minimal media with xylose as the sole carbon source under aerobic conditions or, when O_2 limited, utilize xylose alone or in combination with glucose.[11]

An unusual *Lactobacillus* sp. strain MONT4, isolated from a high-temperature fermenting grape must, is uniquely capable of fermenting L-arabinose to a mixture of D- and L-lactic acids; the organism contains two separate genes encoding lactate dehydrogenase with differing sterochemistries.[12] D-Lactic acid, along with "unnatural" D-acids and amino acids, is contained in a series of bioactive molecules.[13] Chemical transformations of lactic acids can yield a variety of chemical intermediates and feedstocks—notably, acrylic acid and propylene glycol, which are compounds with major existing petrochemical-derived markets.[14]

Few (if any) major industrial-scale processes would fail to utilize individual sugars or mixtures of the carbohydrates emanating from lignocellulosic biomass processing or biodiesel production. Even as atypical a microbe as *Streptomyces clavuligerus* (unable to metabolize free glucose as a carbon source) can use glycerol for fermentations for the medically important β-lactam inhibitor clavulanic acid.[15] The extensive experience of adapting microbes to growing and functioning in lignocellulosic hydrolysates (Chapters 3 and 4) should be readily transferable to industrial strains that already are expected to produce large amounts of high-value products in extremely concentrated media with sugars, oligosaccharides, or plant oils as substrates and with vegetable proteins or high concentrations of ammonium salts as nitrogen sources. Conversely, the genomes of already adapted ethanologens could (with removal of genes for ethanol formation) provide platform hosts for the expression of other biosynthetic pathways.

As an example of a tightly closed circular biorefinery process, consider biodiesel production from a vegetable oil—or, as work in Brazil indicates, intact oil-bearing seeds[16]—using enzyme-catalyzed transesterification rather than alkaline hydrolysis/methanolysis (Chapter 8, Section 8.2.6). The glycerol-containing effluent could be a carbon source for the methylotrophic yeast *Pichia pastoris* (able also to catabolize any contaminating methanol in the feedstock) that is widely used for expressing heterologous proteins, including enzymes.

If the esterase used in the biodiesel process were to be produced on site by a *Pichia* fermentation, the facility could be self-sufficient biotechnologically (Figure 10.3). To maximize biomass substrate utilization, any solid materials from the fermentation, together with any waste plant material, could provide the substrates for solid-state fermentations producing bioinsecticides, biopesticides, and biofertilizers for application to the farmland used to grow the oil crop.[17]

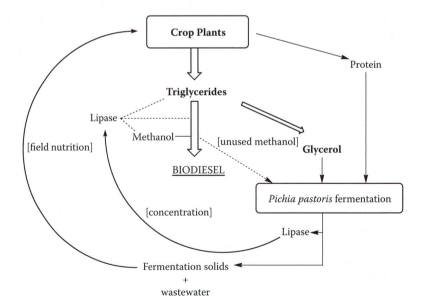

FIGURE 10.3 A "closed loop" integrated biotechnological process for biodiesel production.

Microbial enzymes and single-cell proteins (as well as xylitol, lactic acid, and other fermentation products) have become targets for future innovations in utilizing sugarcane bagasse in Brazil.[18] The bioproduction of bacterial polyhydroxyalkanoate polymers to replace petrochemical-based plastics is another goal of projects aiming to define mass-market uses of agricultural biomass resources in Brazil and elsewhere.[19]

Polyhydroxybutyrate and related polymers have had a long but unsuccessful record of searching for commercialization on a large scale on account of their uncompetitive economics and not always entirely industry-friendly chemical and physical properties. However, high oil prices will increasingly sideline the former while continued research into the vast number of possible biosynthetic structures may alight on unsuspected properties and uses. This also illustrates an important distinction made by the founders of "biocommodity engineering": that replacement of a fossil-resource-derived product by a biomass-derived compound of identical composition is a conservative strategy and that a more radical one of substituting the existing chemical product with a biochemical with equivalent functionalities involves a more protracted transition but could be more promising in the long term.[20]

STEM TOPIC 10.1: COPRODUCTION OF ETHANOL AND OTHER BIOFUELS

One of the many evolving visions of biorefineries is that of coproduction of ethanol and other biofuels to maximize the energy yield from plant biomass feedstocks. The most energy-efficient option may be that of combining ethanol production with the generation of both H_2 and methane (CH_4), requiring only small amounts of the biomass input energy to perform a primary fermentation (to generate some ethanol and H_2) and to trap most of the energy in the form of CH_4[117] (see Figure STEM 10.1).

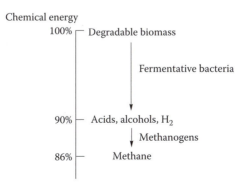

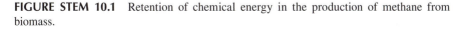

FIGURE STEM 10.1 Retention of chemical energy in the production of methane from biomass.

The key to this high efficiency is that methane formation is often close to the theoretical maximum (3 mol per mole of glucose consumed):

$$C_6H_{12}O_6 \rightarrow 3CO_2 + 3CH_4$$

This chemical equation is a slimmed down version of the Buswell equation, which was derived in 1952 to predict the maximal yields of products from anaerobic digestion processes, most of which are highly malodorous due to the presence of hydrogen sulfide, ammonia, and other noxious gases:

$$C_cH_hO_oN_nS_s + \tfrac{1}{4}(4c-h-2o+3n+2s)H_2O \rightarrow$$

$$\tfrac{1}{8}(4c-h+2o+3n+2s)CO_2 + \tfrac{1}{8}(4c+h-2o-3n-2s)CH_4 + nNH_3 + sH_2S$$

10.3.2 BIOMASS GASIFICATION

The potential of biorefineries based on syngas to generate as wide a collection of chemical feedstocks as are presently derived from petroleum is easily visualized with biologically derived methanol as the starting point (Figure 10.4). Methanol obtained from the gasification of biomass (Chapter 8, Section 8.7) can be

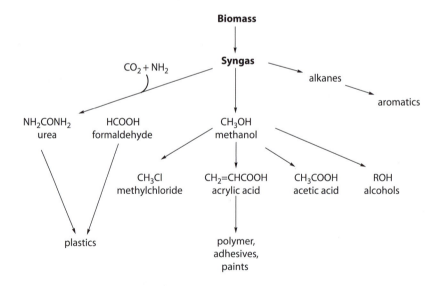

FIGURE 10.4 Chemical production routes for industrial feedstocks from methanol. (Modified from Kamm, B. et al. 2006. In *Biorefineries—Industrial Processes and Products*, vol. 1, ed. Kamm, B., Gruber, P. R., and Kamm, M., chap. 1. Weinheim, Germany: Wiley-VCH Verlag GmbH & Co. KGaA.)

transformed by well-known chemical reactions to form the following, among many other chemicals[22]:

- Formaldehyde (CH_2O) is used in the production of resins, textiles, cosmetics, fungicides, etc. Formaldehyde accounted for 35% of the total worldwide production of methanol in the mid-1990s and is prepared industrially by the oxidation of methanol with atmospheric O_2 using a variety of catalysts.
- Acetic acid (CH_3COOH) is a major acid in the food industry and feed-stock for manufacturing syntheses in the production of some plastics, fibers, etc. Acetic acid is manufactured by the carbonylation of methanol with CO.
- Formic acid (HCOOH) is a preservative.
- Methyl esters of organic and inorganic acids are solvents and methylation reagents and are used in the production of explosives and insecticides.
- Methylamines are precursors for pharmaceuticals.
- Trimethylphosphine is used in the preparation of pharmaceuticals, vitamins, fine chemicals, and fragrances.
- Sodium methoxide is an organic intermediate and catalyst.
- Methyl halides are solvents, organic intermediates, and propellants.
- Ethylene is used for plastics and as an organic intermediate.

STEM TOPIC 10.2: TYPES OF BIOREFINERY

Four major classes of biorefinery are actively being discussed and—however theoretical the exercise—planned and designed:

1. the starch biorefinery
2. the lignocellulosic biorefinery
3. the whole-plant biorefinery
4. the green biorefinery

Designs 1, 2, and 3 do not differ significantly from corn milling, corn grain plus corn stover, or cellulosic processing facilities, respectively, to yield biofuels, solid coproducts, combustibles (e.g., dried lignin), etc.

The green biorefinery would operate on a scientifically different basis and has been best explored in Austria with the agenda of finding new uses for pasture plants (clover, alfalfa, lucerne, etc.) made available by a decrease in dairy farming (Joanneum Research, www.joanneum.at). Extrapolating from traditional silage fermentation of cut grass, the product of this initial fermentation is then processed to protein and lactic acid fractions, residual material can be fermented further to biogas, and fibrous material could be processed for food fiber materials or fiber board (see Figure STEM 10.2).

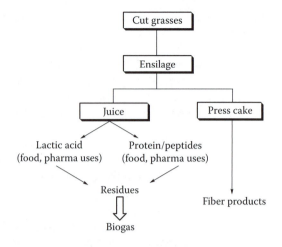

FIGURE STEM 10.2   Material flow in a green biorefinery.

10.4   BIOREFINERY PIVOTAL PRODUCTS

In a reversal of the chemistry used to manufacture industrial alcohol from petrochemical sources, ethanol is readily dehydrated by chemical reactions to ethylene (ethene):

$$C_2H_5OH \rightarrow C_2H_4 + H_2O$$

Ethylene is a key intermediate in organic chemistry for plastics (polyethylenes); in 2005, its worldwide production was 113 million tonnes.[23] From ethylene, the vast hinterland of petrochemical production rapidly opens up by way of

- ethylene oxide (C_2H_4O), a starting material for the manufacture of acrylonitrile, nonionic surfactants, etc. and a ripening agent for fruits
- ethylene glycol ($C_2H_5O_2$), a solvent as well as an intermediate in the synthesis of synthetic fibers
- ethylene chlorohydrin (C_2H_5ClO), another solvent and an intermediate in the production of agrochemicals
- ethyl bromide (C_2H_5Br), an ethylating agent in organic syntheses

Approximately 0.9 million tonne of glycerol are synthesized via chemical routes annually; two thirds of this could be isolated from ethanol fermentation broths as glycerol-rich stillage.[3] Biodiesel wastes and direct fermentations, however, could supply far more glycerol (Chapter 8, Section 8.2.1). Important derivatives of glycerol prepared chemically include[24]:

- glycerol trinitrate (i.e., nitroglycerin for explosives)
- epichlorohydrin, the most important material for the production of epoxy resins (and historically the immediate precursor of glycerol in chemical manufacture)[12]

- oxidation products as intermediates for pharmaceutical synthesis, including glyceric, tartronic, hydroxypyruvic, and mesoxalic acids, and dihydroxyacetone
- *tert*butyl ethers formed by the reaction between glycerol and alkenes such as 2-butene

Approximately 36% of global glycerol production is directed toward such chemical intermediates and products. Most of the remainder find uses in the food and cosmetic industries.

Butanol (Chapter 7, Section 7.2) is a major solvent and finds numerous applications during the manufacture of plastics and textiles.[25] Butanol could be catalytically dehydrated to yield but-1-ene, one of four butene isomers. None of these highly volatile hydrocarbons exists naturally; they are produced from petroleum refineries. The complexity of the composition of the butene-containing C4 streams from crude oil "cracking" has limited the chemical exploitation of butanes as intermediates and feedstocks.[26] The availability of pure but-1-ene in feedstock quantities may expand the horizons for industrial synthetic chemistry.

These examples demonstrate one fundamental feature of biorefineries: the ability to replace petroleum refineries as sources of liquid fuels (ethanol, biobutanol, etc.) and (by catalytic transformation of these compounds) feedstock chemicals. This model implies a short linear chain of sequential biochemical and chemical processes:

Biomass substrate → ethanol, methanol, glycerol, butanol → feedstock chemicals

This could operate on an either/or basis: either producing biofuels for transportation uses or proceeding straight to the production of bulk chemicals. The development of a full infrastructure for biofuels distribution would bring in parallel a means of transporting the liquid biofuels to existing chemical industrial sites, thus reducing the total investment cost associated with the transition from oil dependency to a biobased commodity economy.

Although the economic analysis of biorefineries is poorly developed, the price tag will be high: One estimate puts the cost of a biorefinery capable of processing 2,000 tonnes per day as $500 million.[27] Approximately 500 such facilities would be required to process the billion tons annually required for the mass production of lignocellulosic ethanol in the United States, thus representing a $250 billion total investment.

STEM TOPIC 10.3: CAN LIGNIN BE CHEMICALLY CONVERTED TO HYDROCARBONS?

Lignin is usually portrayed in biorefinery schemes as being combusted as a source of heat. In 2003, however, a patent application was filed that detailed a chemical means of generating C7–C10 alkylbenzenes that could be blended directly with petroleum-derived fuels (US 2003/0115792) (see Figure STEM 10.3).

This piece of chemistry might have gathered dust; however, in September 2009, Chevron Technology Ventures filed two patents for lignin and announced a partnership with Mascoma, Inc., to investigate lignocellulosic lignin as a source of hydrocarbon fuels.

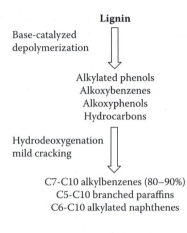

Lignin

Base-catalyzed
depolymerization

Alkylated phenols
Alkoxybenzenes
Alkoxyphenols
Hydrocarbons

Hydrodeoxygenation
mild cracking

C7-C10 alkylbenzenes (80–90%)
C5-C10 branched paraffins
C6-C10 alkylated naphthenes

FIGURE STEM 10.3   Chemical conversion of lignin by hydroprocessing.

A more elaborate model for biorefineries entails the dual production of biofuels and fermentation products other than fuels.[28] The logistical basis for this design is the multiple nature of lignocellulosic and whole-plant carbohydrate streams—that is, hemicellulose-derived pentoses, cellulose-derived glucose and oligoglucans, and starch-derived glucose if grains are processed. Given the wide spectrum of naturally occurring and genetically engineered ethanologens (Chapters 3 and 4), ethanol could be produced entirely from one of these carbohydrate streams, leaving the others as substrates for different types of fermentations—potentially as wide a choice as that of fermentations known or already used for industrial production.

To narrow the field substantially, 12 building block chemicals that can be produced from sugars via biological or chemical conversions and can be subsequently converted to a number of high-value bio-based chemicals or materials have been identified in a report prepared for the U.S. Department of Energy.[29] These building block chemicals are molecules with multiple functional groups that can be transformed into new families of useful molecules:

- 1,4-diacids (succinic, fumaric, and malic)—all intermediates of the tricarboxylic acid cycle and easily bioproduced by microbes
- 2,5-furan dicarboxylic acid—chemically produced by the oxidative dehydration of C6 sugars
- 3-hydroxypropionic acid—a microbial product
- aspartic acid—an amino acid biosynthesized by all Earth's organisms
- glutamic acid—another amino acid and one of the major products of industrial fermentations for fine chemicals (as monosodium glutamate, MSG)
- glucaric acid—chemically produced by the nitric acid oxidation of starch
- itaconic acid—a tricarboxylic acid manufactured on an industrial scale by the fungus *Aspergillus oryzae*
- levulinic acid—chemically produced by the acid-catalyzed dehydration of sugars

- 3-hydroxybutyrolactone—chemically produced by the oxidative degradation of starch by hydrogen peroxide
- glycerol
- sorbitol—a sugar alcohol derived from glucose (chemically by hydrogenation) but also known as an enzyme-catalyzed product of glucose metabolism
- xylitol/arabinitol (Figure 3.2, Chapter 3)

These compounds were chosen by consideration of the potential markets for building blocks and their derivatives and the technical complexity of their (bio)synthetic pathways. A second-tier group of building blocks, most of which are microbial products, was also identified as viable candidates: gluconic acid; lactic acid; malonic acid; propionic acid; the two triacids, citric and aconitic; xylonic acid; acetoin; furfural; levoglucosan; and the three amino acids, lysine, serine, and threonine (Table 10.1).

The building block compounds were clearly differentiated from two other potential biorefinery products: direct product replacements and novel products. The former might include acrylic acid manufactured from lactic acid rather than from fossil-derived propylene. Markets already existed for the compound, and the cost structures and growth potential of these markets were well understood, thus reducing the risks involved in devising novel production routes. On the other hand, novel products such as PLA (Section 10.2) had no competing routes from fossil reserves, had unique properties (thus rendering cost issues less crucial), and were intended to meet new markets.

In contrast, the building block compounds were envisaged as being the starting points for diverse portfolios of products—replacing existing fossil-based compounds as well as offering novel intermediates for chemical syntheses. This combination has three advantages:

1. The market potential is expanded.
2. Multiple possible markets can reduce risks.
3. Capital investment can be spread across different industrial sectors.

To illustrate these points, the 12 building-block biochemicals could give rise to many derivatives that would find immediate or short-term uses in fields as diverse as transportation (polymers for automobile components and fittings, anticorrosion agents, and oxygenates), recreation (footgear, golf equipment, and boats), and health and hygiene (plastic eyeglasses, suntan lotions, and disinfectants). Three of them will now be considered in greater detail to explore possible key features and likely variables in the development of biorefineries.

10.4.1 CASE STUDY 1: SUCCINIC ACID

This C4 dicarboxylic acid is one of the key intermediates of glucose catabolism in aerobic organisms (including *Homo sapiens*), but it can also be formed anaerobically in fermentative microbes (Figure 10.5). In either case, CO_2 is required to be fixed into organic chemicals; in classical microbial texts, this is termed "anaplerosis," acting to replenish the pool of di- and tricarboxylic acids when individual

TABLE 10.1
Microbial Routes Known for Future Biobased Chemical Building Blocks

Building Block	Aerobic Fermentation	Anaerobic Fermentation
Three-Carbon Compounds		
Glycerol	Yeast/fungal and bacterial	Yeast/fungal and bacterial
Lactic acid	Yeast/fungal	Commercial bacterial process
Propionic acid	None	Bacterial
Malonic acid	Yeast/fungal	None
3-Hydroxypropionic acid	Yeast/fungal and bacterial	None
Serine	Commercial bacterial process	None
Four-Carbon Compounds		
3-Hydroxybutyrolactone	None	None
Acetoin	Yeast/fungal and bacterial	Bacterial
Aspartic acid	Yeast/fungal and bacterial	None
Fumaric acid	Yeast/fungal and bacterial	None
Malic acid	Yeast/fungal and bacterial	None
Succinic acid	Yeast/fungal and bacterial	Bacterial
Threonine	Commercial bacterial process	None
Five-Carbon Compounds		
Arabitol	Yeast/fungal	Yeast/fungal
Xylitol	Yeast/fungal	Yeast/fungal and bacterial
Furfural	None	None
Glutamic acid	Commercial bacterial process	None
Itaconic acid	Commercial fungal process	None
Levulinic acid	None	None
Six-Carbon Compounds		
2,5-Furan dicarboxylic acid	None	None
Aconitic acid	Yeast/fungal	None
Citric acid	Commercial fungal process	None
Glucaric acid	Yeast/fungal and bacterial	None
Gluconic acid	Commercial fungal process	None
Levoglucosan	None	None
Lysine	Commercial fungal process	None
Sorbitol	Yeast/fungal and bacterial	None

Source: Data from Werpy, T., and Petersen, G., eds. 2004. *Top Value-Added Chemicals from Biomass. Volume I—Results of Screening for Potential Candidates from Sugars and Synthesis Gas.* Pacific Northwest National Laboratory and National Renewable Energy Laboratory, August 2004 (available electronically at http://www.osti.gov/bridge).

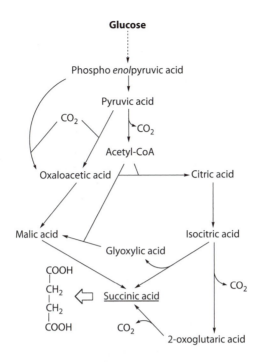

FIGURE 10.5 Succinic acid: known biosynthetic routes from glucose.

compounds (including the major industrial products citric, itaconic, and glutamic acids) are abstracted from the intracellular cycle of reactions and accumulated in the extracellular medium.

Under anaerobic conditions and given the correct balance of fermentation products, a net "dark" fixation of CO_2 can occur. It is this biological option that has been most exploited in the development of modern biosynthetic routes:

- The rumen bacterium *Actinobacillus succinogenes* was discovered at Michigan State University and commercialized by MBI International (Lansing, Michigan); succinate yields as high as 110 g/L have been achieved from glucose.[30,31]
- At the Argonne National Laboratory (ANL; Argonne, Illinois), a mutant of *E. coli* unable to ferment glucose because of inactivation of the genes encoding lactate dehydrogenase and pyruvate formate lyase spontaneously gave rise to a chromosomal mutation that reestablished glucose fermentative capacity but with an unusual spectrum of products: 1 mol of succinate and 0.5 mol each of acetate and ethanol per mole of glucose consumed.[32,33] The second mutation was later mapped to a glucose uptake protein that, when inactivated or impaired, led to slow glucose transport into the cells and avoided any repression of genes involved in this novel fermentation.[34] The result is a fermentation in which redox equivalents are balanced by a partition of carbon between the routes to succinate and that to acetate and

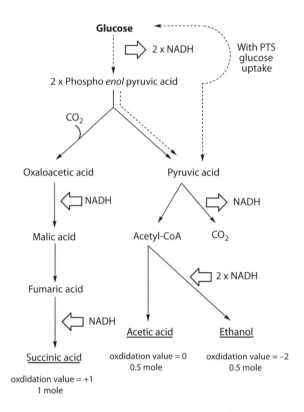

FIGURE 10.6 Redox balance in the fermentation of glucose to succinic acid by *Escherichia coli.*

ethanol (in equal measures); pyruvate is oxidatively decarboxylated rather than split by pyruvate formate lyase activity or reduced to lactic acid—both routes lost from wild-type *E. coli* biochemistry in the parental strain (Figure 10.6). The maximum conversion of glucose to succinate by this route is 1 mol/mol, which is a carbon conversion of 67%; succinate titers have reached 75 g/L. Because *E. coli* is only facultatively anaerobic, biomass in the fermentation can be generated rapidly and to a high level under aerobic conditions; O_2 entry is then restricted to transform the process to one of anaerobic metabolism.[35] The same organism can successfully utilize both glucose and xylose in acid hydrolysates of corn straw and generate succinate as a fermentation product.[36]

- In wild-type *E. coli,* glucose fermentations produce complex mixtures of acid and nonacidic products in which succinate may be only a minor component; nevertheless, the succinate titer can be greatly increased by process optimization, reaching 24 g/L within 30 h with laboratory media and 17 g/L in 30 h in a fermentor with an economical medium based on corn steep liquor and cane sugar molasses.[37,38] Succinate productivity with *Bacteroides fragilis,* another inhabitant of the human gut and intestine but an obligately anaerobic species, has also been explored as a producing organism.[39,40]

- An *aerobic* system for succinate production was designed, with a highly genetically modified *E. coli,* using the same glucose transport inactivation described previously but also inactivating possible competing pathways and expressing a heterologous (*Sorghum vulgare*) gene encoding PEP carboxylase, another route for anaplerosis.[41,42] A succinate yield of 1 mol per mol of glucose consumed was demonstrated, with a high productivity (58 g/L in 59 h) under fed-batch aerobic reactor conditions. The biochemistry involved in this production route entails directing carbon flow via anaplerotic reactions to run backward though the tricarboxylic acid cycle—that is, in the following sequence:

$$PEP \rightarrow oxaloacetate \rightarrow malate \rightarrow fumarate \rightarrow succinate$$

Succinate can be produced in the forward direction by blocking the normal workings of the cyclic pathway (with a necessary loss of carbon as CO_2) and the activation of a pathway (the glyoxylate shunt) normally only functioning when *E. coli* grows on acetate as a carbon source:

$$Citrate \rightarrow isocitrate \rightarrow succinate + glyoxylate$$

The enzymes catalyzing the final two steps in the pathway from PEP—malate dehydratase and fumarate reductase—can be overexpressed in bacterial species and were the subjects of patent applications.[43,44]

To return to anaerobic rumen bacteria, *Anaerobiospirillum succiniciproducens* had a short but intense history as a candidate succinate producer.[45–47] Fermenting glucose in a medium containing corn steep liquor as a cost-effective source of nitrogen and inorganic nutrients, succinate titers reached 18 g/L from 20.2 g/L of glucose—equivalent to a conversion efficiency of 1.35 mol per mole.[48] A novel rumen bacterial species, *Mannheimia succiniciproducens,* now has a complete genomic sequence and a detailed metabolic network.[49–51] Mutants of this microbe can produce succinate with much reduced amounts of other acids and can anaerobically ferment xylose and wood hydrolysate to succinate.[52,53]

A prediction has been that overexpressing genes for anaplerotic pathway enzymes would enhance succinate production (and, in other genotypes or fermentation conditions, the accumulation of other acids of the tricarboxylic cycle); experimental evidence amply confirms this prediction.[54–59] With the capabilities to perform metabolic computer-aided pathway analysis with known gene arrays, comparison of succinate producers and nonproducers and between different species would be expected to accelerate greatly progress toward constructing the ideal microbial cell factory.

Comparison of *E. coli* and *M. succiniciproducens* suggested five target genes for inactivation, but combinatorial inactivation did not result in succinate overproduction in *E. coli.* Two of the identified genes—*ptsG* (the glucose transport system) and *pykF* (encoding pyruvate kinase, the enzyme interconverting PEP and pyruvic acid), together with the second pyruvate kinase gene (*pykA*)—increased succinate accumulation by more than sevenfold, although succinate was still greatly outweighed by the other fermentation products (formate, acetate, etc.).[60] Eliminating the glycolytic pathway below PEP will clearly aid succinate production, but the producing cells will then be highly dependent on organic nutrients (including many amino acids) for growth and maintenance.

The combination of several related technologies is particularly appealing for commercial production of succinic acid:

- optimization of the ANL strains for succinate production by the Oak Ridge National Laboratory (Oak Ridge, Tennessee)
- innovations for succinate recovery from the fermentation broth at ANL
- an improved succinic acid purification process[61]
- the development of catalytic methods for converting succinic acid to 1,4-butanediol and other key derivatives at Pacific Northwest National Laboratory (Richmond, Washington)[62]

These advances have moved biologically derived succinic acid (BDSA) close to commercialization as a component of the first genuine biorefinery (www.mbi.org).

10.4.2 CASE STUDY 2: XYLITOL AND RARE SUGARS AS FINE CHEMICALS

Xylitol was a significant biochemical feature of the metabolic routes for the xylose presented to ethanologenic cultures in hydrolysates of the hemicelluloses from lignocellulosic biomass. However, with an organoleptic sweetness to human taste approximately equivalent to sucrose, it is a fine chemical product in its own right as a low-calorie sweetener; xylose sugars are not metabolized by the human consumers of xylitol-containing chewing gums (Figure 3.2, Chapter 3). Other envisaged uses include[29]:

- production of anhydrosugars (as chemical intermediates) and unsaturated polyester resins
- manufacture of propylene and ethylene glycols as antifreeze agents and unsaturated polyester resins
- oxidation to xylonic and xylaric acids to produce novel polymers (polyesters and nylon-type structures).

The production of xylitol for use as a building block for derivatives essentially requires no technical development and, if the xylose feedstock is inexpensive (as a product of biomass processing), then the production of xylitol could be done for very low cost.

The accumulation of xylitol during ethanologenesis from lignocellulosic substrates is unwanted and quite undesirable. Viewed as an economically valuable product, however, xylitol formation and production acquire a different biotechnological perspective, and patenting activity has recently been intense (Table 10.2).

Biochemical efforts also continue to locate and exploit enzymes for bioprocessing hemicelluloses and hemicellulosic waste streams, for example:

TABLE 10.2
Patents for Xylitol Production Technologies

Date	Title	Assignee	Patent
8/7/2001	Yeast strains for the production of xylitol	Xyrofin Oy (Finland)	US 6,271,007
4/20/2004	Manufacture of xylitol using recombinant microbial hosts	Xyrofin Oy (Finland)	US 6,723,540
1/25/2005	Process for the simultaneous production of xylitol and ethanol	Xyrofin Oy (Finland)	US 6,846,657
4/17/2006	Fermentation process for production of xylitol from *Pichia* sp.	Council of Scientific and Industrial Research (India)	US 6,893,849
5/17/2005	Process for the production of xylitol	Danisco Sweeteners Oy (Finland)	US 6,894,199
6/28/2005	Process for the production of xylitol	Danisco Sweeteners Oy (Finland)	US 6,911,565
8/2/2005	Xylitol dehydrogenase of acetic acid bacteria and gene thereof	Ajinomoto Co., Inc. (Japan)	US 6,924,131
9/19/2005	Process for the simultaneous production of xylitol and ethanol	Danisco Sweeteners Oy (Finland)	US 7,109,005

- An L-xylulose reductase identified from the genome sequence of the filamentous mold *Neurospora crassa* has been heterologously produced in *E. coli* for the production of xylitol.[63]
- A β-xylosidase from *Taloromyces emersonii* has been shown to be superior to the enzyme from the industrial fungus *Hypocrea jecorina* in releasing xylose from vinasse, the solid-waste material from ethanol fermentations.[64]
- Xylan residues in hemicelluloses can be variably esterified with acetyl, feruloyl, and *p*-coumaryl residues; hemicellulose deacetylating esterases have been characterized from fungal species and shown to be highly effective only when mixed in multiplexes of enzymes capable of using all the possible structures as substrates for enzyme action.[65]
- Feruloyl esterases have recently been designed as novel chimeric forms with cellulose and hemicellulose binding proteins to improve their efficiencies with plant polymeric substrates.[66,67]
- Ferulic acid has a number of potential commercial applications as an antioxidant, food preservative, anti-inflammatory agent, photoprotectant, and food flavor precursor. Major sources—brewer's spent grain, wheat bran, sugar beet pulp, and corn cobs—make up 1 or 2% of the daily output of the global food industry, and ferulic acid can be released from brewer's grain and wheat bran by feruloyl esterases from the thermophilic fungus *Humicola insolens.*[68]

A hemicellulose hydrolysate contains a variety of hexoses and pentoses; however, yeast highly suitable for the production of xylitol from the xylose present in the mix may preferentially utilize glucose. One solution to this problem is to remove the rapidly utilized hexoses before removing the cells and replacing them with a purposefully xylose-grown cell batch, as was demonstrated with dilute acid hydrolysates of corn fiber.[69] With *Saccharomyces cerevisiae* expressing the *Pichia stipitis* gene for xylose reductase, the presence of glucose inhibited xylose uptake and biased the culture toward ethanol production; controlling the glucose concentration by feeding the fermentation to maintain a high xylose/glucose ratio resulted in a near-quantitative conversion of xylose to xylitol, reaching a titer of 105 g/L of xylitol.[70]

With a double recombinant strain of *S. cerevisiae* carrying the xylose reductase genes from both *P. stipitis* and *Candida shehatae,* quite different microbial biochemistry occurred in a more process-friendly formation (at close to theoretical levels) of xylitol from a mixture of xylose (the major carbon source) and glucose, galactose, or mannose as the cosubstrate. Indeed, the presence of the cosubstrate was mandatory for continued metabolism of the pentose sugar.[71]

In a gene-disrupted mutant of *C. tropicalis,* with no measurable xylitol dehydrogenase activity, glycerol proved to be the best cosubstrate, allowing cofactor regeneration and redox balancing, with a xylitol yield that was 98% of the maximum possible.[72] Bioprocess engineering for xylitol production is straightforward; aerated cultures with pH regulation and operated at 30°C appear common. With Brazilian sugarcane bagasse as the source of the hemicellulosic sugars, xylitol from the harvested culture broth was readily recovered at up to 94% purity crystallization from a clarified and concentrated broth.[73]

Xylitol dehydrogenase enzymes catalyze reversible reactions. The catabolism of xylitol proceeds via the formation of D-xylulose (Figure 3.2, Chapter 3). Fungal pathways of L-arabinose utilization can include L-xylulose, a much less common pentulose sugar, as an intermediate.[74–77] Such rare sugars are of increasing interest to metabolic biochemists because it has long been appreciated that many naturally occurring antibiotics and other bioactives contain highly unusual sugar residues—sometimes highly modified hexoses derived by lengthy biosynthetic pathways (e.g., erythromycins).

Many secondary metabolites elaborated by microbes may exhibit only weak antibiotic activity but are or can easily be converted into chemicals with a wide range of biologically important effects: antitumor, antiviral, immunosuppressive, anticholesterolemic, cytotoxic, insecticidal, or herbicidal.[78] To the synthetic chemist, therefore, the availability of such novel carbohydrates in large quantities offers new horizons in development of novel therapeutic agents. For example, L-xylulose offers a promising route to inhibiting the glycosylation of proteins, including those of viruses.[79]

Equally, such unconventional chemicals may have very easily exploited properties as novel fine chemicals. The hexose D-tagatose is an isomeric form of the commonly occurring sugar D-galactose (a hexose present in hemicelluloses; see Figure 2.1 in Chapter 2) and has attracted interest for commercial development.[80] Because of its very rare occurrence in the natural world, a structure such as D-tagatose presents a "tooth-friendly" metabolically intractable sugar to human biochemistry.

Fortuitously, the enzyme L-arabinose isomerase includes D-galactose among its spectrum of possible substrates; the enzyme can be found in common bacteria with well advanced molecular genetics and biotechnologies, including *E. coli, Bacillus subtilis,* and *Salmonella typhimurium* and, when expressed in suitable hosts, can convert the hexose into D-tagatose with a 95% yield.[81] Even more promising for industrial use is the efficient bioconversion of D-galactose to D-tagatose using the immobilized enzyme, which is more active than free L-arabinose isomerase and stable for at least 7 days.[82] Both the enzyme and recombinant L-arabinose isomerase-expressing cells can be used in packed-bed bioreactors, and the cells are particularly adaptable to long production cycles.[83,84]

Synthetic chemistry offers only expensive and poorly yielding routes to the rare sugars, but uncommon tetroses, pentoses and hexoses can all be manufactured with whole cells or extracted enzymes acting on cheap and plentiful carbon sources, including hemicellulose sugars. Bioproduction strategies use an expanding tool kit of enzymes, including D-tagatose 4-epimerase, aldose isomerase, and aldose reductase.[85–87] Together, the rare sugars offer new markets for sugars and sugar derivatives of at least the same magnitude as that for high-fructose syrups manufactured with xylose (glucose) isomerase.

10.4.3 CASE STUDY 3: GLYCEROL

Glycerol is an obligatory coproduct of biodiesel production, forming 10% by weight of the triglycerides that act as substrates for chemical or enzymic transesterification (Figure 8.1, Chapter 8). The waste stream, however, is highly impure, with the glycerol mixed with alkali, unreacted methanol, free fatty acids, and chemical degradation products. At the cost of considerable expenditure of energy in gas sparging and flash distillation, the glycerol can be recovered at >99.5% purity.[88] It is much simpler to mix the crude glycerol with methanol and substitute waste petroleum oil and heavy fuel oil as a direct fuel.[89]

Glycerol represents a valuable chemical resource as a potential feedstock and building block chemical.[29] Research into value-added utilization options for biodiesel-derived glycerol has therefore ranged widely in the search for commercial applications using both chemical and biotechnological methods (Table 10.3).[78] For

TABLE 10.3
Chemical and Biotechnological Transformations of Glycerol

Product Formed From Glycerol	Chemical Route	Fermentation Route
1,3-Propanediol	Selective hydroxylation	*Clostridium butyricum, Klebsiella pneumoniae*
1,2-Propanediol	Hydrogenolysis	None
Dihydroxyacetone	Selective catalytic oxidation	*Gluconobacter oxydans*
Succinic acid	None	*Anaerobiospirillum succiniciproducens*
Hydrogen	Catalytic reforming	*Enterobacter aerogenes*
Polyesters	Catalyzed esterification with acids	None
Polyglycerols	Selective etherification	None
Polyhydroxyalkanoates	None	Various osmophilic microbial species

Source: Data from Pachauri, N., and He, B. 2006. 2006 ASABE Annual International Meeting, paper no. 066223.

the production of 1,3-propanediol for polymer manufacture, glycerol as source via microbial fermentation was more expensive than either glucose or chemical routes from ethylene oxide or acrolein 5 or 6 years ago.[3] Those economics have significantly altered now, and glycerol represents the shortest, most direct route for bioproduction of 1,3-propanediol—a two-reaction sequence comprising an enzyme-catalyzed dehydration followed by a reduction:

$$\text{Glycerol} \rightarrow \text{3-hydroxypropionaldehyde} \rightarrow \text{1,3-propanediol}$$

Microbial studies have focused on the fermentative production of 1,3-pro-pandiol by clostridial species, but this is greatly complicated by the multiplicity of other products, including *n*-butanol, ethanol, and acids.[91] Bioproduction of a nutraceutical fatty acid derivative (marketed as a dietary supplement) by microalgae is a route to a higher value product than bulk chemicals.[92] As examples of chemical engineering, recent industrial patent documents have disclosed methods for converting the biodiesel waste glycerol into dichloropropanol and of an alcohol-permeable membrane to generate a purified mixture of esters, glycerol, and excess alcohol.[93,94]

For many years, glycerol was not considered fermentable by *E. coli,* but rather only by a limited number of related bacterial species; however, a landmark publication in 2006 reported that *E. coli* could efficiently ferment glycerol to ethanol (and a small amount of succinic acid), provided high pH in the culture was avoided. This is a crucial point because growth from glycerol requires an anaplerotic step involving CO_2; the CO_2 is generated by pyruvate formate lyase whose activity is reduced by high pH in the growth medium (Figure 10.7).[95] The application of genetically engineered *E. coli* with superior ethanologenic potential (Chapter 4, Section 4.2.2.1), combined with further manipulations to eliminate competing pathways of acid accumulation, could result in ethanol productivity approaching the theoretical molar production (mole per mole) from glycerol.

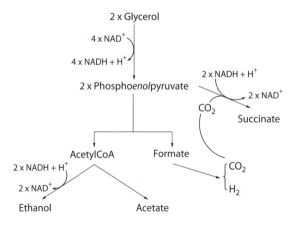

FIGURE 10.7 The anaerobic fermentation of glycerol to ethanol and succinic acid by *Escherichia coli.* (Modified from Dharmadi, Y., Murarka, A., and Gonzalez, R. 2006. *Biotechnology and Bioengineering* 94:821.)

Glycerol-containing biodiesel waste appears at first glance to be an unlikely micro-bial fermentation medium: brown-black, viscous, and with a high inorganic salt con-centration. Diluted and supplemented with nitrogen sources, however, such wastes can be converted with high efficiency (0.85 mol per mol of glycerol) to ethanol—together with the simultaneous evolution of H_2—using immobilized cells of *Enterbobacter aerogenes*.[96] With successful genetic manipulation of this bacterium to maximize H_2 production, the development of biodiesel-based biorefineries may be approaching.[97]

10.5 CENTRAL SUBSTRATES AND BIOREFINERY FLEXIBILITY

In 2008 and 2009, start-up biotech companies began seriously investigating how to retrofit corn ethanol factories left idle by the recession for the fermentative produc-tion of advanced fuels or fine chemicals; this underlined how flexible bioprocess technology is. Multinational companies have often transferred industrial fermen-tations around the world to already established sites with previous uses, with or without extensive reengineering. Future biorefineries will be able to take advantage of the intrinsic flexibility provided by microorganisms utilizing a cluster of key substrates—most obviously glucose derived from starch or cellulose, but also fatty acids and glycerol, both of which are feedstocks for industrial antibiotic production (Figure 10.8).

The implication is that biorefineries can produce biofuels, fine chemicals, and bulk intermediates in response to market demands. To the process chemist's eyes, biore-fineries are entirely analogous to petrochemical refineries in that a limited range of major feedstock chemicals—biomass hydrolysates, vegetable oils, pyrolytic bio-oils, lignin, etc. (in biorefineries) and oils and diesel fuel (in a petrochemical refinery)—are convertible by fermentation, catalytic cracking, hydrotreating or hydrocracking to a range of fuels (biofuels *or* gasoline, diesel, aviation fuel, and liquid petroleum gas) and chemicals (synthetic building blocks *or* olefins, etc.).[98]

Extending this concept further, even if the inputs to a biorefinery are derived biologically rather than thermochemically (via biomass gasification), an interface to chemical processing is provided by applications of chemical catalysis to the pro-duction of fuels and other value-added chemicals from biomass-derived oxygenated

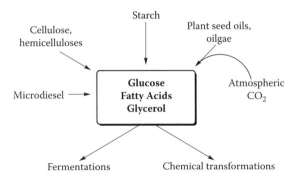

FIGURE 10.8 Glucose, fatty acids, and glycerol as central substrates for biorefinery operations.

feedstocks that have opened up routes involving well-defined reactions of organic chemistry (dehydration, dehydrogenation, aldol condensation, etc.).[99]

Central feedstocks could be provided from food wastes; this biotechnology exists and could produce biofuels or use the same outputs (acetone, butanol, and ethanol) as on-site substrates for other fermentations.[100–102] The most important novel feedstock use, however, will be the microbial biosequestration of atmospheric CO_2 using photosynthetic microorganisms (microalgae and cyanobacteria) to provide the carbon feedstocks for the biorefinery.[103] While plants use the Calvin–Benson cycle to fix CO_2 into organic carbon compounds (initially sugar phosphates), CO_2 incorporation by "dark" metabolic reactions is widely distributed.

The most recently discovered such pathway was defined in 1989 in an archeon, a type of single-celled microbe that is like bacteria but evolved as an ancient line quite separately from the eubacteria and blue-green algae; its members usually inhabit extreme environments (Figure 10.9).[104] Immobilized enzymes using CO_2 as a substrate would be one experimental approach to achieve net CO_2 fixation without photosynthesis, using substrates from the hydrolysis of cellulosic plant material as cosubstrates. Another would be use of immobilized cells in a nutritional environment to maintain viability but avoid respiratory activity of the cell population.

The extensive use of photobioreactors anywhere but in climates and locations with long guaranteed daily hours of intense sunlight is inefficient; however,

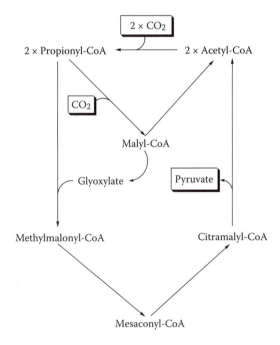

FIGURE 10.9 Schematic of a bicyclic CO_2 fixation pathway in *Chloroflexus aurantiacus* and other archeons. (Modified from Ishii, M. et al. 2004. *Applied Microbiology and Biotechnology* 64:605.)

light-independent bacterial bioprocesses avoid such limitations.[105] Applying biotechnology to use accumulated CO_2 as a process input would achieve far more than might presently envisioned chemical routes. Carbon capture and sequestration (CCS) has been devised mostly for removing or reducing CO_2 emissions from fossil fuel power stations; however, trapping CO_2 with amine-based systems has been questioned as to its economics and its energy demand.[106,107] Pumping trapped CO_2 into sandstone deposits underneath the North Sea (as practiced by the Norwegian oil industry) is of doubtful legality as well as requiring constant monitoring; it has been described as a risky Faustian bargain with uncertain implications for the future despite its technical feasibility.[108]

In contrast, putting the CO_2 already in the atmosphere and that which inevitably will be added in the coming decades is a safer and more productive use of advanced technology (Figure 10.8).

10.6 WHEN WILL THE BIOBASED ECONOMY BE POSSIBLE AND WHEN WILL IT BE UNAVOIDABLE?

Investment is slowly beginning to appear in start-up biorefineries in both the United States and Europe. In April 2008, the U.S. Department of Energy announced up to $86 million to support the development of three small-scale cellulosic biorefineries in Maine, Tennessee, and Kentucky that processed up to 700 tons of feedstock per day. Industrial biorefineries could therefore be operational before 2020. There is one reason for believing that time is important.

Fossil fuels (oil, natural gas, and coal) have often intruded into the discussions of previous chapters—as drivers for biofuels policies as well as benchmarks for estimating production costs and market impacts. Fossil fuels are synonymous in many minds with global warming and atmospheric pollution, and hard truths about the global energy future will be (or already are) unavoidable. The role of biomass and other renewables in the emerging technological mix is a key issue.[109]

The long-term impact of biofuels on greenhouse gas emissions may prove highly disappointing (Chapter 9, Section 9.2.2). Energy sources such as wind, solar, and even nuclear power are genuine low-carbon alternatives, and fuel cells remain highly probable at some point in the present century as developed entities for domestic and personal use.[110] None of these options, however, can substitute for fossil fuels in providing a vast source of *carbon* for industrial use; in other words, the global petrochemical industry (and its myriad industrial product outputs) will remain dependent on oil. Is that a secure future?

The concept of ultimate cumulative production—now usually referred to as estimated ultimate recovery (EUR)—of global oil was introduced in 1956.[111] With a finite quantity of oil in the Earth's crust, production could only reach that fixed amount; the oil extraction rate is mathematically fitted by a curve function with a distinct maximum (i.e., the "peak oil" theory). By 1956, historical oil production rate maxima were well known: for the Ohio oil field before 1900 and for the Illinois oil field in 1940. Projecting forward, the U.S. peak production rate was predicted to occur between 1970 and 1975 and a world peak production rate around the year 2000. In contrast, world coal production would have a much delayed peak rate (approximately 2150), and the total fossil fuel family (oil, coal, and natural gas) might last until 2400–2500.

Although U.S. oil production did peak a little before 1975, the global picture has remained unclear; peak oil rates have been predicted to occur at any date between 2010 and 2120, with a mean value of 2040.[112] The great variability in these rival estimates derives from multiple uncertainties, including those of the extent of future discoverable oil reserves and their timing. The onset of irreversible decline could also be influenced by developments in oil-producing regions; in particular, increasing domestic oil consumption might eliminate over time the ability of some countries to export oil to net consumers. This would result in reducing the number of net exporters from 35 to between 12 and 28 (another large uncertainty) by 2030. In the inverse oil security scenario, Middle Eastern oil exporters attempt to withhold or restrict oil extraction if faced by a concerted attempt on the part of the Organization for Economic Cooperation and Development to reduce dependency on Middle Eastern oil.[113]

An independent perspective is provided by calculating the quotient of the known oil reserves and the actual consumption rate, accepting that "proven" oil reserves may be over- or underestimated, and putting to one side any possible (but unproven) reserves to be substantiated in the future (Figure 10.10). For the past 18 years, this estimate has changed little, after increasing markedly between 1979 and 1998 as new discoveries were made; the average time until exhaustion of the supply has been 41.5 years (with a standard deviation of ±1.0 year) since 1998. The dilemma lies in interpreting the detailed trend line: Is the mean life expectancy of oil reserves now decreasing?

FIGURE 10.10 Crude oil supply longevity as calculated from yearly consumption and estimates of reserves. (Data from *BP Statistical Reviews*.)

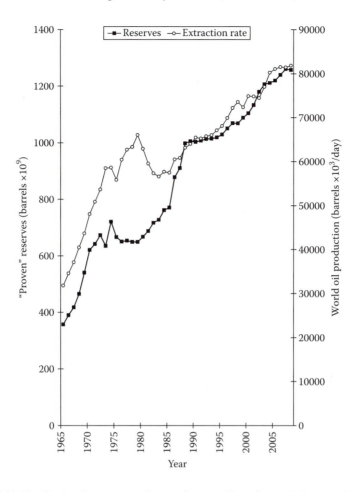

FIGURE 10.11 Crude oil reserves and extraction rate. (Data from *BP Statistical Reviews*.)

After reaching parity in the late 1980s, the rate of discovery has been overtaken by the consumption rate since 2003; if that relative imbalance persists, the original Hubbert prediction will prove to have been accurate (Figure 10.11). There certainly is no sign of the time to eventual exhaustion having increased over the past 20 years; individual years of optimism have been followed by a succession of years that fit better with a static or decreasing trend. So many major production fields are past their peak output that oil supply could decrease consistently to 39 Mb/d by 2030, with all regions (apart from Africa) showing reduced production rates (i.e., "peak oil is now").[114] Table 10.4 collects data of the historical sequence of individual national peaks of production since 1955 and compares these dates with the dwindling rate of new oil field discoveries after the 1960s.

TABLE 10.4
Peak Oil Years and Trend in Oil Discoveries

Geographical Source	Year of Peak	Time Period	Average Oil Discoveries (Gb/Year)
Austria	1955	1950–1959	41.2
Germany	1967	1960–1969	55.4
United States (lower 48)	1971	1970–1979	38.8
Canada	1974		
Romania	1976		
Indonesia	1977		
Alaska	1989	1980–1989	20.9
Egypt	1993	1990–1999	15.1
India	1995		
Syria	1995		
Gabon	1997		
Malaysia	1997		
Argentina	1998		
Venezuela	1998		
Colombia	1999		
Ecuador	1999		
United Kingdom	1999		
Australia	2000	2000–2001	17.0
Oman	2001		
Norway	2001		
Yemen	2001		
Denmark	2004	2004–2005	12.0
Mexico	2004		

Source: Data from Zittel, W., and Schindler, J. 2007. Crude Oil. The Supply Outlook. EWG series no. 3/2007, October 2007. Available from www.energywatchgroup. org.

STEM TOPIC 10.4: DID PEAK OIL CAUSE THE OIL PRICE RISE OF 2004–2008?

Oil depletion causing oil scarcity was one of three hypotheses investigated and rejected in a conference presentation in 2008[115]:

1. Peak oil is some years off—OPEC's production capacities are limited by insufficient investment and not by a lack of resources in the ground.
2. Fast demand growth was discounted because demand was not growing at an unusually high rate in the years in question.
3. Financial speculation would not have caused such a sustained rise in oil price over 4 years.

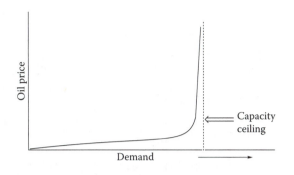

FIGURE STEM 10.4   Schematic of price/demand curve with a capacity limit.

The explanation proposed was that OPEC's lack of overcapacity will cause wild swings in oil prices; that is, with OPEC working close to capacity and unable to rapidly respond positively, the price versus production graph shows a severe "catastrophe" event (see Figure STEM 10.4).

The actual price of removing oil from the ground is only $1–15 per barrel, with many Middle Eastern oil fields at the low end of the scale. Regional producers there would tacitly welcome any mechanism causing oil prices to spiral, however short lived that phenomenon might be.

The implication is that the oil price spike of 2008 will recur.

With natural gas supplies, the horizon of exhaustion remains farther away (i.e., with a mean value of over 66 years for estimates made after 1988) (Figure 10.12). Integrated over those 19 years, however, the outlook does not offer much promise of natural gas supplies having an increased longevity. If anything, the prospect appears to be one now of rapidly dwindling stocks if the trends from 2001 onward prove to be consistent. As with oil, therefore, the era of discovery of large and accessible reserves may be over. Unless fuel economy is radically boosted by technological changes and popular take-up of those choices, price pressures on oil products caused by a dwindling or static supply (and an expected increase in demand from expanding Asian economies) will act to maintain high oil and gasoline prices.

Although there are known novel fossil resources to substitute for petroleum products capable of extending liquid hydrocarbon fuel usage by a factor of up to 10-fold, their estimated production costs lie between two- and sevenfold that of conventional gasoline and oil products (Figure 10.13).[116] Unavoidably, their production from oil shale, tar sands, natural gas, and coal deposits would add massively to total greenhouse gas emissions. Although it is also possible that technological innovations will enable oil to be extracted with high efficiency from such nonconventional sources as oil shale and tar sands and push the limits of geographical and geological possibilities for neglected or undiscovered deep-ocean oil deposits, these too will be costly.

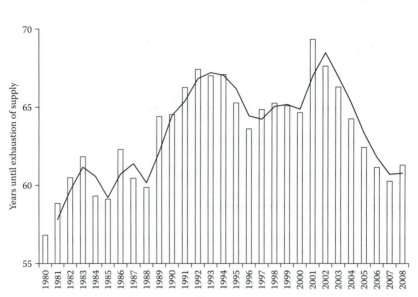

FIGURE 10.12 Natural gas supply longevity as calculated from yearly consumption and estimates of reserves. (Data from *BP Statistical Reviews*.)

Given the highly uncertain time line of global oil supply and the lack of any other substitute for fossil fuel carbon (*not* energy) other than terrestrial and marine plant biomass, the unavoidable conclusion is that—sooner or later (and possibly already)—the onus on scientists (chemists as well as biologists) is to find a biotechnology for when the oil runs out.

This could well happen during the lifetime of a reader of this book.

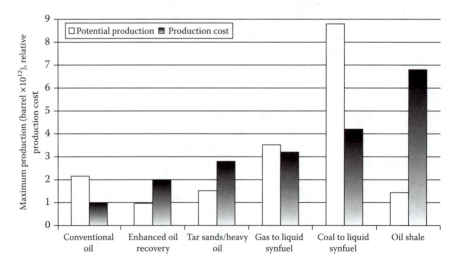

FIGURE 10.13 Potential future global supply of liquid hydrocarbons from fossil resources. (Data from Farrell, A. E., and Brandt, A. R. 2006. *Environmental Research Letters* 1:014004.)

10.7 SUMMARY

Biorefineries use biomass feedstocks as inputs to generate biofuels and chemical outputs in varying proportions. The concept is analogous to oil refineries with the substitution of oil by plant materials supplied on an industrial scale from available resources: whole plant, seed grains, grasses, wood, and even food wastes.

The technological basis for a biorefinery can be either fermentation processes or chemical transformations based on syngas production. Feedstocks are processed to a small number of key chemical substrates—principally glucose, but including other sugars and fatty acids and glycerol from triglycerides. Chemistry and biotechnology will synergize to make the maximum use of the chemical energy and carbon present in the inputs.

Biofuel outputs from biorefineries include ethanol and advanced biofuels (both liquid and gaseous and including hydrogen and methane). Chemical outputs will be a flexible mix of building-block chemicals for synthetic organic chemistry: acids, carbohydrates, amino acids, alcohols, etc. Research and development is active with chemical entities such as succinic acid, rare sugars such as D-tagatose, and derivatives of glycerol.

Biorefineries will be central to the emergence of the biobased (or biocommodity) economy in the course of the twenty-first century. Initial funding has been provided for pilot-scale facilities prior to full commercialization by 2020. If oil supplies begin to be seriously depleted—the "peak oil" effect—by 2050 or even earlier, biorefinery products will offer the only substitutes for the petrochemicals that are the basis of the modern chemical industry worldwide.

REFERENCES

1. Kamm, B. et al. 2006. Biorefinery systems—An overview. In *Biorefineries—Industrial processes and products,* vol. 1, ed. Kamm, B., Gruber, P. R., and Kamm, M., chap. 1. Weinheim, Germany: Wiley-VCH Verlag GmbH & Co. KGaA.
2. Gruber, P., Henton, D. E., and Starr, J. 2006. *Biorefineries—Industrial processes and products,* vol. 2, ed. Kamm, B., Gruber, P. R., and Kamm, M., chap. 14. Weinheim, Germany: Wiley-VCH Verlag GmbH & Co. KGaA.
3. Zeng, A.-P., and Biebl, H. 2002. Bulk chemicals from biotechnology: The case of 1,3-propanediol production and the new trends. *Advances in Biochemical Engineering/Biotechnology* 74:239.
4. U.S. Department of Energy 1997. *Energy, environment and economics (E3) handbook,* 1st ed. September 1997, http://www.oit.doe.gov/e3handbook.
5. National Renewable Energy Laboratory (NREL). 2009. http://www.nrel.gov/biomass/biorefinery.html.
6. Clark, J. 2002. Introduction. In *Handbook of green chemistry and technology,* ed. Clark, J. and MacQuarrie, D. J., chap. 1. Oxford, England: Blackwell Science.
7. Kamm, B., Gruber, P. R., and Kamm, M. 2007. Biorefineries—Industrial processes and products. In *Ullmann's encyclopedia of industrial chemistry,* e-edition. Weinheim, Germany: Wiley-VCH Verlag GmbH & Co. KGaA.
8. Senior, P. J., and Windass, J. 1980. The ICI single cell protein process. *Biotechnology Letters* 2:205.
9. Otero, J. M., Panagiotou, G., and Olsson, L. 2007. Fueling industrial biotechnology growth with bioethanol. *Advances in Biochemical Engineering/Biotechnology* 108:1.

10. Koutinas, A. A. et al. 2007. Cereal-based biorefinery development: Integrated enzyme production for cereal flour hydrolysis. *Biotechnology and Bioengineering* 97:61.

11. Kawaguchi, H. et al. 2006. Engineering of a xylose metabolic pathway in *Corynebacterium glutamicum. Applied and Environmental Microbiology* 72:3418.

12. Weekes, J., and Yüksel, G. U. 2004. Molecular characterization of two lactate dehydrogenase genes with a novel structural organization on the genome of *Lactobacillus* sp. strain MONT4. *Applied and Environmental Microbiology* 70:6290.

13. Yanai, K. et al. 2004. *Para*-position derivatives of fungal anthelmintic cyclodepsipeptides engineered with *Streptomyces venezuelae* antibiotic resistance genes. *Nature Biotechnology* 22:848.

14. Varadarajan, S., and Miller, D. J. 1999. Catalytic upgrading of fermentation-derived organic acids. *Biotechnology Progress* 15:845.

15. Mousdale, D. M. 2006. Metabolic analysis and optimization of microbial and animal cell bioprocesses. In *Fermentation microbiology and biotechnology,* 2nd ed., ed. El-Mansi, E. M. T. et al., chap. 6. Boca Raton, FL: CRC Press.

16. Khalil, C. N., and Leite, L. C. F. 2006. Process for producing biodiesel fuel using triglyceride-rich oleagineous seed directly in a transesterification reaction in the presence of an alkaline alkoxide catalyst. US Patent 7,112,229, September 26, 2006.

17. Pandey, A., Soccol, C. R., and Mitchell, D. 2000. New developments in solid state fermentation. I. Bioprocesses and products. *Process Biochemistry* 35:1153.

18. Pessoa, A. et al. 2005. Perspectives on bioenergy and biotechnology in Brazil. *Applied Biochemistry and Biotechnology* 121–124:59.

19. Solaiman, D. K. et al. 2006. Conversion of agricultural feedstocks and co-products into poly(hydroxyalkanoates). *Applied Microbiology and Biotechnology* 71:783.

20. Lynd, L. R., Wyman, C. E., and Gerngross, T. U. 1999. Biocommodity engineering. *Biotechnology Progress* 15:777.

21. Ahring, B., and Westermann, P. 2007. Coproduction of bioethanol with other biofuels. *Advances in Biochemical Engineering/Biotechnology* 108:290.

22. Fiedler, E. et al. 2005. Methanol. In *Ullmann's encyclopedia of industrial chemistry,* e-edition. Weinheim, German: Wiley-VCH Verlag GmbH & Co. KGaA.

23. Zimmermann, H., and Walzl, R. 2007. Ethylene. In *Ullmann's encyclopedia of industrial chemistry,* e-edition. Weinheim, German: Wiley-VCH Verlag GmbH & Co. KGaA.

24. Christoph, R. et al. 2006. Glycerol. In *Ullmann's encyclopedia of industrial chemistry,* e-edition. Weinheim, German: Wiley-VCH Verlag GmbH & Co. KGaA.

25. Sienel, G., Rieth, R., and Rowbottom, K. T. 2005. Epoxides. In *Ullmann's encyclopedia of industrial chemistry,* e-edition. Weinheim, German: Wiley-VCH Verlag GmbH & Co. KGaA.

26. Hahn, H.-D., Dämbkes, G., and Ruppich, N., et al. 2005. Butanols. In *Ullmann's encyclopedia of industrial chemistry,* e-edition. Weinheim, German: Wiley-VCH Verlag GmbH & Co. KGaA.

27. Obenhaus, F., Droste, W., and Neumeister, J. 2005. Butenes. In *Ullmann's encyclopedia of industrial chemistry,* e-edition. Weinheim, German: Wiley-VCH Verlag GmbH & Co. KGaA.

28. Clements, L. D., and Van Dyne, D. L. 2006. The lignocellulosic biorefinery—A strategy for returning to a sustainable source of fuels and industrial organic chemicals. In *Biorefineries—Industrial processes and products,* vol. 1, ed. Kamm, B., Gruber, P. R., and Kamm, M., chap. 5. Weinheim, Germany: Wiley-VCH Verlag GmbH & Co. KGaA.

29. Werpy, T., and Petersen, G., eds. 2004. Top value added chemicals from biomass. Volume I—Results of screening for potential candidates from sugars and synthesis gas. Pacific Northwest National Laboratory and National Renewable Energy Laboratory, August 2004 (available electronically at http://www.osti.gov/bridge).

30. Guettler, M. V., Jain, M. K., and Soni, B. K. 1998. Process for making succinic acid, microorganisms for use in the process and methods of obtaining the microorganisms. US Patent 5,723,322, March 3, 1998.

31. Zeikus, J. G., Jain, M. K., and Elankovan, P. 1999. Biotechnology of succinic acid production and markets for derived industrial products. *Applied Microbiology and Biotechnology* 51:545.

32. Donnelly, M. I. et al. 1998. A novel fermentation pathway in an *Escherichia coli* mutant producing succinic acid, acetic acid, and ethanol. *Applied Biochemistry and Biotechnology* 70–72:187.

33. Donnelly, M. I., Millard, C. S., and Stols, L. 1998. Mutant *E. coli* strain with increased succinic acid production. US Patent 5,770,435, June 23, 1998.

34. Chatterjee, R. et al. 2001. Mutation of the *ptsG* gene results in increased production of succinic acid in fermentation of glucose by *Escherichia coli*. *Applied and Environmental Microbiology* 67:148.

35. Nghiem, N. P. et al. 1999. Method for the production of dicarboxylic acids. US Patent 5,869,301, February 9, 1999.

36. Donnelly, M. I., Sanville-Millard, C. S., and Nghiem, N. P. 2004. Method to produce succinic acid from raw hydrolysates. US Patent 6,743,610, June 1, 2004.

37. Isar, J. et al. 2006. A statistical method for enhancing the production of succinic acid from *Escherichia coli* under anaerobic conditions. *Bioresource Technology* 97:1443.

38. Agarwal, L. et al. 2006. A cost effective fermentative production of succinic acid from cane molasses and corn steep liquor by *Escherichia coli*. *Journal of Applied Microbiology* 100:1348.

39. Isar, J. et al. 2006. Succinic acid production from *Bacteroides fragilis:* Process optimization and scale up in a bioreactor. *Anaerobe* 12:231.

40. Isar, J. et al. 2007. A statistical approach to study the interactive effects of process parameters on succinic acid production from *Bacteroides fragilis*. *Anaerobe* 13:50.

41. Lin, H., Bennett, G. N., and San, K.-Y. 2005. Fed-batch culture of a metabolically engineered *Escherichia coli* strain designed for high-level succinate production and yield under aerobic conditions. *Biotechnology and Bioengineering* 90:775.

42. San, K.-Y., Bennett, G. N., and Sanchez, A. 2006. Mutant *E. coli* strain with increased succinic acid production. US Patent application 2006/0046288, March 2, 2006.

43. Lee, S. Y. et al. 2007. Novel gene encoding fumarate dehydratase C and method for preparing succinic acid using the same. US Patent application 2007/0042477, February 22, 2007.

44. Murase, M. et al. 2006. Process for producing succinic acid. US Patent application 2006/0205048, September 14, 2006.

45. Lee, P. C. et al. 1999. Succinic acid production by *Anaerobiospirillum succiniciproducens:* Effects of the H_2/CO_2 supply and glucose concentration. *Enzyme and Microbial Technology* 24:549.

46. Lee, P. C. et al. 1999. Effects of medium components on the growth of *Anaerobiospirillum succiniciproducens* and succinic acid production. *Process Biochemistry* 35:49.

47. Lee, P. C. et al. 2000. Batch and continuous fermentation of succinic acid from whey by *Anaerobiospirillum succiniciproducens*. *Applied Microbiology and Biotechnology* 54:23.

48. Lee, P. C. et al. 2000. Fermentative fermentation of succinic acid from glucose and corn steep liquor by *Anaerobiospirillum succiniciproducens*. *Biotechnology and Bioprocess Engineering* 5:379.

49. Lee, P. C. et al. 2002. Isolation and characterization of a new succinic acid-producing bacterium, *Mannheimia succiniciproducens* MBEL55E, from bovine rumen. *Applied Microbiology and Biotechnology* 58:663.

50. Hong, S. H. et al. 2004. The genome sequence of the capnophilic rumen bacterium *Mannheimia succiniciproducens*. *Nature Biotechnology* 22:1275.

51. Kim, T. Y. et al. 2007. Genome-scale analysis of *Mannheimia succiniciproducens* metabolism. *Biotechnology and Bioengineering* 97:657.

52. Lee, S. Y., and Lee, S. J. 2005. Novel rumen bacteria variants and process for preparing succinic acid employing the same. International patent application WO 2005/052135, June 9, 2005.

53. Kim, D. Y. et al. 2004. Batch and continuous fermentations of succinic acid from wood hydrolysate by *Mannheimia succiniciproducens* MBEL55E. *Enzyme and Microbial Technology* 35:648.

54. Millard. C. S. et al. 1996. Enhanced production of succinic acid by overexpression of phosphoenolpyruvate carboxylase in *Escherichia coli*. *Applied and Environmental Microbiology* 62:1808.

55. Stols, L., and Donnelly, M. 1997. Production of succinic acid through overexpression of NAD+-dependent malic enzyme in an *Escherichia coli* mutant. *Applied and Environmental Microbiology* 63:2695.

56. Stols, L. et al. 1997. Expression of *Ascaris suum* malic enzyme in a mutant *Escherichia coli* allows production of succinic acid from glucose. *Applied Biochemistry and Biotechnology* 63–65:153.

57. Hong, S. H., and Lee, S. Y. 2001. Metabolic flux analysis for succinic acid production by recombinant *Escherichia coli* with amplified malic enzyme activity. *Biotechnology and Bioengineering* 74:89.

58. Kim, P. et al. 2004. Effect of overexpression of *Actinobacillus succinogenes* phosphoenolpyruvate carboxylase on succinate production in *Escherichia coli*. *Applied and Environmental Microbiology* 70:1238.

59. Lin, H., Bennett, G. N., and San, K.-Y. 2005. Metabolic engineering of aerobic succinate production systems in *Escherichia coli* to improve process productivity and achieve the maximum theoretical succinate yield. *Metabolic Engineering* 7:116.

60. Lee, S. Y. et al. 2005. Metabolic engineering of *Escherichia coli* for enhanced production of succinic acid based on genome comparison and in silico gene knockout simulation. *Applied and Environmental Microbiology* 71:7880.

61. Yedur, S., Berglund, K. A., and Dunuwila, D. D. 2001. Succinic acid production and purification. US Patent 6,265,190, July 24, 2001.

62. Production of chemicals from biologically derived succinic acid (BDSA). Available from www.pnl.gov/biobased/docs/succinic.pdf

63. Nair, N., and Zhao, H. 2007. Biochemical characterization of an L-xylulose reductase from *Neurospora crassa*. *Applied and Environmental Microbiology* 73:2001.

64. Rasmussen, L. E. et al. 2006. Mode of action and properties of the β-xylosidases from *Taloromyces emersonii* and *Trichoderma reesei*. *Biotechnology and Bioengineering* 94:869.

65. Tenkanen, M. 1998. Action of *Trichoderma reesei* and *Aspergillus oryzae* esterases in the deacetylation of hemicelluloses. *Biotechnology and Applied Biochemistry* 27:19.

66. Lavasseur, A. et al. 2004. Design and production in *Aspergillus niger* of a chimeric protein associating a fungal feruloyl esterase and a clostridial dockerin domain. *Applied and Environmental Microbiology* 70:6984.

67. Lavasseur, A. et al. 2006. Production of a chimeric enzyme tool associating the *Trichoderma reesei* swollenin with the *Aspergillus niger* feruloyl esterase A for release of ferulic acid. *Applied Microbiology and Biotechnology* 73:872.

68. Faulds, C. H. et al. 2004. Arabinoxylan and mono- and dimeric ferulic acid release from brewer's grain and wheat bran by feruloyl esterases and glycosyl hydrolases from *Humicola insolens*. *Applied Microbiology and Biotechnology* 64:644.

69. Leathers, T. D., and Dien, B. S. 2000. Xylitol production from corn fiber hydrolysates by a two-stage fermentation process. *Process Biochemistry* 35:765.

70. Lee, W.-J., Ryu, Y.-W., and Seo, J.-H. 2000. Characterization of two-substrate fermentation processes for xylitol production using recombinant *Saccharomyces cerevisiae* containing xylose reductase gene. *Process Biochemistry* 35:1199.
71. Govinden, R. et al. 2001. Xylitol production by recombinant *Saccharomyces cerevisiae* expressing the *Pichia stipitis* and *Candida shehatae XYL-1* genes. *Applied Microbiology and Biotechnology* 55:76.
72. Ko, B. S., Kim, J., and Kim, J. H. 2006. Production of xylitol from D-xylose by a xylitol dehydrogenase gene-disrupted mutant of *Candida tropicalis. Applied and Environmental Microbiology* 72:4207.
73. Martinez, E. A. et al. 2007. Downstream process for xylitol produced from fermented hydrolysate. *Enzyme and Microbial Technology* 40:1193.
74. Verho, R. et al. 2004. A novel NADH-linked L-xylulose reductase in the L-arabinose catabolic pathway of yeast. *Journal of Biological Chemistry* 279:14746.
75. Poonperm, W. et al. 2007. Production of L-xylulose from xylitol by a newly isolated strain of *Bacillus pallidus* Y25 and characterization of its relevant enzyme xylitol dehydrogenase. *Enzyme and Microbial Technology* 40:1206.
76. Suzuki, T. et al. 2005. Cloning and expression of NAD$^+$-dependent L-arabinitol 4-dehydrogenase gene (*ladA*) of *Aspergillus oryzae. Journal of Bioscience and Bioengineering* 100:472.
77. Aarnikunnas, J. S. et al. 2006. Cloning and expression of a xylitol-4-dehydrogenase gene from *Pantoea ananatis. Applied and Environmental Microbiology* 72:368.
78. Strohl, W. R. 1997. Industrial antibiotics: Today and the future. In *Biotechnology of antibiotics,* ed. Strohl, W. R., chap. 1. New York: Marcel Dekker.
79. Muniruzzaman, S. et al. 1996. Inhibition of glycoprotein processing by L-fructose and L-xylulose. *Glycobiology* 8:795.
80. Kim, P. 2004. Current studies on biological tagatose production using L-arabinose isomerase: A review and future perspective. *Applied Microbiology and Biotechnology* 65:243.
81. Roh, H. J et al. 2000. Bioconversion of D-galactose into D-tagatose by expression of L-arabinose isomerase. *Biotechnology and Applied Biochemistry* 31:1.
82. Kim, P. et al. 2001. High production of D-tagatose, a potential sugar substitute, using immobilized L-arabinose isomerase. *Biotechnology Progress* 17:208.
83. Kim, H. J. et al. 2003. A feasible enzymatic process for D-tagatose production by an immobilized thermostable L-arabinose isomerase in a packed-bed bioreactor. *Biotechnology Progress* 19:400.
84. Jung, E. S., Kim, H. J., and Oh, D. K. 2005. Tagatose production by immobilized recombinant *Escherichia coli* cells containing *Geobacillus stearothermophilus* L-arabinose isomerase mutant in a packed-bed bioreactor. *Biotechnology Progress* 21:1335.
85. Doten, R. C., and Mortlock, R. P. 1985. Production of D- and L-xylulose by mutants of *Klebsiella pneumoniae* and *Erwinia uredovora. Applied and Environmental Microbiology* 49:158.
86. Granström, T. B. et al. 2004. Izumoring: A novel and complete strategy for bioproduction of rare sugars. *Journal of Bioscience and Bioengineering* 97:89.
87. Menavuvu, B. T. et al. 2006. Novel substrate specificity of D-arabinose isomerase from *Klebsiella pneumoniae* and its application to production of D-altrose and D-psicose. *Journal of Bioscience and Bioengineering* 102:436.
88. Aitken, J. E. 2006. Purification of glycerin. US Patent 7,126,032, October 24, 2006.
89. Pech, C. W., and Hopp, J. R. 2007. Method for manufacture and use of the waste stream from biodiesel production (crude glycerin) as a commercial fuel. US Patent application 2007/0113465, May 24, 2007.
90. Pachauri, N., and He, B. 2006. Value-added utilization of crude glycerol from biodiesel production: A survey of current research activities. 2006 ASABE Annual International Meeting, paper no. 066223.

91. Biebl, H. 2001. Fermentation of glycerol by *Clostridium pasteurianum*—Batch and continuous fermentations. *Journal of Industrial Microbiology and Biotechnology* 27:18.

92. Pyle, D., and Wen, Z. 2007. Production of omega-3 polyunsaturated fatty acid from biodiesel—Waste glycerol by microalgal fermentation. 2007 ASABE Annual International Meeting, paper no. 077028.

93. Bournay, L., and Baudot, A. 2006. Process for producing fatty acid alkyl esters and glycerol of high purity. US Patent 7,138,536, November 21, 2006.

94. Frafft, P. et al. 2007. Process for producing dichloropropanol from glycerol, the glycerol coming eventually from the conversion of animal fats in the manufacture of biodiesel. US Patent application 2007/0112224, May 17, 2007.

95. Dharmadi, Y., Murarka, A., and Gonzalez, R. 2006. Anaerobic fermentation of glycerol by *Escherichia coli:* A new platform for metabolic engineering. *Biotechnology and Bioengineering* 94:821.

96. Ito, T. et al. 2005. Hydrogen and ethanol production from glycerol-containing wastes discharged after biodiesel manufacturing process. *Journal of Bioscience and Bioengineering* 100:260.

97. Lu, Y. et al. 2009. Expression of NAD^+-dependent formate dehydrogenase in *Enterobacter aerogenes* and its involvement in anaerobic metabolism and H_2 production. *Biotechnology Letters* 31:1525.

98. Huber, G. W., and Corma, A. 2007. Synergies between bio- and oil refineries for the production of fuels from biomass. *Angewandte Chemie International Edition* 46:7184.

99. Chheda, J. N., Huber, G. W., and Dumesic, J. A. 2007. Liquid-phase catalytic processing of biomass-derived oxygenated hydrocarbons to fuels and chemicals. *Angewandte Chemie International Edition* 46:7164.

100. López-Contreras, A. M. et al. 2000. Utilization of saccharides in extruded domestic organic waste by *Clostridium acetobutylicum* ATCC 824 for production of acetone, butanol and ethanol. *Applied Microbiology and Biotechnology* 54:162.

101. Claassen, P. A., Budde, M. A., and López-Contreras, A. M. 2000. Acetone, butanol and ethanol production by solventogenic clostridia. *Journal of Molecular Microbiology and Biotechnology* 2:39.

102. Zverlov, V. V. et al. 2006. Bacterial acetone and butanol production by industrial fermentation in the Soviet Union: Use of hydrolyzed agricultural waste for biorefinery. *Applied Microbiology and Biotechnology* 71:587.

103. Li, P., Harding, S. E., and Liu, Z. 2001. Cyanobacterial exopolysaccharides; their nature and potential biotechnological applications. *Biotechnology and Genetic Engineering Reviews* 18:404.

104. Ishii, M. et al. 2004. Occurrence, biochemistry and possible biotechnological application of the 3-hydroxypropionate cycle. *Applied Microbiology and Biotechnology* 64:605.

105. Bosma, R. 2007. Prediction of volumetric productivity of an outdoor photobioreactor. *Biotechnology and Bioengineering* 97:1108.

106. Rao, A. B. et al. 2006. Evaluation of potential cost reductions from improved amine-based CO_2 capture systems. *Energy Policy* 34:3765.

107. Rubin, E. S., Chen, C., and Rao, A. B. 2007. Cost and performance of fossil fuel power plants with CO_2 capture and storage. *Energy Policy* 35:4444.

108. Spreng, D., Marland, G., and Weinberg, A. M. 2007. CO_2 capture and storage: Another Faustian bargain? *Energy Policy* 35:850.

109. National Petroleum Council. 2007. Facing the hard truths about energy: A comprehensive view to 2030 of global oil and natural gas, July 2007 (www.npc.org).

110. Asazwa, K. et al. 2007. A platinum-free zero-carbon-emission easy fuelling direct hydrazine fuel cell for vehicles. *Angewandte Chemie International Edition* 46:8024.

111. Hubbert, M. K. 1956. Nuclear energy and the fossil fuels. Publication no. 95, Shell Development Company (Exploration and Production Research Division), Houston, Texas, June 1956.

112. United States Government Accountability Office. 2007. Crude oil: Uncertainty about future oil supply makes it important to develop a strategy for addressing a peak and decline in oil production. GAO-07-283 Washington, D.C., February 2007.

113. Hallock, J. L. et al. 2004. Forecasting the limits to the availability and diversity of global conventional oil supply. *Energy* 29:1673.

114. Zittel, W., and Schindler, J. 2007. Crude oil. The supply outlook. EWG series no. 3/2007, October 2007. Available from www.energywatchgroup.org.

115. Brochmann, B., and Schulte, J. 2008. "Peak oil" and the high price of oil. Second International Association for Energy Economics (IAEE) Asian Conference: Energy Security and Economic Development under Environmental Constraints in the Asia-Pacific Region, Perth, Western Australia, November 5–7, 2008.

116. Farrell, A. E., and Brandt, A. R. 2006. Risks of the oil transition. *Environmental Research Letters* 1:014004.

Index

421